Remote Sensing Applications for Agriculture and Crop Modelling

Remote Sensing Applications for Agriculture and Crop Modelling

Special Issue Editor

Piero Toscano

MDPI • Basel • Beijing • Wuhan • Barcelona • Belgrade

Special Issue Editor
Piero Toscano
Institute of BioEconomy-IBE,
National Research Council-CNR
Italy

Editorial Office
MDPI
St. Alban-Anlage 66
4052 Basel, Switzerland

This is a reprint of articles from the Special Issue published online in the open access journal *Agronomy* (ISSN 2073-4395) from 2018 to 2019 (available at: https://www.mdpi.com/journal/agronomy/special_issues/remote-sensing-agriculture-model).

For citation purposes, cite each article independently as indicated on the article page online and as indicated below:

LastName, A.A.; LastName, B.B.; LastName, C.C. Article Title. *Journal Name* **Year**, *Article Number*, Page Range.

ISBN 978-3-03928-226-5 (Pbk)
ISBN 978-3-03928-227-2 (PDF)

Cover image courtesy of Piero Toscano

Contents

About the Special Issue Editor

Piero Toscano is a meteorologist, ecologist, Research Technologist at the Institute of BioEconomy - National Research Council (https://www.ibe.cnr.it/en/) experienced in proximal and remote sensing, meteorological and micrometeorological observations, analysis and data processing. He has been running the project "Delphi", a comprehensive study on the worldwide durum wheat production forecast, since 2009. He specifically focuses on issues where remote sensing, ground network monitoring and field measurements can activate both a deeper mechanistic understanding of the underlying processes, and a more accurate estimation of land characterization, especially crop-specific agricultural mapping, area estimates, production estimates and productivity estimates at local and national scales. Dr. Toscano has more than 50 publications (ISI) to his credit and has given presentations at international/national congresses, conferences, symposia and workshops. He has been awarded by the Union of Italian Academies for Science applied to Agriculture, Food and Safety (UNASA) for significant research paper publication in the field of agronomy and crop production. Dr. Toscano is also leading the AgroSat project (https://agrosat.it/), a free-to-use and open-source platform to support and foster precision and smart farming in Italy

Preface to "Remote Sensing Applications for Agriculture and Crop Modelling"

Crop models and remote sensing techniques have been combined and applied in agriculture and crop estimation on local and regional scales, or worldwide, based on the simultaneous development of crop models and remote sensing. The literature shows that many new remote sensing sensors and valuable methods have been developed for the retrieval of canopy state variables and soil properties from remote sensing data for assimilating the retrieved variables into crop models. At the same time, remote sensing has been used in a staggering number of applications for agriculture. This book sets the context for remote sensing and modelling for agricultural systems as a mean to minimize the environmental impact, while increasing production and productivity. The eighteen papers published in this Special Issue, although not representative of all the work carried out in the field of Remote Sensing for agriculture and crop modeling, provide insight into the diversity and the complexity of developments of RS applications in agriculture. Five thematic focuses have emerged from the published papers: yield estimation, land cover mapping, soil nutrient balance, time-specific management zone delineation and the use of UAV as agricultural aerial sprayers. All contributions exploited the use of remote sensing data from different platforms (UAV, Sentinel, Landsat, QuickBird, CBERS, MODIS, WorldView), their assimilation into crop models (DSSAT, AQUACROP, EPIC, DELPHI) or on the synergy of Remote Sensing and modeling, applied to cardamom, wheat, tomato, sorghum, rice, sugarcane and olive.

The intended audience is researchers and postgraduate students, as well as those outside academia in policy and practice.

Piero Toscano
Special Issue Editor

Article

Integrating Modelling and Expert Knowledge for Evaluating Current and Future Scenario of Large Cardamom Crop in Eastern Nepal

Sajana Maharjan, Faisal Mueen Qamer *, Mir Matin, Govinda Joshi and Sanjeev Bhuchar

International Centre for Integrated Mountain Development (ICIMOD), Kathmandu 44700, Nepal
* Correspondence: fqamer@icimod.org; Tel.: +977-01-500-3222

Received: 29 May 2019; Accepted: 18 July 2019; Published: 26 August 2019

Abstract: Large Cardamom (*Amomum subulatum* Roxb.) is one of the most valuable cash crop of the Himalayan mountain region including Nepal, India, and Bhutan. Nepal is the world's largest producer of the crop while the Taplejung district contributes a 30%–40% share in Nepal's total production. Large cardamom is an herbaceous perennial crop usually grown under the shade of the Uttis tree in very specialized bioclimatic conditions. In recent years, a decline in cardamom production has been observed which is being attributed to climate-related indicators. To understand the current dynamics of this under-canopy herbaceous crop distribution and its future potential under climate change, a combination of modelling, remote sensing, and expert knowledge is applied for the assessment. The results suggest that currently, Uttis tree cover is 10,735 ha in the district, while 50% (5198 ha) of this cover has a large cardamom crop underneath. When existing cultivation is compared with modelled suitable areas, it is observed that the cultivatable area has not yet reached its full potential. In a future climate scenario, the current habitat will be negatively affected, where mid elevations will remain stable while lower and higher elevation will become infeasible for the crop. Future changes are closely related to temperature and precipitation which are steadily changing in Nepal over time.

Keywords: large cardamom; remote sensing; species modelling; habitat assessment; climate change

1. Introduction

Large cardamom (*Amomum subulatum* Roxb.) is mostly being cultivated in the Himalayan mountain region of Nepal, India, and Bhutan [1]. There is an increasing demand of the spices from local to global markets which has fascinated farmers in its commercial cultivation [2]. The high-value crop cultivation has substantially improved the livelihood of farmers residing in these mountain regions. However, several studies have highlighted threats of cardamom farming which include problems in disease management, change in climatic conditions, and human activities such as infrastructure development in cardamom growing areas [2,3]. Alongside this, comprehensive information on its current distribution and potentially suitable areas for cultivation is unavailable, which is essential for crop planning and management. Further, for its distribution and current suitability, information on the impacts of climate change is also vital for future planning as climate has a significant impact on the geographical distribution of plant species and also alters habitat conditions [4,5].

Large cardamom is an herbaceous perennial crop usually grown under shade. Uttis (*Alnus nepalensis*) trees provide excellent shade, supply a good amount of litter from twigs and leaves, and nitrogen from the root nodules to understory cardamom when they are young. The crop is generally grown at an altitude of 700 to 2000 m above sea level in humid conditions, with temperature ranges between 4–20 °C and annual precipitation around 2000–2500 mm [6].

Although satellite remote sensing data can be used for accurate, timely, and consistent information on the agricultural productivity at local and regional scales [7,8], detection of understory plants is inhibited due to canopy cover, canopy gap shadowing, and terrain variability [9]. Differentiating signals of understory vegetation from overstory canopy is still challenging due to complex interactions between overstory and understory vegetation [10]. Several studies have been conducted to distinguish understory vegetation using remote sensing approaches. In this regard, evergreen understory vegetation were identified from Landsat images of leaf off-season in deciduous forest [11,12]. Phenological difference between overstory and understory vegetation was also used to detect understory vegetation using Landsat images [13,14]. However, such studies require multi-temporal data which are generally available in coarser spatial resolutions. Leduc et al. [15] have mapped wild leek which grows on forest floors using low flying Unmanned Aerial Vehicle (UAV). Other studies have used data from active sensor, mainly LIDAR data, to map understory plants in boreal forests [16,17]. Combinations of LIDAR data and hyperspectral images were used to map understory invasive species in tropical forests [9], and the use of LIDAR data and high-resolution IKONOS imagery to identify understory plant invasion in urban forests [18]. Nevertheless, there are limitations on the use of these data due to its high cost and low availability, mainly in developing countries [19].

Understory vegetation are usually hard to identify only from remote sensing, thus a substantial efforts have been made to identify these vegetation through the integration of several approaches. For instance, Wang et al. [20] have mapped understory bamboo by integrating neural network and Geographic Information System (GIS) expert system, and Tuanmu et al. [10] have detected understory vegetation using phenology metrics derived from time series of Moderate Resolution Imaging Spectroradiometer (MODIS) data together with a suitability model like the maximum entropy (Maxent) model. The Maxent model is a species distribution model (SDM) that provides an approach for making predictions from existing distribution and a set of predictors [21]. Such models predicting the potential distribution of species are useful for several applications in conservation biology [22,23]. Other approaches, such as object-based image analysis (OBIA) on multispectral and hyperspectral data, were found to be effective in identifying understory plant species [24] due to its characteristics of using contextual relationships together with spectral information. In addition, expert knowledge [25] and other ancillary data such as elevation, slope, and aspect can also be used to improve classification accuracy. Moreover, the combination of participatory mapping with remote sensing technique can further improve the accuracy as the two methods complement and validate each other [26]. Participatory mapping is an effective tool to obtain accurate baseline data of the field [27], which can be integrated in remote sensing analyses to produce more accurate maps [28].

Several species distribution models such as the genetic algorithm for rule-set production (GARP), ecological niche factor analysis (ENFA), bioclimatic modeling (BIOCLIM), CLIMEX (climate change experiment), domain environmental envelope (DOMAIN), and Maxent (maximum entropy) have been used for species distribution prediction. Among these, Maxent is widely used as it can perform better even with small sample sizes compared to other modelling methods [29–31]. In addition, it has other merits, such as: it requires only presence data of the species and environmental variables for the study area, it can use both categorical and continuous data and can incorporate interactions among different variables, it can produce spatially explicit habitat suitability maps, and the importance of each environmental variable on the model can be evaluated using the built-in jackknife test.

This study aims to evaluate existing spatial distribution, potential suitable areas for cultivation under various scenarios, and core distributional shift with future climate condition in the Taplejung district of Nepal. The study uses a combination of expert knowledge and species modelling approach to capture all aspects of habitat conditions in the complex environmental conditions in the mountain region.

2. Materials and Methods

2.1. Study Area

The study area, the Taplejung district, lies in eastern Nepal. The district ($27°57'10''$–$27°16'5''$ N and $87°26'40''$–$88°12'6''$ E) with an area of 3646 km^2 physiographically lies in the mid-hill to high-himal region. It lies at an elevation of 498 m to 8464 m. Forest is the major land cover of the district followed by agriculture, bare land, grass, snow, shrub, water body, and built-up, respectively. The district is the major producer of large cardamom in the country, which is the major source of income for farmers in the district. In Nepal, an estimated 12,000 ha of land in over 40 mid-hills districts are under cardamom cultivation with estimated annual production of 6000 metric tons. While the Taplejung district contribute around 4500 ha of land and yearly production of around 2600 metric tons [32].

The study broadly followed two broad approaches for Cardamom habitat mapping including species modelling- and expert knowledge-based assessment (Figure 1).

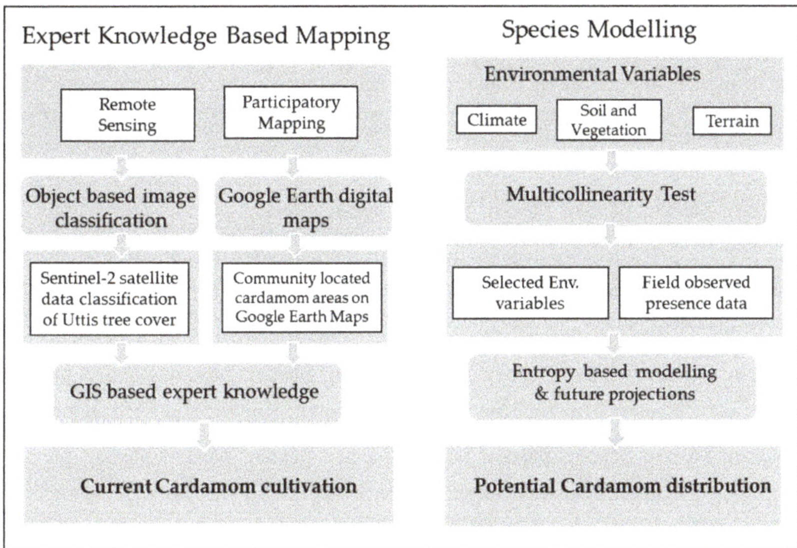

Figure 1. Methodology flow diagram.

2.2. Expert Knowledge-Based Mapping

2.2.1. Participatory Mapping

Participatory mapping with the community in developing countries is found to be effective for understanding local level phenomena [33]. A two-day exercise was conducted with a community representative to map existing large cardamom farming areas on the high-resolution data of Google Earth images available in digital format as well as on the printed maps. During the exercise, large clusters of cardamom cultivated fields were identified. However, detailed delineation of cardamom fields including small patches was not possible. The data was gathered to get existing farming areas on the district which was also cross-checked with outputs obtained from remote sensing data. A part of the pockets identified by the farmers in the participatory exercise was further verified through ground validation exercise. This validation also provided GPS samples of cardamom field points along with other vegetation in the area.

2.2.2. Uttis (*Alnus nepalensis*) Mapping Using High-Resolution Satellite Data

During the participatory mapping and the field activity, it was observed that the Uttis is the major tree species grown for shade to cultivate large cardamom. Since the understory crop could not be directly mapped using remote sensing data, delineation of Uttis cover was consider as proxy to map the large cardamom crop. Our focus was to map Uttis in the entire district, which can provide information on prospective cultivation of large cardamom. Uttis tree cover was mapped by using Sentinel-2A Level 1C product image acquired in February 2016. The image has very high-resolution spectral coverage that includes 12 bands (coastal aerosol, blue green, red, 3 vegetation red edge, NIR, vegetation red edge water vapor, SWIR Cirrus, and 2 SWIR respectively). The spatial resolution of blue green, red and NIR is 10 m. The resolution of the vegetation red edge band is 20 m and for the rest of the bands it is 60 m. These features of the sensor are suited for agricultural monitoring systems [34]. Level 1C product is a Top of Atmosphere (TOA) product for which atmospheric correction had to be done to get reflectance values of the image so that the image can be used for mapping. SNAP (Sentinel toolbox) software (version 5.0.0) was used for atmospheric correction of the image.

Spectral separability of forest types such as coniferous forest, broadleaf forest, Uttis, and shrub was studied before the classification. The mean pixel values of the abovementioned forest types and shrub were plotted against eight bands of Sentinel-2A image to evaluate the potential of image spectral separability before the classification. The object-based image analysis (OBIA) classification approach was adopted for Sentinel-2A image classification. The technique uses spectral and contextual information in an integrative way [25]. The fundamental technique of OBIA is the segmentation of satellite images which overcome the salt and pepper effect [35]. In this study, the chessboard and multi-resolution segmentation algorithm in eCognition software (version 8.7,) was used to develop image objects [36] using scale parameter = 80, shape = 0.1, and compactness = 0.8.

After segmentation, Assign Class algorithm was used to classify general classes (agriculture, conifer forest, broadleaf forest, shrub, water, and snow). This was done to filter the land cover which were not the focus of the study. Several features such as NDVI, brightness, slope, elevation, and field information were used in this step. For the rest of the unclassified image objects, Nearest Neighbor Classification was applied, which is a powerful approach [37] to map Uttis trees. Uttis trees grown between the elevation range of 800–2200 m and slope up to 45 degrees were mapped. This is also the elevation and the slope range where large cardamom is cultivated [38]. Seventy percent (70%) of field data were used to train the samples while the remaining 30% were used as test data for accuracy assessment. Further, in order to reduce error and improve classification accuracy, interactive visual analysis was done on a classified map using Google Earth images [39].

The remote sensing-derived Uttis cover and participatory mapping-based identification of cardamom crop areas were overlaid to get the existing large cardamom farming area. The obtained large cardamom maps in the Taplejung district were further analyzed based on elevation, slope, aspect, and Village Development Committees (VDCs). The analysis was done to understand the pattern of cultivation and environment suitability conditions in the district and the information could be useful in further planning and management of large cardamom farming in the district. Elevation range of the study area was categorized into 9 ranges (below 800 m, 800–1000 m, 1000–1200 m, 1200–1400 m, 1400–1600 m, 1600–1800 m, 1800–2000 m, 2000–2200 m, and above 2200 m). In order to comprehend the farming area slope-wise, the slope was categorized into four gradient levels (0°–15°, 15°–30°, 30°–45°, 45° above).

2.3. Species Modelling

2.3.1. Environmental Variables and Species Occurrence Records

The habitat suitability model was primarily developed based on the variables related to climate, soil, terrain, and vegetation type. Initially, 24 variables were selected to model the current distribution pattern. These comprised of 19 bioclimatic variables with 30 arc seconds (~1 km) spatial resolution from WorldClim dataset (http://www.worldclim.org/), 12.5 m resolution digital elevation model (DEM) generated by The Japan Aerospace Exploration Agency (JAXA) using ALOS PALSAR RTC products [40], slope and aspect layers generated from the DEM using spatial analyst tools in ArcGIS 10.6.1 soil pH with the resolution of 250 m was downloaded from https://soilgrids.org/#!/?layer=TAXNWRB_250m&vector=1 and Uttis cover thematic layer. All bioclimatic variables, topographic layers, and pH were resampled into 20 m spatial resolution to make them compatible with the resolution of Uttis cover. This was done in ArcGIS 10.6.1 with the nearest neighbor resampling technique. For future scenarios (year 2050), we initially selected DEM, slope, aspect, and 19 bioclimatic variables for RCP2.6 (the minimum greenhouse gas emission scenario), and RCP8.5 (the maximum greenhouse gas emission scenario) as adopted by the Intergovernmental Panel on Climate Change (IPCC) in its fifth assessment report (AR5). The climatic variables were also resampled into 20 m to make the variables uniform at one resolution.

Multicollinearity between predictor variables was tested as it can lead to inaccurate prediction by excluding significant explanatory variables [41]. The test was conducted calculating Pearson's Correlation Coefficient (r) to assess the cross-correlation and the one variable from any pair of variables with a cross-correlation coefficient value of $> \pm 0.8$ was excluded [42]. Variables were chosen based on biological relevancy to the species. For example, pH was highly correlated with temperature seasonality (BIO4) ($r = 0.88$), from this pair BIO4 (temperature seasonality) was removed as pH plays a crucial role in cardamom plantation [2]. In addition, the variation inflation factor (VIF) was also used to check collinearity among the variables in R software (version 3.6). Variables with VIF values greater than 10 were excluded for modelling [43]. Out of 24 variables, 8 environmental variables (Uttis cover, pH, slope, aspect, isothermality, maximum temperature of warmest month, minimum temperature of coldest month, and precipitation of wettest month) were selected for the current scenario. For the future scenario of RCP2.6-2050 and for RCP8.5-2050, 6 variables (slope, aspect, isothermality, maximum temperature of warmest month, minimum temperature of coldest month, and precipitation of wettest month) were selected for modelling. As part of the species presence data, a total of 102 occurrence records (GPS coordinates of points) of large cardamom were collected randomly during the field visit conducted in May 2016.

2.3.2. Spatial Modelling and Statistical Analysis

The maximum entropy modelling approach using Maxent software (version 3.3.3k) was applied in this study for predicting habitat suitability of the large cardamom. Maxent is a machine-learning method with a simple and precise mathematical formulation. It uses a maximum entropy algorithm to produce a model that shows the probability of presence of the species that varies from 0 to 1, i.e., from the lowest to the highest probability [21]. We selected 70% of the data for training and the remaining 30% for testing. Area under the ROC (receiver operating characteristic) curve (AUC) was used to evaluate the model performance which ranges from 0 to 1. The curve is plotted with True Positive Rate (sensitivity) at the vertical axis and False Positive Rate (1-specificity) at the horizontal axis. The jackknife was used to evaluate the importance of the variables on the model. The model used logistic format. The final distribution maps have values ranges from 0 to 1 which were grouped into four classes of suitable habitat viz., unsuitable (<0.2), marginally suitable (0.2–0.4), moderately suitable (0.4–0.6), and highly suitable (>0.6). Further, analysis of existing farming area in comparison to current habitat suitability map was performed to understand the gaps and opportunities. In order to assess the change between current and future suitable areas, we quantified the areas (ha) of classes of habitat suitability under different scenarios across different elevation ranges.

3. Results and Discussions

3.1. Participatory Mapping of Large Cardamom

The spatial location of large cardamom fields in the Taplejung District was recorded on the basis of local people knowledge. Out of 50 VDCs in the district, 49 VDCs were found cultivating (Figure 2) large cardamom. However, the farming area varies from VDC to VDC. Participatory mapping has given a general overview of farming area in the district.

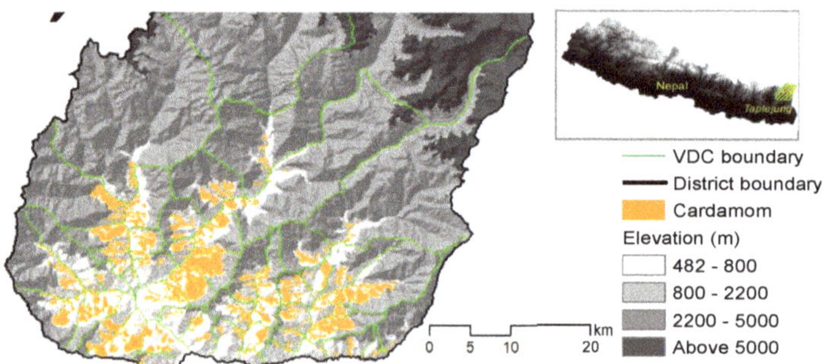

Figure 2. Mapping of large cardamom crop clusters across the Taplejung district based on participatory mapping.

3.2. Uttis (Alnus nepalensis) Cover Mapping and Delineation of Accurate Large Cardamom Map

Spectral separability of major vegetation of the study area using Sentinel-2A image shows that the vegetation are largely separable from each other in NIR and Red Edge bands (Figure 3a). In this study, integration of ancillary data with Sentinel-2A image was applied to map Uttis using OBIA (Figure 3b,c). The classification approach in eCognition has helped to use expert knowledge in differentiating trees within agricultural land and forests, resulting in better classification accuracy. During post classification improvement, small tree patches of Uttis within agricultural land were separated. Shadow in high mountain areas was affecting the classification which was improved during post classification using Google Earth images. There are several studies done on forest type classification [44,45] and land cover classification [39], however, these studies do not include separation of tree species within the forest cover classes.

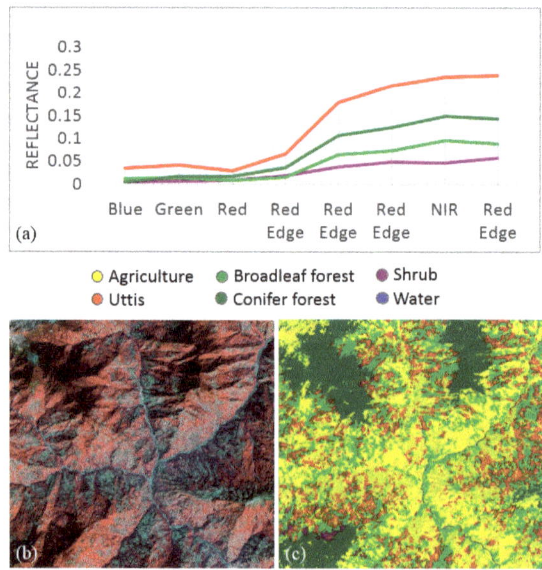

Figure 3. (**a**) Spectral separability of vegetation, (**b**) Sentinel-2A satellite image (False color), (**c**) object-based classification of satellite data.

Uttis cover in the Taplejung district is about 10,735 ha (Figure 4a). Overall, accuracy of the classification was 80% with producer's and user's accuracies at 89% and 84%, respectively (Table 1). The obtained Uttis cover and participatory maps were overplayed to get fine resolution large cardamom farming area (Figure 4b). The overlay produced 5198 ha of existing farming area in the district.

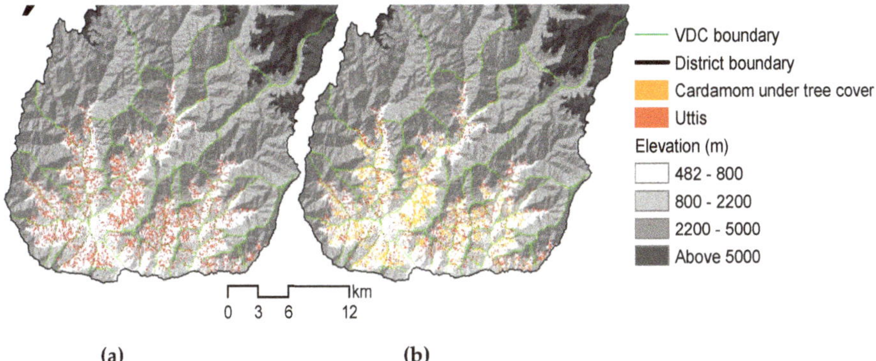

Figure 4. (**a**) Spatial distribution of Uttis tree cover, (**b**) spatial distribution of cardamom under Uttis tree and standalone Uttis tree cover.

Table 1. Contingency matrix for accuracy assessment.

		Observed Vegetation Classes					Grand Total	User's Accuracy
		Agriculture	Conifer	Other Broadleaf	Shrubs	Uttis		
	Agriculture	15	0	0	3	1	19	79
Mapped	Conifer	0	16	2	0	1	19	84
Vegetation	Other Broadleaf	0	2	12	0	2	16	75
Classes	Shrubs	2	0	3	16	0	21	76
	Uttis	2	0	3	1	25	31	81
Grand Total:		19	18	20	20	29	106	
Producer's Accuracy:		79	89	60	80	86		
Overall accuracy: 80%								

The highest cultivation area is found at the elevation range of 1600–1800 m followed by 1800–2000 m, 1400–1600 m, 1200–1400 m, 2000–2200 m, 1000–1200 m, and 800–1000 m, respectively (Figure 5a). Although appropriate elevation for large cardamom is 800–2200 m [38], the farming is also observed below 800 m and above 2200 m. However, there are several species of large cardamom which can be farmed based on altitude [2]. The optimal productivity is possible if large cardamom species is chosen based on elevation.

The largest cultivation area is found in the range of 15°–30° slope while the least is found in the range of 45° and above (Figure 5b). The cultivation area is nearly doubled on the slope of 30°–45° compared to the 0°–15° slope. Several literatures [46,47] stated that the best aspect for cardamom cultivation is North and North-East facing slope. The field data collected for the study showed that cardamom is being cultivated in all aspects. There is no such strict constraint in the selection of aspect for farming. The result shows that the farming area is primarily found in South-West followed by West, North, East, South, North-East, South-East, and North-West (Figure 5c). This indicates that large cardamom can be grown in all kinds of aspects. An evaluation on productivity of large cardamom cultivated at various aspects is essential to identify the best aspect for the crop farming.

Out of 50 VDCs of the Taplejung district, Dhungesanghu, Hangdewa, Hangpang, Phurumbu, Sikecha, and Thukinba are the top six VDCs which have the highest cultivation area (Figure 5d). Among those VDCs, Phurumbu and Hangdewa consist of more than 400 ha of farming area whereas

Dhungesanghu and Hangpang contain approximately 285 ha to 305 ha of cultivation area. Sikecha and Thukimba hold around 200 ha of cardamom field.

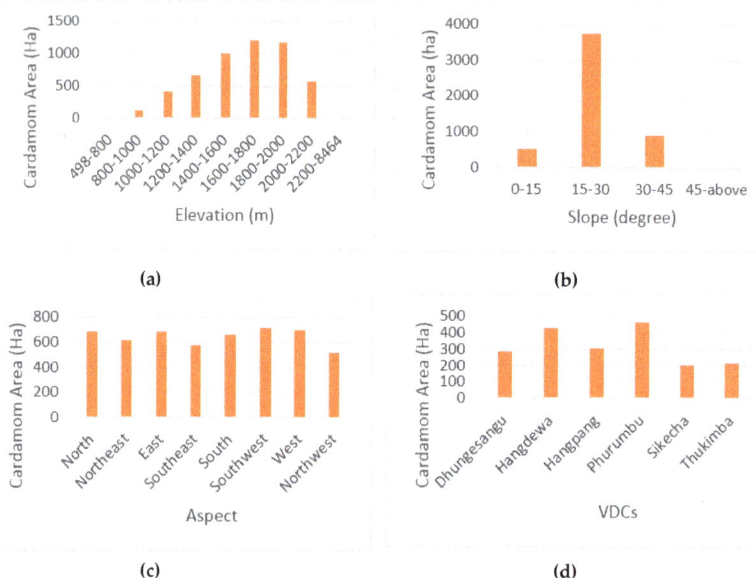

Figure 5. Analysis of large cardamom cultivation area based on elevation, slope, aspect, and VDCs. (**a**). Cardamom cultivation area with elevation; (**b**). Cardamom cultivation area with slope; (**c**). Cardamom cultivation area in various aspects; (**d**). Top 6 VDCs with the highest cultivation area.

3.3. Current Suitable Habitat

The Maxent model provided a comprehensive understanding of the distribution of large cardamom (Figure 6a). Currently, the highly suitable area in the district is about 13,679 ha and the moderately suitable habitat is about 27,778 ha. The most suitable habitat for large cardamom was predicted in the southern part of Taplejung and its distribution is almost continuous. Suitability decreases with an increase in altitude. The maxent-predicted model had high accuracy with an AUC value of 0.941 for training data and 0.934 for test data. Jackknife results showed that Bio6 (minimum temperature of coldest month) among the eight variables considered for the model had the highest predictive power. The Bio5 (maximum temperature of warmest month) is the second most important variable followed by Bio13 (precipitation of wettest month), pH, Uttis cover, and Bio3 (Isothermality) (Figure 6b).

SDM are influenced by several factors, such as data quality [48,49] and decisions taken during the model fitting [50], sample size [51], multicollinearity [52], and selection of independent variables [53]. Despite these, SDM are increasingly used for various purposes, such as species conservation planning [42] and risk analysis [50]. In this study, we have dealt with some of these issues, such as multicollinearity by removing highly correlated variables, selection of important variables for the species, and considered default settings in Maxent as it provided the best model. Maxent performs best among other modelling methods and even performs better with small sample sizes compared to other modelling methods [30,54,55].

3.4. Current Cardamom Cultivation and Habitat Suitability Analysis

Almost 33% of the current farming area is found in the highly suitable class, 44% of the area is found in the moderately suitable class, and nearly 23% of the area is in the marginally suitable class. This indicates that the cultivation has not yet reached its full potential range. Therefore, the farmers need sensitization on potential areas for the farming which would eventually improve productivity (Figure 7).

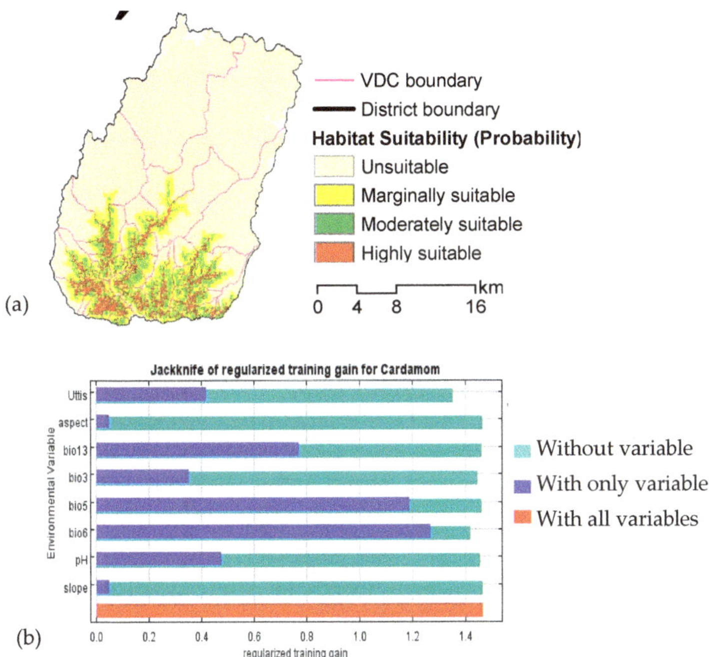

Figure 6. (**a**) Potential distribution of cardamom in the district, (**b**) Relative predictive power of different contributing variables based on the jackknife of regularized training gain in the Maxent model for large cardamom.

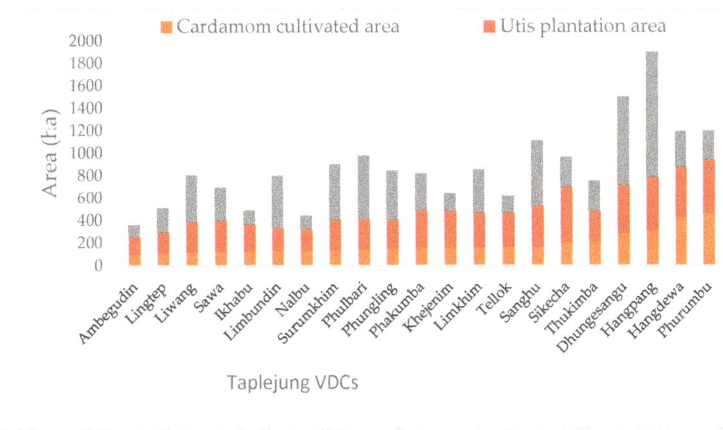

Figure 7. VDC-wise statistics to identify current utilization of the potential habitat.

The Uttis map, existing cultivated area, and current habitat suitability of large cardamom provided the means to look at the gaps and opportunities for cardamom cultivation at the VDC level. Figure 7 demonstrates that the current cultivation area of large cardamom is much less in most of the VDCs, though there are more Uttis areas and highly suitable areas. The cardamom cultivated area and Uttis area in Ambegudin, Hangdewa, Phurumbu, and Thukimba VDCs are much closer, whereas current highly suitable areas in Hangdewa, Ikhabu, Nalbu, Phurumbu, and Tellok VDCs are less than the existing cultivated area. The statistics describing gaps and opportunities for cardamom crop cultivation will support management plans across the administrative area.

Suitability maps help in selecting the areas for better success in cultivating any particular crop. However, productivity of a crop is not solely dependent on suitable site but also heavily depend on many other factors, such as infrastructure and investments (like irrigation, fertilization, and farm management), biological (insect, pests, and disease), and governance (farmers' training on growth management). A map on Large Cardamom suitability classes and crop yield data, of year 2011, across the VDCs of Taplejung District is given in the Supplementary Materials S1 and S2.

3.5. Suitable Habitat under Climate Change Scenerio

The predicted future habitat suitability for large cardamom in the district under RCP2.6-2050 and RCP8.5-2050 is shown in Figure 8. The model predicted under RCP2.6-2050 has accuracy with an AUC value of 0.94 and 0.914 for training and test data, respectively. Under RCP8.5-2050, the AUC values are 0.939 and 0.916 for training and test data, respectively.

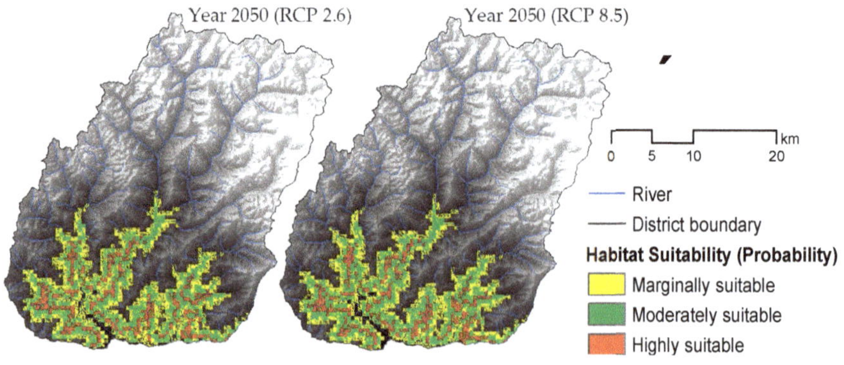

Figure 8. Distribution of large cardamom in Taplejung, Nepal, under future scenarios.

3.6. Projected Changes in the Suitable Habitat Area

Compared to the currently suitable areas, the total areas tend to decrease in both the future scenarios. However, there is no major difference in total areas of the highly suitable class (2%) in RCP 2.6 and a relatively high change in RCP 8.5 (12%). Elevation-wise, we can see there is loss of highly suitable area from the current scenario to both future scenarios, which for the lower is 80% and in higher elevation it is 94% (Table 2). The area tends to maintain a nearly steady-state at elevation range 1000–2000 m, which is ± 0.2%. In conclusion, total potentially suitable areas maintain a nearly steady-state under medium elevation range (1000–2000 m) for all scenarios, whereas the areas are decreasing under future scenarios for the low (500–1000 m) and high elevation (2000–3000 m) (Table 2).

This is the first study to explore current crop cultivation as well as the climate change impacts on the potential distribution of large cardamom crop. Models with AUC values more than 0.75 are considered robust, acceptable, and potentially useful for ecological niche model interpretation [56,57]. Our models obtained AUC values greater than 0.9 which shows our models are satisfactory. The spatial distribution trend under climate change may vary with species causing shifts, contraction, or expansions [58–61]. Our prediction showed suitable habitat for the crop is diminishing under future conditions compared to the current scenario.

Table 2. Areas (ha) of habitat suitability under different scenarios with elevation ranges.

	Suitability	Current Cultivation (2016)	Current				RCP2.6 (2050)				RCP8.5 (2050)			
			Marginal	Moderate	High	Total	Marginal	Moderate	High	Total	Marginal	Moderate	High	Total
Elevation Ranges (m)	500–1000	726	2673	1415	483	4571	2700	667	102	3469	2300	513	107	2920
	1000–2000	17,400	16,699	21,038	12,866	50,602	15,952	22,471	13,374	51,797	15,521	23,318	11,300	50,139
	2000–3000	4603	16,686	5325	331	22,342	11,626	1075	18	12,719	12,266	1051	20	13,336
	Total	22,729	56,058	27,778	13,679	77,515	30,278	24,213	13,494	67,985	30,087	24,882	11,426	66,395

Future changes in suitable habitat area are closely related to temperature and precipitation, which are steadily changing in Nepal over time. It is really important to understand the dynamics of these environmental variables. Stations data showed that the maximum temperature in Nepal increased at the rate of 0.056 °C/year for the period of 1975–2014 [62]. The rate was recorded higher at stations placed at higher altitude, whereas very wet and extremely wet days are diminishing significantly in the northern districts of Nepal [63]. Climate projections for temperature developed for entire Nepal by the Ministry of Forest and Environment (MoFE) [63] showed that the temperature trend is higher in high mountains than other regions in both medium- (2016–2045 for RCP 4.5 and RCP 8.5) and long-term (2036–2065 for RCP 4.5 and RCP 8.5) scenarios. The study found central and western parts of the country will be wetter than the eastern region despite the fact that average annual precipitation change tends to increase by 2.1% in the medium-term period and increase by 7.9% in the long-term period. As large cardamom needs cool and humid climatic conditions [39], most of the current habitat will be negatively affected in future considering the climate projection and data recorded in the past. Particularly, lower altitude would not be favorable for the crop as the temperature tends to reach higher degrees in these elevations. Although annual precipitation change increases in the country, warmer temperature in High Mountain would lead to it becoming drier as it would stimulate more evapotranspiration [58]. These projected drier conditions will negatively influence the habitat suitability of cardamom in the study area. Further, Gudade et al. [64] observed that declining soil nutrition and soil moisture are key drivers in reduction in crop productivity both in India and Nepal.

4. Conclusions

Mapping understory crops and respective habitat conditions is challenging particularly in a complex mountain terrain. The study demonstrated using a combination of expert knowledge-based mapping and species modelling based on bio-climatic factors to understand crop dynamics in a complex mountain terrain. The work applied object-based image analysis on freely available high-resolution satellite data of Sentinel-2A to identify Uttis tree cover in the study area. Use of textures and related indices made it possible to identify a particular tree species within the land cover map. The methodology developed in this study can be used for Uttis cover mapping in other areas of Nepal related to cardamom mapping.

This is the first study to demonstrate potential distribution of large cardamom in the region and the impact of climate change on the distribution. The study showed that crop habitat distribution patterns of understory crop could be modelled using Maxent and coupled with expert knowledge-based mapping to understand the environmental constraints and dynamics. The model under the current scenario showed potentially suitable areas for large cardamom, which has created scope to fill the gaps between farmers' practice in selection of farming land, and planners to develop sustainable land use planning and management. Current cultivation area is shown the highest in the medium suitability class. Farmers can change the farming area on high suitability area. The planning agencies can use the cardamom map and other information produced in this work for land use planning and crop management to optimize the crop sown area. Models under future scenarios showed that the overall suitable habitat is shrinking to mid-hill elevations (1000–2000 m). Therefore, in response to the projected loss of potential distribution, appropriate spatial planning for crop management and livelihood strategies needs to be developed for large cardamom farmers.

Supplementary Materials: The following are available online at http://www.mdpi.com/2073-4395/9/9/481/s1, Supplementary Materials S1: Relative contributions of the environmental variables to the Maxent model; Supplementary Material S2: Map of Suitability map and crop yield.

Author Contributions: S.M. performed the assessment and writing of the manuscript. F.M.Q. designed the study, performed field data collection, data analysis and writing of the manuscript. G.J. performed field data collection and data analysis. M.M. contributed in data analysis and manuscript writing. S.B. contributed to the writing discussion of results.

Acknowledgments: This research was supported by the Rural Livelihoods and Climate Change Adaptation in the Himalayas (Himalica) Programme, which is implemented by the International Center for Integrated Mountain Development (ICIMOD), and is funded by the European Union. The opinions expressed in the paper are those of the authors and do not necessarily reflect the views of their respective institutions.

Conflicts of Interest: The authors declare no conflict of interest.

References

1. Sharma, E.; Sharma, R.; Singh, K.K.; Sharma, G. A Boon for Mountain Populations: Large Cardamom Farming in the Sikkim Himalaya. *Mt. Res. Dev.* **2000**, *20*, 108–111. [CrossRef]
2. Sharma, G.; Joshi, S.R.; Gurung, M.B.; Chilwal, H.C. Package of Practices for Promoting Climate Resilient Cardamom Value Chains in Nepal. In *Training Manual*; International Centre for Integrated Mountain Development (ICIMOD): Kathmandu, Nepal, 2017.
3. Vijayan, A.K. Climate change and its impact of an productivity of large cardamom (*Amomum subutatum* Roxburgh). In Proceedings of the Stakeholders Consultation Workshop on Large Cardamom Development in Nepal, Pakhribas, Nepal, 20 April 2015; pp. 16–27.
4. Lenoir, J.; Gegout, J.C.; Marquet, P.A.; de Ruffray, P.; Brisse, H. A Significant Upward Shift in Plant Species Optimum Elevation during the 20th Century. *Science* **2008**, *320*, 1768–1771. [CrossRef]
5. Bertrand, R.; Lenoir, J.; Piedallu, C.; Riofrío-Dillon, G.; de Ruffray, P.; Vidal, C.; Pierrat, J.C.; Gégout, J.C. Changes in plant community composition lag behind climate warming in lowland forests. *Nature* **2011**, *479*, 517–520. [CrossRef]
6. Rijal, S.P. Impact of climate change on large cardamom-based livelihoods in Panchthar District, Nepal. *Third Pole J. Geogr. Educ.* **2013**, *13*, 33–38. [CrossRef]
7. Justice, C.O.; Becker-Reshef, I. *Developing a Strategy for Global Agricultural Monitoring in the Framework of Group on Earth Observations (GEO) Workshop Report*; FAO: Rome, Italy, 2007.
8. Soares, J.; Williams, M.; Jarvis, I.; Bingfang, W.; Leo, O.; Fabre, P.; Huynh, F.; Kosuth, P.; Lepoutre, D.; Parilar, J.S.; et al. *Strengthening Global Agricultural Monitoring—Improving Sustainable Data for Worldwide Food Security & Commodity Market Transparency.* 2011, pp. 1–16. Available online: http://www.earthobservations.org/documents/cop/ag_gams/201106_g20_global_agricultural_monitoring_initiative.pdf (accessed on 23 July 2019).
9. Asner, G.P.; Knapp, D.E.; Kennedy-Bowdoin, T.; Jones, M.O.; Martin, R.E.; Boardman, J.; Hughes, R.F. Invasive species detection in Hawaiian rainforests using airborne imaging spectroscopy and LiDAR. *Remote Sens. Environ.* **2008**, *112*, 1942–1955. [CrossRef]
10. Tuanmu, M.N.; Viña, A.; Bearer, S.; Xu, W.; Ouyang, Z.; Zhang, H.; Liu, J. Mapping understory vegetation using phenological characteristics derived from remotely sensed data. *Remote Sens. Environ.* **2010**, *114*, 1833–1844. [CrossRef]
11. Hall, R.J.; Peddle, D.R.; Klita, D.L. Mapping conifer understory within boreal mixedwoods from Landsat TM satellite imagery and forest inventory information. *For. Chron.* **2000**, *76*, 887–902. [CrossRef]
12. Johnston, S.E.; Henry, M.C.; Gorchov, D.L. Using Advanced Land Imager (ALI) and Landsat Thematic Mapper (TM) for the Detection of the Invasive Shrub *Lonicera maackii* in Southwestern Ohio Forests. *GIScience Remote Sens.* **2012**, *49*, 450–462. [CrossRef]
13. Resasco, J.; Hale, A.N.; Henry, M.C.; Gorchov, D.L. Detecting an invasive shrub in a deciduous forest understory using latefall Landsat sensor imagery. *Int. J. Remote Sens.* **2007**, *28*, 3739–3745. [CrossRef]
14. Becker, R.H.; Zmijewski, K.A.; Crail, T. Seeing the forest for the invasives: Mapping buckthorn in the Oak Openings. *Biol. Invasions* **2013**, *15*, 315–326. [CrossRef]
15. Leduc, M.B.; Knudby, A.J. Mapping wild leek through the forest canopy using a UAV. *Remote Sens.* **2018**, *10*, 70. [CrossRef]
16. Korpela, I.S. Mapping of understory lichens with airborne discrete-return LiDAR data. *Remote Sens. Environ.* **2008**, *112*, 3891–3897. [CrossRef]
17. Peckham, S.D.; Ahl, D.E.; Gower, S.T. Bryophyte cover estimation in a boreal black spruce forest using airborne lidar and multispectral sensors. *Remote Sens. Environ.* **2009**, *113*, 1127–1132. [CrossRef]
18. Singh, K.K.; Davis, A.J.; Meentemeyer, R.K. Detecting understory plant invasion in urban forests using LiDAR. *Int. J. Appl. Earth Obs. Geoinf.* **2015**, *38*, 267–279. [CrossRef]

19. Vierling, K.T.; Vierling, L.A.; Gould, W.A.; Martinuzzi, S.; Clawges, R.M. Lidar: Shedding new light on habitat characterization and modeling. *Front. Ecol. Environ.* **2008**, *6*, 90–98. [CrossRef]
20. Wang, T.J.; Skidmore, A.K.; Toxopeus, A.G. Improved understorey bamboo cover mapping using a novel hybrid neural network and expert system. *Int. J. Remote Sens.* **2009**, *30*, 965–981. [CrossRef]
21. Phillips, S.B.; Aneja, V.P.; Kang, D.; Arya, S.P. Modelling and analysis of the atmospheric nitrogen deposition in North Carolina. *Int. J. Glob. Environ. Issues* **2006**, *6*, 231–252. [CrossRef]
22. Graham, C.H.; Ferrier, S.; Huettman, F.; Moritz, C.; Peterson, A.T. New developments in museum-based informatics and applications in biodiversity analysis. *Trends Ecol. Evol.* **2004**, *19*, 497–503. [CrossRef]
23. Barik, S.K.; Adhikari, D. Predicting the geographical distribution of an invasive species (*Chromolaena odorata* L. (king) & H.E. Robins) in the Indian subcontinent under climate change scenarios. In *Invasive Alien Plants: An Ecological Appraisal for the Indian Subcontinent*; CABI: Wallingford, UK, 2011.
24. Niphadkar, M.; Nagendra, H.; Tarantino, C.; Adamo, M.; Blonda, P. Comparing Pixel and Object-Based Approaches to Map an Understorey Invasive Shrub in Tropical Mixed Forests. *Front. Plant Sci.* **2017**, *8*, 1–18. [CrossRef]
25. Blaschke, T. Object based image analysis for remote sensing. *ISPRS J. Photogramm. Remote Sens.* **2010**, *65*, 2–16. [CrossRef]
26. Mialhe, F.; Gunnell, Y.; Ignacio, J.A.F.; Delbart, N.; Ogania, J.L.; Henry, S. Monitoring land-use change by combining participatory land-use maps with standard remote sensing techniques: Showcase from a remote forest catchment on Mindanao, Philippines. *Int. J. Appl. Earth Obs. Geoinf.* **2015**, *36*, 69–82. [CrossRef]
27. Robiglio, V.; Mala, W.A. Integrating local and expert knowledge using participatory mapping and GIS to implement intergrated forest management options in Akok, Cameroon. *For. Chron.* **2005**, *81*, 392–397. [CrossRef]
28. Lauer, M.; Aswani, S. Integrating indigenous ecological knowledge and multi-spectral image classification for marine habitat mapping in Oceania. *Ocean Coast. Manag.* **2008**, *51*, 495–504. [CrossRef]
29. Elith, J.; Graham, C.H.; Anderson, R.P.; Dudík, M.; Ferrier, S.; Guisan, A.; Hijmans, R.J.; Huettmann, F.; Leathwick, J.R.; Lehmann, A.; et al. Novel methods improve prediction of species' distributions from occurrence data. *Ecography* **2006**, *29*, 129–151. [CrossRef]
30. Pearson, R.G.; Raxworthy, C.J.; Nakamura, M.; Townsend Peterson, A. Predicting species distributions from small numbers of occurrence records: A test case using cryptic geckos in Madagascar. *J. Biogeogr.* **2007**, *34*, 102–117. [CrossRef]
31. Kumar, S.; Stohlgren, T.J. Maxent modeling for predicting suitable habitat for threatened and endangered tree *Canacomyrica monticola* in New Caledonia. *J. Ecol. Nat. Environ.* **2009**, *1*, 094–098.
32. Shrestha, J.; Prasai, H.K.; Timilsina, K.P.; Shrestha, K.P.; Pokhrel, D.; Paudel, K.; Yadav, M. *Large Cardamom in Nepal: Production Practice and Economics, Processing and Marketing*; National Commercial Agriculture Research Program, Nepal Agriculture Research Council: Pakhribas, Dhankuta, Nepal, 2018.
33. McCall, M.K.; Minang, P.A. Assessing Participatory GIS for Community-Based Natural Resource Management: Claiming Community Forests in Cameroon. *Geogr. J.* **2005**, *171*, 340–356. [CrossRef]
34. Drusch, M.; Del Bello, U.; Carlier, S.; Colin, O.; Fernandez, V.; Gascon, F.; Hoersch, B.; Isola, C.; Laberinti, P.; Martimort, P.; et al. Sentinel-2: ESA's Optical High-Resolution Mission for GMES Operational Services. *Remote Sens. Environ.* **2012**, *120*, 25–36. [CrossRef]
35. Blaschke, T.; Lang, S.; Lorup, E.; Strobl, J.; Zeil, P. Object-Oriented Image Processing in an Integrated GIS/Remote Sensing Environment and Perspectives for Environmental Applications. *Umweltinf. Plan. Polit. Öffentl.keit Environ. Inf. Plan. Polit. Public* **2000**, *2*, 555–570.
36. Hay, G.J.; Blaschke, T.; Marceau, D.J.; Bouchard, A. A comparison of three image-object methods for the multiscale analysis of landscape structure. *ISPRS J. Photogramm. Remote Sens.* **2003**, *57*, 327–345. [CrossRef]
37. Belgiu, M.; Drăguţ, L. Comparing supervised and unsupervised multiresolution segmentation approaches for extracting buildings from very high resolution imagery. *ISPRS J. Photogramm. Remote Sens.* **2014**, *96*, 67–75. [CrossRef]
38. National Spice Crop Development Programme. *Annual Report of National Spice Crop Development Programme*; National Spice Crop Development Programme, Ministry of Agriculture Development, Government of Nepal: Khumaltar, Kathmandu, Nepal, 2009.
39. Uddin, K.; Shrestha, H.L.; Murthy, M.S.R.; Bajracharya, B.; Shrestha, B.; Gilani, H.; Pradhan, S.; Dangol, B. Development of 2010 national land cover database for the Nepal. *J. Environ. Manag.* **2015**, *148*, 82–90. [CrossRef]

40. Tadono, T.; Ishida, H.; Oda, F.; Naito, S.; Minakawa, K.; Iwamoto, H. Precise Global DEM Generation by ALOS PRISM. *ISPRS Ann. Photogramm. Remote Sens. Spat. Inf. Sci.* **2014**, *II–4*, 71–76. [CrossRef]

41. Micheal, H. Graham Confronting Multicollinearity in Ecological Multiple Regression. *Ecology* **2003**, *84*, 2809–2815.

42. Yang, X.Q.; Kushwaha, S.P.S.; Saran, S.; Xu, J.; Roy, P.S. Maxent modeling for predicting the potential distribution of medicinal plant, *Justicia adhatoda* L. in Lesser Himalayan foothills. *Ecol. Eng.* **2013**, *51*, 83–87. [CrossRef]

43. Naimi, B.; Hamm, N.A.S.; Groen, T.A.; Skidmore, A.K.; Toxopeus, A.G. Where is Positional Uncertainty a Problem for Species Distribution Modelling? Available online: https://onlinelibrary.wiley.com/doi/abs/10.1111/j.1600-0587.2013.00205.x (accessed on 11 July 2019).

44. Maharjan, S. Estimation and Mapping Above Ground Woody Carbon Stocks Using Lidar Data And Digital Camera Imagery in The Hilly Forest of Gorkha, Nepal. Master's Thesis, University of Twente, Enschede, The Netherlands, 2012.

45. Karna, Y.K.; Hussin, Y.A.; Gilani, H.; Bronsveld, M.C.; Murthy, M.S.R.; Qamer, F.M.; Karky, B.S.; Bhattarai, T.; Aigong, X.; Baniya, C.B. Integration of WorldView-2 and airborne LiDAR data for tree species level carbon stock mapping in Kayar Khola watershed, Nepal. *Int. J. Appl. Earth Obs. Geoinf.* **2015**, *38*, 280–291. [CrossRef]

46. Baniya, N.; Böehme, M.; Baniya, S. Physical land suitability assessment for the large cardamom amomum subulatum roxb. Cultivation in hills of Kathmandu valley. *Chin. J. Popul. Resour. Environ.* **2009**, *7*, 59–63. [CrossRef]

47. ECDF. *Alaichi Kheti and Prashodhan Hate Pustika*; Environment Conservation and Development Forum: Taplejung, Mechi Zone, Nepal, 2008.

48. Coro, G.; Magliozzi, C.; Ellenbroek, A.; Pagano, P. Improving data quality to build a robust distribution model for *Architeuthis dux*. *Ecol. Model.* **2015**, *305*, 29–39. [CrossRef]

49. Aubry, K.B.; Raley, C.M.; McKelvey, K.S. The importance of data quality for generating reliable distribution models for rare, elusive, and cryptic species. *PLoS ONE* **2017**, *12*, 1–17. [CrossRef]

50. Kumar, S.; Graham, J.; West, A.M.; Evangelista, P.H. Using district-level occurrences in MaxEnt for predicting the invasion potential of an exotic insect pest in India. *Comput. Electron. Agric.* **2014**, *103*, 55–62. [CrossRef]

51. Stockwell, D.R.B.; Peterson, A.T. Effects of sample size on accuracy of species distribution models. *Ecol. Model.* **2002**, *148*, 1–13. [CrossRef]

52. Dormann, C.F.; Elith, J.; Bacher, S.; Buchmann, C.; Carl, G.; Carré, G.; Marquéz, J.R.G.; Gruber, B.; Lafourcade, B.; Leitão, P.J.; et al. Collinearity: A review of methods to deal with it and a simulation study evaluating their performance. *Ecography* **2013**, *36*, 027–046. [CrossRef]

53. Evangelista, P.H.; Kumar, S.; Stohlgren, T.J.; Jarnevich, C.S.; Crall, A.W.; Norman, J.B.; Barnett, D.T. Modelling invasion for a habitat generalist and a specialist plant species. *Divers. Distrib.* **2008**, *14*, 808–817. [CrossRef]

54. Elith, J.; Graham, C.H.; Anderson, R.P.; Dudík, M.; Guisan, A.; Hijmans, R.J.; Huettmann, F.; Leathwick, J.R.; Lehmann, A.; Li, J.; et al. Novel Methods Improve Prediction of Species' Distributions from OccurrenceData. *Ecography.* **2006**, *29*, 129–151. [CrossRef]

55. Ortega-Huerta, M.A.; Peterson, A.T. Modeling ecological niches and predicting geographic distributions: A test of six presence-only methods. *Rev. Mex. De Biodivers.* **2008**, *79*, 205–216.

56. Pearce, J.; Ferrier, S. An evaluation of alternative algorithms for fitting species distribution models using logistic regression. *Ecol. Model.* **2000**, *128*, 127–147. [CrossRef]

57. Elith, J. *Quantitative Methods for Modeling Species Habitat: Comparative Performance and An Application to Australian Plants*; Springer: New York, NY, USA, 2000.

58. Li, R.; Xu, M.; Wong, M.H.G.; Qiu, S.; Sheng, Q.; Li, X.; Song, Z. Climate change-induced decline in bamboo habitats and species diversity: Implications for giant panda conservation. *Divers. Distrib.* **2014**, *21*, 379–391. [CrossRef]

59. Lamsal, P.; Kumar, L.; Aryal, A.; Atreya, K. Invasive alien plant species dynamics in the Himalayan region under climate change. *Ambio* **2018**, *47*, 697–710. [CrossRef]

60. Maclean, I.M.D.; Wilson, R.J. Recent ecological responses to climate change support predictions of high extinction risk. *Proc. Natl. Acad. Sci. USA* **2011**, *108*, 12337–12342. [CrossRef]

61. Ponce-Reyes, R.; Nicholson, E.; Baxter, P.W.J.; Fuller, R.A.; Possingham, H. Extinction risk in cloud forest fragments under climate change and habitat loss. *Divers. Distrib.* **2013**, *19*, 518–529. [CrossRef]

62. *DHM Observed Climate Trend Analysis in the Districts and Physiographic Zones of Nepal (1971–2014)*; Government of Nepal Ministry of Population and Environment Department of Hydrology and Meteorology: Kathmandu, Nepal, 2017.
63. *MoFE Climate Change Scenarios for Nepal for National Adaptation Plan (NAP)*; Ministry of Forests and Environment: Kathmandu, Nepal, 2019.
64. Gudade, B.A.; Harsha, K.N.; Vijayan, A.K.; Chhetri, P.; Deka, T.N.; Babu, S.; Singh, R. Effect of soil application of Zn, Mn and Mg on growth and nutrient content of large cardamom, Amomum subulatum Roxb at Sikkim. *Int. J. Farm Sci.* **2015**, *5*, 51–55.

Article

A Precision Agriculture Approach for Durum Wheat Yield Assessment Using Remote Sensing Data and Yield Mapping

Piero Toscano [1], Annamaria Castrignanò [2], Salvatore Filippo Di Gennaro [1,*], Alessandro Vittorio Vonella [2], Domenico Ventrella [2] and Alessandro Matese [1]

[1] Institute of BioEconomy (IBE), National Research Council (CNR), Via Caproni 8, 50145 Florence, Italy
[2] Council for Agricultural Research and Economics, Research Centre for Agriculture and Environment (CREA-AA), Via Celso Ulpiani 5, 70125 Bari, Italy
* Correspondence: salvatorefilippo.digennaro@cnr.it; Tel.: +39-055-3033711

Received: 19 July 2019; Accepted: 3 August 2019; Published: 8 August 2019

Abstract: The availability of big data in agriculture, enhanced by free remote sensing data and on-board sensor-based data, provides an opportunity to understand within-field and year-to-year variability and promote precision farming practices for site-specific management. This paper explores the performance in durum wheat yield estimation using different technologies and data processing methods. A state-of-the-art data cleaning technique has been applied to data from a yield monitoring system, giving a good agreement between yield monitoring data and hand sampled data. The potential use of Sentinel-2 and Landsat-8 images in precision agriculture for within-field production variability is then assessed, and the optimal time for remote sensing to relate to durum wheat yield is also explored. Comparison of the Normalized Difference Vegetation Index(NDVI) with yield monitoring data reveals significant and highly positive linear relationships (r ranging from 0.54 to 0.74) explaining most within-field variability for all the images acquired between March and April. Remote sensing data analyzed with these methods could be used to assess durum wheat yield and above all to depict spatial variability in order to adopt site-specific management and improve productivity, save time and provide a potential alternative to traditional farming practices.

Keywords: yield mapping; remote sensing; durum wheat; precision agriculture

---

1. Introduction

Durum wheat (*Triticum durum*, Desf.), although it represents only 8% of global wheat production, is one of the most common cereal crops in the Mediterranean basin, traditionally grown under rainfed conditions using conventional tillage [1–3]. Climate variability, price volatility and socio-economic factors are the main sources of uncertainty and concern for farmers in durum wheat cultivation [4,5]. For climate variability, it was shown how, in rainfed conditions, these affected both the quality and quantity of durum wheat production [6]. For other crops, evidence was given by Bowman and Zilberman [7] of how both price volatility and socio-economic factors might influence the agronomic techniques adopted. In light of this, it is increasingly urgent to provide information to optimize crop management and estimate crop yields before harvest for a sustainable agricultural income and ensuring food security. Precision agriculture (PA) has been used for ≥25 years to optimize the use of farm inputs such as fertilizers and herbicides [8,9], and thus maximize profit and minimize negative environmental impacts [10] by addressing spatial variability.

Yield mapping is one of the most widely-used precision agriculture techniques [11–13]. Most of these datasets are characterized by a non-normal distribution due to the presence of errors and outliers and can be misleading if used for decision making processes. Over the last 25 years, several studies have analyzed the sources of errors that cause this non-normality and proposed different processing techniques to reduce their effect [11,14,15]. However, among the studies highlighting accuracy issues associated with the use of the yield monitoring systems, only one compared yield monitoring with hand sampled data [16].

During the same period many studies have been published using satellite imagery to estimate crop parameters and yields [17–19], many of these using empirical relationships between yields and various vegetation indices (VIs) with limited applicability to different areas or years [9], especially in the recent era of prolific satellite data availability [20].

In the last years many studies have focused on the use of medium resolution satellite data (10–100 m) for yield estimation at broader spatial resolution (local, regional, country scales) even for long-term yield series analysis [20–26]. While studies are conducted by the use of very high resolution imagery [27–30] to identify within-field variability of crop growth and yield and for the definition of management zones, few [31] have used Sentinel-2 to provide an insight into field productivity variation for better future management [32–34]. The objectives of this study were to:

- Evaluate the correctness of yield monitoring maps comparing them with hand sampled yield data;
- Evaluate the ability of the most commonly used VI (NDVI) calculated from Landsat-8 and Sentinel-2 satellite platforms to understand within-field variability;
- Understand the optimal time for NDVI acquisition for better yield evaluation;
- Evaluate the relations between NDVI and yield for four durum wheat crop seasons with different climatic conditions and yield performance.

For the last two points, given the scarcity of satellite images, a durum wheat simulation model [5] was used to reconstruct the crop growth variables and analyze in detail the differences that emerged in the yield-NDVI correlation for the four crop seasons.

2. Materials and Methods

2.1. Study Site and Field Trial

The research was performed at the Menichella Experimental Farm of CREA-AA (Council for Agricultural Research and Economics—Research Centre for Agriculture and Environment), located in the Foggia countryside (Southern Italy, 41°27′05.9″ N, 15°30′43.6″ E; 88 m a.s.l.), within the study area of the JECAM site (http://jecam.org/studysite/italy-apulian-tavoliere/), during the 2013–2014, 2014–2015, 2015–2016 and 2016–2017 crop seasons. This study was conducted on a 5 ha field cropped with rainfed durum wheat (*Triticum durum*, Desf., cv Claudio) under conventional management and continuous cultivation.

The field is in a flat area called 'Apulian Tavoliere' and the soil is silty-clay Vertisol of alluvial origin classified as Fine Mesic Typic Cromoxerert by Soil Taxonomy USDA [35].

The soil in the upper 60 cm layer has a good availability of total nitrogen (0.12 g 100 g^{-1}), organic matter (2.07 g 100 g^{-1}) and 41 mg kg^{-1} of available phosphorus (P$_2$O$_5$). In summer 4–5 cm wide cracks frequently appear from the surface to about 50 cm depth.

The climate is classified as Mediterranean subtropical with a thermic soil temperature regime. Rainfall, unevenly distributed throughout the seasons and with a long-term annual average of 550 mm, is mostly concentrated in the winter months, while the dry period is from May to September [36]. Daily weather parameters (air temperature, relative humidity, global solar radiation, rainfall and wind speed), were recorded at the agro-meteorological station of the CNR-IBE weather station network (Foggia, 41°30′00.4″ N, 15°30′46.4″ E, 69 m a.s.l.). The field has been cultivated with a common agronomic management, applying 36 kg ha^{-1} of N as diammonium phosphate (18–46) before sowing and 68.4 kg ha^{-1} as ammonium nitrate (34.2) as top dressing at the end of tillering stage.

Sowing date varied between November and December due to the weather conditions (12 December 2013, 20 November 2014, 19 November 2015 and 29 November 2016). In each cropping season, the sowing density was of 350 germinable seeds/m^2 with 15 cm row spacing.

2.2. Hand Yield Samplings and Yield Map Monitoring

At maturity stage of each season aboveground biomass was collected over 1 m^2 areas in proximity to the 104 sampling points at the nodes of a 20 × 20 m cell-grid (Figure 1). The points were georeferenced in UTM coordinates WGS 84 using a TOPCON GPS, differentially corrected with an accuracy of less than one meter. All measurements were repeated at the same points over the years.

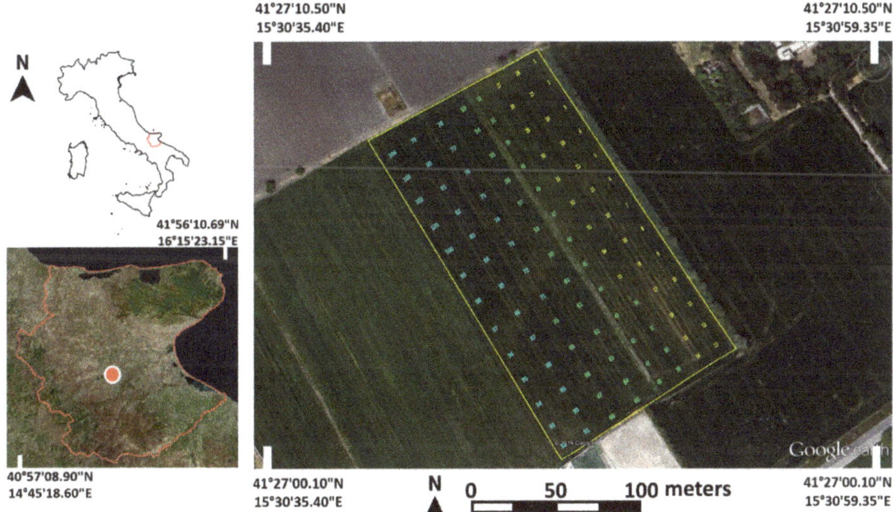

Figure 1. Location of study site and 104 sampling points.

At harvesting in 14 July 2014, 29 June 2015 and 12 July 2016, yield data were recorded by external services provider with a John Deere T670i combine (Deere & Company, Moline, USA) equipped with a yield monitor system (grain mass flow and moisture sensors). The data were recorded every second, which produced a support (footprint) of 6 × 1 m^2 depending on the forward speed of the machine.

For 2017, the last year of analysis, a yield monitoring map was unavailable. The data were measured only at the 104 sampling points.

2.3. Satellite Data

Remote sensing images acquired by the Operational Land Imager (OLI) instrument aboard the Landsat-8 satellite and by the Multi-Spectral Instrument (MSI) aboard the Sentinel-2A satellite were used in the study. Landsat-8/OLI captures images of the earth's surface in nine spectral bands at 30 m spatial resolution (15 m for panchromatic band) while Sentinel-2A/MSI captures images in 13 spectral bands at 10 m, 20 m and 60 m spatial resolution. After cloud and shadow screening, a total of 11 Landsat-8 and five Sentinel-2A (Table 1 images of the study area from 1 March 2013 to 1 June 2017 were selected.

Table 1. Acquired dates of Landsat-8 and Sentinel-2 (2014–2017).

	LANDSAT-8	**SENTINEL-2**
2014	19 March, 20 April	NA
2015	14 April, 30 April	NA
2016	13 March, 9 April, 18 May, 27 May	23 May
2017	2 March, 12 April, 30 May	9 March, 29 March, 8 April, 18 May

The Landsat-8/OLI images were downloaded using USGS (earthexplore.com) that provides data corrected from atmospheric effects.

Sentinel-2A/MSI images were atmospherically corrected for surface reflectance using the European Space Agency's (ESA) Sen2Cor algorithm (http://step.esa.int/main/third-party-plugins-2/sen2cor), which processes ESA's Level-1C top-of-atmosphere reflectance to atmospherically-corrected bottom-of-atmosphere (BoA) reflectance (Level-2A).

Lastly, NDVI [37] was calculated for Landsat-8 images using band 5 (NIR) and band 4 (RED), and for Sentinel-2A images using band 8 (NIR) and band 4 (RED) according to the formula:

$$NDVI = (NIR - RED)/(NIR + RED) \tag{1}$$

Providing two estimates with different support: 30×30 m^2 for Landsat-8 and 10×10 m^2 for Sentinel-2A.

2.4. Data Analysis

Yield map data were firstly normalized to 13% grain moisture content (hereinafter referred to as raw data) and then processed following the Vega et al. [13] protocol. The geographic coordinates of each dataset were converted into UTM Cartesian coordinates, specifying the zone (33, north) and the ellipsoid (WGS84). The 3 years maps in shapefile format (SHP) were pre-processed following a workflow using GeoDa software [38] for Moran index calculation for outliers identification, QGIS [39] for vegetation indices (VIs) calculation, Vesper software [40] for geostatistical interpolation, Matlab [41] for data statistical analysis.

Yield monitoring data underwent a pre-processing procedure in order to automatically identify and delete incorrect values through the two following steps suggested by Vega et al. [13].

Step 1: A threshold was applied by removing the yield values of less than 0.1 t/ha and then yield data points up to 10 m from the edge were removed in order to avoid edge effects. Lastly, yield data out of mean ±3 SD were automatically detected and deleted. This filter was used to prevent changes in inflation as a result of an incorrect estimate of very low data.

Step 2: Moran's local index of spatial autocorrelation and Moran's plot were applied to detect spatial outliers [42,43].

Lastly, the yield map for each year and dataset was assessed separately (raw data, Step1 and Step1 + Step2) by means of ordinary kriging, evaluating spatial variability using a semivariogram of the variable (yield monitoring).

Vesper software was used and an exponential model was fitted to the experimental variogram and model parameters: Nugget (micro-scale variation or measurement error), sill (asymptotic value approximately corresponding to sample variance) and effective range (distance at which 95% sill is reached) were estimated.

Yield values from the prediction map based on raw yield data and Step1 + Step2 yield data were then extracted in the neighborhood (3 m radius) of 104 yield sampled grid points and compared with them.

Step1 + Step2 yield monitoring data were interpolated using block kriging over a block of 10×10 m^2 or 30×30 m^2 to assess the spatial relationship between yield and the two types of remote sensing data (Sentinel-2A and Landsat-8, respectively) and to report on the optimal timing at which spectral measurements should be taken in durum wheat to maximize the correlation with yield (2014, 2015, 2016). The same procedure was adopted for hand yield samplings (block of 10×10 m^2 or 30×30 m^2) to assess the spatial relationship between yield and remote sensing data for 2017.

The Pearson coefficient, which is an index that measures the degree of correlation between linearly related variables and ranges between -1 and $+1$, was used to assess the linear relationship between yield monitoring data and yield sampling data, and between the NDVI and yield monitoring data.

For each regression analysis the correlation coefficient, regression coefficients (intercept and slope) and their corresponding probability levels were estimated, to test the statistical significance (null hypothesis equal to zero).

The performance of the protocol for automating error removal from yield monitoring data was evaluated using the root-mean-square error (RMSE).

2.5. Modelling

To better understand the correlations between remote sensing data and yield and go into detail about crop development and spectral signature, we used the Delphi crop growth model to perform field-scale simulations. The Delphi model was chosen due to the recent validations of its ability to simulate crop growth, yield and product quality conducted in the same study area [1,5]. The Delphi model is based on a FORTRAN-based mechanistic model [44] calibrated for durum wheat in Mediterranean conditions. Plant transpiration and soil evaporation, water and nitrogen soil-plant cycle are incorporated in the model. Input weather data at daily time scale are: Air temperature (maximum, minimum and average), global shortwave radiation, rainfall, wind speed (average) and relative humidity (average). Input data of the main physiological parameters of the durum wheat cultivar, sowing date and number of seeds/m^2, the soil hydrological profile, soil total nitrogen content profile, agronomic data on quality and quantity of nitrogen and roots growth data are also required.

Due to the strong correlation between leaf area index (LAI) and aboveground dry biomass [45], because aboveground dry biomass of plants generally determines LAI, the Delphi model was implemented to calculate LAI for each crop season. The model also predicted the heading, anthesis, maturity dates and length of time between these phases.

These pieces of information were used for detailed analysis of interannual variability and to better understand the different relationships between yield and NDVI over the years.

The weather data input to perform the Delphi simulation were acquired by the weather station located at Foggia, 41°30′00.4″ N, 15°30′46.4″ E, 69 m a.s.l., while sowing date, number of seeds/m^2 and nitrogen fertilization application data were set according to the data reported in Section 2.1. No changes were made to the Delphi model, as it had already been calibrated, validated and tested over 11 crop seasons for this region [5].

3. Results and Discussion

3.1. Yield Map and Yield Sample

A yield map is the basis for understanding yield variability within a field, analyzing its causes and improving management to increase profit [46].

A number of errors may be associated with common yield data collection: The yield monitor may not shut off at the field end and will register 0 value until harvestable crop again moves into the combine; the combine grain-flow system may plug temporarily, especially if the crop has lodged or weeds interfere with continuous grain flow; a time lag can occur between the time the crop is cut and the time its yield is measured in the grain flow [14,15,47]. Researchers have reported that 10% to 50% of observations reveal measurement errors [13,48].

The raw yield data of all three crop seasons used in this study showed high positive skewness coefficient (Table 2 and Figure 2). However, after Step1 and Step1 + Step2, the yield probability distributions were practically symmetric and the statistics were not biased by the presence of atypical data. After removing statistical outliers, the variable distribution tended to be more symmetric, without affecting the spatial structuring.

Table 2. Statistical features of yield monitoring datasets for uncleaned and cleaned yield data. Yield monitoring vs. yield sampling based on interpolated data.

Year	Number of Yield Monitoring Data			Skewness			Yield Monitoring vs. Yield Sampling	
	RAW	Step1	Step2	RAW	Step1	Step2	RAW	Step1 + Step2
2014	6369	4703 −26.2%	4641 −1.32%	17.72	2.72	2.71	$r = 0.11$ p-value = 0.25 RMSE = 1.14 t/ha	$r = 0.40$ p-value < 0.0001 RMSE = 1.05 t/ha
2015	6351	4737 −25.4%	4648 −1.88%	21.58	0.65	0.44	$r = 0.37$ p-value < 0.0001 RMSE = 0.68 t/ha	$r = 0.50$ p-value < 0.0001 RMSE = 0.59 t/ha
2016	6699	5084 −24.1%	4967 −2.30%	23.94	0.39	0.38	$r = 0.43$ p-value < 0.0001 RMSE = 0.84 t/ha	$r = 0.49$ p-value < 0.0001 RMSE = 0.82 t/ha

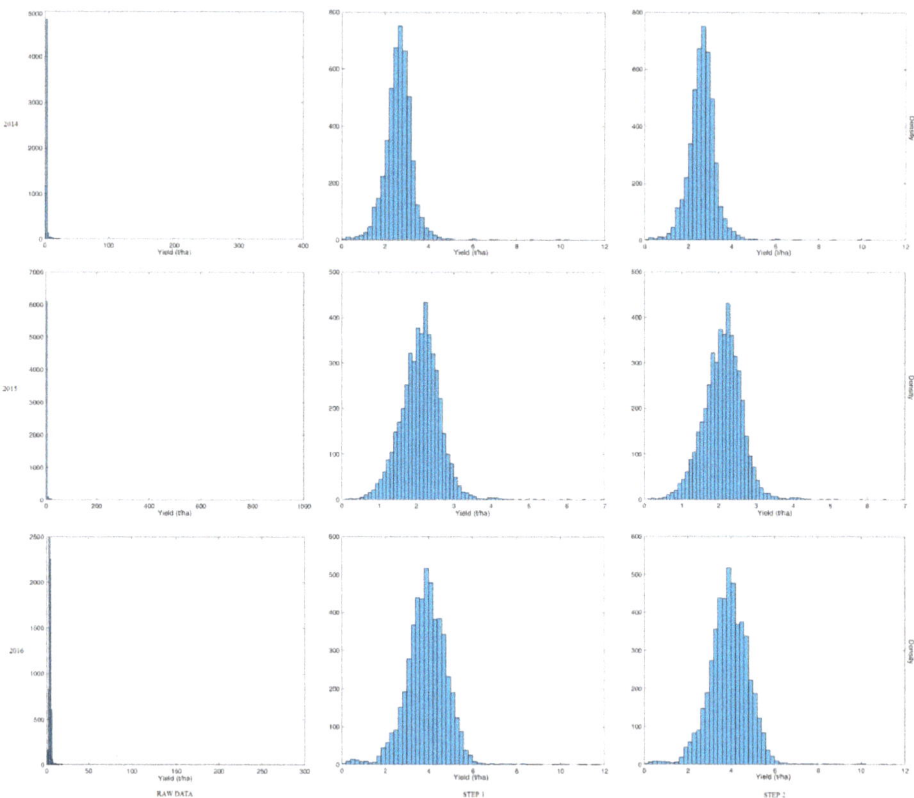

Figure 2. Distribution of yield monitoring datasets for uncleaned and cleaned yield data (data not interpolated).

In fact, after Step1 + Step2 the skewness had values close to zero for 2015 and 2016 crop seasons (Table 2), while for 2014 the final skewness of 2.71 was due to a longer tail on the right side of the data distribution. After Step1 between 24.1% and 26.2% of points were removed, whilst after Step2 a further 1.3% to 2.3% of the dataset was removed. Following the Step1 + Step2 protocol, the percentage of points removed was from 25.9% to 27.1%, close to the range previously reported in the literature [13,47–50].

Figure 3 presents a visual analysis of the location of data removed from each year after both Step1 and Step2: First of all, the highest quantity of removed points came from the filtering of edges, secondly all the data points with overlapping coordinates and lastly, data points identified as outliers through local Moran's index of spatial autocorrelation (Step2).

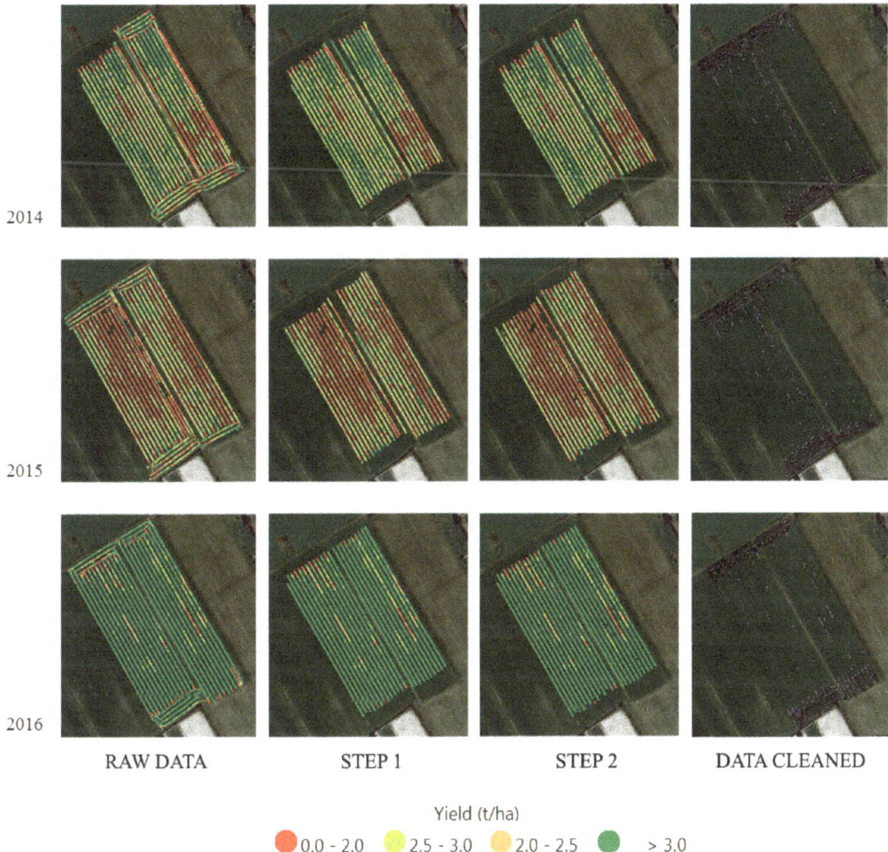

Figure 3. Maps showing yield monitoring datasets: Raw data (left), Step1 and Step2 data (center), data cleaned (right). Raw and Steps 1–2 maps are colored according to quartiles of yield distributions (data not interpolated).

Certainly, the cleaning protocol did not affect the main patterns present in the raw data and allowed both comparison with the sampled yield data and, after interpolation, comparison with the data observed by Landsat-8 (at 30 m of resolution) and Sentinel-2 (at 10 m).

For all three crop seasons, the correlation, its significance and RMSE improved in passing from raw yield monitoring vs. sampled yield to the comparison Step1 + Step2 yield monitoring vs. sampled yield (Table 2). The best improvement was achieved for 2014, where from being non-significant, we found a significant correlation and with a reduction of the RMSE which, however, remained the highest compared to other years. For both 2015 and 2016, the initial correlations between raw data and sampled data were significant and with low RMSE. In any case, the data cleaning procedures improved the performances in terms of both correlation and RMSE.

The different behavior of the 2014 season compared to the other two may be due to intense weeding during the first year of the trial, which caused a large within-field variability of yield.

Given the nature of the comparison between two sets of data that do not refer strictly to the same harvested plants, the worst results obtained for 2014 seem to be clearly linked to the greater variability of within-field yield, unlike crop seasons with uniformly low within-field yield (2015) or uniformly high yield (2016) (Table 3). This corresponds well with Arslan and Colvin [51], who reported that high accuracy cannot be achieved by spot measurements; however, the overall yield trend can be determined. The correlations between spatial data are strongly scale-dependent [52] and of course depend on the coincidence of location between the sampled and monitored data, that in our study is not possible since the hand yield sampling was destructive (before yield mapping).

Table 3. Comparison between yield (t/ha) sampled and monitoring data (mean, max, min and std) (2014–2017).

	2014		2015		2016		2017	
	Sample (t/ha)	Monitor (t/ha)	Sample (t/ha)	Monitor (t/ha)	Sample (t/ha)	Monitor (t/ha)	Sample (t/ha)	Monitor (t/ha)
Mean	3.07	2.65	2.31	2.11	3.88	3.90	5.00	N/A
Min	0.42	0.18	0.87	0.15	2.15	0.12	3.23	N/A
Max	5.69	10.48	3.64	6.90	6.01	11.62	7.43	N/A
Std	1.01	1.12	0.57	0.53	0.71	0.91	0.94	N/A

The correlation coefficients found for all three crop seasons (ranging from 0.40 to 0.50) were in general lower but similar to the findings of Ingeli et al. [16]. The latter is the only published paper to have compared two sources of yield data, the hand sampled data as independent variable and yield monitoring data as dependent variable. For five different crop seasons, the authors found correlations ranging between 0.3 to 0.9 but reporting on a small number of hand sampled data (18 + 3 replications) spread over a larger field area (16 ha) unlike our case study with 104 samples in a smaller area (5 ha).

3.2. Yield and NDVI

For the whole period, few Landsat-8 and Sentinel-2 images were used because most of those acquired were useless due to the presence of clouds. For the first two years (2014 and 2015) the images are limited to two in a fairly short time window and in any case always before crop heading stage. In 2016 there are four useful images spread over a much wider period (March–May), while three images are available in 2017 for the same period.

The scientific and operational life of Sentinel-2 started in July 2015, so the useful passages only relate to 2016 and 2017. In 2016 it is possible to use only one image and it was taken at the end of May. In 2017 there are four useful images in a time window similar to that of the Landsat-8 images for 2017 (Table 3).

A comparison of the NDVI with yield monitoring values (Step1 + Step2) (for 2014–2015–2016, Figures 4–7) reveals significant positive linear relationships (r ranging from 0.54 to 0.74) explaining most of the within-field variability in 2014 with the image acquired in April ($R^2 = 0.55$) and in 2016 with the image acquired in March ($R^2 = 0.55$). In all other cases, although the correlations are significant, R^2 are lower than 0.5.

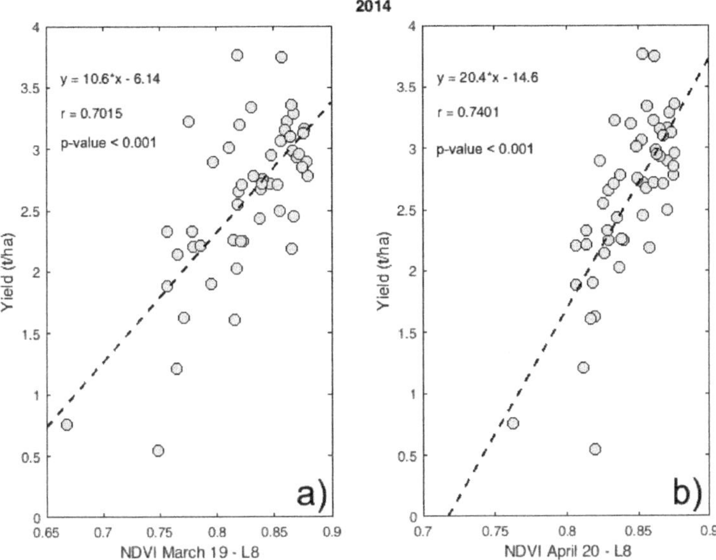

Figure 4. Relationship between observed yield (yield monitor) and NDVI (Landsat-8) for 2013–2014 crop season in (a) 19 March and (b) 20 April.

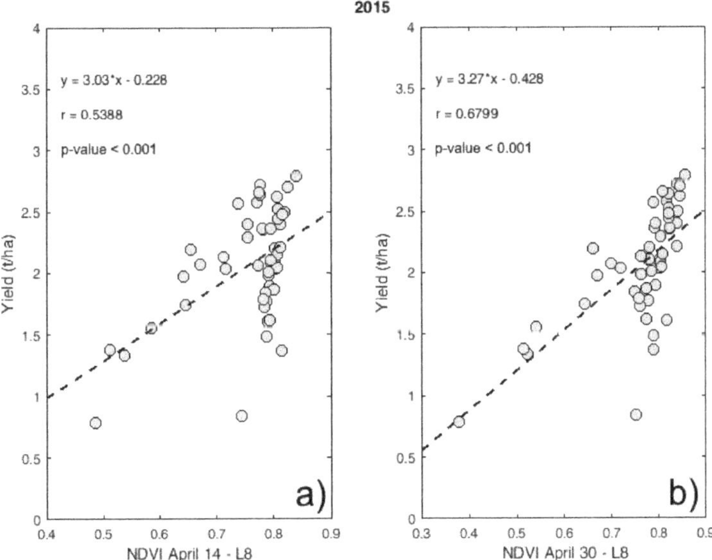

Figure 5. Relationship between observed yield (yield monitor) and NDVI (Landsat-8) for 2014–2015 crop season in (a) 14 April and (b) 30 April.

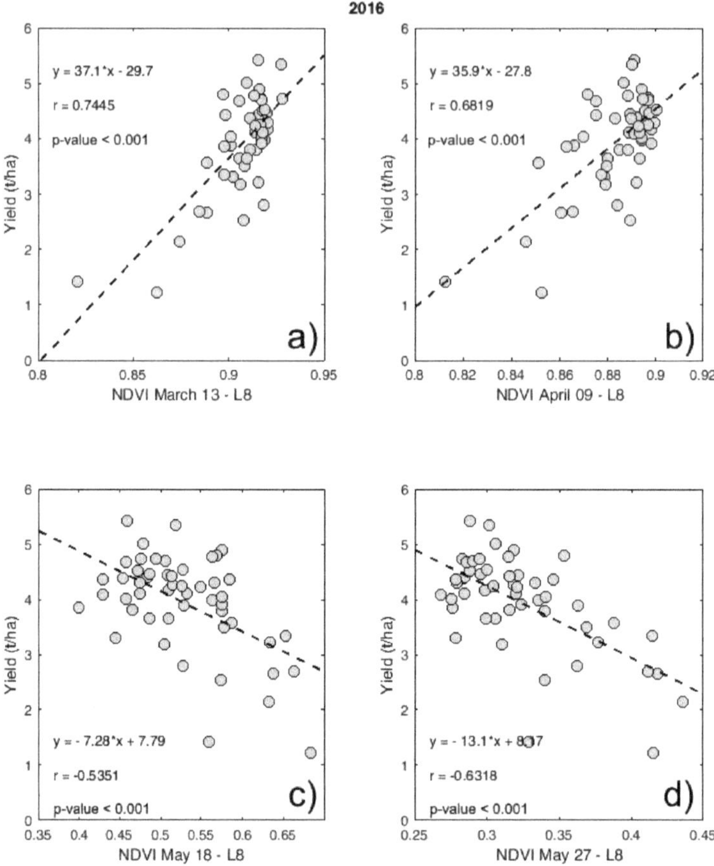

Figure 6. Relationship between observed yield (yield monitor) and NDVI (Landsat-8) for 2015–2016 crop season in (**a**) 13 March, (**b**) 9 April, (**c**) 18 May and (**d**) 27 May.

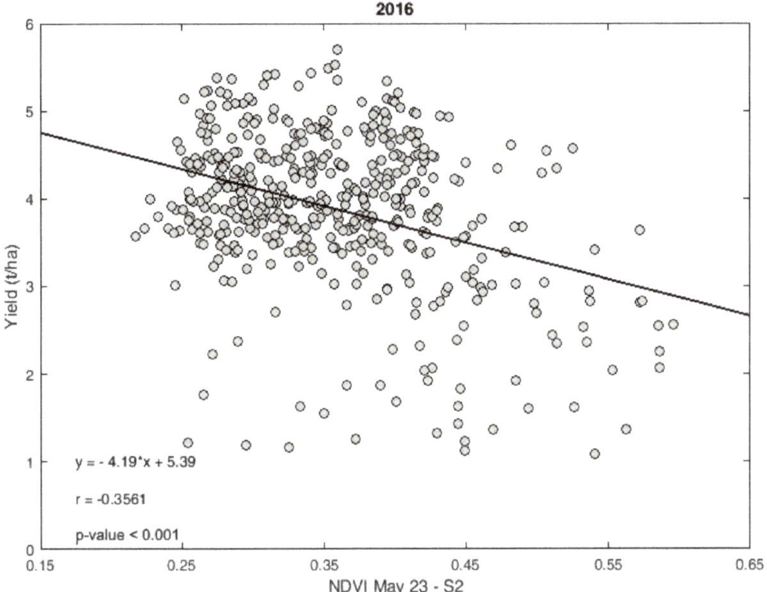

Figure 7. Relationship between observed yield (yield monitor) and NDVI (Sentinel-2) for 2015–2016 crop season.

These results, in terms of both correlations and timing, are in line with Mahey et al. [53]. Freeman et al. [54] found NDVI and wheat grain yield to be highly correlated, establishing the potential to predict yield with remotely sensed data as reported subsequently in several studies for a variety of crop types [55–58].

Freeman et al. [54] also indicated that yield estimates for wheat may be made two months prior to harvest.

Instead, only for 2016, from post-flowering to grain filling, we report weaker but significant negative correlations between NDVI and yield. This is for the Landsat-8 (18 and 27 May, Figure 6) images and the only one available for Sentinel-2 (23 May, Figure 7).

Although negative correlations between NDVI and crop yield are reported in the literature for potato late in the season [55] and for canola, after bolting and once the plants start transitioning to the reproductive stages [59], there are few similar findings for cereal crops when analyzing single or multi cultivars [60–63]. All these latter authors found a negative correlation under severe stress conditions, such as high temperature and drought, during grain filling.

Conversely, in our case study and for 2015–2016 crop season, this unique behavior of NDVI that from strongly positively correlation swings negative to more than −0.6 late in the season is mainly due to opposite climatic conditions (cool-moist) that characterized crop development and above all the period from heading to maturity (Table 4).

Table 4. Climate conditions, phenological length for heading to maturity (H to M) and anthesis to maturity (A to M), leaf area index (LAI) function derivative rate of change for the four crop seasons from anthesis to maturity.

		Number of Days	Total Rainfall (mm)	Air Temperature (°C)	LAI Rate of Change
2014	H to M	79	125.4	20.51	−0.099
	A to M	57	89.6	22.33	
2015	H to M	59	64.0	21.34	−0.142
	A to M	41	64.0	21.64	
2016	H to M	83	129.2	18.88	−0.085
	A to M	64	81.8	19.62	
2017	H to M	67	104.6	19.46	−0.131
	A to M	46	62.4	20.58	

In fact, the longest duration of heading to maturity and anthesis to maturity was observed in 2016 (Table 4) as well as the highest amount of rainfall from heading to maturity (129.2 mm) and lowest mean air temperature both from heading to maturity and anthesis to maturity (18.88 °C and 19.62 °C respectively). These conditions resulted in a general delayed leaf senescence and prolonged late grain filling as a sort of stay green effect [1,64] that is confirmed by the lowest decline rate of LAI in 2016 (Figure 8, Table 4). The derivatives of the modelled LAI function from heading to maturity has a rate of change for 2016 of −0.085, so is lower compared to other years. In the absence of water-stress, as for 2016, stay green is not always correlated with yield (in wheat [65,66] and in sorghum [67]) and can even be associated with reduced yield. For instance, in irrigated wheat and in rice in China, stay green was associated with slow export of leaf carbohydrate to the grain, increased lodging, and harvest difficulties due to delayed ripening, all of which can contribute to reduce yield [68,69]. In our case study, it is likely that this did not occur uniformly due to site-specific soil plant interactions, and the areas within the field that exhibited higher NDVI values during maturation then had translocation problems to the grains. This is confirmed by the fact that the areas with the lowest production (highest NDVI) in 2016 fall in the same low production areas as 2014 and 2015 (Figure 3). Contrary to what happened in 2014, the presence of weeds was not reported for 2016.

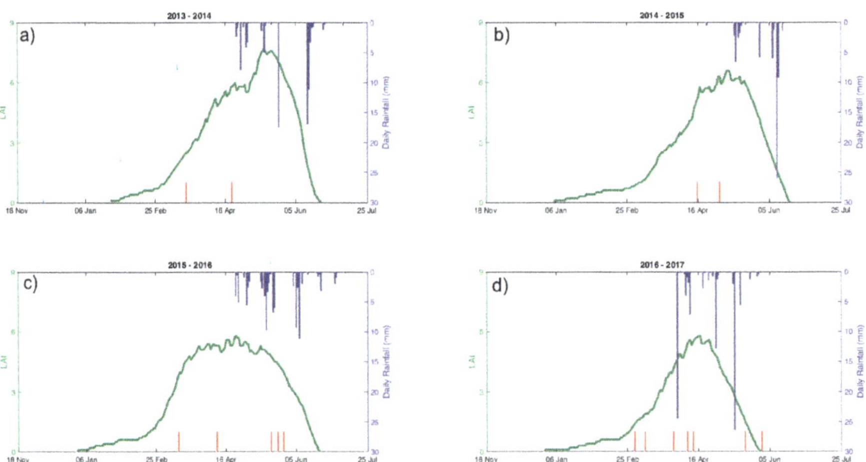

Figure 8. Modelled LAI (green), observed rainfall after heading (blue) and satellite observation (red) for (**a**) 2013–2014, (**b**) 2014–2015, (**c**) 2015–2016 and (**d**) 2016–2017 seasons.

For 2017, having no yield monitoring data, it was possible to compare NDVI data only with hand sampled data (Figures 9 and 10). Also, in this case the highest correlations are observed in the months of March and April, then tend to decrease in May (Sentinel-2, Figure 10) or become non-significant for the Landsat-8 passage in late May (Figure 9). The low and non-significant relationship in late May is probably due to the fact that the drying process had already started unevenly in some areas of the field [33].

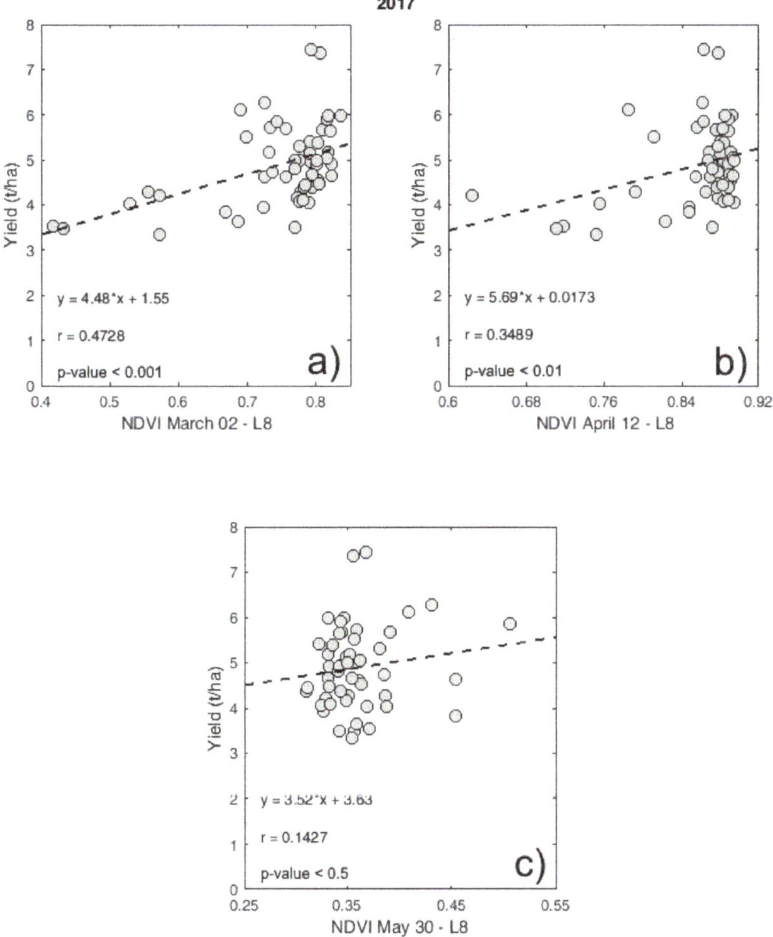

Figure 9. Relationship between observed yield (sampled) and NDVI (Landsat-8) for 2016–2017 crop season in (**a**) 2 March, (**b**) 12 April and (**c**) 30 May.

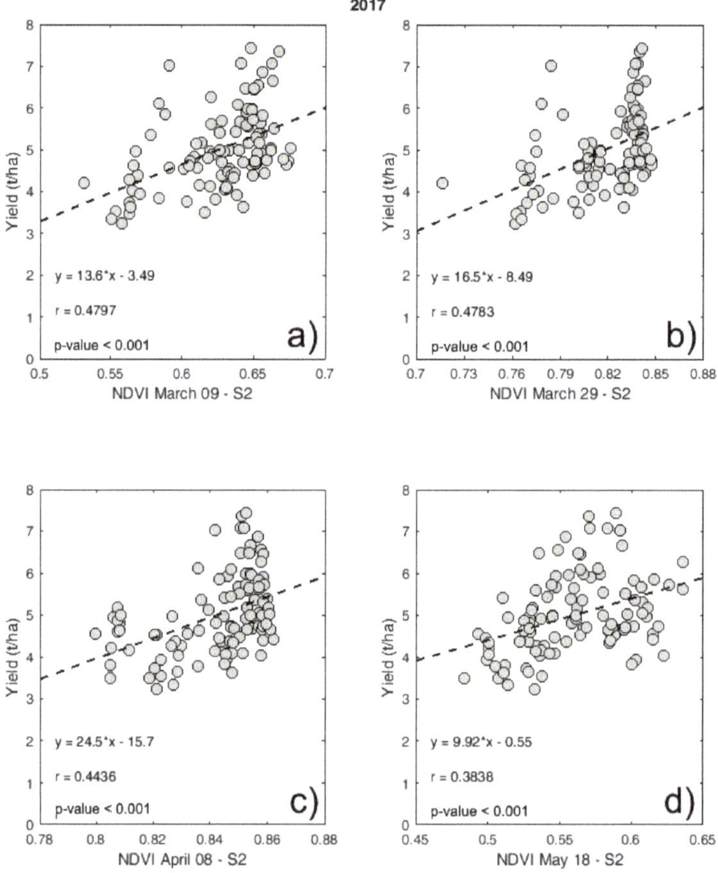

Figure 10. Relationship between observed yield (sampled) and NDVI (Sentinel-2) for 2016–2017 crop season in (**a**) 9 March, (**b**) 29 March, (**c**) 8 April and (**d**) 18 May.

Unfortunately, all the empirical relationships determined over the whole study period cannot be applied elsewhere, since a universal conversion from vegetation indices to yield values does not exist, as pointed out by Georgi et al. [9]. Many efforts have been made to determine this relationship [17,70,71], with results indicating that replicability is mostly limited by crop type and climate zone, confirming our case study findings. Our results highlight the potential use of remote sensing imagery (Sentinel-2 and Landsat-8) for within-field and interannual durum wheat yield assessment under Mediterranean conditions. Although it is not possible to retrieve absolute yield values, the results show the capacity of the NDVI to describe within-field yield levels providing objective criteria, also in terms of potential performance, on which to base nutrient management zones for soil sampling and variable-rate nutrient application, especially thanks to the availability of multiple years of data. This is also facilitated by the fact that, even if multiple surveys are done during crop development, NDVI and yield are strongly correlated at stem elongation and heading stages, which are among the most important for agronomic management to support and improve durum wheat yield and quality.

4. Conclusions

The first part of the study is mainly practice oriented, testing a state of the art protocol for error removal from yield monitoring data and comparing the cleaning map with hand field sampling data. The cleaning process improved measurement accuracy of spatial variability, which is key for adopting precision farming techniques to make daily fieldwork more efficient and increase agricultural productivity. In light of this, in the second part of the study, the usefulness of remote sensing information collected during the optimal period for characterizing within-field spatial variability of durum wheat productivity has been assessed.

Findings suggest that the best time to relate NDVI to durum wheat yields under rainfed conditions in the Mediterranean area is the period leading up to 90–60 days before harvest (March–April). At the same time the results, based on a four year yield dataset, support the conclusion that a unique NDVI-yield relationship cannot be achieved and applied to different years or environments, but year by year can suggest the best management approach while taking farmers' requirements into account.

Additional research is needed in the future to: (i) test different methods of comparing heterogeneous data (different supports, spatial resolution), (ii) address the performance of other VIs and promote them among end users. Furthermore, in case of long periods between satellite images due to cloud cover, the use of a crop simulation model has proved to be of fundamental importance to simulate the crop stage and growth conditions and better understand differences underlying correlations between yield and VIs.

Author Contributions: A.V.V. and A.C. designed the experiment. P.T., A.C., S.F.D.G., A.V.V., D.V. and A.M. formulated the research methodology and wrote the manuscript. P.T., S.F.D.G. and A.M. provided necessary data analysis. All authors reviewed and edited the draft.

Funding: This research was funded by Italian P.O.N. "Ricerca e competitività" 2007–2013 for convergence Regions (grant number: CTN01_00230_450760, D.D. 257/Ric/2012 of "Ministero dell'Istruzione, dell'Università e della Ricerca"): "Sostenibilità della Filiera Agroalimentare Italiana" (SO.FI.A.).

Conflicts of Interest: The authors declare no conflict of interest.

References

1. Toscano, P.; Gioli, B.; Genesio, L.; Vaccari, F.P.; Miglietta, F.; Zaldei, A.; Crisci, A.; Ferrari, E.; Bertuzzi, F.; La Cava, P.; et al. Durum wheat quality prediction in Mediterranean environments: From local to regional scale. *Eur. J. Agron.* **2014**, *61*, 1–9. [CrossRef]
2. Royo, C.; Nazco, R.; Villegas, D. The climate of the zone of origin of Mediterranean durum wheat (Triticum durum Desf.) landraces affects their agronomic performance. *Genet. Resour. Crop Evol.* **2014**, *61*, 1345–1358. [CrossRef]
3. Tedone, L.; Alhajj Ali, S.; Verdini, L.; De Mastro, G. Nitrogen management strategy for optimizing agronomic and environmental performance of rainfed durum wheat under Mediterranean climate. *J. Clean. Prod.* **2018**, *172*, 2058–2074. [CrossRef]
4. Soddu, A.; Deidda, R.; Marrocu, M.; Meloni, R.; Paniconi, C.; Ludwig, R.; Sodde, M.; Mascaro, G.; Perra, E. Climate Variability and Durum Wheat Adaptation Using the AquaCrop Model in Southern Sardinia. *Procedia Environ. Sci.* **2013**, *19*, 830–835. [CrossRef]
5. Toscano, P.; Ranieri, R.; Matese, A.; Vaccari, F.P.; Gioli, B.; Zaldei, A.; Silvestri, M.; Ronchi, C.; La Cava, P.; Porter, J.R.; et al. Durum wheat modeling: The Delphi system, 11 years of observations in Italy. *Eur. J. Agron.* **2012**, *43*, 108–118. [CrossRef]
6. Giuliani, M.M.; Giuzio, L.; de Caro, A.; Flagella, Z. Relationships between nitrogen utilization and grain technological quality in durum wheat: II. Grain yield and quality. *Agron. J.* **2011**, *103*, 1668–1675. [CrossRef]
7. Bowman, M.S.; Zilberman, D. Economic factors affecting diversified farming systems. *Ecol. Soc.* **2013**, *18*, 33. [CrossRef]
8. Mulla, D.J. Twenty five years of remote sensing in precision agriculture: Key advances and remaining knowledge gaps. *Biosyst. Eng.* **2013**, *114*, 358–371. [CrossRef]
9. Georgi, C.; Spengler, D.; Itzerott, S.; Kleinschmit, B. Automatic delineation algorithm for site-specific management zones based on satellite remote sensing data. *Precis. Agric.* **2018**, *19*, 684–707. [CrossRef]
10. Thilakarathna, M.; Raizada, M. Challenges in Using Precision Agriculture to Optimize Symbiotic Nitrogen Fixation in Legumes: Progress, Limitations, and Future Improvements Needed in Diagnostic Testing. *Agronomy* **2018**, *8*, 78. [CrossRef]

11. Ping, J.L.; Dobermann, A. Processing of yield map data. *Precis. Agric.* **2005**, *6*, 193–212. [CrossRef]
12. Thöle, H.; Richter, C.; Ehlert, D. Strategy of statistical model selection for precision farming on-farm experiments. *Precis. Agric.* **2013**, *14*, 434–449. [CrossRef]
13. Vega, A.; Córdoba, M.; Castro-Franco, M.; Balzarini, M. Protocol for automating error removal from yield maps. *Precis. Agric.* 2019. [CrossRef]
14. Lyle, G.; Bryan, B.A.; Ostendorf, B. Post-processing methods to eliminate erroneous grain yield measurements: Review and directions for future development. *Precis. Agric.* **2014**, *15*, 377–402. [CrossRef]
15. Arslan, S. A grain flow model to simulate grain yield sensor response. *Sensors* **2008**, *8*, 952–962. [CrossRef]
16. Ingeli, M.; Galambošová, J.; Macák, M.; Rataj, V. Study on Correlation of Data from Yield Monitoring System and Hand Samples. *Acta Technol. Agric.* **2015**, *18*, 10–13. [CrossRef]
17. Doraiswamy, P.C.; Hatfield, J.L.; Jackson, T.J.; Akhmedov, B.; Prueger, J.; Stern, A. Crop condition and yield simulations using Landsat and MODIS. *Remote Sens. Environ.* **2004**, *92*, 548–559. [CrossRef]
18. Funk, C.; Budde, M.E. Phenologically-tuned MODIS NDVI-based production anomaly estimates for Zimbabwe. *Remote Sens. Environ.* **2009**, *113*, 115–125. [CrossRef]
19. Johnson, D.M. An assessment of pre- and within-season remotely sensed variables for forecasting corn and soybean yields in the United States. *Remote Sens. Environ.* **2014**, *141*, 116–128. [CrossRef]
20. Gao, F.; Anderson, M.; Daughtry, C.; Johnson, D. Assessing the variability of corn and soybean yields in central Iowa using high spatiotemporal resolution multi-satellite imagery. *Remote Sens.* **2018**, *10*, 1489. [CrossRef]
21. Skakun, S.; Vermote, E.; Roger, J.-C.; Franch, B. Combined Use of Landsat-8 and Sentinel-2A Images for Winter Crop Mapping and Winter Wheat Yield Assessment at Regional Scale. *AIMS Geosci.* **2017**, *3*, 163–186. [CrossRef] [PubMed]
22. Azzari, G.; Jain, M.; Lobell, D.B. Towards fine resolution global maps of crop yields: Testing multiple methods and satellites in three countries. *Remote Sens. Environ.* **2017**, *202*, 129–141. [CrossRef]
23. Jin, Z.; Azzari, G.; Lobell, D.B. Improving the accuracy of satellite-based high-resolution yield estimation: A test of multiple scalable approaches. *Agric. For. Meteorol.* **2017**, *247*, 207–220. [CrossRef]
24. Zhong, L.; Yu, L.; Li, X.; Hu, L.; Gong, P. Rapid corn and soybean mapping in US Corn Belt and neighboring areas. *Sci. Rep.* **2016**, *6*, 36240. [CrossRef] [PubMed]
25. Lambert, M.J.; Traoré, P.C.S.; Blaes, X.; Baret, P.; Defourny, P. Estimating smallholder crops production at village level from Sentinel-2 time series in Mali's cotton belt. *Remote Sens. Environ.* **2018**, *216*, 647–657. [CrossRef]
26. Newton, I.H.; Tariqul Islam, A.F.M.; Saiful Islam, A.K.M.; Tarekul Islam, G.M.; Tahsin, A.; Razzaque, S. Yield Prediction Model for Potato Using Landsat Time Series Images Driven Vegetation Indices. *Remote Sens. Earth Syst. Sci.* **2018**, *1*, 29–38. [CrossRef]
27. López-Lozano, R.; Casterad, M.A.; Herrero, J. Site-specific management units in a commercial maize plot delineated using very high resolution remote sensing and soil properties mapping. *Comput. Electron. Agric.* **2010**, *73*, 219–229. [CrossRef]
28. Nawar, S.; Corstanje, R.; Halcro, G.; Mulla, D.; Mouazen, A.M. Delineation of Soil Management Zones for Variable-Rate Fertilization: A Review. In *Advances in Agronomy*; Academic Press: Cambridge, MA, USA, 2017; pp. 175–245.
29. Han, J.; Wei, C.; Chen, Y.; Liu, W.; Song, P.; Zhang, D.; Wang, A.; Song, X.; Wang, X.; Huang, J. Mapping above-ground biomass of winter oilseed rape using high spatial resolution satellite data at parcel scale under waterlogging conditions. *Remote Sens.* **2017**, *9*, 3. [CrossRef]
30. Dong, T.; Shang, J.; Liu, J.; Qian, B.; Jing, Q.; Ma, B.; Huffman, T.; Geng, X.; Sow, A.; Shi, Y.; et al. Using RapidEye imagery to identify within-field variability of crop growth and yield in Ontario, Canada. *Precis. Agric.* **2019**, 1–20. [CrossRef]
31. Maes, W.H.; Steppe, K. Perspectives for Remote Sensing with Unmanned Aerial Vehicles in Precision Agriculture. *Trends Plant Sci.* **2019**, *24*, 152–164. [CrossRef]
32. Al-Gaadi, K.A.; Hassaballa, A.A.; Tola, E.; Kayad, A.G.; Madugundu, R.; Alblewi, B.; Assiri, F. Prediction of potato crop yield using precision agriculture techniques. *PLoS ONE* **2016**, *11*, e0162219. [CrossRef] [PubMed]
33. Escolà, A.; Badia, N.; Arnó, J.; Martínez-Casasnovas, J.A. Using Sentinel-2 images to implement Precision Agriculture techniques in large arable fields: First results of a case study. *Adv. Anim. Biosci.* **2017**, *8*, 377–382. [CrossRef]

34. Jeppesen, J.H.; Jacobsen, R.H.; Jørgensen, R.N.; Halberg, A.; Toftegaard, T.S. Identification of High-Variation Fields based on Open Satellite Imagery. *Adv. Anim. Biosci.* **2017**, *8*, 388–393. [CrossRef]

35. Soil Survey Staff. *Soil Taxonomy: A Basic System of Soil Classification for Making and Interpreting Soil Surveys*, 2nd ed.; USDA-Soil Survey Staff, Ed.; United States Department of Agriculture: Washington, DC, USA, 1999.

36. Ventrella, D.; Stellacci, A.M.; Castrignanò, A.; Charfeddine, M.; Castellini, M. Effects of crop residue management on winter durum wheat productivity in a long term experiment in Southern Italy. *Eur. J. Agron.* **2016**, *77*, 188–198. [CrossRef]

37. Deering, D.W. Rangeland reflectance characteristics measured by aircraft and spacecraft sensors. *Diss. Abstr. Int. B* **1979**, *39*, 3081–3082.

38. Anselin, L.; Syabri, I.; Kho, Y. GeoDa: An introduction to spatial data analysis. *Geogr. Anal.* **2006**, *38*, 5–22. [CrossRef]

39. QGIS. QGIS Geographic Information System. Open Source Geospatial Foundation Project. 2017. Available online: https://qgis.org/en/site/ (accessed on 8 August 2019).

40. Minasny, B.; McBratney, A.B.; Whelan, B.M. *VESPER Version 1.62*, Australian Centre for Precision Agriculture, The University of Sydney: Sydney, Australia, 2005.

41. Mathworks Inc. *MATLAB and Statistics Toolbox Release*, Mathworks Inc.: Natick, MA, USA, 2016.

42. Amelin, L. Local Indicators of Spatial Association-LISA. *Geogr. Anal.* **1995**, *27*, 93–115.

43. Anselin, L. The Moran Scatterplot as an ESDA Tool to Assess Local Instability in Spatial Association. In *Spatial Analytical Perspectives on GIS*; Fischer, M., Scholten, H., Unwin, D., Eds.; Taylor and Francis: London, UK, 1996; pp. 111–125.

44. Porter, J.R.; Porter, J.R. AFRCWHEAT2 A model of the growth and development of wheat incorporating responses to water and nitrogen. *Eur. J. Agron.* **1993**, *2*, 69–82. [CrossRef]

45. Su, L.; Wang, Q.; Wang, C.; Shan, Y. Simulation models of leaf area index and yield for cotton grown with different soil conditioners. *PLoS ONE* **2015**, *10*, e0141835. [CrossRef] [PubMed]

46. Wang, M.H. Field Information Collection and Process Technology. *Agric. Mech.* **1999**, *7*, 22–24.

47. Leroux, C.; Jones, H.; Clenet, A.; Dreux, B.; Becu, M.; Tisseyre, B. A general method to filter out defective spatial observations from yield mapping datasets. *Precis. Agric.* **2018**, *19*, 789–808. [CrossRef]

48. Sudduth, K.A.; Drummond, S.T. Yield editor: Software for removing errors from crop yield maps. *Agron. J.* **2007**, *99*, 1471–1482. [CrossRef]

49. Simbahan, G.C.; Dobermann, A.; Ping, J.L. Site-specific management—Screening yield monitor data improves grain yield maps. *Agron. J.* **2004**, *96*, 1091–1102. [CrossRef]

50. Sun, W.; Whelan, B.; McBratney, A.B.; Minasny, B. An integrated framework for software to provide yield data cleaning and estimation of an opportunity index for site-specific crop management. *Precis. Agric.* **2013**, *14*, 376–391. [CrossRef]

51. Arslan, S.; Colvin, T.S. Grain yield mapping: Yield sensing, yield reconstruction, and errors. *Precis. Agric.* **2002**, *3*, 135–154. [CrossRef]

52. Young, L.J.; Gotway, C.A. Linking spatial data from different sources: The effects of change of support. *Stoch. Environ. Res. Risk Assess.* **2007**, *21*, 589–600. [CrossRef]

53. Mahey, R.K.; Singh, R.; Sidhu, S.S.; Narang, R.S. The use of remote sensing to assess the effects of water stress on wheat. *Exp. Agric.* **1991**, *27*, 423–429. [CrossRef]

54. Freeman, K.W.; Raun, W.R.; Johnson, G.V.; Mullen, R.W.; Stone, M.L.; Solie, J.B. Late-season prediction of wheat grain yield and grain protein. *Commun. Soil Sci. Plant Anal.* **2003**, *34*, 1837–1852. [CrossRef]

55. Johnson, D.M. A comprehensive assessment of the correlations between field crop yields and commonly used MODIS products. *Int. J. Appl. Earth Obs. Geoinf.* **2016**, *52*, 65–81. [CrossRef]

56. Kogan, F.; Kussul, N.; Adamenko, T.; Skakun, S.; Kravchenko, O.; Kryvobok, O.; Shelestov, A.; Kolotii, A.; Kussul, O.; Lavrenyuk, A. Winter wheat yield forecasting in Ukraine based on Earth observation, meteorologicaldata and biophysical models. *Int. J. Appl. Earth Obs. Geoinf.* **2013**, *23*, 192–203. [CrossRef]

57. Mkhabela, M.S.; Bullock, P.; Raj, S.; Wang, S.; Yang, Y. Crop yield forecasting on the Canadian Prairies using MODIS NDVI data. *Agric. For. Meteorol.* **2011**, *151*, 385–393. [CrossRef]

58. Becker-Reshef, I.; Vermote, E.; Lindeman, M.; Justice, C. A generalized regression-based model for forecasting winter wheat yields in Kansas and Ukraine using MODIS data. *Remote Sens. Environ.* **2010**, *114*, 1312–1323. [CrossRef]

59. Weber, J.; Gazali, I. 2017 Annual Report for the Agriculture Demonstration of Practices and Technologies (ADOPT). 2017. Available online: https://sa241e23dba898335.jimcontent.com/download/version/1526570262/module/10558155383/name/20150375%20ADOPT%20Report%20Nitrogen%20Benefits%20of%20adapted%20grain%20legumes%20to%20succeeding%20crops%20in%20NW%20SK.pdf (accessed on 8 August 2019).
60. Kyratzis, A.C.; Skarlatos, D.P.; Menexes, G.C. Assessment of Vegetation Indices Derived by UAV Imagery for Durum Wheat Phenotyping under a Water Limited and Heat Stressed Mediterranean Environment. *Front. Plant Sci.* **2017**, *8*, 1–14. [CrossRef] [PubMed]
61. Lopes, M.S.; Saglam, D.; Ozdogan, M.; Reynolds, M. Traits associated with winter wheat grain yield in Central and West Asia. *J. Integr. Plant Biol.* **2014**, *56*, 673–683. [CrossRef]
62. Corresponding, J.B.; Casadesus, J.; Nachit, M.M.; Araus, J.L. Factors affecting the grain yield predicting attributes of spectral reflectance indices in durum wheat: growing conditions, genotype variability and date of measurement. *Int. J. Remote Sens.* **2005**, *26*, 2337–2358.
63. Rutkoski, J.; Poland, J.; Mondal, S.; Autrique, E.; Pérez, L.G.; Crossa, J.; Reynolds, M.; Singh, R. Canopy Temperature and Vegetation Indices from High-Throughput Phenotyping Improve Accuracy of Pedigree and Genomic Selection for Grain Yield in Wheat. *Genes Genomes Genet.* **2016**, *6*, 2799–2808. [CrossRef]
64. Spano, G.; Di Fonzo, N.; Perrotta, C.; Platani, C.; Ronga, G.; Lawlor, D.W.; Napier, J.A. Physiological characterization of 'stay green' mutants in durum wheat. *J. Exp. Bot.* **2003**, *54*, 1415–1420. [CrossRef]
65. Lopes, M.S.; Reynolds, M.P. Stay-green in spring wheat can be determined by spectral reflectance measurements (normalized difference vegetation index) independently from phenology. *J. Exp. Bot.* **2012**, *63*, 3789–3798. [CrossRef]
66. Gregersen, P.L.; Culetic, A.; Boschian, L.; Krupinska, K. Plant senescence and crop productivity. *Plant Mol. Biol.* **2013**, *82*, 603–622. [CrossRef] [PubMed]
67. Jordan, D.R.; Hunt, C.H.; Cruickshank, A.W.; Borrell, A.K.; Henzell, R.G. The relationship between the stay-green trait and grain yield in elite sorghum hybrids grown in a range of environments. *Crop Sci.* **2012**, *52*, 1153–1161. [CrossRef]
68. Gong, Y.-H.; Zhang, J.; Gao, J.-F.; Lu, J.-Y.; Wang, J.-R. Slow Export of Photoassimilate from Stay-Green Leaves during Late Grain-Filling Stage in Hybrid Winter Wheat (Triticum aestivum L.). *J. Agron. Crop Sci.* **2005**, *191*, 292–299. [CrossRef]
69. Yang, J.; Zhang, J. Grain filling of cereals under soil drying. *New Phytol.* **2006**, *169*, 223–236. [CrossRef] [PubMed]
70. Shanahan, J.F.; Schepers, J.S.; Francis, D.D.; Varvel, G.E.; Wilhelm, W.W.; Tringe, J.M.; Schlemmer, M.R.; Major, D.J. Use of remote-sensing imagery to estimate corn grain yield. *Agron. J.* **2001**, *93*, 583–589. [CrossRef]
71. Dalla Marta, A.; Grifoni, D.; Mancini, M.; Orlando, F.; Guasconi, F.; Orlandini, S. Durum wheat in-field monitoring and early-yield prediction: Assessment of potential use of high resolution satellite imagery in a hilly area of Tuscany, Central Italy. *J. Agric. Sci.* **2015**, *153*, 68–77. [CrossRef]

Article

Integrating Sentinel-2 Imagery with AquaCrop for Dynamic Assessment of Tomato Water Requirements in Southern Italy

Anna Dalla Marta [1,*], **Giovanni Battista Chirico** [2], **Salvatore Falanga Bolognesi** [3], **Marco Mancini** [1], **Guido D'Urso** [2], **Simone Orlandini** [1], **Carlo De Michele** [3] and **Filiberto Altobelli** [4]

[1]   Department of Agriculture, Food, Environment and Forestry, University of Florence, Piazzale delle Cascine 18, 50144 Firenze, Italy

[2]   Department of Agricultural Sciences, University of Naples "Federico II", Via Università 100, 80055 Portici (NA), Italy

[3]   Ariespace s.r.l., Spin off Company, University of Napoli "Federico II", Centro Direzionale IS A3, 80143 Napoli, Italy

[4]   CREA Research Centre for Agricultural Policies and Bioeconomy, Via Po 14, 00198 Roma, Italy

*   Correspondence: anna.dallamarta@unifi.it; Tel.: +39-055-275-5741

Received: 1 June 2019; Accepted: 18 July 2019; Published: 21 July 2019

Abstract: A research study was conducted in an open field tomato crop in order to: (i) Evaluate the capability of Sentinel-2 imagery to assess tomato canopy growth and its crop water requirements; and (ii) explore the possibility to predict crop water requirements by assimilating the canopy cover estimated by Sentinel-2 imagery into AquaCrop model. The pilot area was in Campania, a region in the south west of Italy, characterized by a typical Mediterranean climate, where field campaigns were conducted in seasons 2017 and 2018 on processing tomato. Crop water use and irrigation requirement were estimated by means of three different methods: (i) The AquaCrop model; (ii) an irrigation advisory service based on Sentinel-2 imagery known as IRRISAT and (iii) assimilating the canopy cover estimated by Sentinel-2 imagery into AquaCrop model Sentinel-2 imagery proved to be effective for monitoring canopy growth and for predicting irrigation water requirements during mid-season stage of the crop, when the canopy is fully developed. Conversely, the integration of the Sentinel-2 imagery with a crop growth model can contribute to improve the irrigation water requirement predictions in the early and development stage of the crop, when the soil evaporation is not negligible with respect to the total evapotranspiration.

Keywords: fractional cover; irrigation; satellite; crop simulation model; AquaCrop

---

1. Introduction

Worldwide significant progress has been made to utilize precision agriculture for irrigation as a mean to increase water use efficiency or decrease the water footprint in irrigated agriculture [1,2]. The progress is mainly restricted to advances at the plot scale and individual systems such as installations for drip irrigation or central pivots. As well known, actual crop evapotranspiration (ET) is a major term of water budget in agriculture and it is the main variable used to determine crops water requirement. Beside their massive progress during the recent years, accurate field measurements (soil moisture, plant-based sensors, etc.) are very scarce because sensors and measuring devices are expensive, their use requires specific expertise and complex maintenance and are typically limited to experimental stations. For this reason, many attempts have been made for developing indirect ET estimation methods based on crop data at a field scale easily available across large regions. Concordantly, the use of remotely sensed data has become more common to monitoring and controlling activities at different spatial and temporal scales including precision farming. Surface energy balance methods based on satellite observations in the thermal band [3] have been developed and applied in many areas with the aim of determining actual evapotranspiration and assessing the water balance of irrigated areas and the corresponding water accounting practices [4,5].

Sentinel-2 mission, launched by European Space Agency as part of the Copernicus program (http://www.copernicus.eu/) [6], has been a great step forward for continuous crop monitoring. Sentinel-2 carries a sensor, which captures data at 10, 20 and 60-m spatial resolution over 13 spectral bands and with a very high temporal resolution of five days at the equator. Thanks to the characteristics of these new technologies, commercial services started to develop for operational applications. In this context, Irrigation Advisory Services (IASs) for optimizing water management rapidly grow. Among them, IRRISAT is a fully operative satellite-based IAS provided in Campania Region, which combines Copernicus Sentinel-2A data with daily weather data for estimating irrigation water requirements. Since 2007 it has been active over the area, providing evidence of high efficiency for water saving (up to 30%) [7].

Moreover, the state of the art suggests that significant progresses in saving irrigation volumes at the farm level can be probably attained by assessing crop water requirements through an optimal combination of crop satellite images with a crop growth model since satellite images provide information concerning the current state of the crop canopy and a crop model is able to simulate the biophysical processes of the growing crop.

Ultimately, examples are implemented to derive crop water requirements from satellite estimates of biophysical parameters assimilated into agro-meteorological models [8] to monitor the nitrogen status and to apply fertilizer with variable rates or to derive agronomic variables [9].

Since the rapid development and availability of products with different spatial and temporal resolutions, the integration of remote sensing data into crop growth models has increased in recent years. Among the existing crop models, AquaCrop has been widely used for assessing water requirements and optimal irrigation scheduling for different crops and environments. AquaCrop, in fact, is a crop water productivity model developed by the Land and Water Division of FAO (Food and Agriculture Organization) in 2009 [10,11]. It simulates yield response to water of crops and it is mainly used to increase water efficiency practices in agricultural production

Linker and Ioslovich [12] have successfully shown the possibility of assimilating ground measured canopy cover data (using digital images above the canopy) during the growing season within the AquaCrop model for potato and cotton, based on the Extended Kalman Filter algorithm. However, this innovative procedure requires field data collection during the season, which is budget-and time-consuming but today, it can be facilitated and improved using Earth Observation data, especially from Sentinel-2 satellite sensors whose temporal and spatial resolution is adapted for parcel monitoring. Jin et al. [13] used spectral-based biomass values computed with on field spectral measurement data to calibrate the AquaCrop model with a particle swarm optimization (PSO) algorithm for winter wheat in China.

The same PSO algorithm was used for winter wheat by Silvestro et al. [14] and Jin et al. [15] to assimilate optical and radar satellite data into the AquaCrop model. Several coupling methods have been categorized and detailed by Jin et al. [16]. One of these methods is by substituting crop growth models variables with remotely sensed data [17]. A similar approach using remote sensing data was used to calibrate the AquaCrop model with the leaf area index from MODIS (Moderate Resolution Imaging Spectroradiometer) for winter wheat in Italy by Trombetta et al. [18]. It was only applied to large agricultural parcels, due to MODIS's limited spatial resolution. However, their use in real/operational scenarios is generally limited by data availability, i.e., crop initial conditions, planting date and application of inputs. Under this scenario, a possible alternative for assisting the models in reproducing the actual processes in the field is the use of algorithms relating to remote sensing data and key canopy biophysical parameters in the crop growth models.

This study aims to evaluate:

The capability of Sentinel-2 imagery to assess canopy growth of a tomato open field crop and the corresponding irrigation water requirement by means of Sentinel-2 imagery;

The possibility to predict crop water requirements by assimilating the fractional cover estimated by Sentinel-2 imagery into the AquaCrop model.

2. Materials and Methods

2.1. Test Site

The research was carried out in Frignano, located in Caserta province (Campania, Italy) (Figure 1). The area is characterized by a warm-temperate climate with an average annual temperature of 15.2 °C and 900 mm of annual cumulated precipitation, with the rainiest period occurring in winter and driest period in summer.

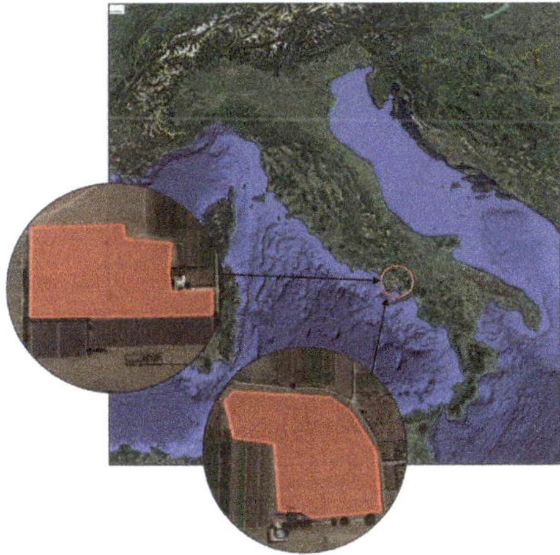

Figure 1. Localization of the pilot area and the two experimental fields.

Processing tomato (*Solanum lycopersicum* L.) was cultivated in years 2017 (41°00′26.33″ N 14°10′13.93″ E) and 2018 (41°01′39.60″ N 14°10′32.87″ E) in two parcels of 4 ha each. Soil samples (three samples were averaged for analysis) were examined for assessing the main physical–chemical properties, including gravel (%), soil texture, bulk density (t/m^3), pH and organic matter (%), at a depth of 15 cm. In 2018 soil characteristics were obtained by the soil map of Region Campania (pedological maps 1:50,000; http://agricoltura.regione.campania.it/pedologia/suoli.html). SPAW software (Soil–Plant–Air–Water; v6.02.75, United States Department of Agriculture-USDA, Washington, DC, USA) was used for assessing soil hydraulic properties, such as soil water retention, soil hydraulic conductivity, field capacity and plant available water.

Soil was ploughed at 40 cm depth, and tomato seedlings were transplanted on April 9 in continuous double rows with 33.5–40.0 cm space between plants, 50 cm between rows, 1.10–1.20 cm between double rows, with a final plant density of 32,000 and 33,500 plants/ha in year 2017 and 2018, respectively.

Fertilization was applied before transplanting (broadcast) with 120 kg/ha of diammonium phosphate (18-46-0) and 400 kg/ha of organic-mineral fertilizer (10-5-12); at transplanting (localized) with 160 kg/ha of diammonium phosphate and 20 kg/ha of seed sprint H5 (12-43-0).

Meteorological daily data of maximum and minimum temperature (°C), relative humidity (%), wind speed (m s^{-1}) and precipitation (mm) were collected at a complete weather station located in the study area for the two cropping seasons.

Plants were watered by light driplines, with 30 cm dripper spacing and 2 L/h flow rate at 1 bar.

Based on the farmer best knowledge, a total volume of 130 L/plant of water, corresponding to 4160 m³/ha, and 120 L/plant corresponding to 4020 m³/ha of irrigation was given throughout the growing period in 2017 and 2018, respectively. Such volumes were then considered as reference values against which the different estimation methods (AquaCrop, IRRISAT, integration of Sentinel 2A imagery into AquaCrop model) were tested.

2.2. Satellite-Based Assessment of Crop Water Requirement: IRRISAT

The remote sensing technique employed is the one employed by IRRISAT, a satellite-based irrigation advisory service developed in Italy and operational since 2007 in Southern Italy [7]. The service aims at providing farmers and water managers with real time information on crop water needs, which are estimated by combining high resolution data from Earth Observation satellites with weather data for the calculation of crop water requirements. More recently, IRRISAT has been combined with numerical weather predictions for forecasting crop water needs up to five days in advance [19,20]. Information is delivered in near-real time (24 h) to users (farmers, Water User Association and water agencies) by means of a dedicated WebGIS accessible from PC, tablets and smartphones (https://www.irrisat.com/en/). In 2016, the operational irrigation advisory service of the Campania region has reached about 2000 farmers with a total irrigation area larger than 80,000 hectares: The achieved water saving has been estimated to be larger than 30% [7]. In 2016 the Italian Ministry of Agriculture has listed IRRISAT among the applicable methodologies for the estimation of the irrigation volume, complying with the EU Water Framework Directive.

The IRRISAT methodology is summarized in Figure 2. Crop potential evapotranspiration (ETp) is computed with the Penman–Monteith equation, with crop parameters albedo (α) and leaf area index (LAI) derived from processing Sentinel-2A/B images in the visible and infrared regions [7] while assuming fixed values for the stomatal resistance (sr $\approx$ 100 sm^{-1}) and crop height (hc = 0.4 m) for herbaceous crop [8]. Following this approach, the calculation of ETp requires standard meteorological data (daily air temperature, relative humidity, solar radiation, wind speed and precipitation), LAI and surface α.

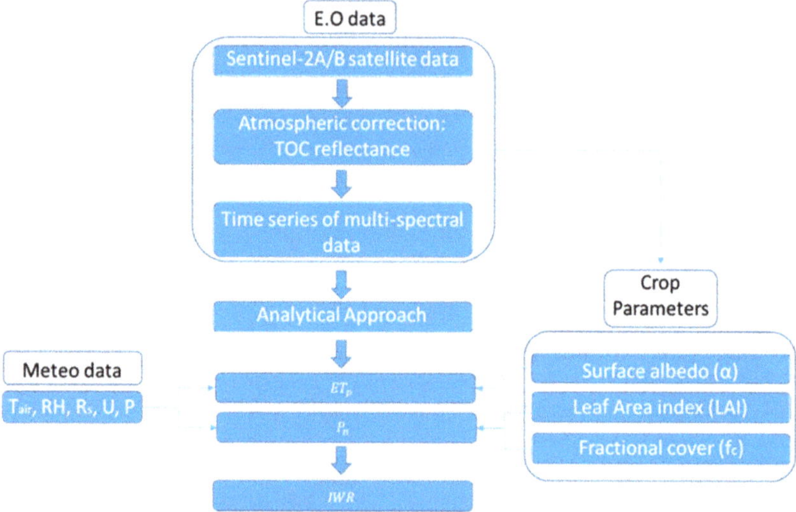

Figure 2. Flowchart shows dataset and processing procedure required to estimate crop potential evapotranspiration (ETp), and irrigation water requirements (IWR), Earth Observation-based direct FAO-56 Penman-Monteith equation method. Legend: Ta = air temperature, RH = relative humidity; Rs = solar radiation; U = wind speed; P = precipitation.

The irrigation water requirement (*IWR*) is then calculated as the difference between *ETp* and the crop effective rainfall (*Pn*), according to the following equation:

$$\text{IWR} = \text{ETp} - \text{Pn} \tag{1}$$

It is assumed that capillary rise does not contribute to root zone soil moisture in the summer season, as usually occurs in the southern European regions. Runoff and deep drainage are assumed to be negligible considering the low amount of rainfall during the two growing seasons. *Pn* is obtained by reducing the precipitation above canopy (P) by a quantity that depends on canopy development, according to an empirical function of the LAI and the fractional vegetation cover fc, as reported in Vuolo et al. [7].

In this study, 10 and 21 multispectral high-resolution images from Sentinel-2A and 2B have been acquired during 2017 and 2018, respectively. The multi spectral instrument (MSI) on board of Sentinel-2A/2B captures data at 10, 20 and 60 m of spatial resolution over 13 spectral bands with a very high temporal resolution of five days at the equator. Individual Sentinel-2 granules Level-1C (processed at the top-of-atmosphere reflectance) were acquired from Copernicus Open Access Hub (https://scihub.copernicus.eu/), already ortho-rectified in UTM/WGS84 (image tiles of 100×100 km^2). The information gathered by Sentinel-2 system (orbit, attitude, date accuracy and viewing directions of all detectors) are exploited for geolocating all Sentinel-2 pixels with an accuracy of about 11 m for about 97% of the cases, which is about the size of one Sentinel-2 pixel. The standard need for multi-temporal registration errors is 0.3 pixels, and the current performances show that for more than 50% of the cases, the performance does not meet that requirement. The resolution is estimated to be three times the registration error, thus the resolution Sentinel-2 time series is around 30 m.

Level-1C products were processed into Level-2A-Bottom-of-Atmosphere (BoA) reflectance-data using the ESA's Sen2Cor v2.5.5 tool (http://step.esa.int/main/third-party-plugins-2/sen2cor/sen2cor_v2-5-5/). Sen2Cor tool performs the atmospheric, terrain and cirrus correction of Top-Of-Atmosphere Level 1C input data, and creates Bottom-Of-Atmosphere, optionally terrain and cirrus corrected reflectance images; additional, aerosol optical thickness, water vapor, scene classification maps and quality indicators for cloud and snow probabilities.

In order to obtain homogeneous and comparable products as time series, all value-added products (LAI, α and fc) are calculated based on atmospherically corrected Level-2A data. LAI and fc are calculated by S2ToolBox [21], an artificial neural network (ANN) algorithm, trained by using radiative transfer simulations from PROSPECT [22] and SAIL [23] models, and tailored for Sentinel-2 data. The algorithm requires eight Sentinel-2 spectral bands (B3–B7, B8a, B11 and B12) at 10 and 20 m (pixel size), which are all resampled to 10 m to derive LAI and fc. Experimental studies have shown the accuracy of this approach for LAI estimation in different environments and crops [24,25]. In this study, average and variance of LAI and fc at parcel scale were assessed by taking a minimum of 50 pixels falling within each parcel, after excluding pixels affected by boundary effects or cloudiness, according to the quality indicator provided by S2ToolBox.

The broadband surface albedo has been calculated, when the observed surface is considered as Lambertian, as the integration of at-surface reflectance across the shortwave spectrum [26], as shown in equation:

$$\alpha = \sum_{bi} \left| \rho_{bi} \cdot \omega_{bi} \right| \tag{2}$$

where α is albedo, ρ_{bi} is surface reflectance for a given band bi at Level-2A Sentinel-2 surface reflectance, ω_{bi} is the weighting coefficient representing the solar radiation fraction derived from the solar irradiance spectrum [26] within the spectral range (spectral response curves) for bands bi and is calculated as equation:

$$\omega_{bi} = \frac{\int_{LO_{bi}}^{UP_{bi}} R_{s\lambda} \cdot d\lambda}{\int_{0.4}^{2.4} R_{s\lambda} \cdot d\lambda} \tag{3}$$

where $R_{s\lambda}$ is extra-terrestrial irradiance for wavelength λ (μm); and UP_{bi} and LO_{bi} are upper and lower wavelength bounds for Sentinel-2A/B band bi, respectively.

2.3. Model-Based Assessment of Crop Water Requirement: AquaCrop

AquaCrop simulates crop yield in four steps: Crop development, crop transpiration, biomass production and yield formation. It calculates the daily soil water balance and divides evapotranspiration into soil evaporation and crop transpiration. AquaCrop describes the foliage development of the crop by the canopy cover (CC), that is formally equivalent to the fractional cover (fc) estimated by Sentinel-2 imagery, i.e., it is the fraction of soil surface covered by the green canopy. Hereinafter, we use the two terms canopy cover (CC) and fractional cover (fc) just to distinguish the two variables, respectively derived with AquaCrop and Sentinel-2 imagery.

Transpiration is a function of CC, while evaporation is proportional to the area of soil not covered by vegetation. The CC is multiplied by reference evapotranspiration (ETo), determined by the FAO Penman–Monteith equation, and the crop coefficient (Kc) to calculate potential crop transpiration. Actual transpiration (Ta) is calculated starting from potential one by accounting for water stress. Then, Ta is used for the calculation of crop biomass though its multiplication with water productivity normalized for climate. By using a harvest index (HI), crop yield is obtained by the biomass. To describe the effect of water stress, the model considers different thresholds of water available to the root zone. The first affects leaf canopy expansion (slowing down); the second threshold affects canopy senescence (quickening); the third is referred to as stomata closure (increase) and so to transpiration. Stress coefficients (Ks) range between 1 (no stress) and 0 (complete stress) and are multiplicative factors of the target process.

Model parameters are grouped into two classes: Conservative and non-conservative. Conservative parameters are not dependent on local and management conditions: Canopy growth (CGC) and canopy decline (CDC) coefficients; full canopy crop transpiration coefficient (Kc); biomass WP and soil water depletion thresholds. Non-conservative parameters vary depending on crop and field management, soil type, and climate (sowing date and density, length of crop cycle and phenological stages, maximum canopy cover, etc.). Such parameters can be either retrieved from AquaCrop literature or calibrated by the user (e.g., by means of field experiments). Main AquaCrop outputs are crop production (biomass and yield) and crop water use.

AquaCrop Implementation

In this study, a limited number of AquaCrop parameters were partly calibrated with field observations, including management information: Transplant dates and densities, flowering date and duration, starting of senescence, maturity, and final yield were used for local calibration of the model. For simulating irrigation, the model was set in net irrigation requirement mode, which estimates the crop water requirement based on a selected threshold of allowed root zone (water) depletion (RZD). In order to reproduce the irrigation method adopted by the farmer, drip irrigation was simulated to ensure that RZD was always above 50% of the readily available water (RAW).

2.4. Assimilation-Based Assessment of Crop Water Requirement

The third method for assessing crop water requirements was based on the integration of Sentinel-2 crop derived data with AquaCrop. The fractional cover (fc) estimated by Sentinel-2 has been sequentially assimilated into AquaCrop, by direct insertion, in place of the canopy cover (CC) simulated by the model. The sequential direct insertion is applied under the assumption that a continuous update of one crop model state based on remote observations can reduce the biases induced by the model simplifications of the processes and environmental conditions influencing the crop growth dynamics.

Crop CC simulated by AquaCrop along the growing season and the fc values measured by satellite were compared and the differences were statistically analyzed by means of the Pearson correlation coefficient (r), root mean square error (RMSE%), Nash–Sutcliffe model efficiency coefficient (EF) and Willmott index of agreement (d).

3. Results and Discussion

3.1. Test Site Characteristics

The parcel cultivated in 2017 had a Vitric Phaeozems (Eutric) soil (WRB classification), with a loamy texture (USDA classification), while soil was Vitric Cambisols with a loamy-sand texture in 2018. Soils were both deep, well drained and with no fertility constraints (Table 1).

Table 1. Main soil characteristics of the two soils.

Parameter	Unit	2017	2018
Texture		Loam	Silty loam
Gravel	(vol%)	<5	<5
Saturation	(vol%)	52	46
Field Capacity	(vol%)	29	33
Wilting Point	(vol%)	10	13
Cation Exchange Capacity	(meq/100g)	32	24
Bulk Density	(t/m^3)	1.1	1.3
pH in H_2O	U.pH	6.9	7.1
Organic Matter	(%)	2.6	2.7

Considering the climatology of the study area, average temperature of 15.2 °C and precipitation of 900 mm (period 1982–2012), year 2017 was relatively dry, with a cumulated rainfall of 579 mm and 16.6 °C of average temperature, whereas 2018 was relatively wet and warm, with 1047 mm of rainfall and an average temperature of 17.25 °C (Figure 3). Observed processing tomato yield (expressed in dry matter) was 7.20 t/ha in 2017 and 7.35 t/ha in 2018.

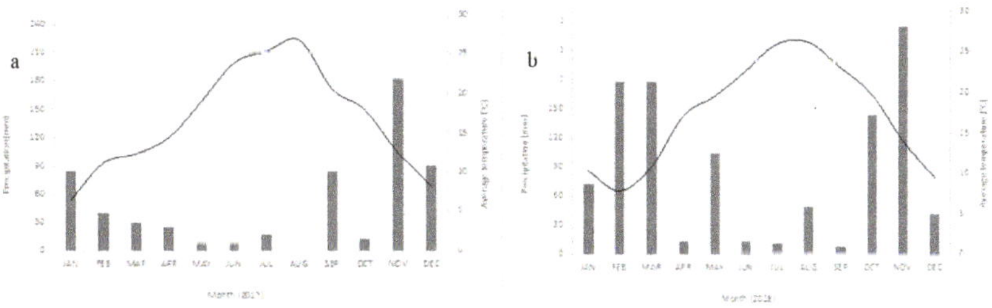

Figure 3. Temperature (line) and precipitation (bars) in the experimental area in year 2017 (**a**) and 2018 (**b**).

3.2. AquaCrop Calibration and Implementation

AquaCrop parameters were calibrated to obtain the best fit between field observations and simulations, both in 2017 and 2018 (Table 2). AquaCrop simulated yields were 7.23 and 7.60 t/ha (dry weight) in 2017 and 2018, with an error of 0.42% and 3.40%, respectively (Table 3). The growth of the green canopy simulated by the model was compared with the fc values observed by IRRISAT. In both seasons the simulated growing curve fitted well with the satellite observations, although an underestimation for the initial canopy cover (late April–early May) and an overestimation during the last part of the growing season (July) was observed (Figure 4). Nevertheless, the statistical analysis (RMSE, EF, d) showed a very good agreement between simulated CC and fc, which correlation was highly significant in both years (α = 0.001; Table 4; Figure 5). In 2017 the spatial variability of the crop canopy within the study parcel was larger than in 2018, as testified by the larger variance of the fc values retrieved in 2017.

Table 2. Input crop file used for AquaCrop simulations (HI: Harvest index, GDD: Growing degree days).

Crop Parameter	Unit	Value
Max canopy cover	GDD	686
Flowering	GDD	612
Flowering duration	GDD	531
Length building up HI	GDD	448
Senescence	GDD	1013
Maturity	GDD	1358
Max rooting depth	GDD	612
Initial canopy cover	%	0.64
Maxi canopy cover	%	75
Reference HI	%	63
Base temperature	°C	5
Upper temperature	°C	30
Canopy expansion Upper		0.15
Canopy expansion Lower		0.55
Stomatal closure		0.5
Early canopy senescence		0.7

Table 3. Crop and water balance variables (Tr: Crop transpiration, E: Soil evaporation, ETp: Potential evapotranspiration, IWR: Irrigation water requirement, WP_{ET}: Evapotranspiration water productivity, WP_{IWR}: Irrigation water requirement water productivity).

		Yield (t/ha)	Tr (mm)	E (mm)	ETp (mm)	IWR (mm)	WP_{ET} (kg/m^3)	WP_{IWR} (kg/m^3)
2017	Observed	7.20				416		1.73
	IRRISAT				450	450		
	AquaCrop	7.23	345	192	537	461	1.35	1.57
	Assimilation	8.23	372	165	537	461	1.52	1.79
2018	Observed	7.35				402		1.82
	IRRISAT				349	298		
	AquaCrop	7.60	291	137	428	332	1.78	2.29
	Assimilation	7.34	273	139	412	317	1.78	2.31

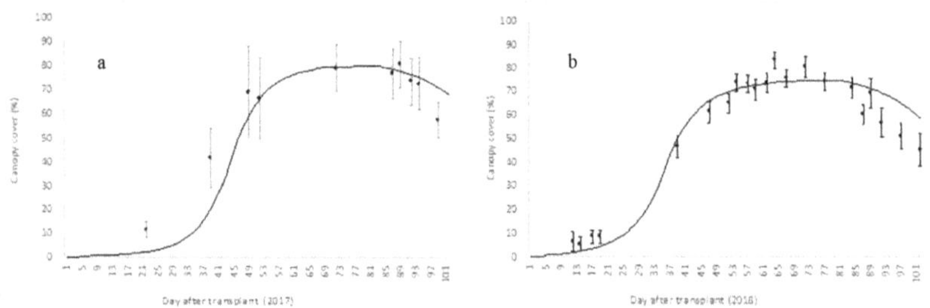

Figure 4. Canopy cover of tomato simulated by AquaCrop, after calibration with field data (line) and corresponding fractional cover values (dots) retrieved by Sentinel-2 imagery during 2017 (**a**) and 2018 (**b**) growing seasons, with corresponding standard deviations.

Table 4. Evaluation of canopy cover simulation results: Number of observations/simulations (n), Pearson correlation coefficient (r), root mean square error (RMSE%), Nash–Sutcliffe model efficiency coefficient (EF) and Willmott index of agreement (d).

	n	*r*	**RMSE**	**EF**	**d**
2017	10	0.95	9.10	0.8	0.96
2018	22	0.97	8.10	0.91	0.98

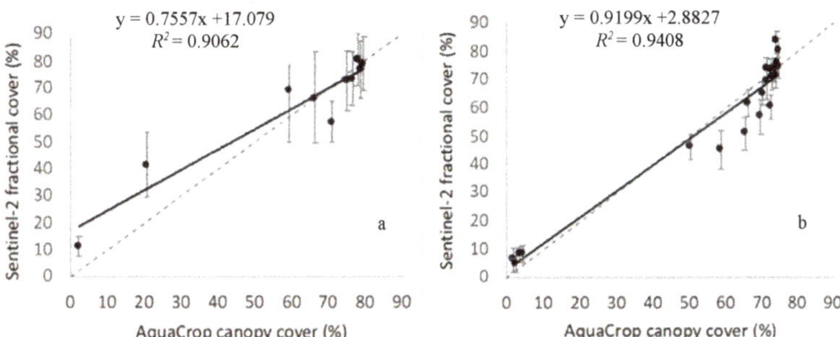

Figure 5. Linear correlation between tomato canopy cover simulated by AquaCrop and fractional cover retrieved by Sentinel-2 imagery in 2017 (**a**) and 2018 (**b**). Standard deviation (bars) and 1:1 line (dashed line) are reported.

3.3. Estimation of Crop Water Irrigation Requirements

During the 2017 crop growing season, spanning from June 9th to July 20th, the accumulated rainfall was 40.4 mm, the estimated reference evapotranspiration (ETo) was 553 mm, while crop transpiration (Tr) and soil evaporation (E) estimated by AquaCrop were 345 and 192 mm, respectively. In such conditions, AquaCrop modeled an irrigation water requirement (IWR) of 461 mm corresponding to 144 L/plants.

In year 2018, wetter conditions occurred, with an accumulated rainfall of 125 mm and an estimated ETo of 439 mm. Under such conditions, both crop Tr (291 mm) and E (137 mm) were lower. Accordingly, a lower IWR, 332 mm, was estimated by the model. Being the final yields very similar, the consequent evapotranspiration (WP$_{ET}$) was higher in 2018 than 2017 (Tab 3).

In Figure 6, three different daily ETp series were compared, respectively produced by calibrated AquaCrop, IRRISAT and AquaCrop after replacing CC with fc estimated by IRRISAT.

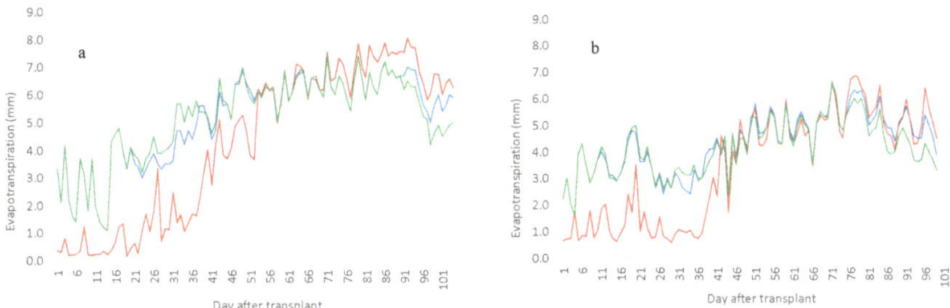

Figure 6. Crop evapotranspiration estimated by AquaCrop (blue), IRRISAT (red) and direct insertion (green) of fc retrieved by Sentinel-2 imagery into AquaCrop, in 2017 (**a**) and 2018 (**b**).

The agreement between AquaCrop and IRRISAT ETp estimates is excellent in the mid-season stage, between about the 50th and 85th day after transplant, when the crop canopy is fully developed. This result is remarkable considering that IRRISAT is fully based on remote observations weather data, and not relying on field or crop data. Moreover, IRRISAT predictions are not based on a continuous assessment of the crop dynamics, rather it provides ETp estimates based on crop parameters retrieved from the last available image, thus with a delay of five or more days, depending on the sky conditions (i.e., cloudiness).

In the initial stage and in the crop development stage (first 50 days after transplant), before canopy is fully developed, the estimation of ETp in AquaCrop is higher than the one determined by IRRISAT. This is due to the conceptually different calculation schemes: Whilst IRRISAT has a resistance-based approach in the Penman–Monteith linked to the LAI at the pixel scale (according to the "big leaf" schematization), AquaCrop has a crop coefficient approach for soil evaporation depending on soil properties. Diversely, with growing LAI and soil cover, the two estimates converge toward very similar values of ETp, hence irrigation requirements.

In the senescence stage, after about the 85th day from transplant, the values of ETp derived by means of the two methods slightly diverge again, with higher ETp for IRRISAT respect to AquaCrop, again due to the difference between the LAI-based approach of the first one and the crop coefficient of the second one. Furthermore, AquaCrop explicitly consider the transpiration reduction associated with the senescence, which cannot be fully predicted by the reduction of the observed LAI applied into the Penman–Monteith equation.

The canopy cover fc maps retrieved from Sentinel-2 imagery can be more effectively employed for constraining AquaCrop predictions to the actual observed crop growth and for assessing its spatial variability. In this study we applied a simple assimilation technique, known in the literature as "direct" insertion [27], consisting in replacing the model state variable CC with the remotely retrieved variable fc. As displayed in Figure 6, AquaCrop ETp estimates after direct insertion are essentially equal to the calibrated AquaCrop, except for the development and senescence stages, when the average fc sensibly deviates from the model calibrated growth curve. As illustrated in Table 3, fc direct insertion does not affect the cumulative ETp, rather it affects its partitioning in transpiration (Tr) an evaporation (E), with a slight increase in Tr and a decrease in E. In 2018, instead, fc direct insertion implies a slight reduction of ETp, mainly associated with a reduction of Tr. Correspondingly, tomato yield was overestimated compared with the observed and the calibrated values in 2017 (+14.3%), while slightly underestimated in 2018 (−1.03%). This result confirmed that sequential assimilation of one state variable does not necessarily improve the prediction performance of all model state variables. In this sense, more complex sequential assimilation techniques are needed, in order to account for the structure of the observation and prediction errors [28], as well of the atmospheric forcing [29].

The impact of the direct insertion in terms of irrigation water requirement (IWR) is null in 2017, while it determines a 5% decrease in 2018. As illustrated in Figure 7, also the temporal pattern of IWR is almost unchanged in the two simulated seasons. A potential water saving of 70 mm could be achieved in 2018 according to AquaCrop, by accounting for the effective contribution of the summer rainfall (125 mm) to the soil water deficit.

Looking again at Table 3, it is interesting to note that IRRISAT, despite the significant reduction of the cumulative ETp both in 2017 (16%) and 2018 (19%) compared with the calibrated AquaCrop, predicts cumulative IWR just 3% smaller in 2017 and 11% smaller in 2018 than the calibrated AquaCrop. As illustrated in Figure 7, this reduction is essentially due to the initial and development stage, when AquaCrop, differently form IRRISAT, accounts for the soil evaporation in the IWR assessment. After the crop is fully developed, the cumulative IWR curves are almost parallel, testifying the good agreement of the remotely assessed daily IWR with AquaCrop.

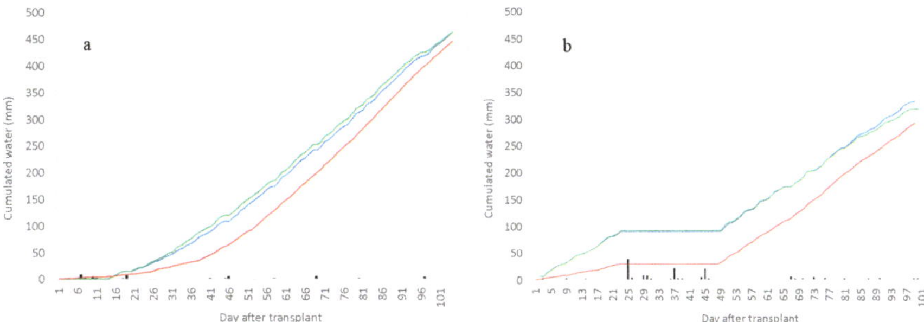

Figure 7. Cumulated irrigation water requirement (mm) estimated by the three methods: AquaCrop (blue), IRRISAT (red) and direct insertion (green) of fractional cover retrieved by Sentinel-2 imagery into AquaCrop, in 2017 (**a**) and 2018 (**b**). Rainfall is also plotted (histograms).

4. Conclusions

Sentinel-2 imagery can be effectively exploited for monitory canopy growth of tomato crops in open field. Irrigation advisory services, such as IRRISAT, which are only based on crop data retrieved by Sentinel-2 and weather data, can provide a reliable assessment of crop water requirements of tomato field crops especially when crop canopy is fully developed. It should be considered that IRRISAT does not require input data concerning the soil or crop phenology, since it is entirely based on crop growth monitoring from space. Hence, integrating Sentinel-2 imagery with a crop growth model such as AquaCrop can be an effective strategy for assessing crop water requirement in the initial and development stage of the crop, as well as for identifying the senescence stage. Further, being the satellite imagery a spatial information, the integration into a crop model can help in assessing crop water requirement at field or higher scales, i.e., at territorial level. Thus, a sequential assimilation can be used to support irrigation planning by Irrigation and Land Reclamation consortia. In this study a simple direct insertion method has been applied for assimilating canopy cover retrieved by Sentinel-2 imagery into AquaCrop, which does not guarantee an optimal model-data integration. Additional studies are required for testing more advanced data assimilation techniques, accounting for the structure of the model state and observation errors.

Author Contributions: Conceptualization, A.D.M., G.B.C. and F.A.; Data curation, S.F.B.; Formal analysis, A.D.M. and G.B.C.; Investigation, A.D.M., M.M. and F.A.; Methodology, A.D.M., G.B.C., S.F.B., M.M., G.D., C.D.M. and F.A.; Software, S.F.B.; Supervision, G.D., S.O. and F.A.; Validation, G.B.C.; Writing—original draft, A.D.M., G.B.C. and F.A.; Writing—review & editing, M.M.

Funding: This study was part of the OPERA project, financed under the ERA-NET Cofund WaterWorks2015 Call.

Acknowledgments: The authors wish to thank OPERA project being carried out within the Water-JPI ERA-NET Cofund WaterWorks2015 Call. We wish to thank also the team of the Irrigation and Land Reclamation Consortium of "Bacino Inferiore del Volturno", and Francesco D'Amore, for their availability, help and support in organizing field campaigns.

Conflicts of Interest: The authors declare no conflict of interest.

References

1. Altobelli, F.; Cimino, O.; Natali, F.; Orlandini, S.; Gitz, V.; Meybeck, A.; Marta, A.D. Irrigated farming systems: Using the water footprint as an indicator of environmental, social and economic sustainability. *J. Agric. Sci.* **2018**, *156*, 711–722. [CrossRef]
2. Tsakmakis, I.D.; Zoidou, M.; Gikas, G.D.; Sylaios, G.K. Impact of Irrigation Technologies and Strategies on Cotton Water Footprint Using AquaCrop and CROPWAT Models. *Environ. Process.* **2018**, *5*, 181–199. [CrossRef]

3. Kustas, W.; Anderson, M. Advances in thermal infrared remote sensing for land surface modeling. *Agric. For. Meteorol.* **2009**, *149*, 2071–2081. [CrossRef]

4. Anderson, M.C.; Norman, J.M.; Mecikalski, J.R.; Otkin, J.A.; Kustas, W.P. A climatological study of evapotranspiration and moisture stress across the continental United States based on thermal remote sensing: 2. Surface moisture climatology. *J. Geophys. Res.* **2007**, *112*, D11112. [CrossRef]

5. Karimi, P.; Bastiaanssen, W.G.M.; Molden, D. Water Accounting Plus (WA+)—A water accounting procedure for complex river basins based on satellite measurements. *Hydrol. Earth Syst. Sci.* **2013**, *17*, 2459–2472. [CrossRef]

6. Drusch, M.; Del Bello, U.; Carlier, S.; Colin, O.; Fernandez, V.; Gascon, F.; Hoersch, B.; Isola, C.; Laberinti, P.; Martimort, P. Sentinel-2: ESA's optical high-resolution mission for GMES operational services. *Remote Sens. Environ.* **2012**, *120*, 25–36. [CrossRef]

7. Vuolo, F.; D'Urso, G.; De Michele, C.; Bianchi, B.; Cutting, M. Satellite-based irrigation advisory services: A common tool for different experiences from Europe to Australia. *Agric. Water Manag.* **2015**, *147*, 82–95. [CrossRef]

8. D'Urso, G. Current Status and Perspectives for the Estimation of Crop Water Requirements from Earth Observation. *Ital. J. Agron.* **2010**, *5*, 107–120.

9. Casa, R.; Varella, H.; Buis, S.; Guerif, M.; de Solan, B.; Frederic, B. Forcing a wheat crop model with LAI data to access agronomic variables: Evaluation of the impact of model and LAI uncertainties and comparison with an empirical approach. *Eur. J. Agron.* **2012**, *37*, 1–10. [CrossRef]

10. Steduto, P.; Hsiao, T.C.; Raes, D.; Fereres, E. Aquacrop-the FAO crop model to simulate yield response to water: I. concepts and underlying principles. *Agron. J.* **2009**, *101*, 426–437. [CrossRef]

11. Raes, D.; Steduto, P.; Hsiao, T.C.; Fereres, E. AquaCropThe FAO Crop Model to Simulate Yield Response to Water: II. Main Algorithms and Software Description. *Agron. J.* **2009**, *101*, 438–447. [CrossRef]

12. Linker, R.; Ioslovich, I. Assimilation of canopy cover and biomass measurements in the crop model AquaCrop. *Biosyst. Eng.* **2017**, *162*, 57–66. [CrossRef]

13. Jin, X.; Kumar, L.; Li, Z.; Xu, X.; Guijun, Y.; Wang, J. Estimation of Winter Wheat Biomass and Yield by Combining the AquaCrop Model and Field Hyperspectral Data. *Remote Sens.* **2016**, *8*, 972. [CrossRef]

14. Silvestro, P.C.; Pignatti, S.; Pascucci, S.; Yang, H.; Li, Z.; Guijun, Y.; Huang, W.; Casa, R. Estimating Wheat Yield in China at the Field and District Scale from the Assimilation of Satellite Data into the Aquacrop and Simple Algorithm for Yield (SAFY) Models. *Remote Sens.* **2017**, *9*, 509. [CrossRef]

15. Jin, X.; Liu, S.; Baret, F.; Hemerlé, M.; Comar, A. Estimates of plant density of wheat crops at emergence from very low altitude UAV imagery. *Remote Sens. Environ.* **2017**, *198*, 105–114. [CrossRef]

16. Jin, X.; Kumar, L.; Li, Z.; Feng, H.; Xu, X.; Guijun, Y.; Wang, J. A review of data assimilation of remote sensing and crop models. *Eur. J. Agron.* **2018**, *92*, 141–152. [CrossRef]

17. Zhao, Y.; Chen, S.; Shen, S. Assimilating remote sensing information with crop model using Ensemble Kalman Filter for improving LAI monitoring and yield estimation. *Ecol. Model.* **2013**, *270*, 30–42. [CrossRef]

18. Trombetta, A.; Iacobellis, V.; Tarantino, E.; Gentile, F. Calibration of the AquaCrop model for winter wheat using MODIS LAI images. *Agric. Water Manag.* **2016**, *164*, 304–316. [CrossRef]

19. Pelosi, A.; Medina, H.; Villani, P.; D'Urso, G.; Chirico, G.B. Probabilistic forecasting of reference evapotranspiration with a limited area ensemble prediction system. *Agric. Water Manag.* **2016**, *178*, 106–118. [CrossRef]

20. Chirico, G.B.; Pelosi, A.; De Michele, C.; Bolognesi, S.F.; D'Urso, G. Forecasting potential evapotranspiration by combining numerical weather predictions and visible and near-infrared satellite images: An application in southern Italy. *J. Agric. Sci.* **2018**, *19*, 702–710. [CrossRef]

21. Garrigues, S.; Lacaze, R.; Baret, F.J.; Morisette, J.T.; Weiss, M.; Nickeson, J.E.; Fernandes, R.; Plummer, S.; Shabanov, N.V.; Myneni, R.B.; et al. Validation and intercomparison of global Leaf Area Index products derived from remotesensing data. *Geophys. Res.* **2008**, *113*. [CrossRef]

22. Jacquemoud, S.; Baret, F. PROSPECT: A model of leaf optical properties spectra. *Remote Sens. Environ.* **1990**, *34*, 75–91. [CrossRef]

23. Verhoef, W. Light scattering by leaf layers with application to canopy reflectance modeling: The SAIL model. *Remote Sens. Environ.* **1984**, *16*, 125–141. [CrossRef]

24. Atzberger, C.; Berger, K. Spatially constrained inversion of radiative transfer models for improved LAI mapping from future Sentinel-2 imagery. *Remote Sens. Environ.* **2012**, *120*, 208–218. [CrossRef]

25. D'Urso, G.; Calera Belmonte, A. Operative approaches to determine crop water requirements from earth observation data: Methodologies and applications. *AIP Conf. Proc.* **2006**, *852*, 14–25.

26. Thuillier, G.; Hersé, M.; Labs, D.; Foujols, T.; Peetermans, W.; Gillotay, D.; Simon, P.C.; Mandel, H. The solar spectral irradiance from 200 to 2400 nm as measured by the solspec spectrometer from the atlas and Eureca missions. *Sol. Phys.* **2003**, *214*, 1–22. [CrossRef]

27. Chirico, G.B.; Medina, H.; Romano, N. Kalman filters for assimilating near-surface observations in the Richards equation. I: Retrieving state profiles with linear and nonlinear numerical schemes. *Hydrol. Earth Syst. Sci.* **2014**, *18*, 2503–2520. [CrossRef]

28. Medina, H.; Romano, N.; Chirico, G.B. Kalman filters for assimilating near-surface observations into the Richards equation—Part 2: A dual filter approach for simultaneous retrieval of states and parameters. *Hydrol. Earth Syst. Sci.* **2014**, *18*, 2521–2541. [CrossRef]

29. Pelosi, A.; Medina, H.; Van den Bergh, J.; Vannitsem, S.; Chirico, G.B. Adaptive Kalman filtering for postprocessing ensemble numerical weather predictions. *Mon. Weather Rev.* **2017**, *145*, 4837–4854. [CrossRef]

 *agronomy*

Article

Discrimination of Tomato Plants (*Solanum lycopersicum*) Grown under Anaerobic Baffled Reactor Effluent, Nitrified Urine Concentrates and Commercial Hydroponic Fertilizer Regimes Using Simulated Sensor Spectral Settings

Mbulisi Sibanda [1,*], Onisimo Mutanga [1], Lembe S. Magwaza [2,3], Timothy Dube [4], Shirly T. Magwaza [2], Alfred O. Odindo [3], Asanda Mditshwa [3] and Paramu L. Mafongoya [1]

[1] Discipline of Geography, School of Agricultural, Earth and Environmental Sciences, University of KwaZulu-Natal, P/Bag X01, Scottsville, Pietermaritzburg 3209, South Africa
[2] Discipline of Horticulture, School of Agricultural, Earth and Environmental Sciences, University of KwaZulu-Natal, P/Bag X01, Scottsville, Pietermaritzburg 3209, South Africa
[3] Discipline of Crop Science, School of Agricultural, Earth and Environmental Sciences, University of KwaZulu-Natal, P/Bag X01, Scottsville, Pietermaritzburg 3209, South Africa
[4] Institute for Water Studies, Department of Earth Sciences, University of the Western Cape, Private Bag X17, Bellville 7535, South Africa
* Correspondence: sibandam3@ukzn.ac.za; Tel.: +27-33-260-5345

Received: 29 May 2019; Accepted: 9 July 2019; Published: 11 July 2019

Abstract: We assess the discriminative strength of three different satellite spectral settings (HyspIRI, the forthcoming Landsat 9 and Sentinel 2-MSI), in mapping tomato (*Solanum lycopersicum* Linnaeus) plants grown under hydroponic system, using human-excreta derived materials (HEDM), namely, anaerobic baffled reactor (ABR) effluent and nitrified urine concentrate (NUC) and commercial hydroponic fertilizer mix (CHFM) as main sources of nutrients. Simulated spectral settings of HyspIRI, Landsat 9 and Sentinel 2-MSI were resampled from spectrometric proximally sensed data. Discriminant analysis (DA) was applied in discriminating tomatoes grown under these different nutrient sources. Results showed that the simulated spectral settings of HyspIRI sensor better discriminate tomatoes grown under different fertilizer regimes when compared to Landsat 9 OLI and Sentinel-2 MSI spectral configurations. Using the DA algorithm, HyspIRI exhibited high overall accuracy (OA) of 0.99 and a kappa statistic of 0.99 whereas Landsat OLI and Sentinel-2 MSI exhibited OA of 0.94 and 0.95 and 0.79 and 0.85 kappa statistics, respectively. Simulated HyspIRI wavebands 710, 720, 690, 840, 1370 and 2110 nm, Sentinel 2-MSI bands 7 (783 nm), 6 (740 nm), 5 (705 nm) and 8a (865 nm) as well as Landsat bands 5 (865 nm), 6 (1610 nm), 7 (2200 nm) and 8 (590 nm), in order of importance, were selected as the most suitable bands for discriminating tomatoes grown under different fertilizer regimes. Overall, the performance of simulated HyspIRI, Landsat 9 OLI-2 and Sentinel-2 MSI spectral bands seem to bring new opportunities for crop monitoring.

Keywords: hydroponic; vegetable monitoring; crop production; spectral simulation; hyperspectral data

1. Introduction

Food insecurity is a large and growing challenge in sub-Saharan Africa [1,2]. It is estimated that at least one out of four people are hungry and undernourished in sub-Saharan Africa. The World Bank estimates that in 2030, nearly 9 in 10 extremely poor people will be living in Sub-Saharan Africa [3]. This is exacerbated by droughts and soil nutrients deficiencies resulting from limited fertilizer applications [1,2]. This is in turn associated with high fertilizer and food prices, amongst other factors. According to FAO [4] the annual food inflation increased from 5% in 2014 to 6% in 2018 whereas in Europe it remained stable and declined in Latin America, Asia, and Oceania. Subsequently, the improvement of crop production which leads to food security has been amongst the principal priorities required to fulfil the goals of sustainable human development as well as the African Union's Agenda 2063 [5]. Furthermore, food demand is anticipated to triple in sub-Saharan Africa after the projected 2.5-fold increase in population increase [6]. Specifically, a 60% increase in agricultural and horticultural production will be required by the increasing population in the light of diminishing water and soil nutrient resources [6]. The major concern is the current dietary transition, which is in favour of vegetables such as tomatoes amongst other crops, which is projected to increase, especially in urban areas, while water and soil nutrients are in decline [6–8].

Tomato fruits have a critical dietary role in providing folate, vitamins A, C and E; as well as antioxidants (lycopene, beta-carotene, gamma-carotene); trace elements of flavonoids; phytosterols and water-soluble vitamins important for human health [7]. To circumvent the challenge of decreasing soil nutrients and increase the production of vegetables (tomatoes) within a small land area, efforts have been exerted towards improving soil fertility and reducing expenses associated with commercial hydroponic fertilizer mix (CHFM) through the use of anaerobic baffled reactor (ABR) effluents and nitrified urine concentrate (NUC) both as a source of soil nutrients and water [8,9]. Smith and Smith [9], for instance, noted that nitrogen recovered from wastewater supported a high increase in tomato (*Solanum lycopersicum*) plant canopy volume, flower and fruit production when compared to plants treated with commercial hydroponic fertilizer mix (CHFM) which contained N, P, K, Ca, Mg and Si. Al-Hamdan, Cruise et al. (2014) in Jordan noted that treatment of tomato crop using waste water facilitated and increased their fruit size by up to 2 cm in diameter, and weight up to 78.7 g in relation to those administered with potable water in their field experiment. However, the challenge that has been lurking in the agricultural sector is the lack of comprehensive spatial explicit frameworks as well as objective criteria for crop growth and productivity monitoring. Spatially explicit data is important in effectively and precisely managing production both in the field and greenhouses to meet the increasing demand of high quality and safe agricultural products such as tomatoes. Information on vegetable crop type, growth, productivity or health status was previously measured in situ or done through routine field surveys which are often time consuming and date lagged [10,11]. Despite the fact that these in situ methods obtained plausible levels of accuracy in characterizing crops, they lacked spatial representativeness. Consequently, there is need for spatial explicit techniques that can be operationally used not only to characterize the crop's areal extent, but also their physiognomies due to lack or excess of soil nutrients. This information can help deduce and understand crop quality, growth and productivity patterns, which are critical in ensuring food security and coming up with well-informed intervention mechanisms or management strategies where necessary.

Meanwhile, earth observation technologies offer spatially explicit non-destructive synoptic views, innovative and economically feasible timely spatial scale means of generating farm scale crop monitoring. Literature shows that crop/plant physiology and structure such as leaf area index, water content plant pigment content, canopy architecture and canopy density are associated with specific key spectral wavebands [12–15]. Subsequently the variations in these biochemical and physical characteristics of crop plants caused by various crop management practices (i.e., different fertilizer regimes) facilitates unique variations in their spectral finger prints. This makes remote sensing and earth observation facilities to be critical reservoirs of spatially and crop explicit information required in ensuring good quality of crop produce. Remotely sensed data is robust and very sensitive to subtle vegetation traits such as those induced by different water and nutrient regimes. Rajah et al. showed that hyperspectral remotely sensed data could discriminate common dry beans that were rain-fed from those that were irrigated. Lu, et al. [16] discriminated tomato crops that were infected with multi-diseases at different phenological stages using hyperspectral data. Their results exhibited a high overall classification accuracy of 100% in discriminating multi-diseases during the early, asymptomatic and late stages of leaves growth. Above all, the advent and advancement of earth observation facilities has unveiled opportunities for assessing previously unresolved crop quality, growth and productivity related questions linked to plant physiognomies such as that induced by different fertilizer regimes on plant spectral characteristics [11]. Despite hyperspectral data's trade-off between cost and accuracy, it remains the most accurate spatial data for monitoring crop growth and productivity. Unlike broadband satellite data, hyperspectral data has numerous contiguous spectral channels with a potential ability to detect and characterize subtle differences in plant traits such as those induced by different fertilizer application regimes in relation. For example, Česonienė, et al. [17] demonstrated that hyperspectral data could discriminate between conventionally and organically grown carrots in Lithuania to Jeffries-Matusita distances ranging between 1.98 and 2.00. They attributed their results to the ability of hyperspectral data to discern on the variations in the cell structure conditions and canopy structure of the carrots grown under different farming methods. Meanwhile, multispectral sensors like Landsat, Satellite Pour l'Observation de la Terre (SPOT), MODerate Resolution Imaging Spectroradiometer (MODIS) are characterized with broad bands making it difficult to discern subtle plant traits as they tend to mask out critical plant information.

Although hyperspectral sensors provide accurate datasets, a number of sensors have been or are being developed with improved sensing capabilities because of the exorbitant acquisition costs associated with it [18]. For example, the earth observation community recently witnessed the launching of Sentinel 2 multispectral imager (MSI) and Landsat 8 OLI etc. Sentinel 2 MSI has been the first freely available sensor with a set of spectral wavebands covering the red-edge section of the electromagnetic spectrum (B5 (705 nm), 6 (740 nm), and 7 (783 nm)) at a relatively fine spatial resolution of 20m. The sensor has a wide swath-width of 290 km, coupled with a high spatial resolution of 10 m as well as a five-day temporal resolution making it a better facility for crop mapping and monitoring. Both sensors (i.e., Sentinel 2 MSI and Landsat 8 OLI) have been tested in various environmental application areas with plausible findings and conclusions [19–22]. However, in some instances they have been reported to experience challenges, especially when applied at farm level monitoring. This has been attributed to the presence of broad wavebands which are perceived to be concealing most important information. As a result, now new sensors such as the proposed Landsat 9 OLI-2 (with improved noise-to-signal ratio), Environmental Mapping and Analysis Program (EnMAP) and Hyperspectral Infrared Imager (HyspIRI) are being developed. The National Aeronautics and Space Administration agency (NASA) is looking forward to launching the state-of-the-art HyspIRI instruments covering the visible and near-infrared section (Vis/NIR) as well as the thermal infrared (TIR). There is a need to evaluate their performance in discriminating subtle plant properties resulting from crop management practices in relation to the available broadband multispectral sensors. These sensors could help by providing a spatially explicit non-destructive method of characterizing crop quality, growth and productivity patterns which are required in the food industry for pricing as well as in agricultural production for ensuring food security.

The upcoming hyperspectral instruments have a potential to supply the much-needed spatially explicit, accurate, consistent information on vegetable crops. Both of these instruments will be spectrometric covering the spectral ranges of 420–2450 nm and 380–2510 nm at different sampling distances of 6.5 nm for EnMAP's VNIR and 10 nm for EnMAP's SWIR section as well as HyspIRI's VSWIR [23,24]. The swath width of HyspIRI will be 185 km at 30 and 60 m spatial resolutions whereas EnMAP have 30-km-wide coverage across-track at a ground-sampling unit of 30m. The temporal resolution of EnMAP will be 4 days at the equator whereas that of HyspIRI will be 5 days. The fine spectral, spatial and temporal resolutions of these sensors make them relatively more suitable for agriculture applications. The major advantage with such instruments is that they will avail quality data at relatively low costs for data scarce regions such as the sub-Saharan Africa where resources are limited. In this regard, there is need to compare the performance of these hyperspectral sensors to the recently launched freely available and forthcoming multispectral sensors (i.e., Sentinel 2 MSI and Landsat 9 OLI-2 with improve spectral settings) so as to ascertain their full potential.

Considering the fact that literature states that hyperspectral data is generally characterized by high collinearity and that there is no specific algorithm that is suitable for discriminating vegetation characteristics at different places and times, this study also examined the performance of discriminant analysis (DA) and partial least squares discriminant analysis (PLS-DA) in discriminating tomato plants grown under different fertilizer regimes. PLS-DA and DA have been widely used in discriminating plants with different characteristics using both hyperspectral and broadband sensors data [25–27]. These Algorithms have been widely used because they offer an opportunity to evaluate and interpret the minute spectral pattern variations in plants, especially those grown under different management regimes. PLS-DA and DA classification ensembles construct a distinctive spectrum that represents the spectral signatures of the plants samples while simplifying the discrimination process when compared with other methods such as K-nearest neighbours [28–30]. Despite the optimal performance of these methods in vegetation discrimination, the robustness and accuracy of these two is yet to be established [31] particularly in discriminating fine spectral variations of vegetable crops (i.e., tomatoes) induced by different fertilizer application regimes. Furthermore, hyperspectral data could be widely renowned for its exceptional performance in literature in relation to broad bands sensors but the performance of these forthcoming sensors remains undocumented. This study therefore, sought to assess the discriminative performance of HyspIRI, Landsat 9 OLI-2 and Sentinel 2 MSI in characterizing tomato (*Solanum lycopersicum*) crops grown under commercial hydroponic fertilizer mix, anaerobic baffled reactor effluent and nitrified urine concentrate as nutrient sources. HyspIRI was selected as a representative of hyperspectral sensors in this study due to the fact that both EnMAP and HyspIRI will have similar spectral and spatial characteristics and that HyspIRI is going to cover a larger spectral portion of the electromagnetic spectrum particularly in the VSWIR sensor in relation to that of EnMAP. In addition, this study also assessed the performance of DA and PLS-DA in discriminating tomato crops grown under different fertilizer regimes using HyspIRI, Landsat 9 OLI-2 and Sentinel 2 MSI spectral data simulated from hyperspectral data.

2. Materials and Methods

2.1. Experimental Set-Up

A pot experiment was conducted in a hydroponic system that was set up in a polyethylene tunnel located at Newlands-Mashu Research Station under eThekwini Municipality, Durban, South Africa (29°46′25.648″ E 30°58′28.329″ S). The hydroponic system was designed to run three nutrients streams namely, anaerobic baffle reactor (ABR) effluents, nitrified urine concentrate (NUC) and commercial hydroponic fertilizer mix (CHFM) as a control. Each hydroponic system consisted of 150 L tank and the nutrient solution for each system was enclosed in a 100 L container stacked on the ground at the foot of each system.

Six-weeks-old, seedlings of 'Monica', a determinate tomato cultivar purchased from a local nursery (Sunshine seedlings, Pietermaritzburg, South Africa) were transplanted to 30 cm polyethylene pots filled with pine sawdust as a growing medium. The nutrient solution for each nutrient source was supplied to the plants using a pressure pump (DAB Model K30/70M, DAB Pumps, MarcoPolo, Mestrino, Italy) via a 20 m irrigation line. A 20 cm drip irrigation emitters (2 L) were placed and irrigation was performed at six intervals of 5 min/duration daily using a timer. The study was arranged using a complete randomised design with three replications of five plants each, giving fifteen experimental units per nutrient source.

Tomato plants of the control treatment were irrigated with a commercial hydroponic fertilizer mix (Hygroponic® and Solu-cal®) at the rate of 800 g + 620 g/1000 L of water as recommended for hydroponic tomato production; NUC, commercial fertilizer application rate was used as a standard as recommended by Jonsson et al. (2004) and ABR effluents with no specified application rate. For the CHFM and NUC treatment, the fertilizer was mixed using municipal tap water whereas for the ABR treatment only effluent from the anaerobic baffle reactor component was used as nutrient and irrigation source. The experiment was allowed to run for 12 weeks before the crop was harvested.

Remotely sensed proximal spectra data was collected at vegetative stage (i.e., four weeks after transplanting) and two weeks after flowering. In this study all requirements for the photosynthetic activity of the tomato plants were the same except for the fertilisation treatments. This was done so as to make sure that any spectral differences between the CHFM, ABR, NUC treated tomato plants were mainly due to the fertiliser applications since all conditions were the same.

2.2. Remotely Sensed Data

The FieldSpec-3 (ASD Inc., Boulder, CO, USA) Analytic Spectral Device (ASD) FieldSpec instrument was used to acquire the spectral reflectance of tomatoes plants receiving ABR effluents, NUC and CHFM. The ASD measured the radiation at 1.4 nm intervals for the 350–1000 nm and 2 nm intervals for the 1000–2500 nm spectral regions. The reflectance measurements were conducted using bare fibre-optic held at nadir position ~0.5m above the tomato canopies resulting in a field of view with a diameter of ~0.225m. This diameter was found to be adequate to capture the reflectance of the tomato canopies. The normalization of spectral measurements was conducted after every 5 to 10 spectral measurements, using a standard white spectralon. A standard spectralon is optically flat to ±4% over the range of 250–2500 nm and ±1% over the photopic region of the spectrum [32,33]. Its diffuse reflectance standards are highly Lambertian [33]. This was done to circumvent the possible changes in weather conditions as well as irradiance from the sun [34]. The spectral measurements were conducted under clear skies during the day between 10:00 and 14:00 since this is the time with maximum net radiation. A total of 900 spectral samples were measured on canopies of tomatoes treated with CHFM ($n = 300$), ABR ($n = 300$) and NUC ($n = 300$). Noise regions between 350 and 399, 1355 and 1420 nm, 1810 and 1940 nm and 2470–2500 nm were cleaned prior to any analysis of the data [35]. The spectrometrically sensed proximal spectra were then resampled into the spectral settings of HyspIRI, Sentinel 2_MSI and Landsat 9 OLI-2 (Table 1). To simulate the HyspIRI VSWIR sampling rate, 21 successive ASD bands were averaged to create one proxy HyspIRI band using a Gaussian averaging window as detailed in Prasad et al. (2009) [36] and Samiappan et al. (2010) [37]. To simulate the ASD data into the spectral settings of broad spectral sensors, the PRISM in IDL-ENVI was used.

Table 1. Properties of HyspIRI, Sentinel 2-MSI and Landsat 9 OLI-2 sensors.

Sensor	Orbital Altitude	Revisit Time (days)	Swath Width (Km)	Spectral Bands	Band Centre	Spatial Resolution
HyspIRI (VSWIR)	626 km	16	185 km	Contiguous (10 nm) 380–2500 nm		30
Landsat 9 OLI-2	705 km	16	185 km	1Coastal/Aerosol	443	30
				2Blue	482	30
				3Green	562	30
				4Red	655	30
				5NIR	865	30
				6SWIR 1	1610	30
				7SWIR 2	2200	30
				8Panchromatic	590	15
				9Cirrus	1375	30
				10Thermal	10,800	100
				11Thermal	12,000	100
Sentinel 2-MSI	786 km	5	280km	1Coastal aerosols	443	60
				2Blue	490	10
				3Green	560	10
				4Yellow	665	10
				5Red edge	705	20
				6Red edge	740	20
				7Red edge	783	20
				8NIR	842	10
				8aNIR	865	20
				9NIR	945	60
				10SWIR	1375	60
				11SWIR	1610	20
				12SWIR	2190	20

2.3. Discriminating Tomato Plants Grown under Different Fertilizer Regimes

Exploratory data analysis was conducted to determine if the data followed a normal distribution curve. Normality test was performed using the Kolmogorov-Smirnov test. Further, we assessed spectral separability as well as administered a pre-filter [38] after hyperspectral data resampling it to HyspIRI, Sentinel-2 MSI and Landsat 8/9 OLI spectral configurations. Resampling was based on the Analysis of Variance test (ANOVA). We then conducted post hoc test to establish the channels that exhibited significant differences between the spectral data of the tomato crops receiving nutrients from ABR effluent, NUC and CHFM.

The other objective of this study was to assess the accuracies of partial least squares discriminant analysis and discriminant analysis algorithms in characterizing tomato crops grown under the three fertilizer regimes. In that regard, we used the PLSDA and the DA to classify the spectral reflectance of tomato crops growing under UNC, UNF and ABR fertilizer treatments. Details about DA and PLS DA are provided in Zhang, et al. [39] and Boulesteix [29]. Prior to conducting PLS-DA and DA, the spectral samples were partitioned into training (70%) and testing (30%) data.

2.4. Classification Accuracy Assessment

The 30 percent of the samples were used for classification accuracy assessment (90 per treatment). Confusion matrices were computed and used to evaluate the classification accuracies of the DA and PLS DA models. We further computed the overall, producer and user accuracies, as well as the kappa statistics for each set of spectral settings as classified by the two algorithms based on the confusion matrices. To compare the performance of the two algorithms, a McNemar's test was conducted as detailed by Manandhar, et al. [40] and de Leeuw, et al. [41].

3. Results

Analysis of variance tests results showed significant differences ($\alpha = 0.05$) between tomato plants treated with different fertilizer combinations based on the spectral settings of HyspIRI, Sentinel-2MSI and Landsat 9 OLI. Figure 1 illustrates spectral signatures (mean spectral reflectance) of tomato plants grown under ABR, CHFM and NUC treatments. Specifically, the spectral signatures of tomatoes grown under NUC exhibited higher reflectance curves when compared to those growing under CHFM and ABR in all the sections of the electromagnetic spectrum across all the sensors. The CHFM spectral signature was the lowest in comparison to the other to fertilizer treatments across all the spectral signatures and sensors. For HyspIRI resampled data, significant differences were observed in the visible, NIR as well as the SWIR portions of the electromagnetic spectrum (Figure 1). Meanwhile the most glaring differences in the reflectance of tomato plants grown under ABR, CHFM and NUC treatments were observed in the NIR portion of the electromagnetic spectrum based on the simulated Sentinel 2 MSI and Landsat 9 OLI-2 data (Figure 1). HyspIRI spectral settings exhibited more potential spectral windows of separability between tomato plants grown under ABR, CHFM and NUC treatments when compared with the broadband sensors. Potential spectral windows of separability exhibited by Sentinel 2 MSI and Landsat 9 OLI-2 were only in the near infrared regions (Figure 1).

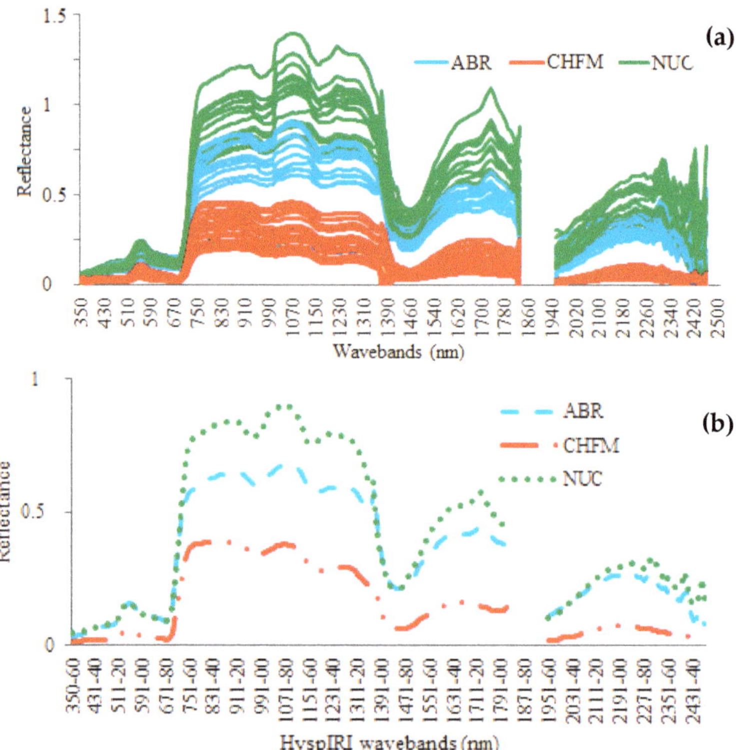

Figure 1. *Cont.*

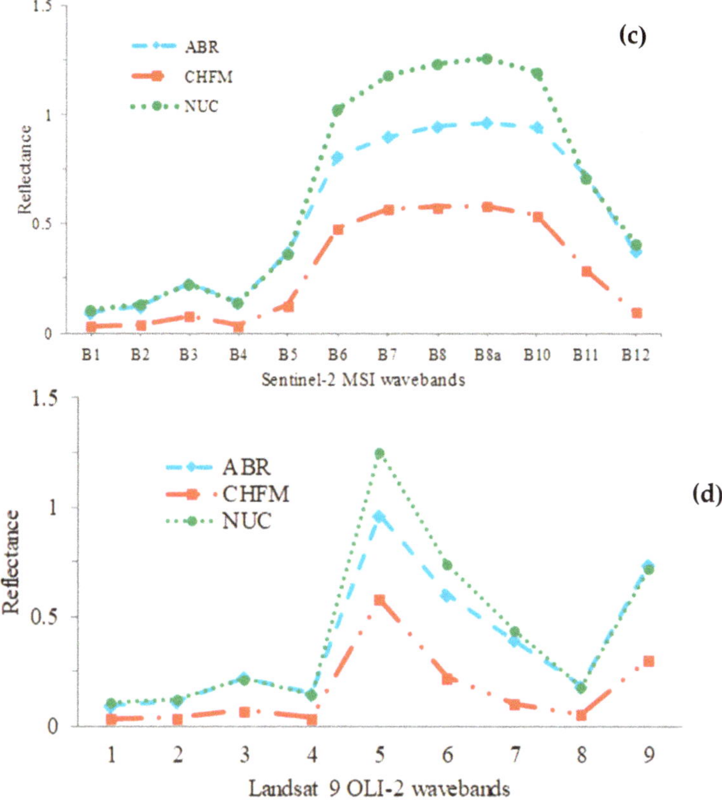

Figure 1. Collected spectra (**a**) and Mean spectral signatures of tomato crops grown under ABR effluent, NUC and CHFM treatments based on (**b**) HyspIRI (**c**) Sentinel -2 MSI and (**d**) Landsat 9 OLI-2 spectral settings.

3.1. Discriminating Tomatoes under ABR Effluent, NUC and CHFM

All sensors better characterized tomato crops administered with ABR when compared to those administered with CHFM and NUC. Specifically, high producer and user accuracies ranging between 91% and 100% were observed in characterizing tomatoes treated with ABR (Table 2). Meanwhile tomatoes administered with ABR and CHFM were characterized with slightly lower accuracies (Table 2). Moreover, HyspIRI, produced high accuracies characterized by kappa statistics of 0.99, whereas the spectral settings of Sentinel-2 MSI and Landsat 9 OLI-2's spectral settings exhibited kappa statistics of 0.85 and 0.79, respectively. HyspIRI spectral settings exhibited high producer accuracies of 100%, 92% and 100% for ABR, CHFM and NUC respectively. Meanwhile Sentinel-2 MSI exhibited slightly lower producer accuracies (ABR = 91%, CHFM = 86% and NUC = 100%). Landsat OLI-2 exhibited producer accuracies that were comparable to those of Sentinel 2 MSI which were 91% for ABR, 83% for CHFM and 100% for NUC. The same trend could be observed on the user accuracies (Table 2). In general, the CHFM treatments had slightly lower-class accuracies across all sensors in relation to the HEDM treatments.

Table 2. Classification accuracies derived using HyspIRI, Sentinel 2 MSI and Lands 9 OLI-2 spectral settings.

Sensor		PLS-DA		DA		PLS-DA		DA	
		PA	UA	PA	UA	OA	Kappa	OA	Kappa
	ABR	100	95	100	100	0.97	0.90	0.99	0.99
HyspIRI	CHFM	92	100	100	100				
	NUC	100	94	100	100				
Sentinel 2 MSI	ABR	91	100	100	100	0.90	0.69	0.95	0.85
	CHFM	86	100	89	100				
	NUC	100	65	100	82				
Landsat 9 OLI-2	ABR	91	95	95	100	0.89	0.63	0.94	0.79
	CHFM	83	100	89	100				
	NUC	100	65	100	76				

3.2. Performance of DA and PLS-DA Algorithms in Discriminating Tomatoes under ABR Effluent, NUC and CHFM

When comparing the performance of algorithms, DA exhibited very high accuracies. For instance, DA's producer accuracies derived using DA across all fertilizer treatments and sensor simulations ranged from a minimum of 89% whereas PLS-DA had a slightly lower minimum of 83% (Table 2). The user accuracies derived using DA ranged from a minimum of 76% whereas those derived using PLS-DA ranged from 65% to 100%. The overall accuracies derived using DA were higher (i.e., 0.94–0.99) when compared to those derived using PLS-DA (i.e., 0.3–0.97). Uniformly, the kappa statistics derived using DA were higher, ranging between 0.79 and 0.99 whereas those derived using PLS-DA were lower ranging between 0.63 and 0.90 (Table 2). Table 3 summarizes potential spectral variables for discriminating tomato plants grown under ABR, CHFM and NUC treatments using HyspIRI, Sentinel 2 MSI and Landsat 9 OLI-2. HyspIRI wavebands 710, 720, 690, 840, 1370 and 2110 nm, Sentinel 2-MSI bands 7, 6, 5 and 8a as well as Landsat bands 5, 6, 7 and 8, in order of importance, were identified as the most optimal spectral variables for discriminating tomato plants grown under different fertilizer regimes. NIR was the most prominent optimal section with a potential of discriminating tomatoes plants grown under different fertilizer regimes across the three sensors, although it did not outperform the red-edge section of the electromagnetic spectrum which was the most influential region.

Table 3. Influential bands in discriminating tomato plants grown under different fertilizer regimes.

Spectrum	Sensor					
	HyspIRI		Sentinel		Landsat	
	DA	PLS-DA	DA	PLS-DA	DA	PLS-DA
Visible	5	4	2	1	2	1
Red edge	12	10	3	3		
NIR	10	10	2	2	1	1
SWIR	8	7	1	1	2	2
M/FWIR	7	7				
Total	42	38	8	7	5	4

NB M/FWIR means Mid/Far wavelength infrared.

4. Discussion

We sought to compare the strength of HyspIRI's spectral configuration in relation to Landsat 9 OLI-2 and Sentinel 2 MSI spectral settings in characterizing tomato (*Solanum lycopersicum*) crops grown under CHFM, ABR and NUC treatment regimes. Results of this study showed that tomatoes that were fertilized using ABR could be optimally discriminated (i.e., Kappa statics ranging 0.79 to 0.99) from those that were administered with CHFM and NUC. The spectral reflectance of HDME fertilized crops was higher than that of those treated with CHFM. This indicates the potential of HEDM fertilizer for sustaining the health of vegetables in a manner comparable to that of conventional chemical fertilizers. This could be attributed to the fact that ABR effluents have nutrient properties favorable for tomato plants, which facilitate excessive vegetative growth, high biomass accumulation, delayed or uneven maturity [42–44]. Tomato crops with excessive vegetative growth, high biomass accumulation, delayed or uneven maturity tend to be easily detected and discriminated by satellite sensors compared to those which are not. Subsequently, tomatoes growing under the ABR treatments have a different spectral signature from those that are fertilized using CHFM and NUC. Literature illustrates that ABR tends to facilitate high biomass accumulation (i.e., increased leaf area index) hence the high classification accuracies exhibited by all remotely sensed data in this study in characterizing the HEDM fertiliser treatments [11,28,29]. For example, Al-Lahham, El Assi and Fayyad [45] illustrated that tomato crops that were administered with high quantities of waste water had big fruit sizes hence high biomass accumulation in relation to those that were administered with potable water. In a related study, Zavadil (2009) noted that primary treated waste water which contained an average of 14-fold nitrogen amounts (70.6 mg/L, which was 89% ammonia form) and also a 3-fold of the total phosphorus, resulted in high biomass accumulation and yields of lettuce salad, radishes and carrots in their study assessing the influence of sewage in relation to potable water on vegetable growth. Subsequently, the increases in biomass accumulation associated with wastewater treated vegetables could explain the discrimination of ABR treated tomato crops in comparison to those administered with CHFM and NUC, in this study. Furthermore, the most influential wavebands that facilitated high classification accuracies in discriminating tomato plants grown under human-excreta derived materials (HEDM) from those growing under CHFM conditions were from the red-edge section of the electromagnetic spectrum which is generally associated with healthy plants. The red-edge section of the electromagnetic spectrum is very sensitive to high chlorophyll, leaf angle distribution, leaf area index levels associated with plants grown in more nutritious environments such as the those exhibited by ABR. As aforementioned, HEDM have a high turnover of N P K plant nutrients which is comparable to that of commercial soluble fertilizers [46–48]. In this regard, the high turnover of plant nutrients such as nitrogen facilitates an increase in biomass, LAD, LAI which in turn makes the canopy spectral signature of those plants treated with HEDM to be discriminable from other plants especially in the red-edge and the NIR sections of the electromagnetic spectrum [49,50]. Subsequently, the red-edge and the NIR regions of the electromagnetic spectrum illustrate that the spectral signatures and the health of plants grown under HEDM are comparable to those grown under the CHFM.

When assessing the performance of sensors, HyspIRI outperformed the two multispectral sensors namely, Sentinel-2 MSI and Landsat OLI spectral settings in discriminating tomato crops grown under different fertilizer regimes. This could be explained by the fact that HyspIRI is a hyperspectral sensor characterized by narrow spectral wavebands that are more sensitive to the spectral reflectance of tomato crops grown under different fertilizer regimes than broadband sensor settings such as those of Landsat which could be masking out those minute tomato crops spectral variations. There is a consistently growing body of literature that supports the claim that hyperspectral sensors are more sensitive to minute vegetation spectral variabilities compared to broadband sensors due to the narrowed bandwidths configuration [51–56]. Specifically, Thenkabail, Smith and De Pauw [55,57] illustrated that narrow bands characterized different crop traits such as yield as well as spectral variations when compared to broadband spectral data. Also, they were able to better characterize wheat from barley using hyperspectral data in relation to the data from broadband sensors. They attributed this to the variation in spectral settings (bandwidths) of the sensors they used. These variations in spectral settings affected the detail that determined the accuracy of their models for plant trait characterization. Meanwhile, Clark [57] noted that there was no significant variation in the performance of HyspIRI and Sentinel-2 MSI as well as Landsat OLI in landcover classification of the San Francisco Bay Area in northern California, USA. However, their results confirmed that HyspIRI exhibited higher classification accuracies in their study.

Results of this study also illustrated that Sentinel 2 MSI and Landsat 9 performed satisfactorily in discriminating tomatoes grown under different fertilizer regimes, although Sentinel-2 MSI outperformed

Landsat OLI. This could be explained by the fact that Sentinel-2 MSI spectral settings cover the red-edge portion of the electromagnetic spectrum which is critical in mapping and detecting various vegetation traits. Moreover, there is a large and growing body of literature illustrating that Sentinel-2 MSI performs better than Landsat OLI in vegetation mapping [22,57,58]. The study by Colkesen and Kavzoglu [58] illustrated that Sentinel-2 MSI outperformed Landsat OLI in discriminating alfalfa, sugar beet and bean in the agricultural lands of the Ferizli district, Turkey. Shoko and Mutanga [22] also illustrated the robustness of Sentinel-2 MSI remotely sensed data in better discriminating C3 *Festuca costata* Nees from the C4 *Themeda triandra* Forssk grasses in a mountainous area in South Africa. They also attributed the optimal performance of Sentinel-2 MSI to the presence of red-edge bands in discriminating *Festuca costata* from *Themeda triandra* grasses.

Even though this study successfully illustrated the robustness of the spectral settings of these sensors in discriminating tomato crops grown under different fertilizer regimes using simulated data, there is still need to assess their satellite remotely sensed data. This study only sought to evaluate the spectral configuration of these sensors, hence there is still need to evaluate their radiometric and spatial resolutions in a similar application setup. Generally, the reflectance of remotely sensed satellite data tends to be affected by numerous factors associated with the sensor's platform, atmospheric related influences etc. [59–61]. Factors that affect satellite remotely sensed data are different from those that affect spectrometric proximally sensed data. Spectrometric proximally sensed data tends to be less affected by atmospheric related influences; hence, it offers a robust dataset suitable for testing the spectral settings of sensors, particularly the forth coming ones. The performance of the same sensors' spectral settings derived using satellite platforms, therefore, still needs to be evaluated.

Although this was not the major objective of the study, DA outperformed PLS-DA in discriminating tomato crops grown under different fertilizer regimes. In this study PLS-DA failed to derive unnecessary variables for characterizing tomato crops grown under different fertilizer regimes. On the other hand, there are numerous studies that have illustrated the optimal performance of DA in dimension reduction as well as feature extraction [62–64]. Our results also indicated that DA selected more spectral bands in relation to PLS-DA as optimal spectral variables for discriminating tomatoes grown under different fertilizer treatment regimes. The optimal performance of DA in relation to PLS-DA could be attributed to the conservative nature of PLS-DA in classifying tomato plants [31].

Implications of the Study'S Findings for Horticultural Crop Production

The optimal performance of HyspIRI's spectral settings in characterizing tomato plants grown under different fertilizer treatments illustrates its great potential in providing additional invaluable spatially explicit information urgently required in horticulture management and decision-making processes particularly for successful site and crop specific management practices. For instance, this study shows that based on remotely sensed data HEDM fertilizers have a great potential of improving vegetable crops health in manner comparable to that of conventional fertilisers. The fact that this study showed that HyspIRI, Sentinel 2 MSI and Landsat 9 OLI-2 could detect the differences between HEDM fertilized plants and those that are grown under conventional fertilizers underscores the potential of these sensors in providing information that could also be used in other applications such as crop inventory, condition, production forecast, assessment of nutrient deficiencies as well as growth and health of tomato plants and other horticultural crops. Furthermore, these sensors offer cheap, fast and reliable spatial explicit information on crop conditions required in the administration of precise and proper fertilizer applications while optimizing resources and increasing net returns especially in sub-Saharan countries with limited data and financial resources for facilitating high production of horticultural crops. However, more and extensive research efforts still need to be exerted in fully ascertaining and exploiting the potential of remotely sensed data in other horticultural crop-management applications.

5. Conclusions

The prime objective of this study was to compare the strength of the forthcoming hyperspectral sensor HyspIRI's spectral settings in the context of characterizing the effects of different fertilizer-treatment regimes on tomato crops. Furthermore, the study assessed the performance of PLS-DA in relation to DA in discriminating tomatoes treated with ABR, NUC and CHFM. Based on the results exhibited by this study we conclude that:

- The forthcoming HyspIRI sensor has the potential to accurately map tomato crops under various fertilizer regimes. Landsat and Sentinel, performed comparably to HyspIRI spectral settings.
- Overall, all the sensors were able to characterise the comparable impact of HEDM in relation to that of CHFM fertilizers on the spectral characteristics of tomato plants as a proxy of their health.
- DA offers optimal accuracies in characterizing tomatoes grown under different fertilizer regimes when compared to PLS-DA.

These findings are a substantial foundation upon which comprehensive precision agricultural assessments initiatives could be formed. These initiatives are required in order to attain sustainable agriculture as well as food security in regions such as sub-Saharan Africa where agricultural crop monitoring is currently hindered by the limited access to robust spatial data sets.

Author Contributions: Conceptualization, (M.S., O.M. and L.S.M.); Data curation, (L.S.M., S.T.M., A.O.O. and A.M.); Formal analysis, (M.S.); Investigation, (M.S., S.T.M. and A.M.); Project administration, (O.M. and P.L.M.); Resources, (O.M. and P.L.M.); Writing–original draft, (M.S.); Writing–review & editing, (M.S. and T.D.).

Funding: This research was funded by the National Research Foundation (NRF) of South Africa (Grant Numbers: 86893 and 84157).

Acknowledgments: The Pollution Research Group (PRG) is acknowledged for technical support and for providing human-excreta derived materials as well as space for experimentation for this study. We also extend our gratitude to the anonymous reviewers for their constructive criticism.

Conflicts of Interest: The authors declare no conflict of interest.

References

1. Mabhaudhi, T.; Chibarabada, T.; Modi, A. Water-Food-Nutrition-Health Nexus: Linking Water to Improving Food, Nutrition and Health in Sub-Saharan Africa. *Int. J. Environ. Res. Public Health* **2016**, *13*, 107. [CrossRef]
2. Van Ittersum, M.K.; Van Bussel, L.G.; Wolf, J.; Grassini, P.; Van Wart, J.; Guilpart, N.; Claessens, L.; de Groot, H.; Wiebe, K.; Mason-D'Croz, D.; et al. Can Sub-Saharan Africa Feed Itself? *Proc. Natl. Acad. Sci. USA* **2016**, *113*, 14964–14969. [CrossRef] [PubMed]
3. The-World-Bank. *Poverty & Equity Data Portal*; The World Bank: Wastington, DC, USA, 2019
4. FAO. *Inflation in Consumer Price Index for Food*; FAO: Rome, Italy, 2019.
5. Conceição, P.; Levine, S.; Lipton, M.; Warren-Rodríguez, A. Toward a Food Secure Future: Ensuring Food Security for Sustainable Human Development in Sub-Saharan Africa. *Food Policy* **2016**, *60*, 1–9. [CrossRef]
6. Nordey, T.; Basset-Mens, C.; De Bon, H.; Martin, T.; Déletré, E.; Simon, S.; Parrot, L.; Despretz, H.; Huat, J.; Biard, Y.; et al. Protected Cultivation of Vegetable Crops in Sub-Saharan Africa: Limits and Prospects for Smallholders. A Review. *Agron. Sustain. Dev.* **2017**, *37*, 53. [CrossRef]
7. Beecher, G.R. Nutrient Content of Tomatoes and Tomato Products. *Proc. Soc. Exp. Biol. Med.* **1998**, *218*, 98–100. [CrossRef]
8. Busari, T.I.; Senzanje, A.; Odindo, A.O.; Buckley, C.A. Evaluating the Effect of Irrigation Water Management Techniques on (Taro) Madumbe (Colocasia Esculenta (L.) Schott) Grown with Anaerobic Filter (Af) Effluent at Newlands, South Africa. *J. Water Reuse Desalin.* **2019**, *9*, 203–212. [CrossRef]
9. Smith, D.P.; Smith, N.T. Recovery of Wastewater Nitrogen for Solanum Lycopersicum Propagation. *Waste Biomass Valorization* **2019**, *10*, 1192–1202. [CrossRef]
10. Al-Hamdan, M.; Cruise, J.; Rickman, D.; Quattrochi, D. Forest Stand Size-Species Models Using Spatial Analyses of Remotely Sensed Data. *Remote Sens.* **2014**, *6*, 9802–9828. [CrossRef]

11. Petropoulos, G.; Srivastava, P.; Piles, M.; Pearson, S. Earth Observation-Based Operational Estimation of Soil Moisture and Evapotranspiration for Agricultural Crops in Support of Sustainable Water Management. *Sustainability* **2018**, *10*, 181. [CrossRef]

12. Peng, Y.; Nguy-Robertson, A.; Arkebauer, T.; Gitelson, A. Assessment of Canopy Chlorophyll Content Retrieval in Maize and Soybean: Implications of Hysteresis on the Development of Generic Algorithms. *Remote Sens.* **2017**, *9*, 226. [CrossRef]

13. Nguy-Robertson, A.L.; Peng, Y.; Gitelson, A.A.; Arkebauer, T.J.; Pimstein, A.; Herrmann, I.; Karnieli, A.; Rundquist, D.C.; Bonfil, D.J. Estimating Green Lai in Four Crops: Potential of Determining Optimal Spectral Bands for a Universal Algorithm. *Agric. For. Meteorol.* **2014**, *192*, 140–148. [CrossRef]

14. Luo, S.; Wang, C.; Xi, X.; Nie, S.; Fan, X.; Chen, H.; Yang, X.; Peng, D.; Lin, Y.; Zhou, G. Combining Hyperspectral Imagery and Lidar Pseudo-Waveform for Predicting Crop Lai, Canopy Height and above-Ground Biomass. *Ecol. Indic.* **2019**, *102*, 801–812. [CrossRef]

15. Zhao, T.; Koumis, A.; Niu, H.; Wang, D.; Chen, Y. Onion Irrigation Treatment Inference Using a Low-Cost Hyperspectral Scanner. In Proceedings of the Paper Presented at the Multispectral, Hyperspectral, and Ultraspectral Remote Sensing Technology, Techniques and Applications VII, Honolulu, HI, USA, 24–26 September 2018.

16. Lu, J.; Ehsani, R.; Shi, Y.; de Castro, A.I.; Wang, S. Detection of Multi-Tomato Leaf Diseases (Late Blight, Target and Bacterial Spots) in Different Stages by Using a Spectral-Based Sensor. *Sci. Rep.* **2018**, *8*, 2793. [CrossRef] [PubMed]

17. Česonienė, L.; Masaitis, G.; Mozgeris, G.; Gadal, S.; Šileikienė, D.; Karklelienė, R. Visible and near-Infrared Hyperspectral Imaging to Describe Properties of Conventionally and Organically Grown Carrots. *J. Elem.* **2019**, *24*, 421–435.

18. Transon, J.; d'Andrimont, R.; Maugnard, A.; Defourny, P. Survey of Hyperspectral Earth Observation Applications from Space in the Sentinel-2 Context. *Remote Sens.* **2018**, *10*, 157. [CrossRef]

19. Korhonen, L.; Packalen, P.; Rautiainen, M. Comparison of Sentinel-2 and Landsat 8 in the Estimation of Boreal Forest Canopy Cover and Leaf Area Index. *Remote Sens. Environ.* **2017**, *195*, 259–274. [CrossRef]

20. Dube, T.; Mutanga, O. Evaluating the Utility of the Medium-Spatial Resolution Landsat 8 Multispectral Sensor in Quantifying Aboveground Biomass in Umgeni Catchment, South Africa. *ISPRS J. Photogramm. Remote Sens.* **2015**, *101*, 36–46. [CrossRef]

21. Ahmadian, N.; Ghasemi, S.; Wigneron, J.P.; Zölitz, R. Comprehensive Study of the Biophysical Parameters of Agricultural Crops Based on Assessing Landsat 8 Oli and Landsat 7 Etm+ Vegetation Indices. *GISci. Remote Sens.* **2016**, *53*, 337–359. [CrossRef]

22. Shoko, C.; Mutanga, O. Examining the Strength of the Newly-Launched Sentinel 2 Msi Sensor in Detecting and Discriminating Subtle Differences between C3 and C4 Grass Species. *ISPRS J. Photogramm. Remote Sens.* **2017**, *129*, 32–40. [CrossRef]

23. Guanter, L.; Kaufmann, H.; Segl, K.; Foerster, S.; Rogass, C.; Chabrillat, S.; Kuester, T.; Hollstein, A.; Rossner, G.; Chlebek, C.; et al. The Enmap Spaceborne Imaging Spectroscopy Mission for Earth Observation. *Remote Sens.* **2015**, *7*, 8830–8857. [CrossRef]

24. Lee, C.M.; Cable, M.L.; Hook, S.J.; Green, R.O.; Ustin, S.L.; Mandl, D.J.; Middleton, E.M. An Introduction to the Nasa Hyperspectral Infrared Imager (Hyspiri) Mission and Preparatory Activities. *Remote Sens. Environ.* **2015**, *167*, 6–19. [CrossRef]

25. Sheik, O.M.; Kabir, P.; Ilaria, G.; Onisimo, M.; Zakariyyaa, O. Detecting Canopy Damage Caused by Uromycladium Acaciae on South African Black Wattle Forest Compartments Using Moderate Resolution Satellite Imagery. *S. Afr. J. Geomat.* **2019**, *8*, 69–83.

26. Huang, H.C.; Liu, S.C.; Wang, C.; Xia, K.B.; Zhang, D.J.; Wang, H.Z.; Zhan, S.Y.; Huang, H.; He, S.Y.; Liu, C.C. On-Site Visualized Classification of Transparent Hazards and Noxious Substances on a Water Surface by Multispectral Techniques. *Appl. Opt.* **2019**, *58*, 4458–4466. [CrossRef] [PubMed]

27. Li, X.Y.; Zhang, F.L.; Jane, Y. Locally Weighted Discriminant Analysis for Hyperspectral Image Classification. *Remote Sens.* **2019**, *11*, 109. [CrossRef]

28. Corbane, C.; Alleaume, S.; Deshayes, M. Mapping Natural Habitats Using Remote Sensing and Sparse Partial Least Square Discriminant Analysis. *Int. J. Remote Sens.* **2013**, *34*, 7625–7647. [CrossRef]

29. Boulesteix, A.L. Pls Dimension Reduction for Classification with Microarray Data. *Stat. Appl. Genet. Mol. Biol.* **2004**, *3*, 1–32. [CrossRef] [PubMed]

30. Bagheri, N.; Mohamadi-Monavar, H.; Azizi, A.; Ghasemi, A. Detection of Fire Blight Disease in Pear Trees by Hyperspectral Data. *Eur. J. Remote Sens.* **2018**, *50*, 1–10. [CrossRef]
31. Prats-Montalbán, J.M.; Ferrer, A.; Malo, J.L.; Gorbena, J. A Comparison of Different Discriminant Analysis Techniques in a Steel Industry Welding Process. *Chemom. Intell. Lab. Syst.* **2006**, *80*, 109–119. [CrossRef]
32. Julia, A. Colour Correction of Underwater Images Using Spectral Data. Ph.D. Thesis, Acta Universitatis Upsaliensis, Uppsala, Sweden, 2005.
33. Springsteen, A. Standards for Reflectance Measurements. *Appl. Spectrosc. A Compact. Ref. Pract.* **1998**, *15*, 247.
34. Abdel-Rahman, E.M.; Mutanga, O.; Odindi, J.; Adam, E.; Odindo, A.; Ismail, R. A Comparison of Partial Least Squares (Pls) and Sparse Pls Regressions for Predicting Yield of Swiss Chard Grown under Different Irrigation Water Sources Using Hyperspectral Data. *Comput. Electron. Agric.* **2014**, *106*, 11–19. [CrossRef]
35. Curran, P.J. Imaging Spectrqmetry-Its Present and Future Rôle in Environmental Research. In *Imaging Spectrometry—A Tool for Environmental Observations*; Springer: Berlin, Germany, 1994; pp. 1–23.
36. Prasad, S.; Lori, M.B.; Hemanth, K. Data Exploitation of Hyspiri Observations for Precision Vegetation Mapping. In Proceedings of the 2009 IEEE International Geoscience and Remote Sensing Symposium, Cape Town, South Africa, 12–17 July 2009.
37. Sathishkumar, S.; Prasad, S.; Bruce, M.L.; Robles, W. Nasa's Upcoming Hyspiri Mission—Precision Vegetation Mapping with Limited Ground Truth. In Proceedings of the 2010 IEEE International Geoscience and Remote Sensing Symposium, Honolulu, HI, USA, 25–30 July 2010.
38. Adelabu, S.; Mutanga, O.; Adam, E.; Sebego, R. Spectral Discrimination of Insect Defoliation Levels in Mopane Woodland Using Hyperspectral Data. Selected Topics in Applied Earth Observations and Remote Sensing. *IEEE J. Sel. Top. Appl. Earth Obs. Remote Sens.* **2014**, *7*, 177–186. [CrossRef]
39. Zhang, H.; Lan, Y.; Suh, C.P.; Westbrook, J.K.; Lacey, R.; Hoffmann, W.C. Differentiation of Cotton from Other Crops at Different Growth Stages Using Spectral Properties and Discriminant Analysis. *Trans. Am. Soc. Agric. Biol. Eng.* **2012**, *55*, 1623–1630. [CrossRef]
40. Manandhar, R.; Odeh, I.; Ancev, T. Improving the Accuracy of Land Use and Land Cover Classification of Landsat Data Using Post-Classification Enhancement. *Remote Sens.* **2009**, *1*, 330–344. [CrossRef]
41. de Leeuw, J.; Jia, H.; Yang, L.; Liu, X.; Schmidt, K.; Skidmore, A.K. Comparing Accuracy Assessments to Infer Superiority of Image Classification Methods. *Int. J. Remote Sens.* **2006**, *27*, 223–232. [CrossRef]
42. Pedrero, F.; Kalavrouziotis, I.; Alarcón, J.J.; Koukoulakis, P.; Asano, T. Use of Treated Municipal Wastewater in Irrigated Agriculture—Review of Some Practices in Spain and Greece. *Agric. Water Manag.* **2010**, *97*, 1233–1241. [CrossRef]
43. Maurer, M.A.; Davies, F.S.; Graetz, D.A. Reclaimed Wastewater Irrigation and Fertilization of Matureredblush'grapefruit Trees on Spodosols in Florida. *J. Am. Soc. Hortic. Sci.* **1995**, *120*, 394–402. [CrossRef]
44. Zavadil, J. The Effect of Municipal Wastewater Irrigation on the Yield and Quality of Vegetables and Crops. *Soil Water Res.* **2009**, *4*, 91–103. [CrossRef]
45. Al-Lahham, O.; El Assi, N.M.; Fayyad, M. Impact of Treated Wastewater Irrigation on Quality Attributes and Contamination of Tomato Fruit. *Agric. Water Manag.* **2003**, *61*, 51–62. [CrossRef]
46. Viskari, E.L.; Grobler, G.; Karimäki, K.; Gorbatova, A.; Vilpas, R.; Lehtoranta, S. Nitrogen Recovery with Source Separation of Human Urine–Preliminary Results of Its Fertiliser Potential and Use in Agriculture. *Front. Sustain. Food Syst.* **2018**, *2*, 32. [CrossRef]
47. Moya, B.; Parker, A.; Sakrabani, R. Challenges to the Use of Fertilisers Derived from Human Excreta: The Case of Vegetable Exports from Kenya to Europe and Influence of Certification Systems. *Food Policy* **2019**, *85*, 72–78. [CrossRef]
48. Moya, B.; Parker, R.S.; Mesa, B. Evaluating the Efficacy of Fertilisers Derived from Human Excreta in Agriculture and Their Perception in Antananarivo, Madagascar. *Waste Biomass Valorization* **2019**, *10*, 941–952. [CrossRef]
49. Bassegio, D.; Santos, R.F.; de Oliveira, E.; Wernecke, I.; Secco, D.; de Souza, S.N.M. Effect of Nitrogen Fertilization and Cutting Age on Yield of Tropical Forage Plants. *Afr. J. Agric. Res.* **2013**, *8*, 1427–1432.
50. Clevers, J.G.; Gitelson, A.A. Remote Estimation of Crop and Grass Chlorophyll and Nitrogen Content Using Red-Edge Bands on Sentinel-2 and-3. *Int. J. Appl. Earth Obs. Geoinf.* **2013**, *23*, 344–351. [CrossRef]
51. Adam, E.; Mutanga, O.; Rugege, D. Multispectral and Hyperspectral Remote Sensing for Identification and Mapping of Wetland Vegetation: A Review. *Wetl. Ecol. Manag.* **2010**, *18*, 281–296. [CrossRef]

52. Mansour, K.; Mutanga, O.; Everson, T. Remote Sensing Based Indicators of Vegetation Species for Assessing Rangeland Degradation: Opportunities and Challenges. *Afr. J. Agric. Res.* **2012**, *7*, 3261–3270.

53. Thenkabail, P.S.; Lyon, J.G. *Hyperspectral Remote Sensing of Vegetation: Knowledge Gain and Knowledge Gap after 50 Years of Research (Conference Presentation)*; Paper Presented at the Hyperspectral Imaging Sensors: Innovative Applications and Sensor Standards 2017; CRC Press: Boca Raton, FL, USA, 2017.

54. Thenkabail, P.S.; Enclona, E.A.; Ashton, M.S.; Legg, C.; De Dieu, M.J. Hyperion, Ikonos, Ali, and Etm+ Sensors in the Study of African Rainforests. *Remote Sens. Environ.* **2004**, *90*, 23–43. [CrossRef]

55. Thenkabail, P.S.; Smith, R.B.; De Pauw, E. Evaluation of Narrowband and Broadband Vegetation Indices for Determining Optimal Hyperspectral Wavebands for Agricultural Crop Characterization. *Photogramm. Eng. Remote Sens.* **2002**, *68*, 607–622.

56. Thenkabail, P.S.; Lyon, J.G. *Hyperspectral Remote Sensing of Vegetation*; CRC Press: Boca Raton, FL, USA, 2016.

57. Clark, M.L. Comparison of Simulated Hyperspectral Hyspiri and Multispectral Landsat 8 and Sentinel-2 Imagery for Multi-Seasonal, Regional Land-Cover Mapping. *Remote Sens. Environ.* **2017**, *300*, 311–325. [CrossRef]

58. Colkesen, I.; Kavzoglu, T. Ensemble-Based Canonical Correlation Forest (Ccf) for Land Use and Land Cover Classification Using Sentinel-2 and Landsat Oli Imagery. *Remote Sens. Lett.* **2017**, *8*, 1082–1091. [CrossRef]

59. Lillesand, T.; Ralph, W.K.; Jonathan, C. *Remote Sensing and Image Interpretation*; John Wiley & Sons: Hoboken, NJ, USA, 2015.

60. Thenkabail, P.S.; Mariotto, I.; Gumma, M.K.; Middleton, M.E.; Landis, R.D.; Huemmrich, F.K. Selection of Hyperspectral Narrowbands (Hnbs) and Composition of Hyperspectral Twoband Vegetation Indices (Hvis) for Biophysical Characterization and Discrimination of Crop Types Using Field Reflectance and Hyperion/Eo-1 Data. *Sel. Top. Appl. Earth Obs. Remote. Sens. IEEE J.* **2013**, *6*, 427–439. [CrossRef]

61. Mariotto, I.; Prasad, S.; Thenkabail, A.H.; Terrence Slonecker, E.; Platonov, A. Hyperspectral Versus Multispectral Crop-Productivity Modeling and Type Discrimination for the Hyspiri Mission. *Remote Sens. Environ.* **2013**, *139*, 291–305. [CrossRef]

62. Pu, R.; Liu, D. Segmented Canonical Discriminant Analysis of in Situ Hyperspectral Data for Identifying 13 Urban Tree Species. *Int. J. Remote. Sens.* **2011**, *32*, 2207–2226. [CrossRef]

63. Karimi Prasher, Y.S.O.; McNairn, H.; Bonnell, R.B.; Dutilleul, P.; Goel, P.K. Classification Accuracy of Discriminant Analysis, Artificial Neural Networks, and Decision Trees for Weed and Nitrogen Stress Detection in Corn. *Trans. ASAE* **2005**, *35*, 1261–1268. [CrossRef]

64. Filella, I.; Serrano, L.; Serra, J.; Penuelas, J. Evaluating Wheat Nitrogen Status with Canopy Reflectance Indices and Discriminant Analysis. *Crop Sci.* **1995**, *35*, 1400–1405. [CrossRef]

Article

Multi-Temporal Agricultural Land-Cover Mapping Using Single-Year and Multi-Year Models Based on Landsat Imagery and IACS Data

Isaac Kyere *, Thomas Astor [ID], Rüdiger Graß and Michael Wachendorf [ID]

Grassland Science and Renewable Plant Resources, Universität Kassel, Steinstraße 19,
D-37213 Witzenhausen, Germany; thastor@uni-kassel.de (T.A.); grass@wiz.uni-kassel.de (R.G.);
mwach@uni-kassel.de (M.W.)
* Correspondence: isaac.kyere@uni-kassel.de; Tel.: +49-5542-98-1338

Received: 26 April 2019; Accepted: 9 June 2019; Published: 12 June 2019

Abstract: The spatial distribution and location of crops are necessary information for agricultural planning. The free availability of optical satellites such as Landsat offers an opportunity to obtain this key information. Crop type mapping using satellite data is challenged by its reliance on ground truth data. The Integrated Administration and Control System (IACS) data, submitted by farmers in Europe for subsidy payments, provide a solution to the issue of periodic field data collection. The present study tested the performance of the IACS data in the development of a generalized predictive crop type model, which is independent of the calibration year. Using the IACS polygons as objects, the mean spectral information based on four different vegetation indices and six Landsat bands were extracted for each crop type and used as predictors in a random forest model. Two modelling methods called single-year (SY) and multiple-year (MY) calibration were tested to find out their performance in the prediction of grassland, maize, summer, and winter crops. The independent validation of SY and MY resulted in a mean overall accuracy of 71.5% and 77.3%, respectively. The field-based approach of calibration used in this study dealt with the 'salt and pepper' effects of the pixel-based approach.

Keywords: agricultural land-cover; multi-spectral; generalized model; machine learning; crop type mapping; Integrated Administration and Control System; remote sensing

---

1. Introduction

The increasing world population coupled with the high demand for agricultural resources [1] require reliable data on agricultural lands for decision making and planning towards the future [2]. The knowledge on available croplands is fundamental to food security [3], sustainable cropping [4] and the maximization of food production [5]. Information about the spatial distribution of crops and the spatial extent of croplands are also essential to ascertain the impact of any human activity on croplands [6].

Reliable and accurate information about agricultural lands requires an efficient and precise approach, which remote sensing (RS) can offer [7–9]. RS-based methods can be used to obtain various crop information, such as crop type [10], biomass [11], or yield [12]. The advent of satellite-based optical RS has revolutionized large-scale cropland mapping and has been used in many local, regional, and global agricultural projects [4,13–15].

The free availability of some of these images adds to the many advantages of satellite-based optical remote sensing in agriculture [16]. Such data, which is also available for historic time periods back to the early 1970s, provides a means to study the present landscapes in relation to how they were in the past. Landsat, which is the oldest running earth monitoring program, provides a 47-year archive of satellite data of the entire earth at a 30-meter resolution. As a result, most of the crop types and other agricultural mapping studies have used Landsat images as the main data source. For instance, Lui et al., [16] used multi-temporal Landsat-8 to successfully map winter wheat in China. Maxwell et al., [17] demonstrated an effective corn classification from Landsat images through an automated process in south-central Nebraska of USA, and Yin et al., [18] used dense Landsat time series data to map agricultural and land abandonment with a high level of accuracy in the Caucasus, covering parts of Russia and Georgia. Many of these studies either employed supervised or unsupervised classification to ascertain the needed land-cover information [19,20], either at the pixel or object-based levels. Despite their accurate performances, they have some limitations [21,22]. Supervised learning always requires field information, also known as training or ground-truth data [23]. Even with the unsupervised learning, knowledge about the study area is required to assign the correct land cover type to the classification results.

A large number of studies on crop type and cropland mapping used field data from the same mapping year [24–26]. This way of mapping is limited in situations, where there are no ground truth data available, or data collection is impossible for the period of interest. Due to the yearly and periodic changes in cultivated crops, continuous collection of ground truth information is necessary to reliably map crop types. However, given the labor-intensive, expensive and difficult nature of ground truth data collection [27], studies such as Botkin et al., [28] and Sonobe et al., [29] have recommended research into the development of training and classification methods, which is applicable to years where field information is not available (i.e., a generalized classifier).

Given the rotation of crops on fields at different seasons and the fast changes in biomass and phenology of crops, the use of temporal information is very crucial in the discrimination of crops. Prediction of land-cover based on multi-temporal data involves the use of data from several different seasons and has proven to be effective in many studies [30,31], as it integrates the varying phenological characteristics among vegetation. Leaf pigment, water, and canopy structure are proven to relate with spectral reflectance of crops but varies at different growing seasons [32]. The use of data from a single date is known to inefficiently capture the differences among the many crops which share similar spectral characteristics [10]. Manfron et al., [33] for instance, analyzed time series of satellite images to efficiently estimate the inter-annual variability of the sowing dates of winter wheat Many other studies have employed vegetation indices such as the normalized difference vegetation index and the enhanced vegetation index to capture the seasonal dynamics of crops and other land-cover characteristics [19,34].

In Europe, there exists a remarkably rich agricultural land cover data body within the Integrated Administration and Control System (IACS), which are regularly collected by farmers as part of the subsidy payment scheme in the common agricultural policy [35]. A similar agricultural data in the United States of America is the reference data collected by the Department of Agriculture (USDA) to produce the annual crop data layer (CDL) [36]. These reference data are not available to the public, which may be the reason why the CDL has served as validation data in many crop type mapping studies [8,37]. Conversely, the IACS data can be freely obtained by scientists and research institutions for scientific purposes upon an official request. However, not much has been done with the IACS data in crop type mapping. Griffiths et al., [38] created a national single-year wall-to-wall land-cover map of Germany and used IACS data as a reference to validate some part of the study area. The study of Vuolo et al., [30] demonstrated how multi-temporal Sentinel-2 data can improve the accuracy of crop prediction when IACS data was used to independently validate the classified map. To the best of the authors' knowledge, there is no research that has used multi-temporal IACS data as training data to develop a generalized model to predict crop types from satellite data at the field level.

Therefore, we hypothesize that the IACS data can be used to train a multi-temporal field-based model, which can predict crop types from a satellite image that is independent of the model's training year. Hence, the calibration data, as well as the data used for testing the models, are from different years. In addressing the stated hypothesis, two different modelling approaches, i.e., multiple-year

(MY) and single-year (SY) calibrations were tested in the present study. While SY models are calibrated using data from just one year, MY modelling involves model training based on data from two or more years. The SY and MY approaches have been applied in some crop mapping studies, e.g., [8], but were done at the pixel level, which is characterized by the problem of 'salt and pepper' effects (i.e., a misclassification of neighbouring pixels despite large similarities). On the other side, the object-based method of land cover classification, that has recently attracted considerable attention [39] as a replacement for the pixel-based [40], suffers from difficulties in the segmentation scale selection. Further, it was shown to depend on the size of the objects being mapped [41] and tend to misclassify small land-cover objects in low to medium satellite images, such as Landsat [22]. Therefore, this study employs a field/polygon-based calibration approach using the exact crop field shapes from the IACS database. Our study addresses the following questions:

(1) How well do models based on a single year's spectral information predict crops when tested on years not included in the model calibration process?
(2) What is the prediction performance of models calibrated on spectral information from multiple years?
(3) Is the accuracy of the classification models affected by field size?

2. Materials and Methods

2.1. Study Area

The study was done in the Northern Hesse region of Germany (Figure 1) comprising the districts of Kassel, Waldeck-Frankenberg, Schwalm-Eder, Hersfeld Rotenburg, and Werra-Meissner. The study area comprises ca. 6900 km^2 and is characterized by diverse landscapes and sites with favorable and less favorable environmental conditions for farming. The favourable arable lands are mostly found in flat valleys and on plateaus with moderate slopes, which are often covered by loess of substantial thickness mainly in the western and northern parts [42]. The less favourable arable sites show shallow soils with less native water and nutrient availability.

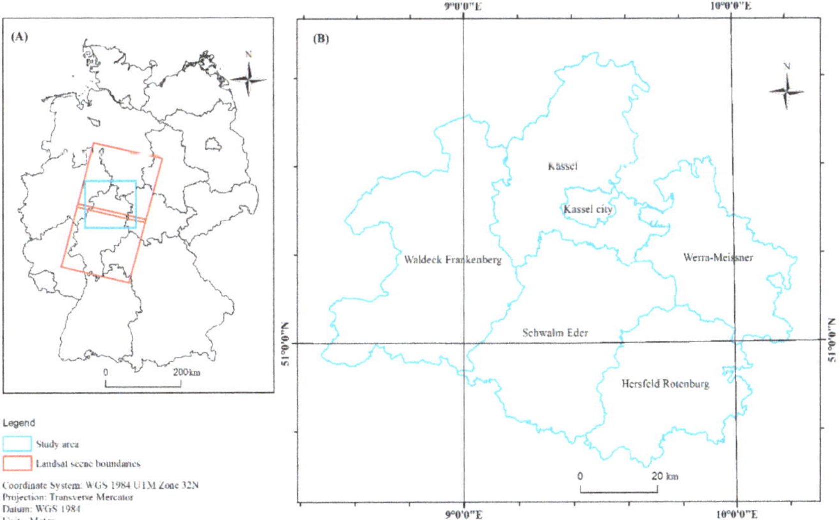

Figure 1. Map of the study area. (**A**) shows a map of Germany and the location of the study area, with the boundaries of the Landsat scenes; (**B**) shows the five districts, where the study was done.

Elevation ranges from 101 to 754 m with mean annual temperatures of 9–10 °C in the lowlands and 5–6 °C in the highlands. The mean annual rainfall ranges from 500–1300 mm [43]. The calendar of the crop types considered in this study can be seen in Table 1.

Table 1. Generalized calendar of the four crop types in the study area.

Crop Types	Sowing Window	Peak Greenness	Harvesting Window
Grassland	Depending on the grassland management system		
Maize	Late April	Mid-August	Mid-late September
Summer crops	Late March-Mid April	Mid-Late June	July-September
Winter crops	September-October	Mid-June	July-August

2.2. Data

2.2.1. Satellite Data

A total of 63 satellite images from the period April to October between 2005 and 2015 were used for this study. Surface reflectance Landsat scenes (Level-2) as summarized in Table 2 were downloaded from USGS's Earth Explorer [44]. Images from only six years were used because of little to no clouds. The images of Landsat 5 TM and 7 ETM+ had been atmospherically corrected using Landsat Ecosystem Disturbances Adaptive Processing System (LEDAPS) by NASA [45]. Surface reflectance of Landsat-8 was produced using Landsat Surface Reflectance Code [46]. Table 3 shows detailed information about the six spectral bands of the Landsat data used. Despite the small differences in the spectral ranges of the Landsat types, which have been well studied [47,48] to be smaller than 1 standard deviation of time-series of the spectral curve had no significant effect on classification results. Additionally, according to USGS [49], the Level-2 product (surface reflectance) of the Landsat images are similar, therefore, the Landsat data was not normalized. The atmospherically corrected Landsat images were accompanied by cloud mask layers. The images were categorized according to the dates when the images were captured, i.e., early summer (ES, April to May) and late summer (LS, July to October). ES and LS seasons cover the growing period of crop types in the study area; hence their use can help capture the different phenology of the crops at different stages of their development.

Table 2. Summary of satellite images used. (TM: Thematic Mapper, ETM+: Enhanced Thematic Mapper plus). The numbers in brackets represent the number of images used per date.

		Date of Image Acquisition	
Year	Satellite	Early Summer	Late Summer
2005	Landsat 5 TM	03-Apr. (1), 21-Apr. (2)	18-Aug. (2)
	Landsat 7 ETM+	4-Apr. (2)	
2007	Landsat 5 TM	02-Apr. (2), 25-Apr. (1), 27-Apr. (1)	16-Jul. (2), 01-Aug. (1), 24-Aug. (2)
	Landsat 7 ETM+	26-Apr (2)	
2009	Landsat 5 TM	07-Apr. (2), 14-Apr. (2), 16-Apr. (2), 02-May (1), 25-May (1)	06-Aug. (2), 20-Aug. (2)
	Landsat 7 ETM+		05-Aug. (1), 22-Sep. (1)
2010	Landsat 5 TM	17- Apr. (2), 19-Apr. (2)	08-Jul. (1), 31-Jul. (1), 07-Aug. (2)
	Landsat 7 ETM+	18-Apr. (2)	
2011	Landsat 5 TM	20-Apr. (1), 22-Apr. (2), 08-May (1)	03-Aug. (1), 15-Oct. (1), 22-Oct. (1)
	Landsat 7 ETM+	21-Apr. (2), 07-May(2)	20-Aug. (1), 03-Sep. (2), 21-Sep (1), 28-Sep. (2)
2015	Landsat 8	24-Apr. (2)	30-Aug. (2)

Table 3. Summary of the six spectral bands of the Landsat 5, 7 and 8. NIR = Near infra-red, SWIR = Shortwave infra-red, TM = Thematic Mapper, ETM+ = Enhanced Thematic Mapper plus.

Landsat 5 TM and 7 ETM+			Landsat 8		
Band Number	**Band Name**	**Wavelength (μm)**	**Band Number**	**Band Name**	**Wavelength (μm)**
Band 1	Blue	0.441–0.514	Band 2	Blue	0.452–0.512
Band 2	Green	0.519–0.601	Band 3	Green	0.533–0.590
Band 3	Red	0.631–0.692	Band 4	Red	0.636–0.673
Band 4	NIR	0.772–0.898	Band 5	NIR	0.851–0.879
Band 5	SWIR-1	1.547–1.749	Band 6	SWIR-1	1.566–1.651
Band 7	SWIR-2	2.064–2.345	Band 7	SWIR-2	2.107–2.294

2.2.2. Reference Data

The IACS data were used as ancillary data in this study. These are spatial data collected by farmers as part of the subsidy support system within the EU. It is made up of the shapes of agricultural fields and the crop types planted in each cropping season. The models were initially developed to predict individual crop species in the study area, but tests (not shown) exhibited incorrect predictions among crops species of similar spectral characteristics and growing periods. Therefore, several crops were grouped into four crop types. They were grassland, maize, summer, and winter crops with their vegetation profiles shown in Figure 2. These vegetation profiles depict the spectral characteristics of the crop types at different stages of their development and show a similar trend across all years. Farmers are not obliged to register their fields; except for farmers who apply for subsidies. Therefore, the reference data used in this study is limited to the declared fields as submitted to the responsible agency. Fallow fields were not considered in this study since the reference data (i.e., the IACS) used for modelling consists of only cultivated fields.

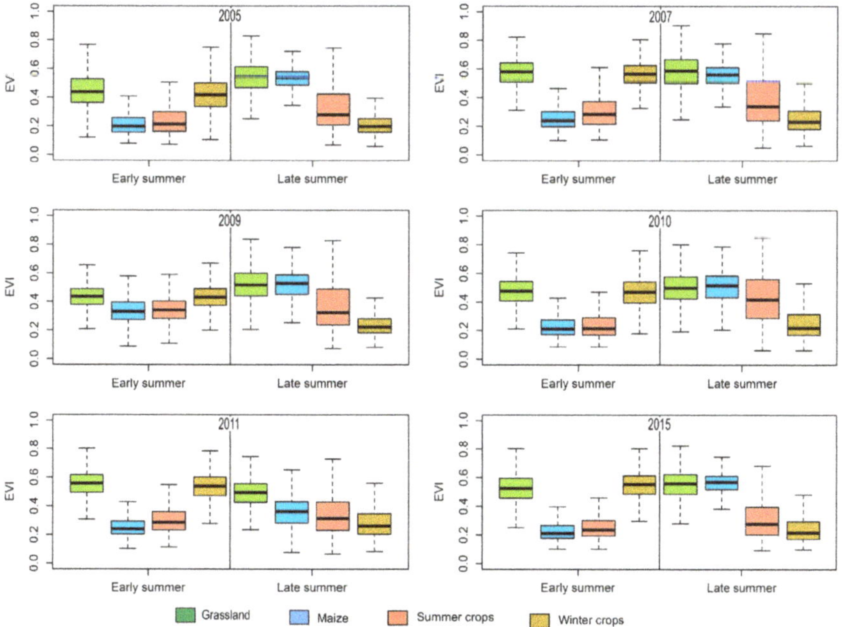

Figure 2. Vegetation profiles of the four crops based on enhanced vegetation index (EVI). Early summer = April to May, Late summer = July to October.

2.3. Data Processing

Image Pre-processing

The different bands (i.e., blue, green, red, near infra-red and shortwave infra-red 1 and 2) of the satellite images were stacked together, based on the years and acquisition time (Figure 3). Some of the images had small areas of clouds since cloud cover of less than 10% was considered appropriate for the study purpose. As a result, the cloud masks that came with the images were used to mask out all clouds. The masked areas were replaced with non-cloudy data from other images of the same area around the same time frame (i.e., May for ES and September, October for LS). With respect to the Landsat 7, the scan lines which resulted from the failure of the scan line corrector of the ETM+ sensor were also replaced with cloud-free satellite data from other images of the same area using the "cover" function [50] from the "raster" package in R software [51]. As our study area includes more than one Landsat image, some images were mosaicked to cover the entire area of interest. Mean values were used for overlapping layers during the mosaicking process. Figure 3 shows a complete workflow of the data analysis of this study.

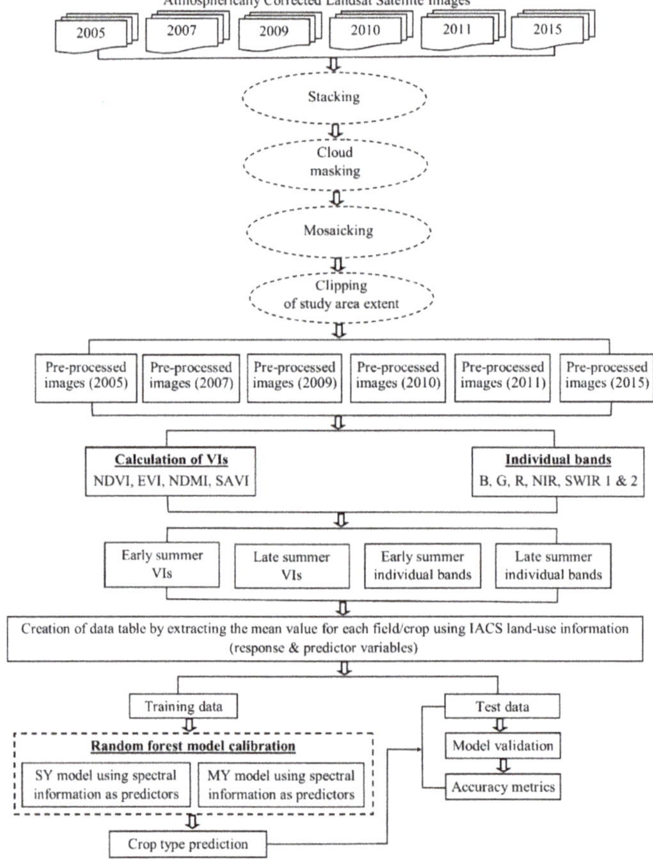

Figure 3. The workflow of the data analysis. SY-Single-year, MY-Multiple-years, VIs-Vegetation Indices, B-Blue, G-Green, R-Red, NIR-Near Infra-red, SWIR-Shortwave Infra-red, NDVI-Normalized Difference Vegetation Index, EVI-Enhanced Vegetation Index, SAVI-Soil Adjusted Vegetation Index, NDMI-Normalized Difference Moisture Index.

2.4. Model Calibration and Validation

Crop type prediction models were built based on random forest (RF) algorithm, which is an ensemble supervised machine learning classifier that creates numerous decision trees for prediction by randomly selecting subsets of the training data through the process of bagging [52]. Higher accuracies have been achieved with RF as compared to other machine learning algorithms in many crop mapping studies [7,8]. It can effectively function with only two main parameters, i.e., the number of trees to grow (*Ntree*) and the number of predictor variables selected for the best splitting of each tree node (*Mtry*) [53]. In this study, *Ntree* was set at 500 for all models since the error steadies before this number is reached, while *Mtry* was set to the square root of the input variables as reviewed by Belgiu and Drăgu [53].

2.4.1. Input Variables Used in the Model

Spectral data obtained from the satellite data and used as predictors in the crop type predictive models consisted of blue, green, red, near infra-red (NIR), shortwave infra-red 1 (SWIR 1) and shortwave infra-red 2 (SWIR 2). Additionally, four widely used spectral vegetation indices (VIs), i.e., normalized difference vegetation index (NDVI) [54], enhanced vegetation index (EVI) [34], soil adjusted vegetation index (SAVI) [55] and normalized difference moisture index (NDMI) [56], were computed from a ratio of different satellite bands (see Equations (1)–(4)) and included as explanatory variables. These VIs capture the dynamics of vegetation like greenness and vigor among others at different phenological stages. The potential of NDVI to assess vegetation dynamics of crops has been demonstrated by a number of studies [10,19]. However, it has shortcomings of sensitivity to saturation, soil background effects, or atmospheric effects. In dealing with these limitations, EVI and SAVI were added. SAVI deals with the soil background effects, while EVI uses the blue band to deal with the atmospheric influences by aerosols. EVI, NDVI, and SAVI, as shown in Equations (1)–(3), respectively, use the NIR and red bands in their computation, and they complement each other when used in vegetation analysis. NDMI, which uses NIR and SWIR for measuring the water content in vegetation, was also included in the analysis (Equation (4)), as it adds some complementary information to the other VIs.

$$NDVI = (NIR - Red)/(NIR + Red) \tag{1}$$

$$EVI = G \times (NIR - Red)/(NIR + C_1 \times Red - C_2 \times Blue + L) \tag{2}$$

$$SAVI = [(NIR - Red)/(NIR + Red + L)] \times (1 + L) \tag{3}$$

$$NDMI = (NIR - SWIR)/(NIR + SWIR) \tag{4}$$

where NIR = Near Infra-red, G = gain factor, C_1 and C_2 are aerosol resistance term coefficients, L in Equation (2) is non-linear canopy background adjustment, L in Equation (3) is soil brightness factor and SWIR = Shortwave Infra − red (values: G = 2.5, C_1 = 6, C_2 = 7.5, L_{EVI} = 1, L_{SAVI} = 0.5).

2.4.2. Field-based Extraction of Spectral Information.

The extraction of the spectral information was done at field base with the exact crop fields as objects. Mean values of each crop's field were extracted from the spectral data, which consisted of six individual bands of the satellite image, as well as four vegetation indices; grouped into early summer (ES) and late summer (LS) spectral information. In all, 20 different spectral information were used as predictors in the RF models. Since the IACS data (Figure 5) represent field information and, crops are cultivated with unequal distribution of fields for each crop type, almost equal numbers of polygon/field samples for each crop type were selected. Thus, crop types with many fields were always undersampled compared to those which were represented by fewer fields. Six different data tables were built for the six respective years, which were later used to calibrate and validate the crop type prediction model, with the spectral information and crop types as predictor and response variables, respectively.

2.4.3. Crop Type Prediction Modelling

Two different modelling approaches called same-year (SY) and multiple-year (MY) training were employed. With respect to SY models, an RF algorithm was trained using only the spectral information of one year and cross-validated using the remaining years as shown in Figure 4A. This was repeated six times, where for each repetition a different year was used to train the model, and the model's performance in predicting crop types was assessed using the remaining single-year data.

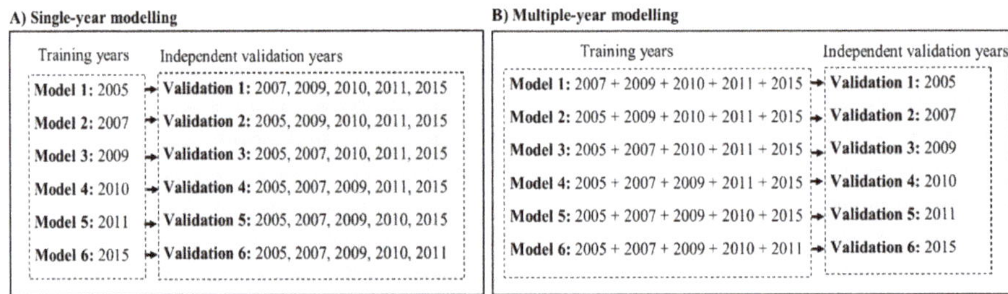

Figure 4. An illustration of the two modelling approaches.

MY models were trained by combining the extracted spectral information from five different years during the training phase and tested on an independent year. The training combination with multiple years was done six times, and with each repetition a single year was left out to validate the efficacy of the MY models (Figure 4B). The contribution of the predictors was assessed based on the internal mean decrease Gini of RF, which is the average of all Gini impurity recorded for each input variable when selected for splitting at each tree node [57]. Graphs showing the six most important predictor variables (based on percentages of the mean decrease Gini Index) of the best modelling method were created for visualization.

2.4.4. Accuracy Assessment

The performance of the models in predicting crop types was independently evaluated at field scale based on a confusion matrix. The independent validation was done by comparing the predicted crop types with the known crops using the reference data. The three most important and widely used metrics namely, overall accuracy (OA), user accuracy (UA) and producer accuracy (PA), resulting from the confusion matrix were calculated. OA assesses the overall performance of a model and is the ratio of correctly predicted crops and the total number of predicted crops. UA evaluates how well the predicted crops agree with the known reference data (i.e., the IACS field data), while PA measures the agreement between the reference data and the prediction. From the confusion matrix, the error of commission (EC) and error of omission (EO) of the respective land cover types can be obtained. Since the performance of each developed SY model was tested for all years, except for the training years, the presented accuracy measures are averages of OA, UA, and PA of the same years. Since accuracies and errors of spatial data are spatially explicit, a map was created that demonstrates our models' ability to visualize correctly and wrongly predicted fields using the best modelling method.

2.5. Relationship between Field Size and Accuracy

To check whether the accuracy of crop type prediction depended on the size of crop fields, the correctly and wrongly predicted fields along with their sizes were extracted for each of the MY models. Field sizes were rounded to the nearest multiples of 1.5 ha (i.e., the average field size) to create field size classes, with the count of the all correct (True) and wrong (False) predictions for each field size category. The percentages of correct predictions for all the field size categories were computed.

3. Results

3.1. Agricultural Land Cover Data

The IACS data used as reference data to calibrate and validate the developed models showed different field numbers, average field size, and area for grassland, maize, summer crops, and winter crops (Figure 5). The average field size for grassland was around 1 ha and did not change significantly over time. Summer crops showed a slightly higher average field size between 1.2 and 1.5 ha for all years under consideration. Maize and winter crops had the biggest average field sizes, ranging between 1.8 ha to 2 ha from 2005 to 2015.

While grassland had the highest number of fields followed by winter crops, the number of maize and summer crops were the lowest. Consequently, winter crops covered the largest area of arable lands in the study area followed by grassland, whereas the area of maize and summer crops was comparatively small.

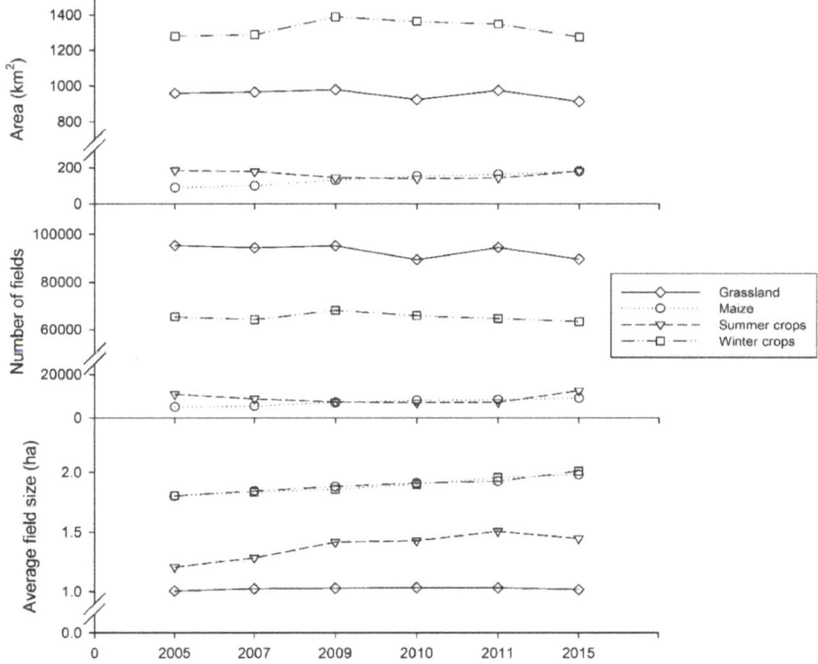

Figure 5. Characteristics of Integrated Administration and Control System (IACS) data used as reference information for the predictive crop type models.

3.2. Assessment of the Modelling Approaches

The general performance of the two modelling approaches as indicated by the OA (Figure 6) ranged from 67.7% to 73.4% with an average value of 71.5% for the SY models. The MY modelling approach showed an OA between 67.1% and 86.1% with an average of 77.3% (Figure 7). The lowest OA for the SY models was observed when they were tested with crops in the year 2015, while the highest OA value was achieved in 2011. Conversely, MY models achieved their highest OA in 2015, whereas, the lowest performance was observed in 2011.

The SY prediction of individual crop types in different years based on user (UA) and producer accuracy (PA) showed a high UA (81.6%) and PA (82.2%) values for grassland, while UA (76.7%) and PA values (67%) for maize were somewhat lower. Prediction of summer crops (UA and PA of 60.6 % and 69.5 % respectively) was less accurate than winter crops (UA 76%, PA 71.1%).

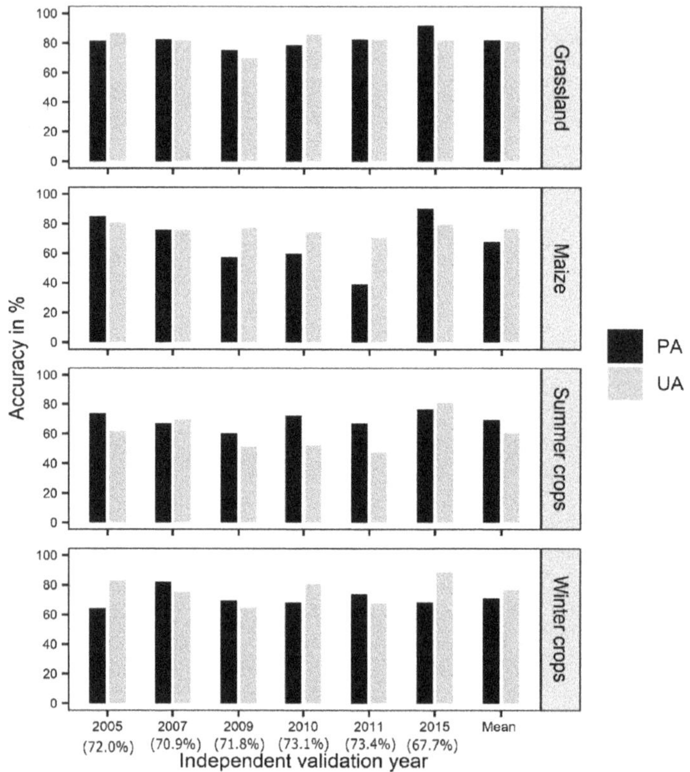

Figure 6. Accuracies of single year models. The mean bars represent average PA and UA, respectively, for each crop across years. Values in brackets represent the average overall accuracies of all models. UA = User accuracy and PA = producer accuracy.

The application of MY models to identify grassland area of different years resulted in a mean UA of 83.1% and a PA of 87.8% (Figure 7). Maize was discriminated with a mean UA and PA of 71.8% and 85.2% respectively, while winter crops were discriminated with a mean accuracy of 79% (UA) and 79.6% (PA). The average accuracy for summer crops was comparatively low (PA = 71.5% and PA = 69.3%), which was mainly a result of the confusion between summer crops and the other crops.

Overall, grasslands always exhibited higher accuracies (UA and PA > 80%) across years when predicted by the two modelling methods, whereas the arable crops were better predicted by the MY models in comparison to the SY models.

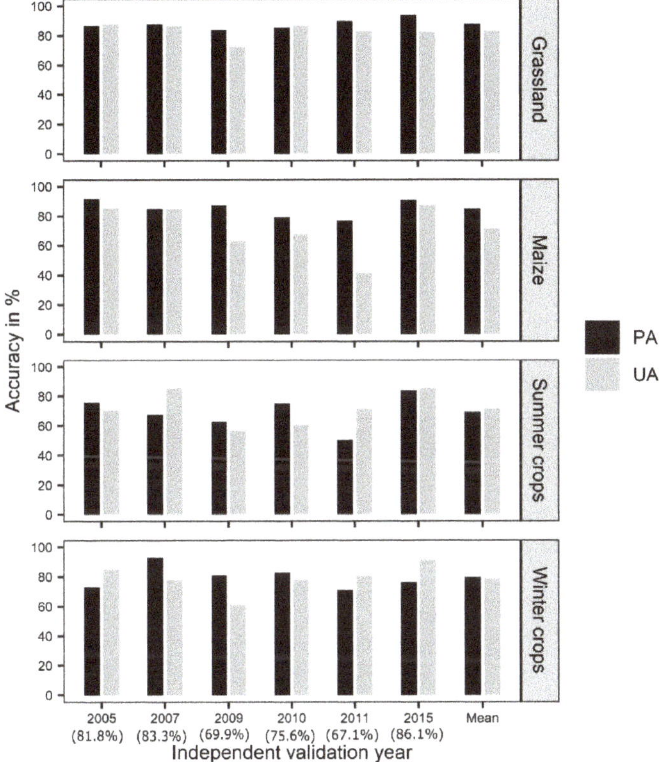

Figure 7. Accuracies of multiple-year models. The mean bars represent average producer accuracies (PA) and user accuracy (UA), respectively, for each crop across years. Values in brackets represent the average overall accuracies of all models.

3.3. Classified Maps Based on Best Modelling Method

Since MY models proved to be the best modelling approach, their capability to create accurate crop-type maps are exhibited in Figure 8 for 2015, which corresponds to an OA of 86.1% (see Appendix A, Table A6 for a confusion matrix). The map was derived with model 6 (Figure 4B), which is an MY model developed based on information from 2005, 2007, 2009, 2010, and 2011. On closer examination, regions at higher altitudes with less favourable growth conditions, which are dominated by grasslands (Figure 8A), can be clearly distinguished from fertile areas, where a multitude of arable crops is grown and where grassland is only interspersed. Moreover, the spatial distribution and patterns of crops, the shape, and edges of fields can clearly be observed. The so-called 'salt and pepper' effects, that characterize most land-cover maps at a pixel base were not experienced with the maps produced in this research, which may be a result of the fact that our models were calibrated with the mean spectral information at field scale based on the IACS field polygons. Maps of the remaining years and their respective confusion matrices are shown in the Appendix A, that is, Figures A1–A5 and Tables A1–A5, respectively.

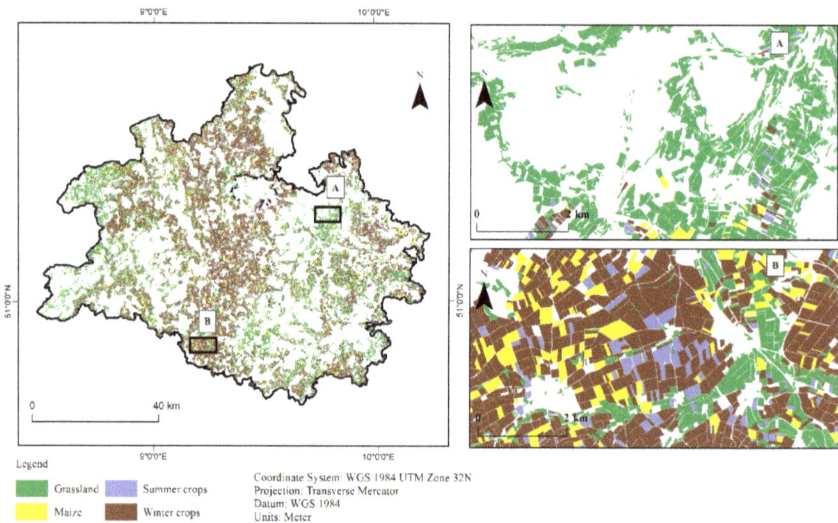

Figure 8. A classified map of 2015 resulting from a multiple-year model based on spectral information of 2005, 2007, 2009, 2010 and 2011. 'A' and 'B' show areas dominated by grassland at higher altitudes and fertile areas dominated by arable crops respectively.

3.4. Relationship between Model Accuracy and Field Size

An increase in mean accuracy from 74% to 87% was observed as the field size increased (i.e., 1.5–9 ha) (Figure 9). Furthermore, a slight decrease in the mean accuracy from 87% to 85% of crop type prediction was seen with increasing field size from 9 ha to 12 ha. As the field size increased further to 13.5 ha, a slight increase in accuracy was observed (from 85% to 88%). However, a marginal reduction to 87% accuracy was observed as the field size increased further. The consistent and marginal rise and fall in accuracy with increasing field sizes indicate that crop type prediction by MY models is independent of field size. But it is important to state that the majority of fields (> 60%) belong to the smallest field category (1.5 ha). The accuracy maps of the MY models can be seen in Figure 10 and the rest in the Appendix A (Figures A6–A10)

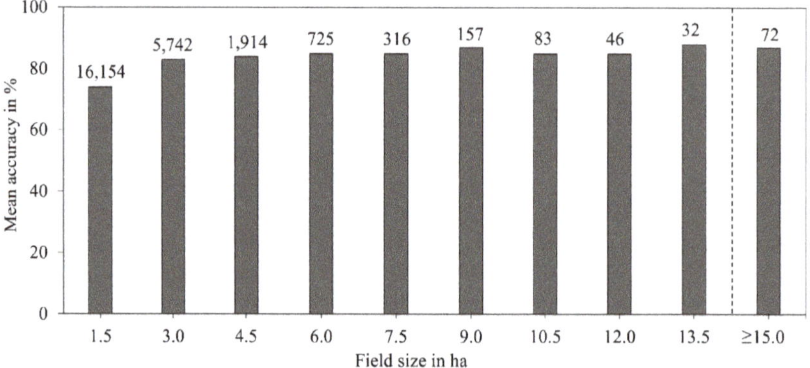

Figure 9. The relationship between the accuracy of crop type prediction and field size based on the multiple-year models. The accuracies are mean values for all years, with the last bin representing the average accuracy of all fields ranging 15–34.5 ha. The value on top of each bar represents the number of fields for each field size group.

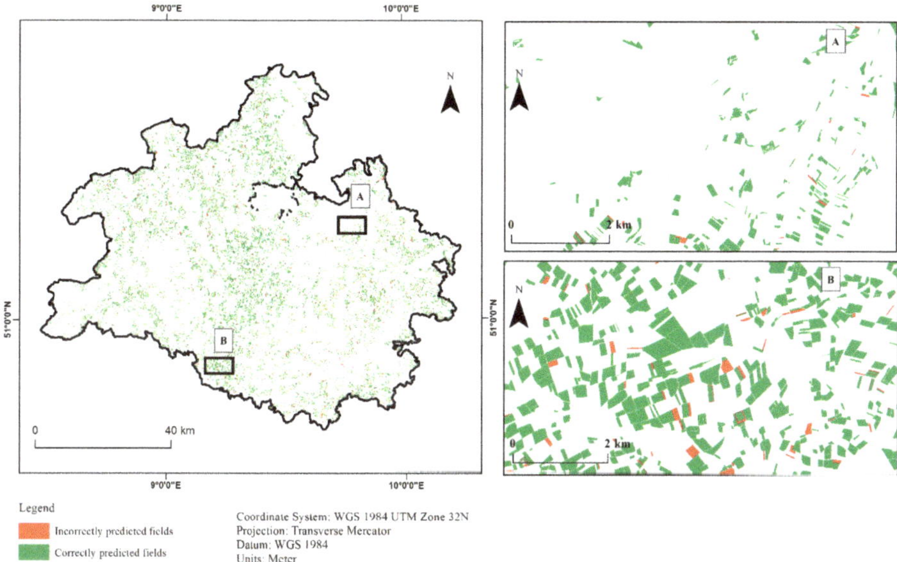

Figure 10. An IACS-based accuracy map of 2015 resulting from a multiple-year model calibrated using spectral data of 2005, 2007, 2009, 2010 and 2011 representing wrongly and correctly predicted fields. 'A' and 'B' show areas dominated by grassland at higher altitudes and fertile areas dominated by arable crops respectively.

3.5. Importance of Predictor Variables

One of the strengths of RF models is the ability to measure and assess the contribution of each predictor variable used. Figure 11 presents the importance values of the first 6 most important input variables used in the MY models. Late summer (LS) NDMI was the first most important predictor variable in most models, and only in one instance the means of early summer (ES) NDVI was ranked first (Figure 11B). Another predictor that seemed to be important across models was the early summer NDVI as shown in Figure 11A,C–F. The contribution of VIs in the prediction of crops was much stronger than the individual bands. Red and green bands are the only bands that appeared among the first six important predictors, with red being the dominant one across all models.

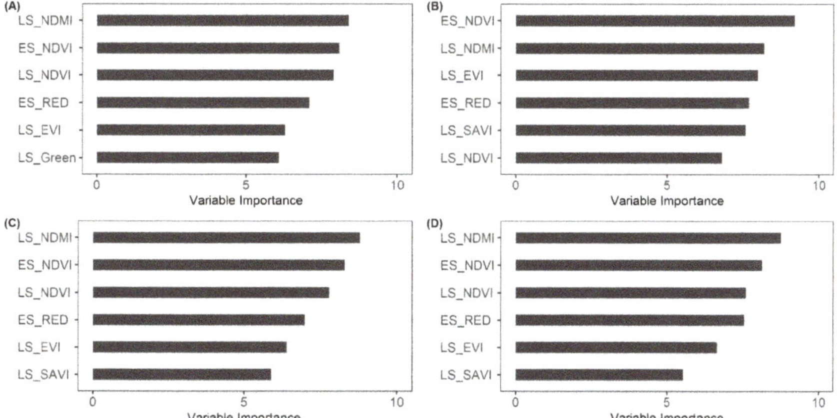

Figure 11. *Cont.*

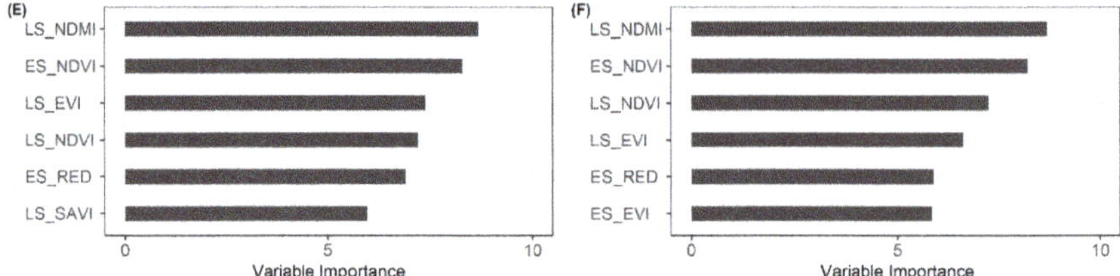

Figure 11. The 6 most important predictors used in the multi-year models expressed in decreasing order of importance from the top of the y-axis. Variable importance on the x-axis is expressed as a percentage of Mean Decrease Gini. (**A–F**) are MY models trained by a combination of (2005 + 2007 + 2009 + 2010 + 2011), (2005 + 2007 + 2009 + 2010 + 2015), (2005 + 2007 + 2009 + 2011 + 2015), (2005 + 2007 + 2010 + 2011 + 2015), (2005 + 2009 + 2010 + 2011 + 2015) and (2007 + 2009 + 2010 + 2011 + 2015) respectively. LS = Late summer, ES = Early summer.

4. Discussion

Crop type mapping in large agricultural landscapes is challenged by the daunting task of periodic training data collection. The traditional satellite-based mapping approach of using reference data from the same year impedes mapping specifically in periods where reference data is not available. The IACS data, which is a field-based crop type data presents reliable reference datasets to deal with the problem of frequent training data collection for satellite-based crop type mapping through the development of a generalized model. The issue of generalized classifiers has been raised and exploited in a few agricultural mapping studies, but less focus was put on specific crop type mapping. Thus, this study aimed at assessing the efficacy of IACS data to be used as reference data for the development of generalized SY and MY crop type models to predict grassland, maize, summer and winter crops as land-cover categories.

The accuracies achieved in our study are similar to the work of [8], where corn and soybeans were predicted based on spectral characteristics using the single-year modelling approach. While our study considers four different crop types with an acceptable average UA and PA for grassland across years (>80%), maize, summer and winter crops were predicted at somewhat lower accuracies. That means that despite the somewhat low performance of SY models in predicting the other crop types, it is able to predict grassland with an acceptable level of accuracy across years.

The overall accuracy of the SY calibration method employed in our study is rather low compared to the traditional method, where calibration and testing data are from the same year. Probable reasons may be different growing dates of the crops in different years, inter-annual differences in climate, image acquisition time as well as variation in image quality between years, as was also suggested by Laborte et al., [58] and Zhong et al., [8]. The performance of MY models in predicting crop types showed higher robustness across years than the SY calibration approach. An average increment of 6% in OA (i.e., 77.3%) was achieved by MY models across years. A similar OA of 73.1% was achieved by Massey et al., [37] when an MY calibrated model was used to predict crop types from MODIS data for an independent year. The higher robustness of MY models might be attributable to the fact that, through the use of spectral information from different years, the interannual climate variability as well as variations in image quality from different years, are reduced to a certain degree [8,58]. Thus, these factors make them generic with high prediction accuracies when applied to predict crops from data not seen by the model. Additionally, MY calibration compensates for the phenological differences of crops among years through the inclusion of many phenological situations using multiple data from different years. This makes MY models more efficient and generalized for satellite-based crop type classification when training data is not available for a period of interest.

The prediction of crops from the spectral-temporal profiles of satellite images can heavily depend on the time and quality of the images used [10,58]. Occasionally, a compromise has to be made between the quality and time of images in the same growing season, which may ultimately have the consequence that not all years will have high accuracies when data from multiple years are used [8]. In this study, for example, the 2009 and 2011 seasons comprised of very late September and October images (Table 2) due to the lack of earlier images. However, in the study area, maize fields are harvested as late as the end of September or early October, whereas summer crops are harvested much earlier to give way for winter crops, which might have developed one or two leaves at this time. Such sources of variation might explain the somewhat lower accuracies of 2009 and 2011 data as compared to the other years.

The data on the assessment of the spectral predictors used in the MY models indicates a paramount contribution of VIs in the prediction of crops as compared to the individual bands. A similar conclusion was drawn by Fletcher [59] in the discrimination of soybean and three weed species. The highest contribution of VIs in the prediction of the crops was expected since their calculation involved two or more bands and as a result, used the unique spectral characteristics of the individual bands to produce a single layer which captured the different phenological dynamics among the crops. Vegetation indices, which are based on SWIR and NIR spectral bands, are known for their contributions to plant separation [59]. Therefore, despite the contributions of the other predictor variables, the late summer NDMI is ranked as the topmost dominant predictor variables for almost all years, followed by early summer NDVI. NDMI uses a normalized ratio of the difference and sum of NIR and SWIR and is known to be sensitive to changes in water content of vegetations canopies [56]. It can, therefore, be inferred that the differences among the four crop types are better captured by the content of moisture in their leaf canopies during late summer.

The subject of object size and prediction accuracy is very crucial in the mapping of agricultural areas [39]. Our results suggest that the prediction of a particular crop type does not necessarily depend on the corresponding size of the field. It was expected that the prediction of bigger fields may be easier than with smaller fields, but the results do not confirm that. The biggest crop field category (≥15 ha) achieved a prediction accuracy of 88%, nonetheless, comparatively smaller field sizes (9 ha) also achieved the same accuracy. Thus, the present study does not support the conclusions of Castilla et al., [41], that the possibility of correct classification of land-cover type decreases with decreasing object size. The reason may be that Castilla et al., [41] employed a segmentation method, which is dependent on the land cover size, whereas the present study used the exact field polygons declared by farmers in the study area as objects for the prediction of the crop types. However, since the present study area is dominated by smaller fields with very few large fields, future research is required to further investigate the relationship between field size and accuracy of crop types prediction in agricultural areas with a relatively even distribution of field sizes.

The uniqueness of this study compared to other studies of generalized classifiers for cropland mapping is the field-based approach employed. This approach deals with some of the challenges associated with the widely used methods. The issue of segmentation scale selection of the other object-based classification [21] is avoided. Moreover, the 'salt and pepper' effects that characterize the pixel-based prediction of land-cover types are equally averted in this study. Hence, our MY modelling approach can be used to map past and present crop types which may be necessary to ascertain the impacts of any agricultural activity (e.g., biogas production) heavily dependent on croplands.

Finally, the hypothesis that IACS data can be used to calibrate models for the prediction of crop types from a satellite image differing from the calibration year has been proven through a field-based SY and MY calibration approach. However, Cai et al., [60] stated that increasing the calibration years to a maximum of 10 years can further increase accuracy. Therefore, our five-year MY models could be improved further by incorporating more years of spectral information, as more satellite data (Sentinel and EnMap) become available in the future.

5. Conclusions

For the first time, this study used a field-based approach to test the usefulness of IACS data in calibrating an RF-based model to predict crop types from satellite images, that are not from the same year as the calibration year. Thus, two modelling methods called SY (i.e., using spectral data from a single year) and MY calibration (i.e., using spectral data from multiple years) were tested in the discrimination of grassland, maize, summer and winter crops.

The results depict a superior performance of the MY approach as compared to SY model. The MY approach included a larger range of inter-annual variability in image quality, climate, and growing dates of crops from different years, thus, contributing to its robustness in predicting crop type from satellite images of different years. The approach employed in this work, unlike other object-based methods, is not dependent on field size. It is, therefore, recommended to use the field-based MY calibration approach for practical crop type mapping, particularly when reference data for the mapping year is not available. This method is useful for practical reasons and can be used to map past and present croplands for comparative analysis. However, the inclusion of soil data and phenological metrics as predictors of MY model may have a potential for future research. This might help improve performance and provide an opportunity for more specific crop type mapping, rather than generic crops like summer and winter crops as used in this study. A combination of data from different satellites like Sentinel or upcoming satellites like EnMap or HyspIRI might further improve the MY modelling approach due to higher revisiting time and thus a denser time series

Author Contributions: T.A., R.G., and M.W. conceived the idea of the research. I.K. processed the data, analyzed the results and wrote the manuscript. T.A., R.G., and M.W. supervised the study and contributed to the writing and revision of the manuscript.

Funding: This study was supported by the Federal Ministry of Education and Research of Germany [grant number 031B0281A].

Conflicts of Interest: The authors declare no conflict of interest. The funding agency had no role in the design of the study; in the collection, analyses, or interpretation of data; in the writing of the manuscript, or in the decision to publish the results.

Appendix A

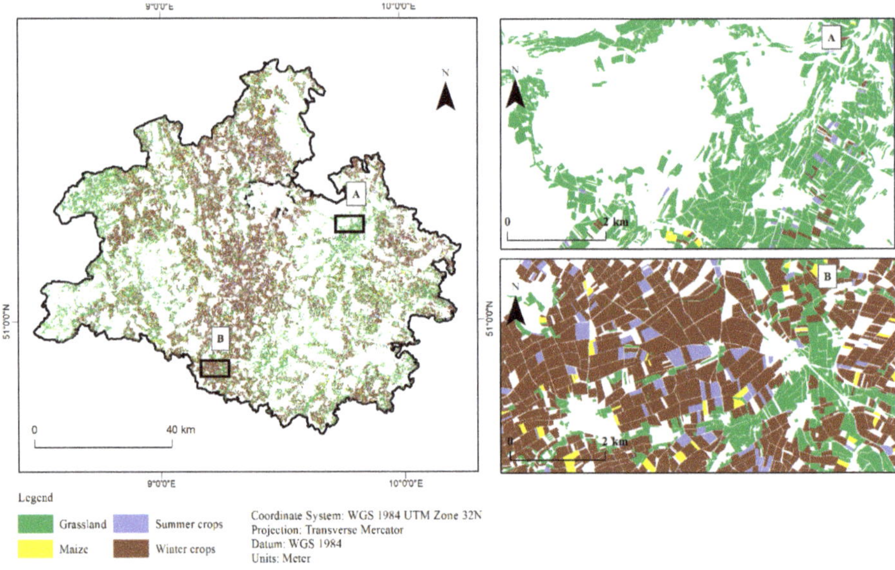

Figure A1. A classified map of 2005 resulting from a multiple-year model based on spectral information of 2007, 2009, 2010, 2011 and 2015.

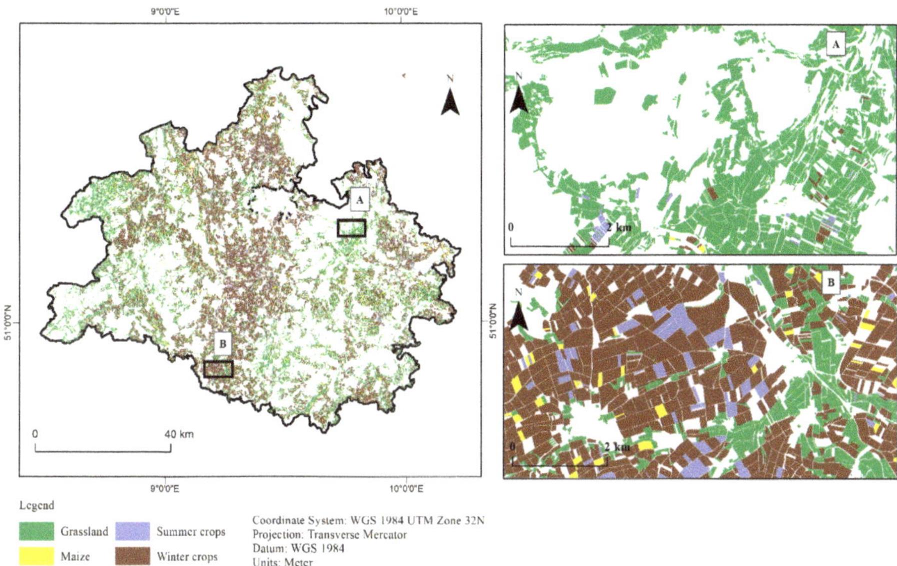

Figure A2. A classified map of 2007 resulting from a multiple-year model based on spectral information of 2005, 2009, 2010, 2011 and 2015.

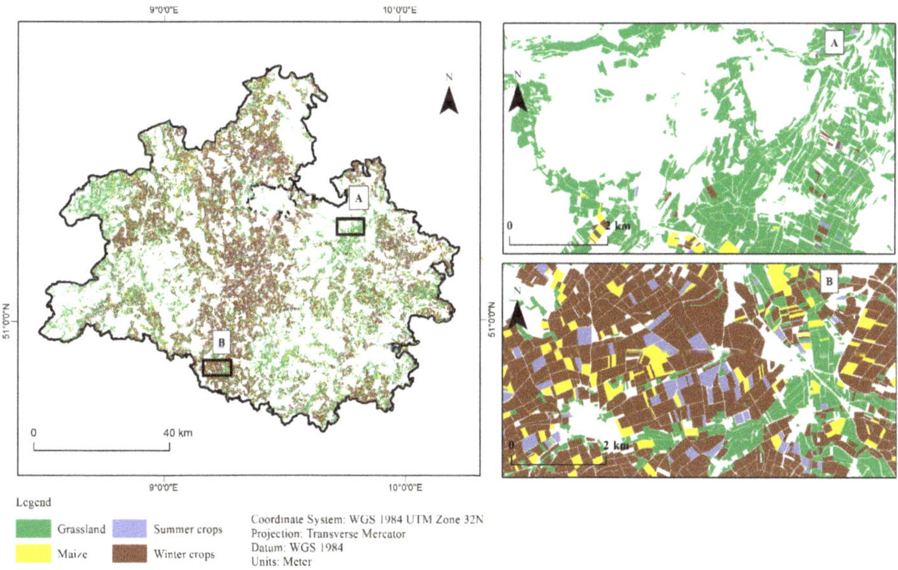

Figure A3. A classified map of 2009 resulting from a multiple-year model based on spectral information of 2005, 2007, 2010, 2011 and 2015.

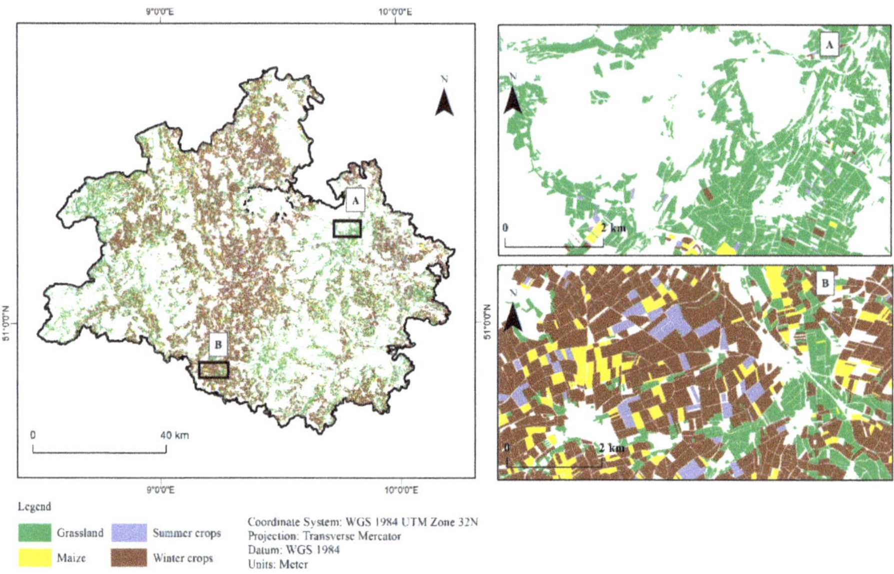

Figure A4. A classified map of 2010 resulting from a multiple-year model based on spectral information of 2005, 2007, 2009, 2011 and 2015.

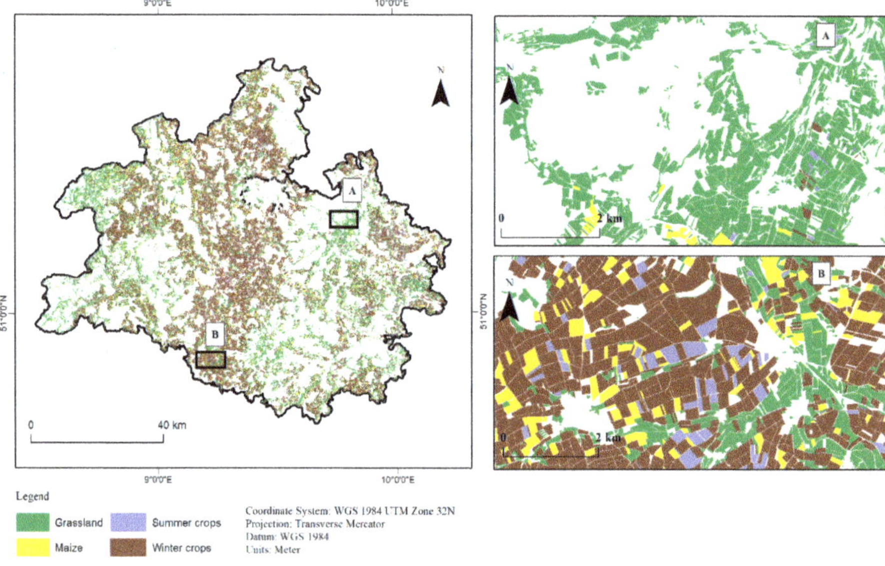

Figure A5. A classified map of 2011 resulting from a multiple-year model based on spectral information of 2005, 2007, 2009, 2010 and 2015.

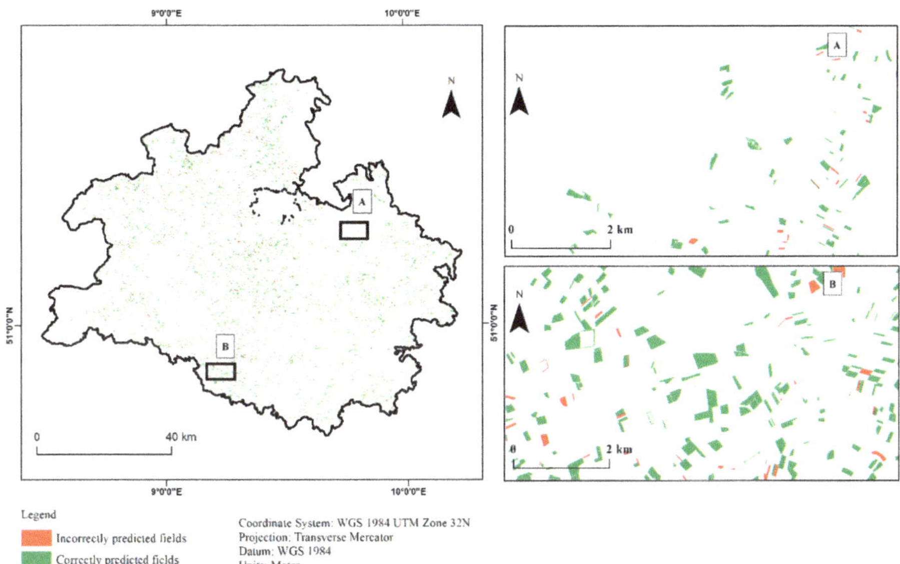

Figure A6. An IACS-based accuracy map of 2005 resulting from a multiple-year model calibrated using spectral data from 2007, 2009, 2010, 2011 and 2015.

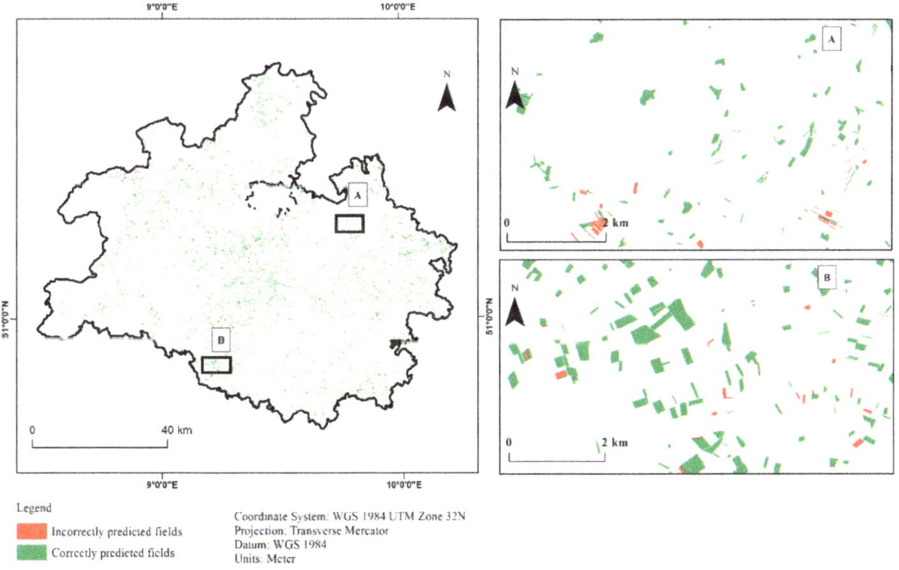

Figure A7. An IACS-based accuracy map of 2007 resulting from a multiple-year model calibrated using spectral data from 2005, 2009, 2010, 2011 and 2015.

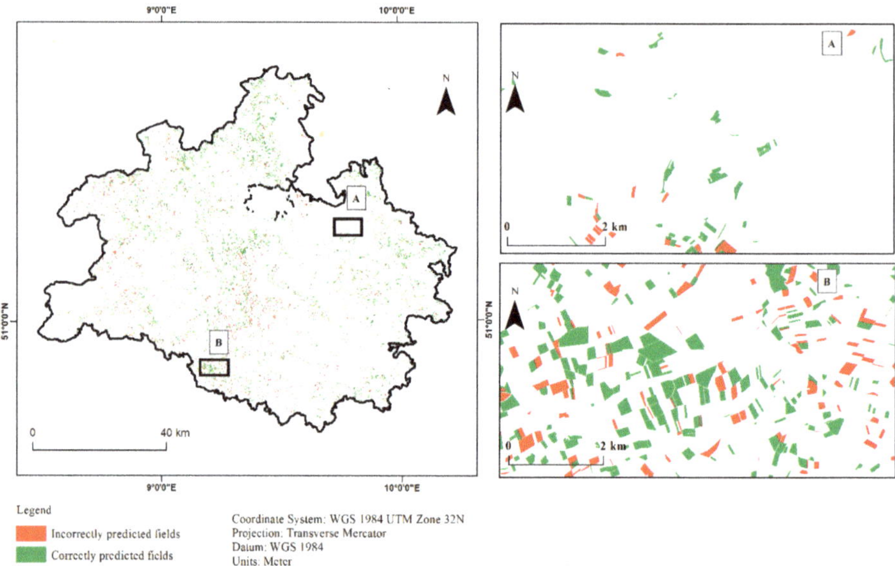

Figure A8. An IACS-based accuracy map of 2009 resulting from a multiple-year model calibrated using spectral data from 2005, 2007, 2010, 2011 and 2015.

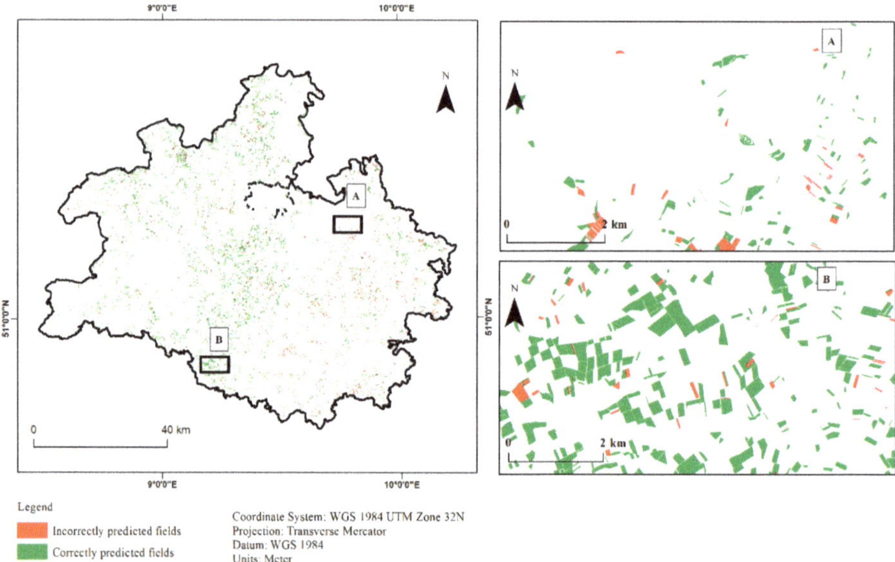

Figure A9. An IACS-based accuracy map of 2010 resulting from a multiple-year model calibrated using spectral data from 2005, 2007, 2009, 2011 and 2015.

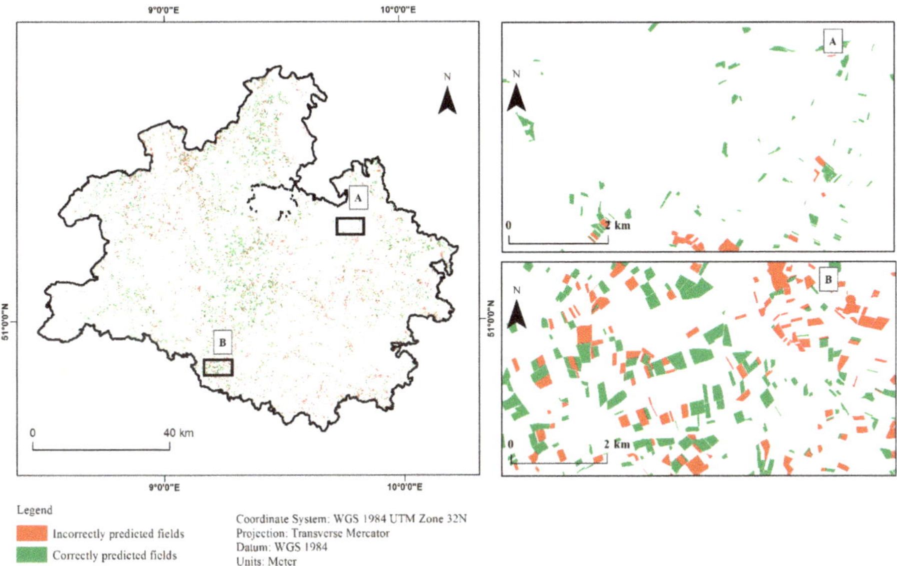

Figure A10. An IACS-based accuracy map of 2011 resulting from a multiple-year model calibrated using spectral data from 2005, 2007, 2009, 2010 and 2015.

Table A1. Confusion matrix of the classified map of 2005 using the multiple-year model. The shaded diagonals represent the number of correctly predicted crops. GL = Grassland, MZ = Maize, SC = Summer crops, WC = Winter crops, UA = User accuracy, PA = Producer accuracy, EC = Error of commission, EO = Error of omission and OA = Overall accuracy.

		Reference						
		GL	**MZ**	**SC**	**WC**	**Total**	**UA (%)**	**CE (%)**
	GL	4339	137	270	214	4960	87.48	12.52
	MZ	292	4527	442	33	5294	85.51	14.49
	SC	264	229	3792	1099	5384	70.43	29.57
Prediction	WC	105	48	496	3654	4303	84.92	15.08
	Total	5000	4941	5000	5000			
	PA (%)	86.78	91.62	75.84	73.08			
	OE (%)	13.22	8.38	24.16	26.92			
	OA (%)	81.8						

Table A2. Confusion matrix of the classified map of 2007 using the multiple-year model. The shaded diagonals represent the number of correctly predicted crops. GL = Grassland, MZ = Maize, SC = Summer crops, WC = Winter crops, UA = User accuracy, PA = Producer accuracy, EC = Error of commission, EO = Error of omission and OA = Overall accuracy.

		Reference						
		GL	**MZ**	**SC**	**WC**	**Total**	**UA (%)**	**CE (%)**
	GL	4380	203	316	156	5055	86.65	13.35
	MZ	158	3738	463	38	4397	85.01	14.99
	SC	68	358	3384	154	3964	85.37	14.63
Prediction	WC	394	90	837	4652	5973	77.88	22.12
	Total	5000	4389	5000	5000			
	PA (%)	87.60	85.17	67.68	93.04			
	OE (%)	12.40	14.83	32.32	6.96			
	OA (%)	83.32						

Table A3. Confusion matrix of the classified map of 2009 using the multiple-year model. The shaded diagonals represent the number of correctly predicted crops. GL = Grassland, MZ = Maize, SC = Summer crops, WC = Winter crops, UA = User accuracy, PA = Producer accuracy, EC = Error of commission, EO = Error of omission and OA = Overall accuracy.

		Reference						
		GL	**MZ**	**SC**	**WC**	**Total**	**UA (%)**	**CE (%)**
	GL	4185	986	430	175	5776	72.45	27.55
	MZ	166	4070	394	26	4656	87.41	12.59
	SC	345	889	3347	748	5329	62.81	37.19
Prediction	WC	304	531	1743	4051	6629	61.11	38.89
	Total	5000	6476	5914	5000			
	PA (%)	83.70	62.85	56.59	81.02			
	OE (%)	16.30	37.15	43.41	18.98			
	OA (%)	69.91						

Table A4. Confusion matrix of the classified map of 2010 using the multiple-year model. The shaded diagonals represent the number of correctly predicted crops. GL = Grassland, MZ = Maize, SC = Summer crops, WC = Winter crops, UA = User accuracy, PA = Producer accuracy, EC = Error of commission, EO = Error of omission and OA = Overall accuracy.

		Reference						
		GL	**MZ**	**SC**	**WC**	**Total**	**UA (%)**	**CE (%)**
	GL	4262	376	179	91	4908	86.84	13.16
	MZ	172	4990	956	185	6303	79.17	20.83
	SC	238	1830	4388	827	7283	60.25	39.75
Prediction	WC	328	171	307	3897	4703	82.86	17.14
	Total	5000	7367	5830	5000			
	PA (%)	85.24	67.73	75.27	77.94			
	OE (%)	14.76	32.27	24.73	22.06			
	OA (%)	75.6						

Table A5. Confusion matrix of the classified map of 2011 using the multiple-year model. The shaded diagonals represent the number of correctly predicted crops. GL = Grassland, MZ = Maize, SC = Summer crops, WC = Winter crops, UA = User accuracy, PA = Producer accuracy, EC = Error of commission, EO = Error of omission and OA = Overall accuracy.

		Reference						
		GL	**MZ**	**SC**	**WC**	**Total**	**UA (%)**	**CE (%)**
	GL	4493	209	264	460	5426	82.81	17.19
	MZ	141	3427	732	151	4451	76.99	23.01
	SC	108	4073	4644	363	9188	50.54	49.46
Prediction	WC	258	507	872	4026	5663	71.09	28.91
	Total	5000	8216	6512	5000			
	PA (%)	89.86	41.71	71.31	80.52			
	OE (%)	10.14	58.29	28.69	19.48			
	OA (%)	67.09						

Table A6. Confusion matrix of the classified map of 2015 using the multiple-year model. The shaded diagonals represent the number of correctly predicted crops. GL = Grassland, MZ = Maize, SC = Summer crops, WC = Winter crops, UA = User accuracy, PA = Producer accuracy, EC = Error of commission, EO = Error of omission and OA = Overall accuracy.

		Reference						
		GL	**MZ**	**SC**	**WC**	**Total**	**UA (%)**	**CE (%)**
	GL	10,313	461	592	1163	12,529	82.31	17.69
	MZ	183	8111	891	69	9254	87.65	12.35
	SC	305	296	9943	1134	11,678	85.14	14.86
Prediction	WC	199	53	464	7634	8350	91.43	8.57
	Total	11,000	8921	11,890	10,000			
	PA (%)	93.75	90.92	83.62	76.34			
	OE (%)	6.25	9.08	16.38	23.66			
	OA (%)	86.1						

References

1. Tilman, D.; Balzer, C.; Hill, J.; Befort, B.L. Global food demand and the sustainable intensification of agriculture. *Proc. Natl. Acad. Sci. USA* **2011**, *108*, 20260–20264. [CrossRef] [PubMed]
2. Nagy, A.; Fehér, J.; Tamás, J. Wheat and maize yield forecasting for the Tisza river catchment using MODIS NDVI time series and reported crop statistics. *Comput. Electron. Agric.* **2018**, *151*, 41–49. [CrossRef]
3. See, L.; Fritz, S.; You, L.; Ramankutty, N.; Herrero, M.; Justice, C.; Becker-Reshef, I.; Thornton, P.; Erb, K.; Gong, P.; et al. Improved global cropland data as an essential ingredient for food security. *Glob. Food Secur.* **2015**, *4*, 37–45. [CrossRef]
4. Gumma, M.K.; Thenkabail, P.S.; Teluguntla, P.; Rao, M.N.; Mohammed, I.A.; Whitbread, A.M. Mapping rice-fallow cropland areas for short-season grain legumes intensification in South Asia using MODIS 250 m time-series data. *Int. J. Digit. Earth* **2016**, *9*, 981–1003. [CrossRef]
5. Funk, C.C.; Brown, M.E. Declining global per capita agricultural production and warming oceans threaten food security. *Food Secur.* **2009**, *1*, 271–289. [CrossRef]
6. Pérez-Hoyos, A.; Rembold, F.; Kerdiles, H.; Gallego, J. Comparison of global land cover datasets for cropland monitoring. *Remote Sens.* **2017**, *9*, 1118. [CrossRef]
7. Tatsumi, K.; Yamashiki, Y.; Canales Torres, M.A.; Taipe, C.L.R. Crop classification of upland fields using Random forest of time-series Landsat 7 ETM+ data. *Comput. Electron. Agric.* **2015**, *115*, 171–179. [CrossRef]
8. Zhong, L.; Gong, P.; Biging, G.S. Efficient corn and soybean mapping with temporal extendability: A multi-year experiment using Landsat imagery. *Remote Sens. Environ.* **2014**, *140*, 1–13. [CrossRef]
9. Castillejo González, I. Mapping of Olive Trees Using Pansharpened QuickBird Images: An Evaluation of Pixel- and Object-Based Analyses. *Agronomy* **2018**, *8*, 288. [CrossRef]
10. Foerster, S.; Kaden, K.; Foerster, M.; Itzerott, S. Crop type mapping using spectral-temporal profiles and phenological information. *Comput. Electron. Agric.* **2012**, *89*, 30–40. [CrossRef]
11. Moeckel, T.; Dayananda, S.; Nidamanuri, R.R.; Nautiyal, S.; Hanumaiah, N.; Buerkert, A.; Wachendorf, M. Estimation of vegetable crop parameter by multi-temporal UAV-borne images. *Remote Sens.* **2018**, *10*, 805. [CrossRef]
12. Wu, B.; Meng, J.; Li, Q.; Yan, N.; Du, X.; Zhang, M. Remote sensing-based global crop monitoring: Experiences with China's CropWatch system. *Int. J. Digit. Earth* **2014**, *7*, 113–137. [CrossRef]
13. Pittman, K.; Hansen, M.C.; Becker-Reshef, I.; Potapov, P.V.; Justice, C.O. Estimating global cropland extent with multi-year MODIS data. *Remote Sens.* **2010**, *2*, 1844–1863. [CrossRef]
14. Waldhoff, G.; Lussem, U.; Bareth, G. Multi-Data Approach for remote sensing-based regional crop rotation mapping: A case study for the Rur catchment, Germany. *Int. J. Appl. Earth Obs. Geoinf.* **2017**, *61*, 55–69. [CrossRef]
15. Habyarimana, E.; Piccard, I.; Catellani, M.; De Franceschi, P.; Dall'Agata, M. Towards Predictive Modeling of Sorghum Biomass Yields Using Fraction of Absorbed Photosynthetically Active Radiation Derived from Sentinel-2 Satellite Imagery and Supervised Machine Learning Techniques. *Agronomy* **2019**, *9*, 203. [CrossRef]

16. Liu, J.; Feng, Q.; Gong, J.; Zhou, J.; Liang, J.; Li, Y. Winter wheat mapping using a random forest classifier combined with multi-temporal and multi-sensor data. *Int. J. Digit. Earth* **2018**, *11*, 783–802. [CrossRef]

17. Maxwell, S.K.; Nuckols, J.R.; Ward, M.H.; Hoffer, R.M. An automated approach to mapping corn from Landsat imagery. *Comput. Electron. Agric.* **2004**, *43*, 43–54. [CrossRef]

18. Yin, H.; Prishchepov, A.V.; Kuemmerle, T.; Bleyhl, B.; Buchner, J.; Radeloff, V.C. Mapping agricultural land abandonment from spatial and temporal segmentation of Landsat time series. *Remote Sens. Environ.* **2018**, *210*, 12–24. [CrossRef]

19. Zheng, B.; Myint, S.W.; Thenkabail, P.S.; Aggarwal, R.M. A support vector machine to identify irrigated crop types using time-series Landsat NDVI data. *Int. J. Appl. Earth Obs. Geoinf.* **2015**, *34*, 103–112. [CrossRef]

20. Waldner, F.; Hansen, M.C.; Potapov, P.V.; Löw, F.; Newby, T.; Ferreira, S.; Defourny, P. National-scale cropland mapping based on spectral-temporal features and outdated land cover information. *PLoS ONE* **2017**, *12*, e0181911. [CrossRef]

21. Phiri, D.; Morgenroth, J. Developments in Landsat land cover classification methods: A review. *Remote Sens.* **2017**, *9*, 967. [CrossRef]

22. Liu, D.; Xia, F. Assessing object-based classification: Advantages and limitations. *Remote Sens. Lett.* **2010**, *1*, 187–194. [CrossRef]

23. Chen, D.; Stow, D. The Effect of Training Strategies on Supervised Classification at Different Spatial Resolutions. *Photogramm. Eng. Remote Sens.* **2002**, *68*, 1155–1162.

24. Li, Q.; Wang, C.; Zhang, B.; Lu, L. Object-based crop classification with Landsat-MODIS enhanced time-series data. *Remote Sens.* **2015**, *7*, 16091–16107. [CrossRef]

25. Zhu, L.; Radeloff, V.C.; Ives, A.R. Improving the mapping of crop types in the Midwestern U.S. by fusing Landsat and MODIS satellite data. *Int. J. Appl. Earth Obs. Geoinf.* **2017**, *58*, 1–11. [CrossRef]

26. Chen, Y.; Lu, D.; Moran, E.; Batistella, M.; Dutra, L.V.; Sanches, I.D.; da Silva, R.F.B.; Huang, J.; Luiz, A.J.B.; de Oliveira, M.A.F. Mapping croplands, cropping patterns, and crop types using MODIS time-series data. *Int. J. Appl. Earth Obs. Geoinf.* **2018**, *69*, 133–147. [CrossRef]

27. Foody, G.M.; Mathur, A.; Sanchez-Hernandez, C.; Boyd, D.S. Training set size requirements for the classification of a specific class. *Remote Sens. Environ.* **2006**, *104*, 1–14. [CrossRef]

28. Botkin, D.B.; Estes, J.E.; MacDonald, R.M.; Wilson, M.V. Studying the Earth's Vegetation from Space. *Bioscience* **1984**, *34*, 508–514. [CrossRef]

29. Sonobe, R.; Yamaya, Y.; Tani, H.; Wang, X.; Kobayashi, N.; Mochizuki, K.I. Mapping crop cover using multi-temporal landsat 8 oli imagery. *Int. J. Remote Sens.* **2017**, *38*, 4348–4361. [CrossRef]

30. Vuolo, F.; Neuwirth, M.; Immitzer, M.; Atzberger, C.; Ng, W.-T. How much does multi-temporal Sentinel-2 data improve crop type classification? *Int. J. Appl. Earth Obs. Geoinf.* **2018**, *72*, 122–130. [CrossRef]

31. Zhao, Y.; Feng, D.; Yu, L.; Wang, X.; Chen, Y.; Bai, Y.; Hernández, H.J.; Galleguillos, M.; Estades, C.; Biging, G.S.; et al. Detailed dynamic land cover mapping of Chile: Accuracy improvement by integrating multi-temporal data. *Remote Sens. Environ.* **2016**, *183*, 170–185. [CrossRef]

32. Xiao, X.; Boles, S.; Liu, J.; Zhuang, D.; Frolking, S.; Li, C.; Salas, W.; Moore, B. Mapping paddy rice agriculture in southern China using multi-temporal MODIS images. *Remote Sens. Environ.* **2005**, *95*, 480–492. [CrossRef]

33. Manfron, G.; Delmotte, S.; Busetto, L.; Hossard, L.; Ranghetti, L.; Brivio, P.A.; Boschetti, M. Estimating inter-annual variability in winter wheat sowing dates from satellite time series in Camargue, France. *Int. J. Appl. Earth Obs. Geoinf.* **2017**, *57*, 190–201. [CrossRef]

34. Huete, A.; Didan, K.; Miura, T.; Rodriguez, E.P.; Gao, X.; Ferreira, L.G. Overview of the radiometric and biophysical performance of the MODIS vegetation indices. *Remote Sens. Environ.* **2002**, *83*, 195–213. [CrossRef]

35. Bill, R.; Nash, E.; Grenzdörffer, G. Integrated Administration and Control System. In *Springer Handbook of Geographic Information System*; Kresse, W., Danko, D.M., Eds.; Springer: Berlin, Germany, 2012; pp. 801–803. ISBN 978-3-540-72678-4.

36. Boryan, C.; Yang, Z.; Mueller, R.; Craig, M. Monitoring US agriculture: The US department of agriculture, national agricultural statistics service, cropland data layer program. *Geocarto Int.* **2011**, *26*, 341–358. [CrossRef]

37. Massey, R.; Sankey, T.T.; Congalton, R.G.; Yadav, K.; Thenkabail, P.S.; Ozdogan, M.; Sánchez Meador, A.J. MODIS phenology-derived, multi-year distribution of conterminous U.S. crop types. *Remote Sens. Environ.* **2017**, *198*, 490–503. [CrossRef]

38. Griffiths, P.; Nendel, C.; Hostert, P. Intra-annual reflectance composites from Sentinel-2 and Landsat for national-scale crop and land cover mapping. *Remote Sens. Environ.* **2019**, *220*, 135–151. [CrossRef]

39. Ma, L.; Li, M.; Ma, X.; Cheng, L.; Du, P.; Liu, Y. ISPRS Journal of Photogrammetry and Remote Sensing A review of supervised object-based land-cover image classification. *ISPRS J. Photogramm. Remote Sens.* **2017**, *130*, 277–293. [CrossRef]

40. Blaschke, T.; Hay, G.J.; Kelly, M.; Lang, S.; Hofmann, P.; Addink, E.; Queiroz Feitosa, R.; van der Meer, F.; van der Werff, H.; van Coillie, F.; et al. Geographic Object-Based Image Analysis—Towards a new paradigm. *ISPRS J. Photogramm. Remote Sens.* **2014**, *87*, 180–191. [CrossRef]

41. Castilla, G.; Hernando, A.; Zhang, C.; McDermid, G.J. The impact of object size on the thematic accuracy of landcover maps. *Int. J. Remote Sens.* **2014**, *35*, 1029–1037. [CrossRef]

42. Wagner, B. Spatial analysis of loess and loess-like sediments in the Weser-Aller catchment (Lower Saxony and Northern Hesse, NW Germany). *Quarernary Sci. J.* **2011**, *60*, 27–46. [CrossRef]

43. HLNUG Hessian Agency for Nature Conservation, Environment and Geology. Available online: Atlas.umwelt.hessen.de (accessed on 9 November 2018).

44. USGS Landsat Surface Reflectance Products. Available online: https://earthexplorer.usgs.gov/ (accessed on 24 July 2018).

45. Masek, J.G.; Vermote, E.F.; Saleous, N.E.; Wolfe, R.; Hall, F.G.; Huemmrich, K.F.; Gao, F.; Kutler, J.; Lim, T.K. A landsat surface reflectance dataset for North America, 1990–2000. *IEEE Geosci. Remote Sens. Lett.* **2006**, *3*, 68–72. [CrossRef]

46. Vermote, E.; Justice, C.; Claverie, M.; Franch, B. Preliminary analysis of the performance of the Landsat 8/OLI land surface reflectance product. *Remote Sens. Environ.* **2016**, *185*, 46–56. [CrossRef]

47. Flood, N. Continuity of reflectance data between landsat-7 ETM+ and landsat-8 OLI, for both top-of-atmosphere and surface reflectance: A study in the australian landscape. *Remote Sens.* **2014**, *6*, 7952–7970. [CrossRef]

48. Li, P.; Jiang, L.; Feng, Z. Cross-comparison of vegetation indices derived from landsat-7 enhanced thematic mapper plus (ETM+) and landsat-8 operational land imager (OLI) sensors. *Remote Sens.* **2014**, *6*, 310–329. [CrossRef]

49. USGS Landsat Science Products-Landsat Surface Reflectance. Available online: https://www.usgs.gov/land-resources/nli/landsat/landsat-surface-reflectance?qt-science_support_page_related_con=0#qt-science_support_page_related_con (accessed on 10 May 2019).

50. Hijmans, R.J.; van Etten, J.; Cheng, J.; Mattiuzzi, M.; Sumner, M.; Greenberg, J.A.; Lamigueiro, O.P.; Bevan, A.; Racine, E.B.; Shortridge, A. Package "raster". *R Package* **2019**, version 2.9 5. 1 60.

51. Team, R.D.C. R Development Core Team R. *R A Lang. Environ. Stat. Comput.* **2015**, *1*, 409.

52. Breiman, L. Random forest. *Mach. Learn.* **2001**, *45*, 5–32. [CrossRef]

53. Belgiu, M.; Drăgu, L. Random forest in remote sensing: A review of applications and future directions. *ISPRS J. Photogramm. Remote Sens.* **2016**, *114*, 24–31. [CrossRef]

54. Tucker, C.J. Red and photographic infrared linear combinations for monitoring vegetation. *Remote Sens. Environ.* **1979**, *8*, 127–150. [CrossRef]

55. Huete, A. A soil-adjusted vegetation index (SAVI). *Remote Sens. Environ.* **1988**, *25*, 295–309. [CrossRef]

56. Gao, B.C. NDWI-A normalized difference water index for remote sensing of vegetation liquid water from space. *Remote Sens. Environ.* **1996**, *58*, 257–266. [CrossRef]

57. Liaw, A.; Wiener, M. Classification and Regression by randomForest. *R News* **2002**, *2*, 18–22.

58. Laborte, A.G.; Maunahan, A.A.; Hijmans, R.J. Spectral Signature Generalization and Expansion Can Improve the Accuracy of Satellite Image Classification. *PLoS ONE* **2010**, *5*, e10516. [CrossRef] [PubMed]

59. Fletcher, R.S. Using Vegetation Indices as Input into Random Forest for Soybean and Weed Classification. *Am. J. Plant Sci.* **2016**, *07*, 2186–2198. [CrossRef]

60. Cai, Y.; Guan, K.; Peng, J.; Wang, S.; Seifert, C.; Wardlow, B.; Li, Z. A high-performance and in-season classification system of field-level crop types using time-series Landsat data and a machine learning approach. *Remote Sens. Environ.* **2018**, *210*, 35–47. [CrossRef]

 *agronomy*

Article

Performance Characterization of the UAV Chemical Application Based on CFD Simulation

Hang Zhu [1],*, Hongze Li [1], Cui Zhang [1], Junxing Li [2] and Huihui Zhang [3],*

[1] School of Mechanical and Aerospace Engineering, Jilin University, Changchun 130022, China;
    lihz18@mails.jlu.edu.cn (H.L.); zhangcui@jlu.edu.cn (C.Z.)
[2] Jilin Academy of Agricultural Machinery, Changchun 130062, China; junxingli_jl@163.com
[3] Water Management and Systems Research Unit, USDA-ARS, Fort Collins, CO 80526, USA
*   Correspondence: zhuhang@jlu.edu.cn (H.Z.); Huihui.Zhang@ars.usda.gov (H.Z.);
    Tel.: +86-1808-866-5997 (H.Z.); +01-970-492-7413 (H.Z.)

Received: 14 May 2019; Accepted: 11 June 2019; Published: 12 June 2019

Abstract: Battery-powered multi-rotor UAVs (Unmanned Aerial Vehicles) have been employed as chemical applicators in agriculture for small fields in China. Major challenges in spraying include reducing the influence of environmental factors and appropriate chemical use. Therefore, the objective of this research was to obtain the law of droplet drift and deposition by CFD (Computational Fluid Dynamics), a universal method to solve the fluid problem using a discretization mathematical method. DPM (Discrete Phase Model) was taken to simulate the motion of droplet particles since it is an appropriate way to simulate discrete phase in flow field and can track particle trajectory. The figure of deposition concentration and trace of droplet drift was obtained by controlling the variables of wind speed, pressure, and spray height. The droplet drifting models influenced by different factors were established by least square method after analysis of drift quantity to get the equation of drift quantity and safe distance. The relationship model, $Y_i(m)$, between three dependent variables, wind speed $X_w(m\ s^{-1})$, pressure $X_p(MPa)$ and spray height $X_h(m)$, are listed as follows: The edge drift distance model was $Y_1 = 0.887X_w + 0.550X_p + 1.552X_h - 3.906$ and the correlation coefficient (R^2) was 0.837; the center drift distance model was $Y_2 = 0.167X_w + 0.085X_p + 0.308X_h - 0.667$ and the correlation coefficient (R^2) was 0.774; the overlap width model was $Y_3 = 0.692x_w + 0.529x_p + 1.469x_h - 3.374$ and the correlation coefficient (R^2) was 0.795. For the three models, the coefficients of the three variables were all positive, indicating that the three factors were all positively correlated with edge drift distance, center drift distance, and overlap width. The results of this study can provide theoretical support for improving the spray quality of UAV and reducing the drift of droplets.

Keywords: UAV chemical application; droplet drift; flat-fan atomizer; simulation analysis; control variables

---

1. Introduction

China is a large agricultural country with the most serious occurrence of crop diseases and pests in the world, and has the largest use of pesticides [1]. Diseases and pests are important factors in agricultural production, and chemical pesticides are the main means to prevent and control crop diseases and pests in China [2,3]. The traditional method of pesticide application is manual control of ground application machinery, but this is strictly limited by the terrain [4]. With the development and implementation of new aerial application technology, the use of unmanned aerial vehicles (UAV) for aerial pesticide application is an inevitable trend for the intelligent development of green agriculture [5,6]. Therefore, it is critical to analyze and evaluate various performance parameters of UAV pesticide application technology. The performance evaluation of a UAV pesticide application system mainly involves an analysis of droplet drift and deposition characteristics under different working conditions [7].

In recent years, research concerning drift deposition characteristics and nozzle performance of UAVs used for the protection of small plants has increased. Due to aerial operation environment and the influence of natural airflow, the drifting and deposition problems of spraying is complicated [8–10]. Chen et al. [11] measured the wind field distribution under the rotor of a multi-rotor electric UAV using a UAV rotor wind field measurement system. The results showed that the vertical wind speed had an impact on the droplet deposition in the effective spray area. Qiu et al. [12] studied the relationship between spray deposition and flight height for unmanned helicopters. The results showed the significant effects of application height on deposition concentration. Shen et al. [13] obtained the flow field characteristics of multi-rotor UAV at different speeds by simulating the flow field of multi-rotor UAV through CFD. Nuyttens et al. [14] established a CFD three-dimensional spray drift model, which considered droplet characteristics, meteorological conditions, chemical characteristics, canopy structure and crop characteristics, etc. They carried out field experiments, verifying that their CFD model was useful for reducing spray drift in the field. Teske et al. [15–17] developed the AGDISP model based on the droplet trajectory model, which covered aircraft models, aircraft vortices, nozzle types, weather factors, and more. Luo et al. [18] carried out gas–solid two-phase flow field simulations for three types of nozzles, acquiring data to assist in the selection of nozzles for specific applications. Chen et al. [19] carried out multi-nozzle atomization field simulations using the UDF method. Their results showed that there was interference between multiple nozzles, and the number and position of nozzles affected the overall atomization effect. There have been many analyses of UAV performance as agricultural aerial sprayers, but relatively fewer analyses concerned with the effect of nozzle characteristics. Thus, this study aimed to simulate and analyze the droplet drift and deposition law of a flat fan nozzle under different working conditions and explore the droplet drift and deposition phenomenon under the influence of different factors.

2. Materials and Methods

2.1. Geometric Model Building

According to the actual spray situation (Figure 1), a cuboid model is established in software ANSYS ICEM CFD15.0 (NASDAQ: ANSS, Canonsburg, PA, USA). The length and width of the simulation calculation area was set as 10 m and 4 m to simulate the spraying area, the height was set as 0.8 m, 1 m and 1.5 m to simulate the spraying heights, for total grid area of 228,000. The left side of the cuboid is an application target area, and the right side is a droplet drift (off-target) area. Grids in the target area are encrypted to have a better spatial resolution. The two nozzles are positioned directly above the center point of the target area (Figure 2).

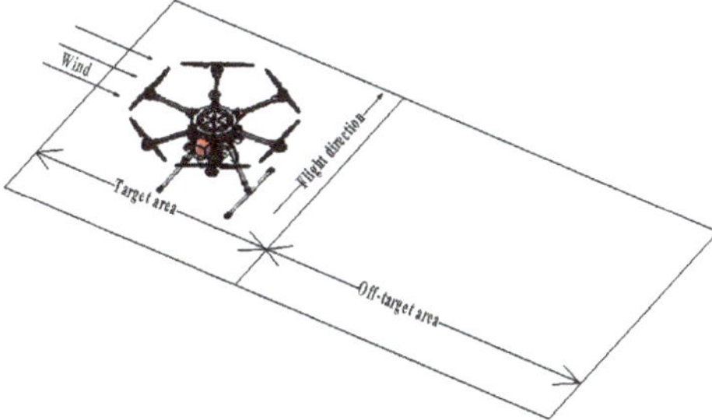

Figure 1. The sketch of UAV, wind and flight direction is shown in diagram to simulate the actual spray picture. The target area is set under the UAV and the off-target area is set down wind relative to the UAV.

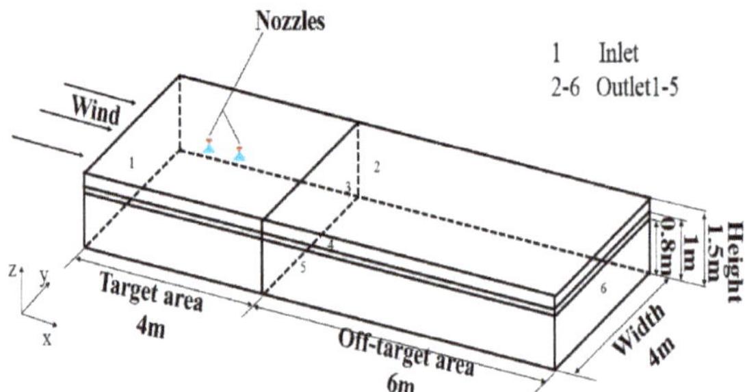

Figure 2. Computational domain and boundary setup are shown. The two nozzles are positioned on top of the domain.

The simulation analysis mainly obtains the droplet deposition rule and its influencing factor indexes. The boundary conditions are set as follows: The left face of the cuboid is the velocity inlet; the right face is the droplet receiving area pressure outlet; the other four surfaces are all pressure outlets; and the outlets are set as boundary escapes. According to the actual spraying operation height of the plant protection UAV, the sprayer height was changed to 0.8 m, 1 m and 1.5 m, respectively, by adjusting the position of the nozzles. The parameter settings are shown in Table 1.

Table 1. Parameters of the flat fan atomizer used to simulate XR8002 nozzle.

Parameters (Unit)	Value
X-Center (m)	1.75/2.25
Y-Center (m)	2
Z-Center (m)	1.494
X-Virtual Center (m)	1.75/2.25
Y-Virtual Center (m)	2
Z-Virtual Center (m)	1.5
X-Fan Normal Vector	0
Y-Fan Normal Vector	−1
Z-Fan Normal Vector	1
Flow Rate (kg s^{-1})	0.01316
Spray Half Angle (deg)	40
Orifice Width (m)	0.00091
Flat Fan Sheet Constant	3
Atomizer Dispersion Angle (deg)	6

In the simulation analysis, the continuous phase substance is air and the discrete uses the parameters of liquid water to simulate chemical. In the steady-state calculation mode, the standard k-ε model (ANSYS, 15.0) [20,21] is selected to simulate the turbulent wind flow. Its transport equations are shown in Equations (1) and (2).

$$\frac{\partial(\rho k)}{\partial t} + \frac{\partial(\rho k u_i)}{\partial x_i} = \frac{\partial}{\partial x_j}\left[(\mu + \frac{\mu_t}{\sigma_k})\frac{\partial k}{\partial x_j}\right] + G_k + G_b - \rho\varepsilon - Y_M + S_k \tag{1}$$

$$\frac{\partial(\rho\varepsilon)}{\partial t} + \frac{\partial(\rho\varepsilon u_i)}{\partial x_i} = \frac{\partial}{\partial x_j}\left[(\mu + \frac{\mu_t}{\sigma_\varepsilon})\frac{\partial\varepsilon}{\partial x_j}\right] + C_{1\varepsilon}\frac{\varepsilon}{k}(G_k + C_{3\varepsilon}G_b) - C_{2\varepsilon}\rho\frac{\varepsilon^2}{k} + S_\varepsilon \tag{2}$$

where k is the turbulence kinetic energy, ε is the turbulence dissipation rate, μ is the dynamic viscosity, μ_t is the turbulence viscosity, G_k is generated by turbulent kinetic energy caused by the average velocity gradient, and G_b is generated by turbulent kinetic energy caused by buoyancy. Y_M is a pulsating expansion term in compressible turbulence, $C_{1\varepsilon}$, $C_{2\varepsilon}$ and $C_{3\varepsilon}$ are empirical constants, Prandtl numbers of σ_k and σ_ε correspond to turbulent kinetic energy k and turbulent dissipation rate respectively, S_k and S_ε is a user-defined source item.

The variation of air density under standard atmospheric pressure and normal temperature is less than 5%, it is regarded as an incompressible fluid, the pressure-based solver type is selected. All the simulations are based on the transient calculation. The convergence criterion is set to 10^{-5}, which means that the converged solutions are reached when the residuals of several significant variables are equal to or less than 10^{-5}.

2.2. Design of Experiment

As is shown in Figure 3, the spray effect is seriously affected by wind speed. The control variates method was used to solve the problem including multivariate by changing one of the factors. This study used different wind speeds, 0 m s^{-1}, 1 m s^{-1}, 3 m s^{-1} and 5 m s^{-1}, to seek the law of droplet drift, and different particle mass flow rates, 0.01083 kg s^{-1}, 0.01316 kg s^{-1} and 0.01516 kg s^{-1}, to control the spray pressure at 0.2 MPa, 0.3 MPa and 0.4 MPa, respectively. The overall control variates of the parameters are shown in Table 2.

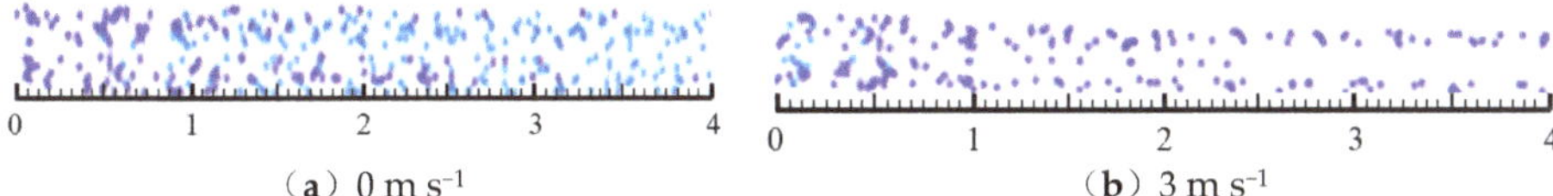

(**a**) 0 m s^{-1} (**b**) 3 m s^{-1}

Figure 3. Droplets deposition in target area at different wind speed (0 m s^{-1} and 3 m s^{-1}). (**a**) The distribution of droplets is ideal at 0 m s^{-1}, (**b**) The droplet was influenced by wind clearly and the distribution of droplets is not satisfied at 3 m s^{-1}.

Table 2. Total variates of spray height (0.8, 1, and 1.5 m), pressure (0.2, 0.3, and 0.4 MPa) and wind speed (0, 1, 3, and 5 m s^{-1}).

Spray Height(m)	0.8			1			1.5		
Pressure (MPa)	0.2	0.3	0.4	0.2	0.3	0.4	0.2	0.3	0.4
Wind Speed (m s^{-1})									
0	A_{11}	A_{12}	A_{13}	B_{11}	B_{12}	B_{13}	C_{11}	C_{12}	C_{13}
1	A_{21}	A_{22}	A_{23}	B_{21}	B_{22}	B_{23}	C_{21}	C_{22}	C_{23}
3	A_{31}	A_{32}	A_{33}	B_{31}	B_{32}	B_{33}	C_{31}	C_{32}	C_{33}
5	A_{41}	A_{42}	A_{43}	B_{41}	B_{42}	B_{43}	C_{41}	C_{42}	C_{43}

The ability of DPM has been shown to accurately simulate particle dispersion and deposition [22–24]. In this study, the flat fan atomizer model of the DPM was selected to simulate the XR8002 nozzle of Teejet Company (Wheaton, IL, USA, 60187). In the DPM model, Euler method is used to describe the continuous phase. Navier-Stokes equation [25] is used to obtain velocity and other parameters. The discrete phase is described by Lagrange method, and its movement is obtained by integrating the motion equations of a large number of particles. Therefore, this model is called Euler-Lagrange model, and its transport equation can be expressed as [26]

$$\frac{du_p}{dt} = \frac{18\mu C_D R_e}{24\rho_p dp^2}(u - u_p) + \frac{g_x(\rho_p - \rho)}{\rho_p} + \frac{1}{2}\frac{\rho}{\rho_p}\frac{d}{dt}(u - u_p) \tag{3}$$

where u is the continuous phase velocity, u_p is the velocity of particle, ρ_p is the density of particle, d_p is the particle diameter, g_x is the acceleration of gravity, R_e is the relative Reynolds number, and C_D is the drag coefficient.

2.3. Method of Analysis

In order to show the influence of three factors (wind speed, spray pressure, and spray height) on the droplet drift distance, deposition center drift distance, and overlap width, the multivariate curve fitting was carried out by the least square method [27,28]. The equation is listed as follows:

$$y_1 = a_0 + a_1 x_w + a_2 x_p + a_3 x_h \tag{4}$$

$$y_2 = a_4 + a_5 x_w + a_6 x_p + a_7 x_h \tag{5}$$

$$y_3 = a_8 + a_9 x_w + a_{10} x_p + a_{11} x_h \tag{6}$$

Satisfy

$$\sum_{i=1}^{n} (y - y_i)^2 = \min \sum_{i=1}^{n} (y - y_i)^2 \tag{7}$$

The model obtained by the least square method can be verified by Equation (8),

$$y_o = a_0 - y_n + y_f \tag{8}$$

where y_o is the overlap width influenced by wind, a_o is the overlap width in nature, y_n is the drift distance of near tuyere edge, and y_f is the drift distance of far tuyere edge. Their relationship can be portrayed in Figure 4.

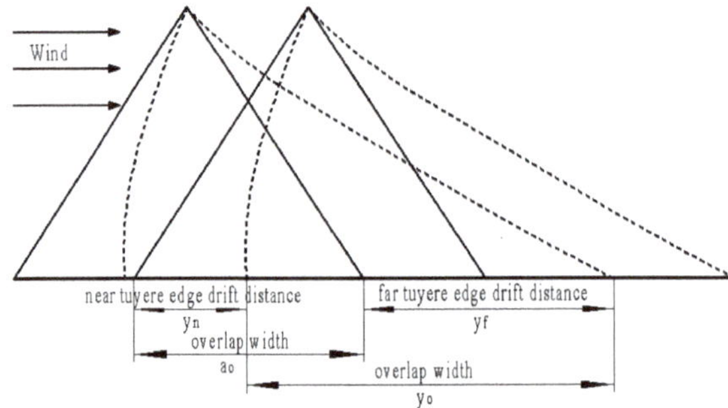

Figure 4. The natural curve of droplets is shown by solid lines and the drift curve influenced by wind is shown by dotted lines, a_0 is overlap width, y_2 is the drift distance of near tuyere edge, y_3 is the drift distance of far tuyere edge, and y_4 is the overlap width influenced by wind.

3. Results and Discussion

3.1. Simulation of Influence of Different Factors on Droplet Drift

3.1.1. Influence of Wind Speed on Droplet Drift

Inlet pressure (0.3 MPa) and spray height (1.5 m) were set to constant and the effect of ambient wind speeds of 0 m s^{-1}, 1 m s^{-1}, 3 m s^{-1}, and 5 m s^{-1} were explored. The droplet deposition density at different target positions is plotted in Figure 5.

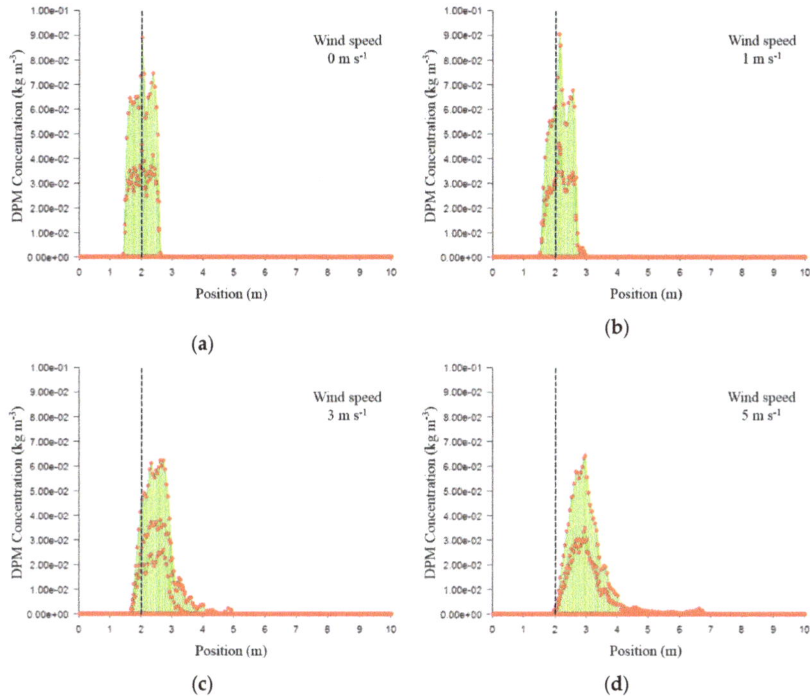

Figure 5. Concentration of droplets deposition at different wind speeds, (**a**) 0 m s^{-1}, (**b**) 1 m s^{-1}, (**c**) 3 m s^{-1}, and (**d**) 5 m s^{-1}. X-axis indicates a different position of target area and off-target area. Y-axis shows DPM concentration (kg m^{-3}) of droplets deposition at different wind speeds, 0 m s^{-1}, 1 m s^{-1}, 3 m s^{-1}, and 5 m s^{-1}. The point means concentration of different position, the dotted line means the droplet concentration deposition center, and the solid line means the biggest concentration of corresponding point.

It is clear that droplet deposition concentration at different target locations varied with wind speed (Figure 5). When the wind speed was 0 m s^{-1}, the droplets did not drift. The droplet concentration deposition center point was 2 m and the drift distance was 0. When the wind speed was 1 m s^{-1}, the droplet had a minor drift phenomenon. The droplet concentration deposition center point was 2.05 m and the drift distance was 0.05 m. When the wind speed was 3 m s^{-1}, the droplet had obvious drift phenomenon. The droplet concentration deposition center point was 2.35 m and the drift distance was 0.35 m. When the wind speed was 5 m s^{-1}, the droplets had a large amount of drift. The droplet concentration deposition center point was 2.85 m and the drift distance was 0.85 m. The three main performance indicators to evaluate drift characteristics, drift distance of deposition center, near tuyere edge, and far tuyere edge, and the relationships between these parameters and wind speed are shown in Table 3 and Figure 6.

Table 3. Droplet drift distance of deposition center, near tuyere edge, far tuyere edge, and overlap width at different wind speed, 0 m s^{-1}, 1 m s^{-1}, 3 m s^{-1} and 5 m s^{-1}.

Wind Speed (m s^{-1})	Deposition Center (m)	Near Tuyere Edge (m)	Far Tuyere Edge (m)	Overlap Width (m)
0	0	0	0	0.20
1	0.05	0.05	0.05	0.65
3	0.45	0.25	2.40	2.35
5	0.85	0.55	4.60	3.95

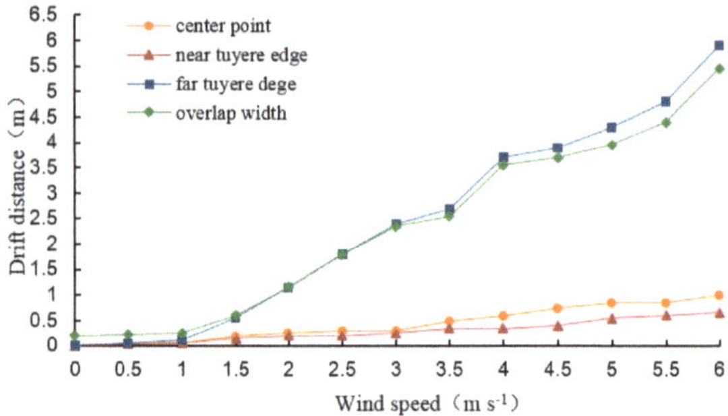

Figure 6. The relationship between wind speed and drift distance (center point, near tuyere edge, far tuyere edge, and overlap width).

The least square method was used to fit the drift distance and overlap width under different wind speeds. The center point drift model was $y_1 = 0.176x_w + 0.091$, near tuyere edge drift model was $y_2 = 0.110x_w - 0.038$, far tuyere edge drift model was $y_3 = 1.016x_w - 0.636$, and the overlap width model was $y_4 = 0.901x_w - 0.382$. In these models, x_w was wind speed (m s^{-1}) and y was distance (m). It can be seen that the wind speed was positively correlated with drift distance and overlap width. For every 1 m s^{-1} increase in wind speed, drift distance of droplet center point was increased by 0.176 m, near tuyere edge was increased by 0.11 m, far tuyere edge was increased by 1.016 m, and overlap width was increased by 0.901 m. The accuracy can be verified by the model obtained above with a_0 representing the droplet overlap width in the natural state. The simulation showed that when $a_0 = 0.2$, $y_4 - (a_0 - y_2 + y_3) = 0.005x + 0.0156$ was about 0, which proved that the model fits the actual situation.

3.1.2. Influence of Inlet Pressure on Droplet Drift

Wind speed (3 m s^{-1}) and spray height (1.5 m) were set to constant and the pressure of 0.2 MPa, 0.3 MPa and 0.4 MPa was explored. The droplet deposition density at different target positions is plotted in Figure 7.

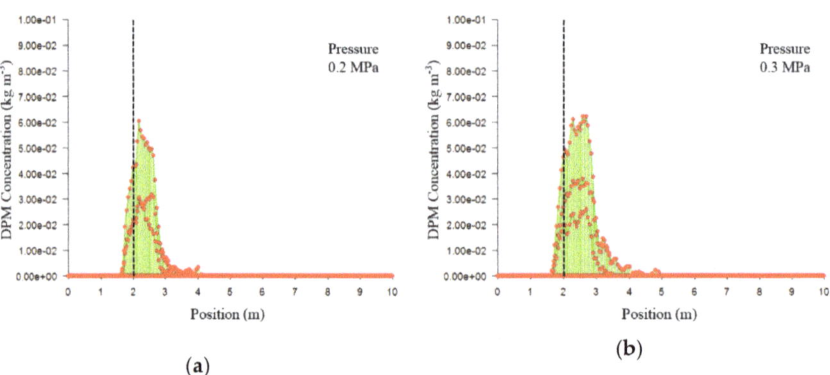

(a)

(b)

Figure 7. *Cont.*

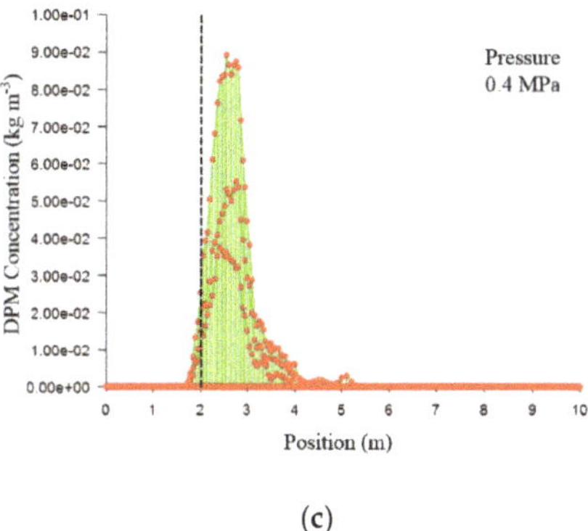

(**c**)

Figure 7. Concentration of droplets deposition under different pressures, (**a**) 0.2 MPa, (**b**) 0.3 MPa, and (**c**) 0.4 MPa. X-axis indicates different position of target area and off-target area, Y-axis shows DPM Concentration (kg m^{-3}) of droplets deposition under different pressure, 0.2 MPa, 0.3 MPa, and 0.4 MPa. The point means concentration of different position, the dotted line means the droplet concentration deposition center, and the solid line means the biggest concentration of corresponding point.

It is clear that droplet deposition concentration at different target locations varies under the influence of pressure (Figure 7). When the pressure was 0.2 MPa, the droplet concentration deposition center point is 2.35 m and the drift distance is 0.35 m. When the pressure was 0.3 MPa, the droplet concentration deposition center point is 2.45 m and the drift distance is 0.45 m. When the pressure was 0.4 MPa, the droplet concentration deposition center point is 2.6 m and the drift distance is 0.6 m. The drift distance of deposition center, the edge of near tuyere, and the edge of far tuyere and the relationships between these parameters and pressure are shown in Table 4 and Figure 8.

Table 4. Droplet drift distance of deposition center, near tuyere edge, far tuyere edge, and overlap width under different pressure, 0.2 MPa, 0.3 MPa, and 0.4 MPa.

Pressure (MPa)	Deposition Center (m)	Near Tuyere Edge (m)	Far Tuyere Edge (m)	Overlap Width (m)
0.2	0.60	0.25	1.70	1.55
0.3	0.45	0.25	2.40	2.35
0.4	0.65	0.35	2.85	2.70

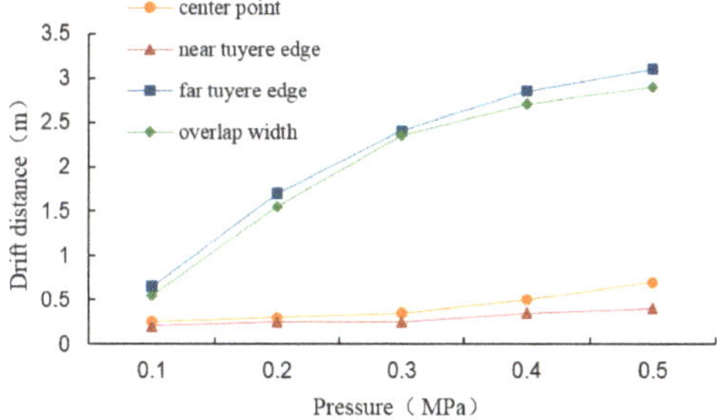

Figure 8. The relationship between pressure and drift distance (center point, near tuyere edge, far tuyere edge and overlap width).

The least square method was used to fit the drift distance and overlap width under different wind speeds. The center point drift model is $y_5 = 0.110x_p + 0.090$, the near tuyere edge drift model is $y_6 = 0.050x_p + 0.140$, the far tuyere edge drift model is $y_7 = 0.605x_p + 0.325$, and the overlap width model is $y_8 = 0.601x_p + 0.600$. In the models, x is inlet pressure (MPa) and y is distance (m). It can be seen that inlet pressure is positively correlated with droplet drift distance and overlap width. For every 0.1 MPa increased in the inlet pressure, the drift distance of deposition center point was increased by 0.11 m, the edge of near tuyere was increased by 0.05 m, the edge of far tuyere was increased by 0.605 m, and the overlap width was increased by 0.6 m.

3.1.3. Influence of Spray Height on Droplet Drift

Wind speed (3 m s^{-1}) and pressure (0.3 MPa) were set to constant and the spray height of 0.8 m, 1 m, and 1.5 m was explored. The droplet deposition density at different target positions is plotted in Figure 9.

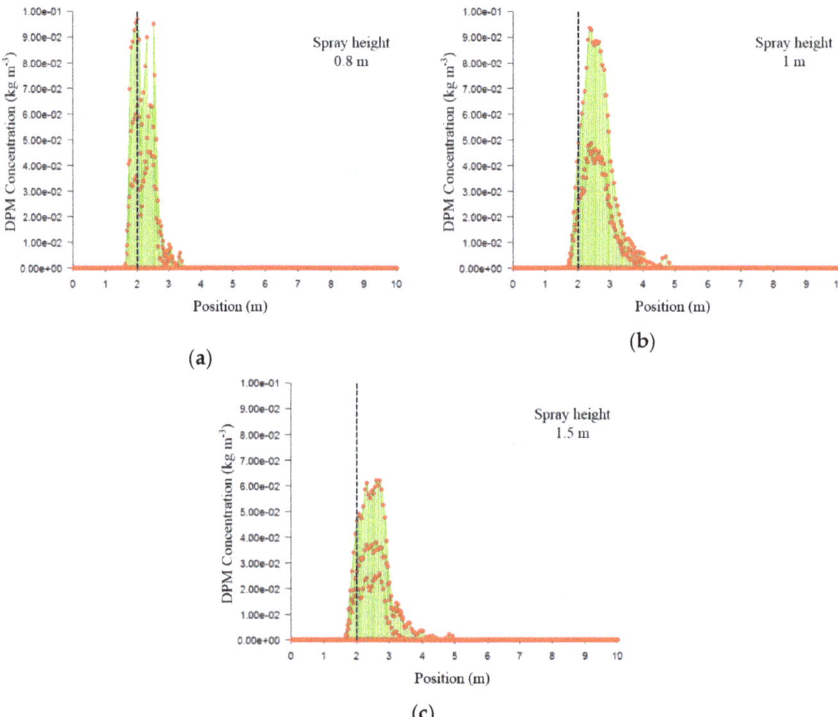

Figure 9. Concentration of droplets deposition in different spray heights, (**a**) 0.8 m, (**b**) 1 m, and (**c**) 1.5 m. X-axis means different position of target area and off-target area, Y-axis means DPM Concentration (kg m^{-3}) of droplets deposition in different spray height, 0.8 m, 1 m, and 1.5 m. The point means concentration of different position, the dotted line means the droplet concentration deposition center and the solid line means the biggest concentration of corresponding point.

It is clear that droplet deposition concentration at different target locations varied with spray height (Figure 9). When the spray height was 0.8 m, the droplet concentration deposition center point is 2.35 m and the drift distance is 0.35 m. When the spray height was 1 m, the droplet concentration deposition center point is 2.45 m and the drift distance is 0.45 m. When the spray height was 1.5 m, the droplet concentration deposition center point is 2.6 m and the drift distance is 0.6 m. The deposition center, near tuyere edge, and far tuyere edge, and the relationship between these parameters and spray height are shown in Table 5 and Figure 10.

Table 5. Droplet drift distance of deposition center, near tuyere edge, far tuyere edge, and overlap width in different spray height, 0.8 m, 1 m, and 1.5 m.

Spray Height (m)	Deposition Center (m)	Near Tuyere Edge (m)	Far Tuyere Edge (m)	Overlap Width (m)
0.8	0.35	0.15	1.55	1.35
1	0.45	0.25	2.40	2.60
1.5	0.60	0.35	3.25	2.75

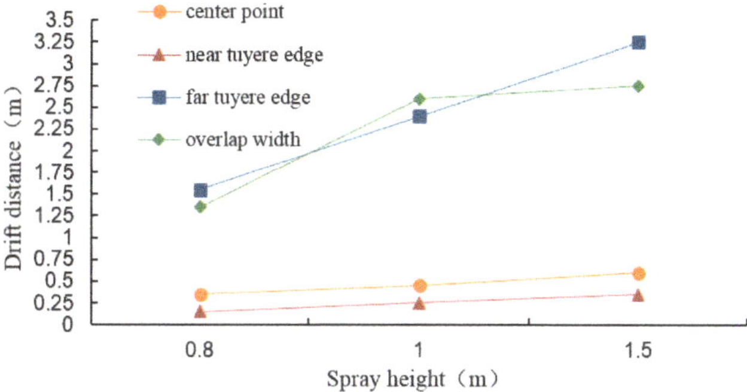

Figure 10. The relationship between spray height and drift distance (center point, near tuyere edge, far tuyere edge, and overlap width).

Again, the least square method was used to fit the drift distance and overlap width under different spray heights. The center point drift model is $y_9 = 0.235x_h + 0.125$, the near tuyere drift model is $y_{10} = 0.115x_h + 0.902$, the far tuyere drift model is $y_{11} = 1.654x_h - 0.081$, and the overlap width is $y_{12} = 1.670x_h + 0.789$. In the models, x is the spray height (m) and y is the distance (m). It can be seen that the spray height is positively correlated with droplet drift distance and overlap width, i.e., for every 1m increase in spray height, the drift distance of deposition center point was increased by 0.235 m, the drift distance of near tuyere increased by 0.115 m, the drift distance of far tuyere was increased by 1.654 m, and the overlap width was increased by 1.67 m.

3.2. Drift Distance Analysis of Fitting Regression Results

The least square method was taken to fit a curve of drift, with y_1 (edge drift distance), y_2 (deposition center drift distance), and y_3 (overlap width) as dependent variables, and x_w (wind speed), x_p (inlet pressure), and x_h (spray height) as independent variables, and C (constant term). The corresponding equations of multivariate linear function groups can be explained as follows:

$$\begin{pmatrix} 27 & 81 & 81 & 29.7 \\ 81 & 315 & 243 & 89.1 \\ 89.1 & 243 & 288 & 89.1 \\ 29.7 & 89.1 & 89.1 & 35 \end{pmatrix} \begin{pmatrix} a_0 \\ a_1 \\ a_2 \\ a_3 \end{pmatrix} = \begin{pmatrix} 57.025 \\ 243.925 \\ 180.975 \\ 66.36 \end{pmatrix} \tag{9}$$

$$\begin{pmatrix} 27 & 81 & 81 & 29.7 \\ 81 & 315 & 243 & 89.1 \\ 89.1 & 243 & 288 & 89.1 \\ 29.7 & 89.1 & 89.1 & 35 \end{pmatrix} \begin{pmatrix} a_0 \\ a_1 \\ a_2 \\ a_3 \end{pmatrix} = \begin{pmatrix} 11.5 \\ 46.5 \\ 36.025 \\ 13.37 \end{pmatrix} \tag{10}$$

$$
\begin{pmatrix}
27 & 81 & 81 & 29.7 \\
81 & 315 & 243 & 89.1 \\
89.1 & 243 & 288 & 89.1 \\
29.7 & 89.1 & 89.1 & 35
\end{pmatrix}
\begin{pmatrix}
a_0 \\ a_1 \\ a_2 \\ a_3
\end{pmatrix}
=
\begin{pmatrix}
51.475 \\ 204.275 \\ 172.5 \\ 60.06
\end{pmatrix}
\tag{11}
$$

and solved by MATLAB 2015a®(MathWorks, MA, USA),

$$
(a_0\ a_1\ a_2\ a_3)^{\mathrm{T}} = (-3.096\ 0.887\ 0.550\ 1.552)^{\mathrm{T}} \tag{12}
$$

$$
(a_0\ a_1\ a_2\ a_3)^{\mathrm{T}} = (-0.667\ 0.167\ 0.085\ 0.308)^{\mathrm{T}} \tag{13}
$$

$$
(a_0\ a_1\ a_2\ a_3)^{\mathrm{T}} = (-3.374\ 0.692\ 0.529\ 1.469)^{\mathrm{T}} \tag{14}
$$

Table 6 shows the variance and regression analysis of the influence of three factors on edge drift distance, deposition center drift distance and overlap width. According to the analysis results (Table 6), the influence of wind speed on edge drift distance, deposition center drift distance and overlap width is very significant, pressure and spray height is significant. The influence of three factors on edge drift distance, deposition center drift distance, and overlap width is also significant, so a linear equation can be established.

Table 6. The variance and regression analysis of the influence of three factors on edge drift distance, deposition center drift distance, and overlap width analysis.

Dependent Variable	Source of Difference	Regression Coefficient	T-Distribution Value	Significance	95% Confidence Interval		R	R^2
					Lower Limit	Upper Limit		
Y_1	X_w	0.887	9.941	**	0.702	1.071	0.915	0.837
	X_p	0.550	3.083	*	0.181	0.919		
	X_h	1.552	3.137	*	0.529	2.576		
	C	−3.096	−4.752		−5.606	−2.206		
Y_2	X_w	0.167	8.196	**	0.125	0.209	0.880	0.774
	X_p	0.085	2.083	*	0.001	0.169		
	X_h	0.308	2.728	*	0.074	0.541		
	C	−0.667	−3.558		−1.054	−0.279		
Y_3	X_w	0.692	8.303	**	0.520	0.865	0.892	0.795
	X_p	0.529	3.173	*	0.184	0.874		
	X_h	1.469	3.176	*	0.512	2.426		
	C	−3.374	−4.391		−4.963	−1.785		

The dependent variables (Y_1, Y_2, Y_3) are edge drift distance, deposition center drift distance, and overlap width. The different sources (X_w, X_p, X_h, C) are wind speed, inlet pressure, spray height and constant term. The significances are the results of significance analysis and the number of stars means the degree of influence by independent variables. The two stars (**) means the influence is very significant, one star (*) is significant, and no star is not significant.

The regression coefficients of the three variables in the regression equation of edge drift distance are 0.887, 0.550, and 1.552, and the constant term (C) is −3.096. Therefore, the relationship model between edge drift distance Y_1 and wind speed X_w (m s^{-1}), pressure X_p (MPa) and spray height X_h (m) is

$$
Y_1 = 0.887X_w + 0.550X_p + 1.552X_h - 3.906\ (R^2 = 0.837) \tag{15}
$$

In this model (15), the coefficients of the three variables are all positive, indicating that the three factors are all positively correlated with droplet drift distance of far tuyere edge. At the same time, this model also provides a reference for safe distance. This model is the drift distance of the droplet far from far tuyere edge, which is also the farthest distance to which the droplet can drift.

The regression coefficients of the three variables in the regression equation of center drift distance are 0.167, 0.085, and 0.308, and the C is −0.667. Therefore, the relationship model between drift distance Y_2 of deposition center and wind speed X_w (m s^{-1}), pressure X_p (MPa) and spray height X_h (m) is

$$
Y_2 = 0.167X_w + 0.085X_p + 0.308X_h - 0.667\ (R^2 = 0.774) \tag{16}
$$

In this model (16), the coefficients of the three variables are all positive, indicating that the three factors are positively correlated with the drift distance of droplet deposition center. At the same time, this model also provides a reference for the selection of spray deposition center. This model is the drift distance of droplet deposition center, which is also the point where droplet deposition concentration is the largest.

The regression coefficients of the three variables in the regression equation of overlap width are 0.692, 0.529, and 1.469, and the C is −3.374. Therefore, the relationship model between overlap width Y_3 and wind speed X_w (m s^{-1}), pressure X_p (MPa), and spray height X_h (m) is

$$Y_3 = 0.692x_w + 0.529x_p + 1.469x_h - 3.374 \ (R^2 = 0.795) \tag{17}$$

In this model (17), the coefficients of the three variables are all positive, indicating that the three factors are positively correlated with the overlap width. At the same time, this model also provides a reference for the selection of nozzle distance and how to get the best droplet overlapping effect.

3.3. Analysis of Droplet Drift Curve Characteristic

Under the influence of different wind speeds, the droplet drift curve of XR8002 (spray height 1.5 m; inlet pressure 0.3 MPa) is shown in Figure 11.

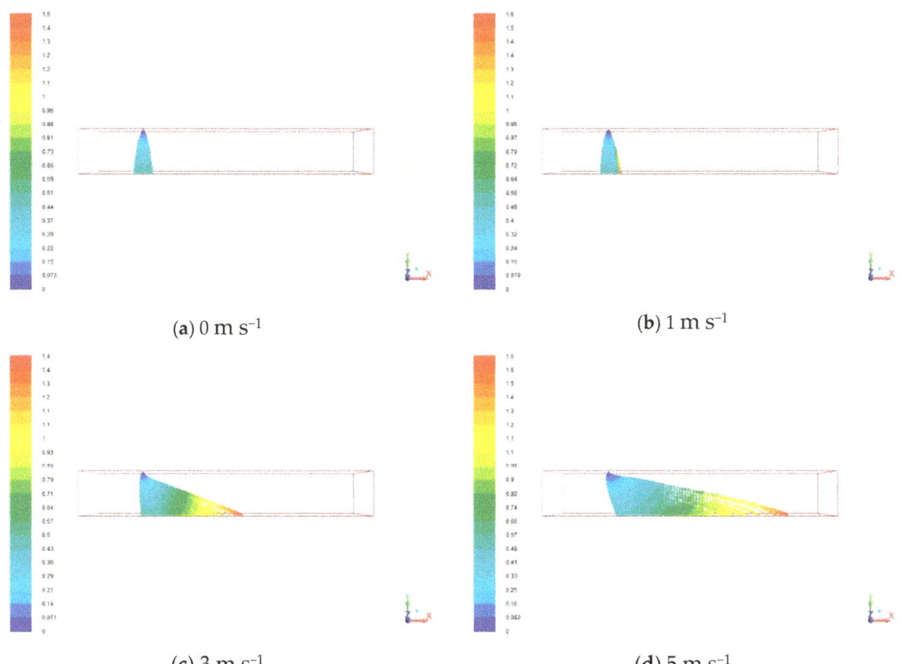

(**a**) 0 m s^{-1} (**b**) 1 m s^{-1}

(**c**) 3 m s^{-1} (**d**) 5 m s^{-1}

Figure 11. Trace of droplet drift curve at wind speed 0, 1, 3, and 5 m s^{-1} when spray height was 1.5 m and inlet pressure was 3 MPa. (**a**) The droplets have no drift at 0 m s^{-1}, (**b**) The droplets hardly drift at 1 m s^{-1}, (**c**) The droplets have obvious drift at 3 m s^{-1}, (**d**) The droplets have serious drift at 5 m s^{-1}.

As can be seen from the above Figure 11, when the wind speed was 1 m s^{-1}, the droplets hardly drift. When the wind speed exceeded 1 m s^{-1}, the droplets drift slightly. When the wind speed exceeded 3 m s^{-1}, the droplets will obviously drift. It is also the reason why UAV could not work in high wind speed. The analysis of droplet drift curves at the wind speeds of 3 m s^{-1} and 5 m s^{-1} were also made in Figure 12.

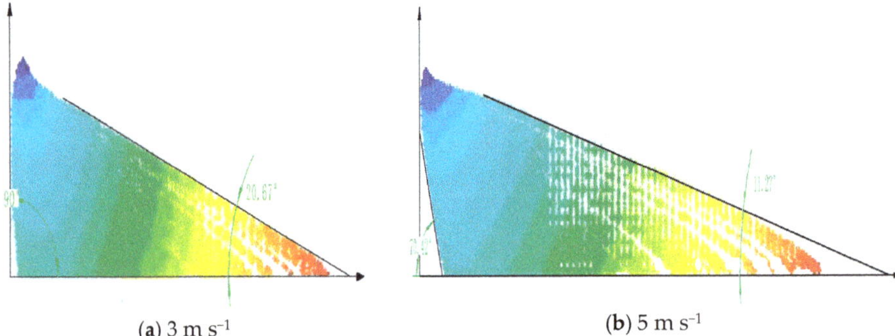

Figure 12. Droplet drift curve of different wind speed (3 and 5 m s^{-1}), X-axis means deposition area of 0 to 10m, and Y-axis means spray height. (**a**) The droplet angle was 90° and 20.67° at 3 m s^{-1}, (**b**) The droplet angle was 70.42° and 11.27° at 5 m s^{-1}.

As is shown in Figure 12, both sides of the droplet drift curve were approximately a straight line. When the wind speed was 3 m s^{-1}, the droplet angle on the left side was 90°and the right side was 20.67°. When the wind speed was 5 m s^{-1}, the droplet angle on the left side was 70.42°and the right side was 11.27°, which showed that the inclination angle of the straight line was negatively related to the wind speed. Results displayed in Figure 12 show trends matching the previously observed behavior of a drift distance increase with a droplet angle decrease. The droplet angle also could be used to estimate the influence of wind, and the drift distance.

4. Conclusions

In this study, CFD simulation method was used to acquire droplet deposition distribution and drift under the influence of different factors. DPM model was used to simulate droplet drift from a double XR8002 nozzle at various wind speeds (0, 1, 3, 5 m s^{-1}), nozzle pressures (0.2, 0.3, 0.4 MPa) and spray heights (0.8, 1, 1.5 m). The information obtained from these simulations provided valuable insight into the characteristic of spray drift. General analysis suggest that the droplet drift curves were influenced by the three factors. Additionally, the influence coefficients of the three factors on the droplet drift distance were calculated. On the basis of analysis of the variance and regression results aimed at the edge drift distance, the center drift distance, and the overlap width, the three models were established. The expressions of three models are important on guiding significance to the practice. The analysis of the droplet drift curves showed that the droplet angle is closely related to drift. Results made from these simulations have provided a tool which can be used to ensure future UAV chemical application can be designed to maximize efficacy, reduce waste, and minimize damage to organisms not being targeted.

Author Contributions: Formal analysis, C.Z.; resources, J.L.; supervision, H.Z. (Hang Zhu); writing—original draft, H.L.; writing—review & editing, H.Z. (Huihui Zhang). All authors have read and approved the final manuscript.

Funding: This work was financially supported by National Key R&D Program of China (No.2016YFD0200701).

Conflicts of Interest: The authors declare no conflict of interest.

References

1. Han, H.Q.; Gao, H.J.; Yang, H.T.; Zhang, X.; Hu, T.; Zhou, Q.Q. Analysis on decoupling between total grain production and pesticide usage in China. *Subtrop. Agric. Res.* **2018**, *14*, 91–98.

2. Cheng, X.B. Current situation and development countermeasures of agricultural extension in China. *Agric. Sci. Technol.* **2018**, *23*, 255–259.
3. Wu, K.M.; Lu, Y.H.; Wang, Z.Y. Advance in integrated pest management of crops in China. *Chin. Bull. Entomol.* **2009**, *46*, 831–836.
4. Wang, G.; Gong, Y.; Zhang, X.; Chen, X.; Liu, D.J.; Chen, W.; Miao, Y.Y. The test of droplet deposition distribution for different kinds of pesticide application equipment in corn field. *J. Agric. Mech. Res.* **2017**, *39*, 177–182.
5. Meng, Y.H.; Zhou, G.Q.; Wu, C.B. Discussion on application and promotion of agricultural plant protection unmanned aerial vehicle in China. *China Plant Prot.* **2014**, *34*, 33–39.
6. Shang, C.Y.; Cai, J.F.; Huang, S.J.; Pan, H.L.; Zhong, F.L. Applications status and prospect analysis of agricultural UAVs in China. *J. Anhui Agric. Sci.* **2017**, *45*, 193–195.
7. Wang, X.N.; He, X.K.; Wang, C.L.; Wang, Z.C.; Li, L.L.; Wang, S.L.; Janes, B.; Andreas, H.; Wang, Z.G. Spray drift characteristics of fuel powered single-rotor UAV for plant protection. *Trans. Chin. Soc. Agric. Eng.* **2017**, *33*, 117–123.
8. Bradley, K.F. Role of atmospheric stability in drift and deposition of aerially applied sprays-preliminary results. *ASAE* **2004**, *58*, 742–755.
9. Kirk, L.W.; Teske, M.E.; Thistle, H.W. What about upwind buffer zones for aerial applications? *J. Agric. Saf. Health* **2002**, *8*, 333–336. [CrossRef]
10. Ivan, W.K.; Bradley, K.F.; Wesley, C.H. Aerial methods for increasing spray deposits on wheat heads. *ASAE* **2004**, *58*, 716–729.
11. Chen, S.D.; Lan, Y.B.; Bradley, K.F. Effect of wind Field below Rotor on Distribution of Aerial Spraying Droplet Deposition by Using Multi-rotor UAV. *Trans. Chin. Soc. Agric. Mach.* **2017**, *48*, 105–113.
12. Qiu, B.J.; Wang, L.W.; Cai, D.L.; Wu, J.H.; Ding, G.R.; Guan, X.P. Effects of flight altitude and speed of unmanned helicopter on spray deposition uniform. *Trans. Chin. Soc. Agric. Mach.* **2013**, *29*, 25–32.
13. Shen, A.; Zhou, S.D.; Wang, M. Simulation and analysis of multi-rotor UAV flow field. *Trans. Chin. Soc. Agric. Mach.* **2018**, *36*, 29–33.
14. Nuyttens, D.; De Schampheleire, M.; Baetens, K.; Brusselman, E.; Dekeyser, D.; Verboven, P. Drift from field crop sprayers using an integrated approach: Results of a 5-year study. *Trans. ASABE* **2011**, *54*, 403–408. [CrossRef]
15. Teske, M.E.; Bird, S.L.; Esterly, D.M.; Curbishley, T.B.; Ray, S.L.; Perry, S.G. AgDrift: A model for estimating near-field spray drift from aerial applications. *Environ. Toxicol. Chem.* **2002**, *21*, 659–671. [CrossRef] [PubMed]
16. Teske, M.E.; Thistle, H.W.; Riley, C.M.; Hewitt, A.J. Laboratory measurements on the sensitivity of evaporation rate of droplets inside a spray cloud. *Trans. ASABE* **2017**, *60*, 361–366.
17. Bilanin, A.J.; Teske, M.E.; Barry, J.W.; Ekblad, R.B. AGDISP: The aircraft spray dispersion model, code development and experimental validation. *Trans. ASABE* **1989**, *32*, 327–334. [CrossRef]
18. Luo, J.; Zeng, G.H.; Li, B.K. Simulation analysis of two-phase flow field inside pass of nozzle base on FLUENT. *Modul. Mach. Tool Autom. Manuf. Tech.* **2016**, *10*, 44–47.
19. Chen, X.; Ge, S.C.; Zhang, Z.W.; Jing, D.J. Numerical simulation and analysis of multi-nozzle interference base on fluent. *Chin. J. Environ. Eng.* **2014**, *8*, 2503–2508.
20. Sun, G.X.; Wang, X.C.; Ding, W.M.; Zhang, Y. Simulation and analysis on characteristics of droplet deposition base on CFD discrete phase model. *Trans. Chin. Soc. Agric. Eng.* **2012**, *28*, 13–19.
21. Zhang, S.F.; Song, L.L. Simulation of Spray characteristics of pressure swirl nozzle. *J. Anhui Agric. Sci.* **2009**, *37*, 8098–8100.
22. Adeniyi, A.A.; Morvan, H.P.; Simmons, K.A. A coupled Euler-Lagrange CFD modelling of droplets-to-film. *Aeronaut. J.* **2017**, *121*, 1897–1918. [CrossRef]
23. Ryan, S.D.; Gerber, A.G.; Holloway, A.G.L. A computational study on spray dispersal in the wake of an aircraft. *Trans. ASABE* **2013**, *56*, 847–868.
24. Zhang, B.; Tang, Q.; Chen, L.P.; Xu, M. Numerical simulation of wake vortices of crop spraying aircraft close to the ground. *Biosyst. Eng.* **2016**, *145*, 52–64. [CrossRef]
25. Yang, F.B.; Xue, X.Y.; Cai, C. Numerical Simulation and analysis on spray drift movement of multirotor plant protection unmanned aerial vehicle. *Energies* **2018**, *11*, 2399. [CrossRef]
26. Roberts, T.W.; Murman, E.M. Solution method for a hovering helicopter rotor using the Euler equations. In Proceedings of the 23rd Aerospace Sciences Meeting, Reno, NV, USA, 14–17 January 1985.

27. Jin, Y.H.; Jae, S.R.; Kyu, H.K. New least squares method with geometric conservation law (GC-LSM) for compressible flow computation in meshless method. *Comput. Fluids* **2018**, *172*, 122–146.
28. Salazar-Campoy, M.M.; Morales, R.D.; Najera-Bastida, A.; Calderon-Ramos, I.; Cedillo-Hernandez, V.; Delgado-Pureco, J.C. A physical model to study the effects of nozzle design on dispersed two-phase flows in a slab mold casting ultra-low-carbon steels. *Metall. Mater. Trans. B* **2018**, *49*, 812–830. [CrossRef]

Article

Combined Use of Low-Cost Remote Sensing Techniques and δ^{13}C to Assess Bread Wheat Grain Yield under Different Water and Nitrogen Conditions

Salima Yousfi [1,2], Adrian Gracia-Romero [1,2], Nassim Kellas [3], Mohamed Kaddour [3], Ahmed Chadouli [3], Mohamed Karrou [4], José Luis Araus [1,2] and Maria Dolores Serret [1,2,*]

[1] Section of Plant Physiology, University of Barcelona, 08028 Barcelona, Spain; yousfisalima@hotmail.com (S.Y.); adriangraciaromero@hotmail.com (A.G.-R.); jaraus@ub.edu (J.L.A.)

[2] AGROTECNIO Center, University of Lleida, 25198 Lleida, Spain

[3] Institut Technique des Grandes Cultures, Alger 16016, Algeria; nassim.gfield@yahoo.fr (N.K.); kaddour.mohamed47@yahoo.fr (M.K.); chadouli_ahmed@yahoo.fr (A.C.)

[4] International Center for Agricultural Research in the Dry Areas (ICARDA), Avenue Hafiane Cherkaoui, Rabat 10112, Morocco; m_karrou@yahoo.fr

* Correspondence: dserret@ub.edu; Tel.: +34-93-4021-963

Received: 11 April 2019; Accepted: 30 May 2019; Published: 31 May 2019

Abstract: Vegetation indices and canopy temperature are the most usual remote sensing approaches to assess cereal performance. Understanding the relationships of these parameters and yield may help design more efficient strategies to monitor crop performance. We present an evaluation of vegetation indices (derived from RGB images and multispectral data) and water status traits (through the canopy temperature, stomatal conductance and carbon isotopic composition) measured during the reproductive stage for genotype phenotyping in a study of four wheat genotypes growing under different water and nitrogen regimes in north Algeria. Differences among the cultivars were reported through the vegetation indices, but not with the water status traits. Both approximations correlated significantly with grain yield (GY), reporting stronger correlations under support irrigation and N fertilization than the rainfed or the no N-fertilization conditions. For N-fertilized trials (irrigated or rainfed) water status parameters were the main factors predicting relative GY performance, while in the absence of N-fertilization, the green canopy area (assessed through GGA) was the main factor negatively correlated with GY. Regression models for GY estimation were generated using data from three consecutive growing seasons. The results highlighted the usefulness of vegetation indices derived from RGB images predicting GY.

Keywords: wheat; canopy temperature depression; NDVI; RGB images; grain yield; δ^{13}C

---

1. Introduction

Bread wheat is one of the most cultivated herbaceous crops in the Mediterranean region [1], with water stress and low nitrogen fertility being the main constraints limiting productivity [2]. These limitations are likely to increase in the future because climatic change is expected to decrease precipitation and increase evapotranspiration in the Mediterranean region [3]. Increasing productivity in these semi-arid environments depends on the efficiency of crop management [2] and breeding [4], where efficient and affordable methodologies to monitor crop performance, or to assess phenotypic variability agr breeding, are needed. Remote sensing techniques at the canopy level have become valuable tools for precision agriculture and high throughput phenotyping [5–7]. Thus, both spectral and thermal approaches have been proposed as potential solutions to identify crop N status and water stress across large areas [8,9]. In this way, these techniques can help farmers to practice more sustainable agriculture, minimizing risks of losing the harvest by providing (whenever possible) the resources (e.g., water and fertilizer) needed to secure yield. However, the adoption of new technologies often requires much up-front investment and is therefore restricted to large-scale production and/or farmers with substantial economic resources. This limitation is particularly evident for smallholder farmers from developing countries. Nevertheless, satellite-derived indices can be used in local management to support farmers' decision making, including the rate of irrigation and fertilizer application, and eventually yield prediction in wheat [10,11] and other crops [12,13]. While satellite images are often freely available, as in the case of Sentinel-2 [14], the resolution (not higher than 100 square meters per pixel), together with the periodicity of image acquisition and weather constraints (e.g., clouds) and the need for computing support and trained staff makes this form of remote sensing is unattainable for smallholder farmers. Different approaches to small-scale-tailored crop management have been proposed. For example, site-specific nitrogen management using leaf color charts has been proposed in irrigated wheat [15,16]. However, the interaction of the water regime with nitrogen status may affect leaf color, making this method impractical for rainfed or deficit-irrigation crops. A more flexible alternative uses optical sensors such as portable spectroradiometers (like, for example, the GreenSeeker) [17,18]. However, the cost of the equipment may limit the uptake of this approach. In this sense, the use of low-cost remote sensing methods to schedule irrigation and fertilization and predict yield, such as digital conventional imagery and/or infrared thermometry [19], may contribute to more sustainable agriculture in arid and semi-arid regions of the Mediterranean where irrigation and fertilization are not optimized in terms of timing and amount. While remote sensing has been regarded as a potentially useful approach in predicting grain yield, an inherent limitation of remote sensing methods is that the relationships between yield and vegetation indices may be site and season specific, changing between sites and years. Thus, for example, in the case of sensor calibration for N management, site-year characteristics have a critical impact [20]. While new methods for sensor-based site-specific N management are probably needed, it is likely that the best approaches will arise from the use of multiple sensors [20,21], therefore increasing the cost of deployment. Even when low cost remote sensing approaches using single sensors have shown great potential in experimental trials with wheat [22], their practical application needs to be proven.

The normalized difference vegetation index (NDVI) is one of the most well-known multispectral vegetation indices. The NDVI has been used extensively to estimate plant biomass [23–25], nitrogen status [26] and yield in wheat and other cereals [27–29]. The leaf chlorophyll content measured with a portable chlorophyll meter, which uses the same principle as the NDVI, but on the basis of the light transmitted through the leaf, has also been used extensively [30]. As an alternative, information derived from conventional digital Red-Green-Blue (RGB) images to formulate canopy vegetation indices is a low-cost and an easy proximal sensing approach to assess grain yield in cereals [31–33], even when limitations related to shadows and changes in ambient light conditions need to be taken into consideration [34]. Information derived from RGB images allows estimation of a wide range of crop traits in durum and bread wheat, such as early vigor, leaf area index, leaf senescence, aerial biomass and grain yield [31,33]. The green area and the greener area are two indices derived from conventional digital images [31]. The first parameter describes the amount of green biomass in the picture, while the second one excludes the more yellowish-green pixels. In fact, greener area is aimed at capturing active photosynthetic area and plant senescence [31]. Such indices are formulated using open access software [19,35].

It has been long recognized that plant temperature may represent a valuable index to detect differences in plant water regimes [36–38]. Reynolds et al. (2007) [39] have reported that wheat canopy temperature is a relative measure of plant transpiration associated with water uptake from the soil. Under water limited conditions, transpiration and its associated evaporative cooling are reduced, resulting in higher leaf temperatures. Given that a major role of transpiration is leaf cooling, canopy temperature and its depression relative to ambient air temperature is an indicator of the degree to which transpiration cools leaves under a demanding environmental load [40]. In that sense, infrared thermometry has been proposed as a low-cost approach in crop management to enable scheduling of support irrigation [41], to assess spatial soil heterogeneity [42], or to evaluate genotypic performance to drought [43]. However, potential interaction effects between N fertilization and water regime should be considered. In particular, how does N fertilization affect the water by temperature relationships and even how does water affect the interaction of N fertilization with vegetation indices. This is not trivial because haying-off, which is the negative effect of nitrogen fertilization on productivity caused by an imbalance between transpired biomass and the available water, is regarded as a potential problem for wheat cultivation in Mediterranean regions [44]. As a consequence, unexpected relationships between remote sensing readings and grain yield may occur.

Similarly, other physiological characteristics related to plant water status, such as stable carbon isotope composition (δ^{13}C; frequently measured as discrimination from the surrounding air, Δ^{13}C) are also often used for evaluating genotypic performance under water stress [40,44] or even to monitor spatial variability and water status [45]. The natural ^{13}C abundance in plant matter provides time-integrated information of the effects of water stress on the photosynthetic carbon assimilation of C3 species, including wheat [46–48]. Conditions inducing stomatal closure (e.g., water deficit or salinity) restrict the CO_2 supply to carboxylation sites, which then decreases the Δ^{13}C (or increases the δ^{13}C) of plant matter [47,49]. Under Mediterranean conditions the δ^{13}C of mature kernels is better correlated with grain yield than the δ^{13}C of other plant parts [50]. The costs of these analyses have decreased throughout the years, making their analysis increasingly feasible.

The objective of this study was to assess the grain yield performance of wheat under a range of water and fertilization conditions in the Mediterranean, using different low-cost remote sensing approaches to assess canopy green biomass (NDVI and vegetation indices derived from conventional RGB images), and characteristics associated with plant water status, (canopy temperature depression), together with additional traits informing on the water status (δ^{13}C of mature grains and the stomatal conductance of the flag leaf). The novelty of the study centers on (i) testing how different low-cost, user-friendly remote sensing techniques may contribute to site-specific wheat management and eventually to the prediction of yield across seasons; and (ii) how interactions between growing conditions (water regime and N fertilization) may affect the predictive strength of these techniques. Moreover, to better explore the potential usefulness of our study for wheat phenotyping we developed a conceptual model of how the combination of these different traits explains genotypic variability in grain yield under different combinations of water regimes and nitrogen fertilization.

2. Materials and Methods

2.1. Plant Material and Growing Conditions

Field trials were conducted during the 2014–2015 crop season at Bir Ould Khelifa in the area of Khemis Miliana commune, approximately 230 km to the south west of Algiers (Algeria) and with geographical coordinates 36°11′50.23″ N and 2°13′17.69″ E. This commune receives an average rainfall of between 400 and 450 mm, and it is characterized by clay-silty fertile soils, with high organic matter content and high levels of total and mineral nitrogen (Table S1). Monthly total accumulated rainfall and temperatures for the study region for the 2014–2015 crop season are presented in Table S2. Four bread wheat (*Triticum aestivum* L.) genotypes were planted on 14 December 2014. The wheat genotypes were "Ain-Abid" and "Arz" (modern varieties) and "Wifak" and "Maaouna" (local varieties). The experimental design was a split-split-plot, with the main-plot factor being water regime, the subplot factor was the N amount and the sub-subplot factor was genotype (Figure S1). A total of 108 plots (four genotypes, three replicates per genotype, three water regimes, and three nitrogen fertilization treatments) each with a size of 10 m × 1.2 m and six rows, 20 cm apart, were studied. The three water regimes consisted of one rainfed and two support irrigation treatments of 30 mm (SI-1, a single amount of supplemental irrigation) and 60 mm (SI-2, a double amount of supplemental irrigation) aimed at providing water in the typical range for the agronomic practices in this area. Supplementary irrigation was applied with sprinklers around the beginning of stem elongation. For SI-1, one irrigation (30 mm) was applied at the beginning of stem elongation (31 Zadoks stage), whereas for SI-2, a second irrigation was also delivered at heading (58 Zadoks stage). Nitrogen fertilization was applied using urea fertilizer, and treatments consisted of no fertilizer (N0) and 60 kg ha^{-1} and 120 kg ha^{-1} of nitrogen fertilizer (N60, N120 respectively). Application of N fertilization was achieved at two growing stages, tillering and jointing (26 and 31 Zadoks stages). Plots were harvested with a sickle after physiological maturity and grain yield was estimated. Thousand kernel weight and the number of kernels per square meter (Kernel m^{-2}) were evaluated.

2.2. Vegetation Indices

Remote sensing measurements were performed on a sunny day around anthesis (26 March 2015) between 10 h and 15 h solar time. The NDVI was determined with a portable spectroradiometer (GreenSeeker handheld crop sensor, Trimble, Sunnyvale, CA, USA). The NDVI is formulated using the following equation: (NIR − R)/(NIR + R), where R is the reflectance in the red band and NIR is the reflectance in the near-infrared band. The distance between the sensor and the plots was kept constant at around 50–60 cm above and perpendicular to the canopy. Additionally, one conventional digital picture was taken per plot, holding the camera about 80 cm above the plant canopy in a zenithal plane and focusing near the center of each plot. The images were acquired with an Olympus E-M10 camera (Olympus Corporation, Tokyo, Japan), using a 14 mm lens, triggered at a speed of 1/125 seconds with the aperture programmed in automatic mode. The size of the images was 4608 × 3456 pixels stored in JPG format using RGB color standard [51]. Pictures were analyzed with the free-access BreedPix 0.2 software, now integrated within the CerealScanner plugin (https://integrativecropecophysiology.com/software-development/cerealscanner/), from the Mediterranean Crop Ecophysiology Group, University of Barcelona [22], which was developed for digital image processing. This software quickly provides digital values from different color properties. The vegetation indices measured were the green area (GA) and the greener area (GGA). GA is the portion (as a %) of pixels with 60 < Hue < 120 from the total amount of pixels, whereas greener area is formulated as the % of pixels with 80 < Hue < 120 [31,52]. GGA is designed to capture the active photosynthetic area excluding senescent leaves. In addition, the leaf chlorophyll content of five flag leaf blades per plot was measured using a Minolta SPAD-502 portable meter (Spectrum Technologies Inc., Plainfield, IL, USA).

2.3. Canopy Temperature Measurements

Canopy temperature (CT) was measured at noon (12 h–14 h), on the same day around anthesis as the vegetation indices, using an infrared thermometer (PhotoTempTM MXSTMTD, Raytek®, Santa Cruz, CA, USA). Measurements were taken above the plants, pointing towards the canopy at a distance of about one meter and having the sun towards the rear. The air temperature was measured simultaneously for each plot with a temperature humidity meter (Testo 177-H1 Logger, Testo, Lenzkirch, Germany) and employed for the calculation of the canopy temperature depression (CTD) as the difference between the ambient and the canopy temperature.

2.4. Stomatal Conductance

Stomatal conductance was measured on the flag leaves on the same days as the remote sensing traits. Two measurements per plot were taken around noon (12 h–14 h), using a Decagon SC-1 Leaf Porometer (Decagon Devices Inc., Pullman, WA, USA).

2.5. Stable Carbon Isotope Composition

Carbon isotope composition was analyzed in mature grains using an Elemental Analyzer (Flash 1112 EA; ThermoFinnigan, Bremen, Germany) coupled with an Isotope Ratio Mass Spectrometer (Delta C IRMS, ThermoFinnigan, Bremen, Germany) operating in continuous flow mode to determine the stable carbon ($^{13}C/^{12}C$) isotope ratios. Samples of about 0.7 mg of dry matter and reference materials were weighed into tin capsules, sealed, and then loaded into an automatic sampler (ThermoFinnigan, Bremen, Germany) prior to EAIRMS analysis. Measurements were carried out at the Scientific Facilities of the University of Barcelona. The $^{13}C/^{12}C$ ratios were expressed in δ notation determined by: $\delta^{13}C = (^{13}C/^{12}C)_{sample}/(^{13}C/^{12}C)_{standard} - 1$ [53], where sample refers to plant material and standard to Pee Dee Belemnite (PDB) calcium carbonate. International isotope secondary standards of known $^{13}C/^{12}C$ ratios (IAEA CH$_7$ polyethylene foil, IAEA CH$_6$ sucrose, and USGS 40 L-glutamic acid) were used for calibration to a precision of 0.1‰.

2.6. Statistical Analysis

Data were subjected to factorial ANOVA to test the effects of the growing conditions (water regime and nitrogen fertilization), genotype, and their interaction. Mean comparisons were performed using Tukey's honestly significant difference (HSD) test. Pearson correlation coefficients between grain yield and all different traits were calculated. Multiple linear regression analysis (stepwise) was used to analyze grain yield under different growing conditions. Data were analyzed using IBM SPSS Statistics 24 (SPSS Inc., Chicago, IL, USA). Figures were created using Sigma-Plot 11.0 for Windows (Systat Software Inc., Point Richmond, CA, USA).

The performance of the different remote sensing traits in predicting yield performance across the seasons was evaluated using data from the present study, together with data already published by our team from the two previous seasons (2012–2013 and 2013–2014) related to grain yield and different remote sensing traits measured in a set of genotypes grown under different water regimes [54]. The equations of the linear relationships between these traits and grain yield (GY) determined in the first crop season (2012–2013) were further tested during the two following crop seasons 2013–2014 and 2014–2015 (the latter of these conducted during the present study). Predicted and measured grain yields were expressed as relative values; grain yields were normalized with regard to the highest yield combination (genotype and growing condition), then the means of the three replicates per genotype were calculated. Finally, we performed path analyses to quantify the relative contributions of direct and indirect effects of water status ($\delta^{13}C$, g$_s$ and CTD) and vegetation indices (GA, GGA) on grain yield. This methodology offers the possibility of building associations between variables on the basis of prior knowledge. Mechanisms that play potential roles in grain yield variation and involving traits that exhibited genotypic differences have been proposed, as detailed in the conceptual model displayed in Figure S2. This model was aimed at understanding grain yield responses to genotypic differences under different levels of nitrogen fertilization (N0, N60, N120) and under different water regimes (supplementary irrigation and rainfed). A model with a comparative fit index (CFI) [42] with values > 0.9 was taken as indicative of a good fit. Data were analyzed using IBM, SPSS, Amos 21 (SPSS Inc., Chicago, IL, USA).

3. Results

3.1. Irrigation and Fertilization Effects on Grain Yield

Both irrigation and nitrogen fertilization significantly affected the grain yield (GY) and the agronomic yield components (Table 1). The doubled amount of supplemental irrigation (SI-2 with all nitrogen fertilization combined) and the highest nitrogen fertilization N120 (120 kg N ha^{-1} of fertilizer with all irrigations combined) were the growing conditions that exhibited the highest grain yield. Significant interactions only existed for grain yield between genotype and water regime ($p = 0.049$).

Table 1. Mean values of genotypes, water regimes, nitrogen fertilization levels for grain yield (GY), thousand kernel weight (TKW) and kernels m^{-2} and the corresponding ANOVA.

	GY (T ha^{-1})	TKW (g)	Kernels m^{-2}
Genotypes			
Ain Abid	2.20 [a]	21.73 [a]	275.9 [a]
Arz	2.35 [ab]	23.70 [b]	298.3 [ab]
Maaouna	2.35 [ab]	24.35 [b]	307.9 [b]
Wifak	2.50 [b]	24.83 [b]	308.6 [b]
Water regime			
RF	2.07 [a]	22.42 [a]	270.6 [a]
SI-1	2.29 [b]	22.72 [a]	308.8 [b]
SI-2	2.68 [c]	25.83 [b]	313.7 [b]
Nitrogen fertilization			
N0	2.05 [a]	22.60 [a]	291.3 [a]
N60	2.40 [b]	23.43 [a]	307.6 [a]
N120	2.60 [c]	24.93 [b]	294.1 [a]
Level of significance			
Genotype (G)	0.026	0.000	0.029
Water regime (WR)	0.000	0.000	0.000
Nitrogen fertilization (N)	0.000	0.001	0.256
G × WR	0.049	0.126	0.388
G × N	0.065	0.141	0.688
N × WR	0.070	0.110	0.840
G × WR × N	0.801	0.050	0.863

Means followed by different letters are significantly different ($p < 0.05$) according to Tukey's honestly significant difference (HSD) test. For more details, including the acronyms for the treatments, see Materials and Methods.

3.2. Vegetation Indices

The water regime significantly affected the NDVI ($p < 0.001$), as well as the green area (GA) ($p < 0.001$) and the greener area (GGA) ($p < 0.001$) indices (Table 2). The values of these three vegetation indices were highest under SI-2 compared to the single amount of supplemental irrigation (SI-1) and rainfed conditions. However, no difference was observed in vegetation indices across fertilization treatments except for GGA ($p = 0.005$). Moreover, leaf chlorophyll content (LC) slightly increased (Table 2) under SI-1 and rainfed conditions compared to SI-2 and under N60 and N120 compared to N0. Interactions were not significant, whatever the combination (genotypes, water regimes and fertilization levels) or variables considered.

Table 2. Mean values of genotypes, water regimes, and nitrogen fertilization levels for the vegetation indices NDVI (Normalized Difference Vegetation Index), GA (Green Area), GGA (Greener Area) and LC (Leaf chlorophyll content) and the corresponding ANOVA. Parameters were measured around anthesis. Means followed by different letters are significantly different ($p < 0.05$) according to Tukey's honestly significant difference (HSD) test.

	NDVI	GA	GGA	LC
Genotypes				
Ain Abid	0.70 [c]	0.88 [c]	0.74 [d]	46.88 [a]
Arz	0.64 [b]	0.82 [b]	0.58 [b]	52.32 [c]
Maaouna	0.57 [a]	0.74 [a]	0.50 [a]	50.04 [b]
Wifak	0.58 [a]	0.81 [b]	0.64 [c]	50.57 [bc]
Water regime				
RF	0.56 [a]	0.73 [a]	0.52 [a]	50.52 [b]
SI-1	0.62 [b]	0.78 [b]	0.60 [b]	50.35 [ab]
SI-2	0.70 [c]	0.93 [c]	0.73 [c]	49.00 [a]

Table 2. *Cont.*

	NDVI	GA	GGA	LC
Nitrogen fertilization				
N0	0.62 [a]	0.80 [a]	0.58 [a]	49.08 [a]
N60	0.63 [a]	0.82 [a]	0.64 [b]	50.75 [b]
N120	0.63 [a]	0.82 [a]	0.63 [b]	50.03 [ab]
Level of significance				
Genotype (G)	0.000	0.000	0.000	0.000
Water regime (WR)	0.000	0.000	0.000	0.037
Nitrogen fertilization (N)	0.699	0.304	0.005	0.038
G × WR	0.071	0.073	0.381	0.144
G × N	0.729	0.423	0.562	0.438
N × WR	0.765	0.721	0.598	0.495
G × WR × N	0.975	0.518	0.492	0.257

3.3. Canopy Temperature Depression, Stable Carbon Isotope Composition and Stomatal Conductance

Water regime affected significantly the canopy temperature depression (CTD) ($p < 0.001$), the stomatal conductance (g_s) ($p < 0.001$) and the stable carbon isotope composition ($\delta^{13}C$) ($p < 0.001$) of mature grains. Rainfed conditions decreased g_s and CTD, whereas $\delta^{13}C$ increased compared to support irrigation conditions (Table 3). Nevertheless, fertilization treatments did not affect any of these three parameters (Table 3). No significant interactions were observed except for the $\delta^{13}C$ between the water regime and nitrogen fertilization ($p = 0.022$).

Table 3. Mean values of genotypes, water regimes, and nitrogen fertilization levels for canopy temperature depression (CTD), stomatal conductance (g_s) of the flag leaves and the stable carbon isotope composition ($\delta^{13}C$) of the mature grains. Means followed by different letters are significantly different ($p < 0.05$) according to Tukey's honestly significant difference (HSD) test.

	CTD (°C)	g_s (μmol CO$_2$ m^{-2} s^{-1})	$\delta^{13}C$ (‰)
Genotypes			
Ain Abid	1.92 [a]	143.71 [a]	−24.07 [b]
Arz	1.48 [a]	119.07 [a]	−24.09 [b]
Maaouna	1.20 [a]	119.51 [a]	−24.43 [a]
Wifak	1.07 [a]	149.90 [a]	−24.64 [a]
Water regime			
RF	−0.49 [a]	71.18 [a]	−23.50 [c]
SI-1	1.68 [b]	95.31 [a]	−24.19 [b]
SI-2	3.07 [c]	232.65 [b]	−25.23 [a]
Nitrogen fertilization			
N0	1.67 [a]	132.37 [a]	−24.28 [a]
N60	1.40 [a]	146.86 [a]	−24.34 [a]
N120	1.19 [a]	119.93 [a]	−24.31 [a]
Level of significance			
Genotype (G)	0.066	0.102	0.000
Water regime (WR)	0.000	0.000	0.000
Nitrogen fertilization (N)	0.269	0.142	0.841
G × WR	0.743	0.689	0.371
G × N	0.307	0.963	0.502
N × WR	0.336	0.950	0.022
G × WR × N	0.658	0.106	0.932

3.4. Genotypic Effect on Grain Yield, Vegetation Indices and Water Status Traits

The genotypic effect was significant ($p < 0.05$) for grain yield (GY) and ($p < 0.001$) for vegetation indices (Tables 1 and 2) under the growing conditions analyzed together, whereas only the δ^{13}C (in the case of water status traits) was significantly ($p < 0.001$) different between genotypes (Table 3). The genotypic difference was also examined within each of the nine growing conditions, resulting from the combination of the three water regimes and the three nitrogen fertilization levels (Table 4). GY only showed a genotypic effect under SI-2 combined with nitrogen fertilization (either N120 or N60). However, under the SI-1 and rainfed conditions, GY did not show genotypic differences, regardless of the N fertilization regime. The vegetation indices, LC content, δ^{13}C and g_s also showed genotypic effects in some specific growing conditions while CTD did not (Table 4).

Table 4. Genotype effect on the Normalized Difference Vegetation Index (NDVI) and the Green Area and the Greener Area (GA and GGA) vegetation indices, the leaf chlorophyll content (LC), the stomatal conductance (g_s) of the flag leaf, the canopy temperature depression (CTD), the stable carbon isotope composition (δ^{13}C) of the mature grains and the grain yield (GY).

Growing Conditions	NDVI	GA	GGA	LC	g_s	CTD	δ^{13}C	GY
SI-2 with N120	0.002 **	0.089 ns	0.010 **	0.037 *	0.061 ns	0.493 ns	0.308 ns	0.048 *
SI-2 with N60	0.000 ***	0.009 **	0.002 **	0.043 *	0.970 ns	0.564 ns	0.216 ns	0.024 *
SI-2 without N	0.003 **	0.000 ***	0.001 ***	0.008 **	0.796 ns	0.111 ns	0.112 ns	0.144 ns
SI-1 with N120	0.127 ns	0.650 ns	0.299 ns	0.118 ns	0.624 ns	0.703 ns	0.292 ns	0.279 ns
SI-1 with N60	0.050 *	0.109 ns	0.006 **	0.071 ns	0.000 ***	0.076 ns	0.205 ns	0.708 ns
SI-1 without N	0.129 ns	0.519 ns	0.455 ns	0.883 ns	0.973 ns	0.610 ns	0.027 *	0.070 ns
Rainfed with N120	0.064 ns	0.150 ns	0.056 ns	0.035 *	0.212 ns	0.738 ns	0.594 ns	0.732 ns
Rainfed with N60	0.034 *	0.005 **	0.034 *	0.002 **	0.416 ns	0.682 ns	0.019 *	0.751 ns
Rainfed without N	0.000 ***	0.002 **	0.001 ***	0.103 ns	0.242 ns	0.519 ns	0.086 ns	0.149 ns

Significance levels—ns, not significant; * $p < 0.05$; ** $p < 0.01$ and *** $p < 0.001$.

3.5. Relationships of the Grain Yield with the Vegetation Indices

Combining both irrigation (SI-2 and SI-1) conditions, the two RGB vegetation indices (GA and GGA) were positively correlated with GY at N120 and only GA at N60, while the NDVI was not correlated with GY (Figure 1). Moreover, under rainfed conditions GA, GGA and the NDVI were correlated with GY at N60 (Figure 1). In the absence of nitrogen fertilization, no correlation was found between GY and the different vegetation indices (Table S3).

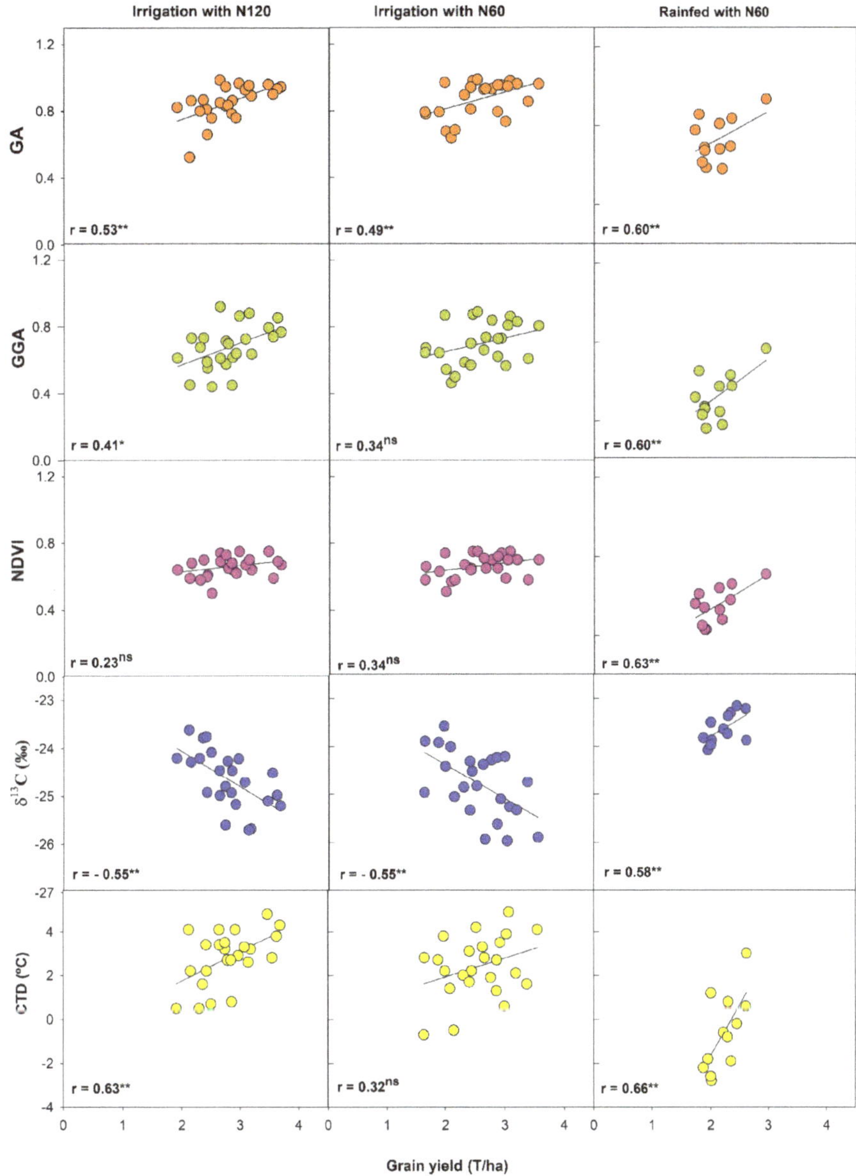

Figure 1. Relationships of grain yield (GY) with NDVI, GA, and GGA, $\delta^{13}C$ of mature grains and the canopy temperature depression (CTD) under (left column) irrigation with N120, (central column) irrigation with N60, and (central column) rainfed conditions with N60. The irrigation data combines the two support-irrigation regimes (SI-1, SI-2). Significance levels—ns, not significant; * $p < 0.05$ and ** $p < 0.01$. Abbreviations for variables and growing conditions are as defined in Tables 1 and 2.

3.6. Relationships of the Grain Yield with the Canopy Temperature Depression and the $\delta^{13}C$

Grain yield was negatively correlated with the $\delta^{13}C$ of mature grains under irrigation at both N120 and N60, and positively correlated with the $\delta^{13}C$ of rainfed conditions at N120 (Figure 1). Moreover, GY was positively correlated with CTD under either irrigation (SI-1 and SI-2 combined) or rainfed conditions at N120, while under N60 no association was found (Figure 1 and Table S3). In the absence of N fertilizer, no correlations were found (Table S3).

3.7. Relationships of Vegetation and Water Status Indices

The NDVI correlated highly and positively with both the GA and GGA vegetation indices when all genotypes, growing conditions, and replicates analyzed were combined (Table 5). Moreover, the CTD was also significantly ($p < 0.01$) correlated with the other water status parameters; negatively with $\delta^{13}C$ and positively with g_s. Likewise, both vegetation and water status indices were significantly ($p < 0.01$) correlated; the NDVI, GA and GGA were positively associated with CTD and g_s and negatively correlated with $\delta^{13}C$ (Table 5).

Table 5. Correlation coefficients of the relationships between different indices used to measure crop biomass and water status. Parameters were measured at anthesis. Treatments and genotypes were analyzed together. Significance levels—** $p < 0.010$; *** $p < 0.000$. Abbreviations of variables as in Tables 1 and 3.

Vegetation Indices	Correlation Coefficients
NDVI vs. GA	0.77 ***
NDVI vs. GGA	0.73 ***
Water status indices	
CTD vs. $\delta^{13}C$	−0.57 **
CTD vs. g_s	0.63 **
Vegetation and water status indices	
NDVI vs. CTD	0.63 **
GA vs. CTD	0.55 **
GGA vs. CTD	0.54 **
NDVI vs. g_s	0.51 **
GA vs. g_s	0.58 **
GGA vs. g_s	0.55 **
NDVI vs. $\delta^{13}C$	−0.43 **
GA vs. $\delta^{13}C$	−0.49 **
GGA vs. $\delta^{13}C$	−0.44 **

3.8. Grain Yield Estimation Using Vegetation Indices and Canopy Temperature

A general model using linear regression of the NDVI, GA and CTD with GY in the 2012–2013 crop season was developed in this study to estimate GY in the two successive crop seasons (2013–2014 and 2014–2015). Only the NDVI, GA and CTD were involved in this model because they were the three variables included in the stepwise models to explain the difference in GY in the present study (see Section 3.9 below) and were significantly ($p < 0.01$) correlated with GY in the first crop season [54]. With all growing conditions and genotypes analyzed together, the predicted and measured grain yields were positively and highly significantly ($p < 0.01$) correlated in 2013-2014 and 2014–2015 (Figure 2) using any of the three parameters studied alone (NDVI, GA, CTD). However, the range of normalized values predicted was smaller, in general, using the NDVI (at 2014–2015 crop season) than either of the other two indices. Predicted and measured GY for the 2013–2014 crop season was also positively and highly correlated using the NDVI, GA and CTD (Figure 3) under irrigation (SI-2 and SI-1 combined) with N fertilization (N120 and N60 combined) and without N fertilization (N0). For the 2014–2015 crop season, predicted and measured grain yields were also highly significantly ($p < 0.01$) correlated using either the NDVI, GA or CTD (Figure 3), but only under irrigation (SI-2 and SI-1 combined) with N fertilizer (N120 and N60 combined) and with no correlation under N0. In the absence of irrigation (rainfed conditions) and regardless of the N fertilization conditions, we did not find any association between the predicted and the measured grain yields (Figure 3). Grain yield estimation was also examined within each of the nine growing conditions resulting from the combination of the three water regimes and the three nitrogen fertilization levels (Figure S3). Under any of the two irrigation conditions with and without N fertilizer and the rainfed conditions with N fertilizer, the predicted GY was positively correlated with the measured GY in both 2013–2014 and 2014–2015 using at least one of the NDVI, GA or CTD parameters (Figure S3). Under rainfed conditions and N0 the predicted and measured GYs were not correlated in either crop season (Figure S3).

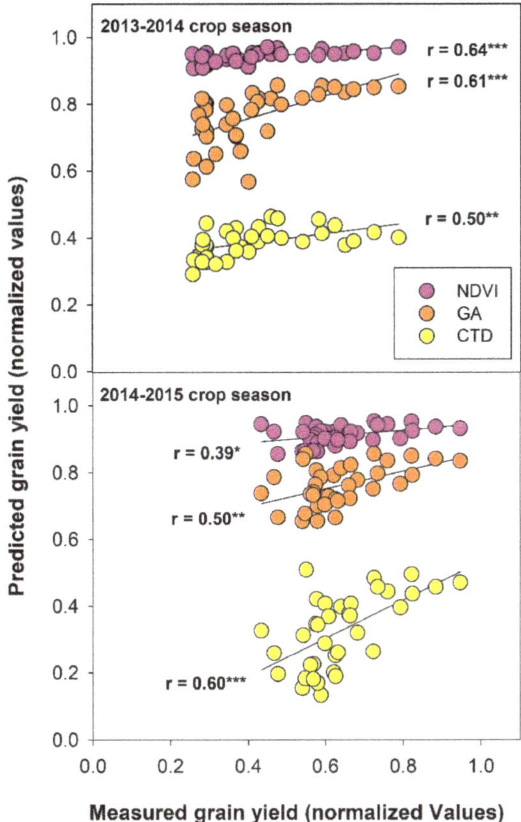

Figure 2. Relationships of the measured versus the predicted grain yields of wheat achieved during two successive crop seasons (2013–2014 and 2014–2015). Predicted grain yield values were calculated using the linear relationships of the grain yield with two vegetation indices, the Normalized Difference Vegetation Index (NDVI), the relative Green Area index (GA), as well as the canopy temperature depression (CTD). All variables were evaluated during the 2012–2013 crop season. Measurements were performed in the same region as the present study. For each crop season, different combinations of wheat genotypes under different water regimes and nitrogen fertilization levels are plotted together. Significance levels—ns, not significant; * $p < 0.05$; *** $p < 0.01$ and *** $p < 0.001$.

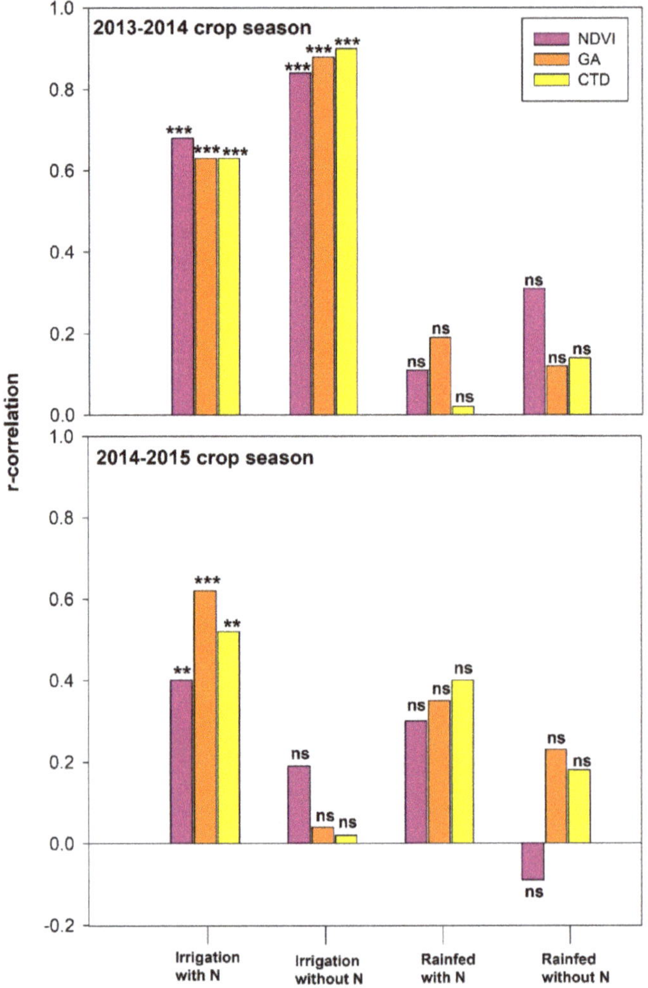

Figure 3. Correlation coefficients of the relationships of the measured versus the predicted grain yields achieved during two successive crop seasons (2013–2014 and 2014–2015). Predicted grain yield values were calculated using the linear relationships of the grain yield with the NDVI, (GA) and CTD evaluated during the 2012–2013 crop season in the same region as the present study. Predicted and measured grain yield were correlated under irrigation and rainfed conditions both with and without N. The irrigation with N values combine both irrigation regimes and the N120 and N60 treatments, irrigation without N combines the two irrigation regimes without N fertilization, and the rainfed conditions with N combine the two levels of N fertilization. Significance levels—ns, not significant; ** $p < 0.01$ and *** $p < 0.001$. Abbreviations of variables are as in Tables 1–3.

3.9. The Combined Effect of Remote Sensing Indices and Physiological Traits in Explaining Grain Yield

Stepwise regressions were performed for the irrigation conditions combined (SI-1 and SI-2) and the rainfed conditions, and considered both groups under each of the three N fertilization levels using grain yield (GY) as the dependent variable and the vegetation indices (NDVI, GA, GGA and LC) and water status traits (CTD, g_s, and $\delta^{13}C$) as independent variables (Table 6). Except for the rainfed conditions with N60, the first trait selected by the model to explain GY was related to the plant water status, while under rainfed conditions the NDVI was the first and only trait chosen by the model (Table 6).

Table 6. Multiple linear regressions (stepwise) explaining grain yield (GY) variation as a dependent variable and the NDVI, GA, GGA, LC, CTD, g_s and $\delta^{13}C$ as independent variables.

Dependent Variable	Growing Conditions	Variable Chosen	Correlation Coefficients	Final Stepwise Model
GY	Irrigation with N120	CTD	0.63 ***	0.24 CTD + 2.14
GY	Irrigation with N60	$\delta^{13}C$	0.55 **	$-0.42\ \delta^{13}C - 7.91$
GY	Irrigation without N	No variables were entered into the equation for this treatment		
GY	Rainfed with N120	CTD	0.66 ***	0.09 CTD + 2.27
GY	Rainfed with N60	NDVI	0.63 **	0.80 NDVI + 2.33
GY	Rainfed without N	No variables were entered into the equation for this treatment		

Data were analyzed under each level of N fertilizer for the irrigation (combined SI-1 and SI-2) and the rainfed treatments. Significance levels—** $p < 0.01$ and *** $p < 0.001$. Abbreviations for variables and growing conditions as defined in Tables 1–3.

Furthermore, the genotypic differences in grain yield within each of the nine growing conditions (resulting from the combination of the three water regimes and the three nitrogen fertilization levels) were assessed through a stepwise model having GY as the dependent variable and any of the vegetation indices and water status traits as independent variables. The model identified at least one trait positively correlated with GY in only four of the nine growing conditions (Table S4).

Additionally, a conceptual model based on a path analysis was proposed (Figure S2) that separated direct acclimation responses in grain yield related to water status traits and vegetation indices (through GA and GGA). The three water status traits were included in the model because they represent different scales: Temporal ($\delta^{13}C$), individual organ (g_s) and canopy (CTD). Concerning the vegetation indices, the NDVI was discarded because GA (whole photosynthetic biomass) and GGA (non-senescent biomass) already tracked the same parameter. The final objective of the model was to dissect how these physiological traits may have directly or indirectly assessed GY performance within different growing conditions. The four path models proposed (Figure 4) provided an acceptable fit to the data (CFI > 0.9 in all cases). Under irrigation and with and without N fertilizer, g_s had a strong and negative association with $\delta^{13}C$ and strong and positive associations with CTD and GA. Significant paths corresponding to a direct (negative) association of $\delta_{13}C$ with GY were also observed in irrigation conditions without N fertilizer and without, while CTD had a positive association with GY only under irrigation with N fertilizer. A direct positive and strong association of GA with GGA was observed under irrigation. GGA was in turn strongly and negatively associated with GY only under irrigation without N fertilizer. Under rainfed conditions, GY was not associated with $\delta^{13}C$, but was positively associated with CTD and GGA with N fertilizer, and negatively associated with GGA without fertilizer.

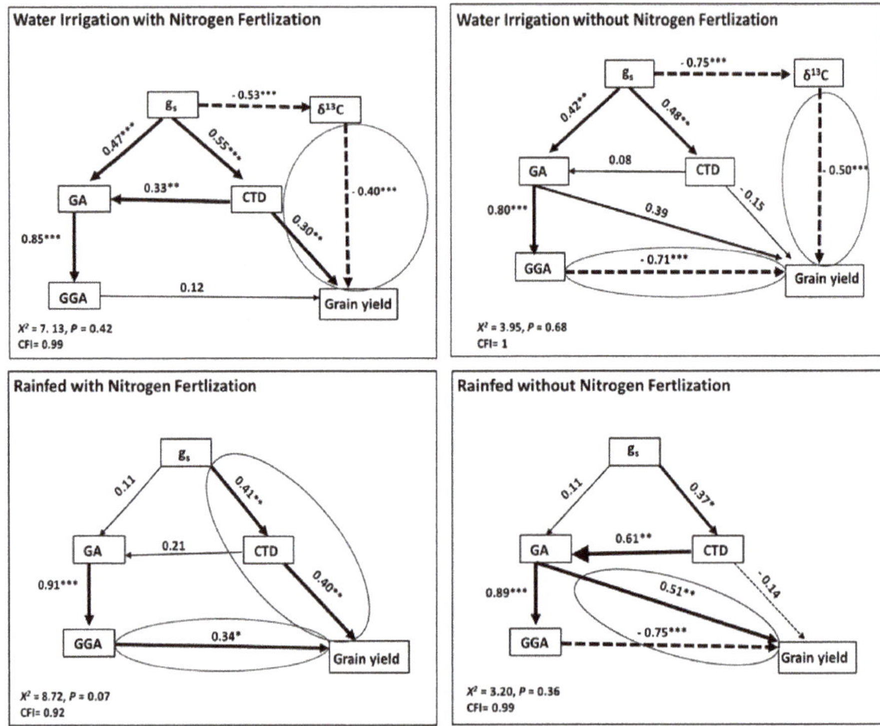

Figure 4. Path analyses of four wheat genotypes grown under different combinations of nitrogen fertilization and water regimes. The irrigation with N values combine both the SI-1 and SI-2 irrigation regimes and the N120 and N60 nitrogen treatments, irrigation without N combines the two irrigation regimes (SI-1 and SI-2) without N fertilization, and the rainfed conditions with N combine the two levels of N fertilization (N60 and N120). Physiological parameters included in the model are: The stomatal conductance (g_s), the stable carbon isotope composition ($\delta^{13}C$) of mature grains, the Relative Green Area (GA) and the Relative Greener Area (GGA), and the canopy temperature depression (CTD). The width of the arrows is proportional to the path coefficient values. Dashed lines indicate negative relationships. CFIs with values > 0.9 are taken as indicative of a good fit. Significance levels—* $p < 0.05$; ** $p < 0.01$ and *** $p < 0.001$.

Path analysis was also examined within each of the nine growing conditions (resulting from the combination of the three water regimes and the three nitrogen fertilization levels (Figure S4). Water irrigation conditions with and without N fertilization indicated the dependence of grain yield on the water status traits $\delta^{13}C$ and CTD (and the latter also under rainfed conditions with N fertilizer). In the absence of N fertilization under both irrigation and rainfed conditions, GY seemed strongly and negatively dependent on the vegetation index GGA.

4. Discussion

The grain yields achieved, even under the best growing conditions (SI-2 with N120) are below 3 tonnes per hectare, and this clearly corresponds with moderate–low yielding conditions in the Mediterranean [44,54]. The vegetation indices tested in this study, generated either spectroradiometrically (NDVI) or derived from RGB images, performed well when assessing differences in water conditions. The efficacy of these indices in capturing differences in growth and senescence in response to water regime has been reported for wheat already [32,33,54]. The choice of anthesis as the phenological stage for the remote sensing measurements was decided based on the results of Yousfi et al. (2016) [54] under similar agro-ecological conditions as in the present study, together with an additional study under Mediterranean conditions where NDVI and RGB indices were measured periodically during the crop cycle [22], with robust correlations between those indices measured at anthesis and grain yield being reported. While the argument may be valid that any action on N management at anthesis is probably too late to significantly affect yield, particularly for fully irrigated wheat, it may still positively affect grain quality. More importantly, scheduling irrigation at anthesis may be fully relevant for cereals under Mediterranean conditions where drought increases progressively during the reproductive stage of the crop. Thus, following previous reports in wheat [54–56], a significant association between the NDVI, GA and GGA vegetation indices and the water status parameters (δ^{13}C, g_s and CTD) was also found in this study. The NDVI, GA and GGA were positively associated with CTD and g_s and negatively associated with δ^{13}C. In this context, Lopes and Reynolds (2012) [57] reported that the relationship observed between chlorophyll retention or 'stay-green' (assessed via the NDVI) and canopy temperature would confirm the functionality of stay-green in terms of gas exchange and would explain a better capacity to use water by the stay-green genotypes under stressful environments related to low fertilization and the lack of water. Our data confirm the close association between vegetation indices and water status parameters and identified the canopy greenness as a good indicator of crop water status and irrigation management.

4.1. Vegetation Indices and Nitrogen Fertilization

Digital images have been used to evaluate the nitrogen status of crops [58,59]. Our results showed that GGA was the only vegetation index exhibiting a significant difference between N treatments, with lower values found under N0 compared with N60 and N120. The absence of N fertilizer limits plant growth, whereas it may accelerate plant senescence during the reproductive stage of the crop, therefore decreasing GGA compared to plants fertilized with nitrogen. In this context, [31] described the greener area index (GGA) derived from RGB images is a good parameter for capturing active photosynthetic area and plant senescence because it is formulated with green pixels alone. Furthermore, the NDVI failed to assess differences under different N treatments. Digital pictures provide information that is not currently acquired through spectral reflectance measurements, such as the portion of yellow leaves in wheat growing under field conditions [31,32,52]. In the case of GA, this index, which takes into account yellow/green pixels, is less stringent in terms of excluding non-senescent parts of the plant. This may explain why GA measured during anthesis was not affected by N fertilization.

4.2. Canopy Temperature and Water Status in Wheat

Many studies have recognized canopy temperature depression (CTD) as an indicator of overall plant water status [36,60] and a potential tool for irrigation management [9,61]. In our study, CTD measured with an infrared thermometer was lower (and even negative) under rainfed conditions compared to support irrigation. A priori, a higher CTD indicates a greater capacity for transpiration, for taking up water from the soil, and therefore for maintaining a better plant water status [36]. In the case of the rainfed trials, the fact that the leaf temperature was higher than the air temperature indicates that rainfed plants were subjected to severe water stress that closed the stomata. In fact, the stomatal conductance measured in the rainfed plants was very low and one third of that measured under the best support-irrigation regime.

In addition, Gutierrez et al. (2010) [62] reported that the association between canopy temperature and the normalized difference water index confirmed that canopy temperature is a good indicator of hydration status. According to this, our results showed highly significant associations of CTD with δ^{13}C (negative) and g_s (positive). Furthermore, CTD seems to be a better indicator of the water status at the crop level than other traits related to water status, such as leaf g_s [63]. In our study, CTD was strongly associated with δ^{13}C ($r = 0.84$ ***) under SI-1 at N60, with the NDVI ($r = 0.67$ **) under rainfed conditions without N, and with grain yield ($r = 0.66$ **) under rainfed conditions at N120, while g_s

was not correlated with any of these parameters. These results confirm the close association between canopy temperature and other water status parameters and identify the canopy temperature as a good indicator of crop water status.

4.3. Relationship of Vegetation Indices and Water Status Traits with GY

As found in previous studies in wheat, the RGB canopy indices measured at flowering were strongly correlated with GY [31,32,64]. For wheat under Mediterranean conditions, the reproductive stage is usually the best period for crop monitoring, since the crop is exposed to increasing stress (drought) conditions during the last part of the crop cycle. Following on from this, the present study revealed a positive relationship between GA and GGA with grain yield under irrigation. Additionally, stepwise analysis reinforced the evidence for the usefulness of the RGB vegetation indices to assess GY. The GA vegetation index was chosen by the model as the first independent variable, explaining 66% of GY variability under SI-2 without N fertilizer. Moreover, various studies have reported that RGB-based indices may perform far better than the NDVI for GY prediction in wheat [32,56,65]. In our study, the NDVI failed to assess GY under irrigation. In contrast, the NDVI was correlated (positively) with GY under rainfed conditions and was also the only variable chosen by the stepwise model, explaining 63% of GY variation under rainfed conditions. In this context, Casadesus et al. 2007 [18] reported that the NDVI measured at anthesis in durum wheat correlated positively with GY under severe water stress conditions, but failed to correlate under well-watered conditions. Verhulst and Govaerts (2010) [66] have also reported that the NDVI has been correlated with long-term water stress. The reason for the low correlation of NDVI against GY under well-watered conditions is because plant canopies during anthesis are very dense and the measured NDVI values become saturated. NDVI is an index based on the strong contrast between the near infrared and the red band reflectance of a vegetation canopy, and this difference becomes wider as the canopy cover increases. Thus, NDVI works better with stress conditions where canopies are sparse and/or early senescence is present [27,31]. In any case, anthesis proved to be the correct phenological stage for remote sensing evaluations when crops under different levels of stress were compared, which may be the case for crops exposed to a range of different combinations of water and nitrogen fertilization conditions.

Furthermore, the water status of plants can also be associated with grain yield. In our study, CTD (positively) and $\delta^{13}C$ (negatively) correlated with GY and both parameters were chosen by the stepwise model as the first variables to explain GY variation under different irrigated and rainfed growing conditions. In this context, previous studies have shown that a higher CTD is associated with increased wheat yield under irrigated, hot environments [38,67], but also under dryland environments [68].

However, in our study, and regardless of the water regime (rainfed or irrigation), neither the vegetation indices (NDVI, GA and GGA) nor the water status indices (CTD and $\delta^{13}C$) were associated with GY in the absence of N fertilizer. A lack of variability in green biomass and grain yield associated with the lack of nitrogen fertilization might explain this outcome.

4.4. Phenotyping Parameters under Different Water and N Supplies

The results of our study have shown the usefulness of vegetation indices with low implementation costs as a means to identify genetic variability under different growing conditions in the field. From nine of the growing conditions studied (resulting from the combination of the three water regimes and the three nitrogen fertilization levels), GY was significantly different between genotypes under only two of the growing conditions (SI-2 at N120 and N60). In contrast, two of the vegetation indices (NDVI, GGA) measured at anthesis were able to distinguish between the genotypes growing under six of the nine growing conditions (including rainfed), regardless of the N fertilization conditions. In this context, multispectral ground-based portable spectroradiometric devices have been used in wheat phenotyping [24,27]. The conventional RGB images have also been proposed as a selection tool for cereal breeding [18–20].

The genotypic differences observed using vegetation indices possibly reflect differences in canopy stay green during the reproductive stage. Lopes and Reynolds (2012) [57] reported that stay-green is regarded as a key indicator of stress adaptation. Thus, our study revealed the usefulness of the vegetation indices to select the most tolerant genotypes in terms of retaining a greener biomass during the last part of the crop cycle. The three vegetation indices assayed were able to identify genotypic differences, even under the most severe growing conditions, such as rainfed with and without N fertilizer and where GY failed to detect differences among genotypes. Phenotyping wheat genotypes for water and N fertilization deficit at anthesis using these vegetation indices should permit the formulation of the best crosses between genotypes. However, by comparison, canopy temperature performed much worse as a phenotyping parameter in our study. It has been reported that CTD is a poor indicator of plant performance when the yield is highly dependent on limited amounts of soil-stored water [69,70]. Moreover, the canopy in these trials, particularly during the reproductive stages, frequently leaves areas of bare soil exposed that may affect the canopy temperature readings. Leaf chlorophyll content measured by a portable device was perfect for distinguishing among genotypes, regardless of the water status (irrigation or rainfed), but only when trials were provided with nitrogen fertilizer. Therefore, for the agronomic conditions of our study, the vegetation indices assessed at the canopy level performed better as phenotyping tools than canopy temperature and chlorophyll content measures.

4.5. Grain Yield Prediction across Crop Seasons Using Low-Cost Remote Sensing Techniques

The results of data combining the growing conditions, genotypes and replicates support the use of different affordable remote sensing techniques to estimate grain yield across crop seasons. However, in agreement with Clevers (1997) [71], estimates of crop growth and yield using crop growth models often lost accuracy as the growing conditions became more stressed. The loss of accuracy may be the consequence of a very narrow range of variability in grain yield associated with stressed growing conditions. Moreover, vegetation indices derived from RGB images performed comparatively better than the NDVI, probably because GA was less saturated than the NDVI. The application timing could have played a critical role here—i.e., saturated NDVI at anthesis is indeed not expected to perform well, while the saturation pattern of RGB indices is less evident. In fact, the acquisition of high-resolution RGB images is fast and its dependence on atmospheric conditions (e.g., sunny versus cloudy days) is minimal [22,32]. Therefore, the availability, cost and practicality of digital cameras make them an ideal tool for the management of crop water and fertilization status [19,22]. However, in agreement with previous reports, the relationships between yield and vegetation were site and season specific [20,21]. In our study models were not able to predict absolute yields, but rather relative differences in yield, which makes the approach unfeasible for yield forecasting, but it is still useful in terms of crop management and even phenotyping. The strong relationships of these vegetation and water status indices with grain yield expressed in relative units support the effectiveness of these low-cost indices in crop management. Nevertheless, whereas the evaluations in the three successive seasons were performed in the same region, crop management conditions (water and fertilization regimes) affected the performance of the models. Hence, to make yield predictions more holistic and effective across different environments, it is necessary to use more robust calibration; for example, incorporating site-year covariates or a multi sensor approach [21].

4.6. An Integrated Model to Predict Grain Yield That Combines Remote-Sensing Traits, Canopy Reflectance Measurements and Grain $\delta^{13}C$

In this study, we performed a path analysis to dissect how the vegetation indices (GA and GGA) and the water status (CTD, $\delta^{13}C$ and g_s) indices directly or indirectly assessed GY performance within each of the growing conditions assayed. The following parameters may provide suitable coverage of the factors affecting GY performance under a given water regime and nitrogen fertilization supply: Vegetation indices as indicators of photosynthetic capacity (GA) and the effect of early senescence (GGA) on the canopy; CTD which informs about the current water status of the canopy; the $\delta^{13}C$

in mature grains as a time-integrated indicator of photosynthetic and transpirative gas exchange of the crop; and the water status at the single organ level (assessed as the g_s of the flag leaf). Under irrigation and nitrogen fertilization, both $\delta^{13}C$ and CTD (indicators of photosynthetic and water status) had a direct association with GY. Better grain yield performance is associated with higher CTD and lower $\delta^{13}C$ under supplementary irrigation. The association of $\delta^{13}C$ (negative) and CTD (positive) with GY is probably due the higher stomatal conductance and transpiration, (which increases CTD), therefore increasing the photosynthetic capacity even at the expense of a lower water use efficiency (and thus $\delta^{13}C$) and the consequent increases in GY [47,72]. Under rainfed conditions with N fertilizer, the transpiration/water status (assessed through CTD) and photosynthetic potential (evaluated through vegetation indices) affected GY. Under rainfed conditions with N120, CTD had a positive effect on GY. We suggest that N fertilization only had a positive effect on grain yield providing that there was water available to maintain transpiration (higher CTD), stomatal conductance and thus photosynthesis in the available canopy (higher vegetation indices).

In the absence of nitrogen fertilization, and despite the water conditions (irrigated or rainfed), the total canopy area (evaluated through GA) has a positive effect on GY. However, an excess of young (not senescing) leaf area (evaluated through GAA) had a strong negative association with GY. Stay-green character may have a negative effect on GY in the absence of nitrogen fertilization because it limits the retranslocation of N to the inflorescences and ultimately affects grain filling. In the case of the rainfed crop fertilized with a limited amount of nitrogen (N60), the active canopy area (evaluated through GGA) had a positive effect on grain yield, which may indicate that the limitation is imposed by the amount of photosynthetic area rather than by the availability of N to reproductive tissues. Under conditions of high nitrogen fertilization (N120) and irrespective of the water regime (irrigated or rainfed), GY is not affected by the size of the canopy or even by its greenness (assessed through GA or GGA), but by the water status of the crop (evaluated through CTD and $\delta^{13}C$).

5. Conclusions

This study demonstrated the potential of low-cost RGB vegetation indices and the canopy temperature for the management of growing conditions (essentially the water and nitrogen regimes) under Mediterranean conditions. Although the models did not predict absolute yields, they are still useful in terms of crop management and even phenotyping. Nevertheless, grain yield estimation performs better under irrigation than under the low-yielding conditions of rainfed cultivation in the absence of nitrogen fertilization and this illustrates one of the potential limitations associated with the remote sensing-based yield predictions; they are affected by specific environmental conditions. Even though a multispectral vegetation index, such as the NDVI is a widely accepted approach to monitor changes in growth under different conditions, in this study we have shown that vegetation indices derived from conventional images like the GA and GAA indices provided a similar if not better prediction of grain yield and at a comparatively lower cost than the NDVI. The use of vegetation indices derived from RGB images to assess GY could be popularized in the near future via apps installed on mobile phones.

The vegetation indices have also proven their suitability for differentiating among genotypes. Furthermore, these traits may contribute through path analysis to develop physiological models for assessing wheat ideotypes best suited to different water and nitrogen regimes. The models showed that for nitrogen fertilized trials, and regardless of the water regime imposed, the water status parameters were the main factors determining GY performance. Moreover, a larger green area at anthesis may also contribute to a larger yield. In the absence of nitrogen fertilization, a large greener canopy area (assessed through GGA) at anthesis is a factor that negatively affects grain yield.

Supplementary Materials: The following are available online at http://www.mdpi.com/2073-4395/9/6/285/s1, Figure S1: Scheme detailing the different plots of the experimental design, Figure S2: Conceptual model of the path analyses quantifying the relative strengths of the direct and indirect relationships of the different physiological traits and grain yield. Physiological parameters included in the model are: The stomatal conductance (g_s), the stable carbon isotope composition (δ^{13}C) of mature grains, the Relative Green Area (GA) and the Relative Greener Area (GGA) indices calculated from digital pictures, and the canopy temperature depression (CTD) measured with an infrared thermometer, Figure S3: Relationships of the measured versus the predicted grain yields of wheat achieved during two successive crop seasons (2013–2014 and 2014–2015). Predicted grain yield values were calculated using the linear relationships of the grain yield with the two vegetation indices, the Normalized Difference Vegetation Index (NDVI) measured with a portable spectroradiometer, and the relative Green Area index (GA) calculated from digital images, and the canopy temperature depression (CTD) measured with an infrared thermometer. All variables were evaluated during the 2012–2013 crop season. Measurements were performed in the same region as the present study. Grain yield and vegetation index data of the two first crop seasons have been reported in Yousfi et al. (2016). For each crop season, the nine different growing conditions (resulting from the combination of the three water regimes and the three nitrogen fertilization levels) were analysed. The different combinations of nitrogen fertilization and water regimes for the 2013–2014 crop season were as follows: Supplementary irrigation (SI-2) with nitrogen fertilization N120 and N60 and without N fertilization; Supplementary irrigation (SI-1) with nitrogen fertilization N120 and N60 and without N fertilization; Rainfed with nitrogen fertilization N120 and N60 and without N fertilization. In addition, for the 2014–2015 crop season the combinations were as follows: Supplementary irrigation (SI-2) with nitrogen fertilization N120 and N60 and without N fertilization; Supplementary irrigation (SI-1) with nitrogen fertilization N120 and N60 and without N fertilization; Rainfed with nitrogen fertilization N120 and N60 and without N fertilization. Significance levels—ns, not significant; * $p < 0.05$; ** $p < 0.01$ and *** $p < 0.001$, Figure S4: Path analyses of four wheat genotypes grown under different combinations of nitrogen fertilization and water regimes. The different combinations of nitrogen fertilization and water regimes are as follows: (A) SI-2 with high fertilization (N120); (B) SI-2 with medium nitrogen fertilization (N60); (C) SI-2 without nitrogen fertilization; (D) SI-I with high nitrogen fertilization (N120); (E) SI-I without nitrogen fertilization (N60); (F) Rainfed with high fertilization (N120); (G) Rainfed with medium nitrogen fertilization (N60); (H) Rainfed without nitrogen fertilization. Physiological parameters included in the model are: The stomatal conductance (g_s), the stable carbon isotope composition (δ^{13}C) of mature grains, the Relative Green Area (GA) and the Relative Greener Area (GGA) indices calculated from digital pictures, and the canopy temperature depression (CTD) measured with an infrared thermometer. The width of the arrows is proportional to the path coefficient values. Dashed lines indicate negative relationships. Overall fit statistics for each path model (chi-squared, the probability and comparative fit index, CFI) are shown at the bottom left of each panel. CFIs with values > 0.9 were taken as indicative of a good fit. Significance levels—* $p < 0.05$; ** $p < 0.01$ and *** $p < 0.001$; Table S1: Soil chemical characteristics at different depths, Table S2: Monthly total accumulated rainfall (PP), minimum air temperature (T min), maximum air temperature (T max) and average air temperature (T aver) for the 2014–2015 crop season. Values were collected at the meteorological station of Khemis Miliana (Algeria), Table S3: Correlation coefficients of the linear relationships of grain yield (GY) with NDVI, GA, GGA, CTD and δ^{13}C under irrigation without N fertilization and rainfed conditions with N120, N60 and without N, Table S4. Multiple linear regressions (stepwise) across genotypes and replicates explaining grain yield (GY) variation as a dependent variable and the NDVI, GA, GGA, LC, CTD, g_s and δ^{13}C as independent variables. Data were analyzed within each level of nitrogen fertilization and water regime. Significance levels—* $p < 0.05$; ** $p < 0.01$ and *** $p < 0.001$. Abbreviations for variables and growing conditions as defined in Tables 1 and 2.

Author Contributions: S.Y., J.L.A., M.D.S., M.K. (Mohamed Karrou), N.K. and A.C. conceived and designed the experiment. N.K., M.K. (Mohamed Kaddour) and A.C. contributed to the experimental work N.K., M.K. (Mohamed Kaddour), A.C., M.K. (Mohamed Karrou), S.Y., J.L.A. and M.D.S. contributed to the experimental measures and field sampling. A.G.-R. and M.D.S. performed stable isotope analyses. S.Y., J.L.A. and M.D.S. contributed to the data analysis and interpreted the results. S.Y. wrote the paper under the supervision of J.L.A. and M.D.S. and all three revised the manuscript. All authors read and approved the final manuscript.

Funding: This study was supported in part by the European project ACLIMAS (EuropeAid/131046/C/ACT/Multi) and the Spanish MINECO project grant No. AGL2016-76527-R).

Acknowledgments: We acknowledge the support of the staff from the ITGC (Institut Technique des Grandes Cultures, Algeria). J. L Araus acknowledges the support of the ICREA Academia Award.

Conflicts of Interest: The authors declare no conflict of interest.

References

1. FAO. Statistical Year Book. 2012. Available online: http://www.fao.org/docrep/015/i2490e/i249e00.htm (accessed on 14 June 2015).

2. Oweis, T.; Zhang, H.; Pala, M. Water use efficiency of rainfed and irrigated bread wheat in a Mediterranean environment. *Agron. J.* **2000**, *92*, 231–238. [CrossRef]

3. Lobell, D.B.; Burke, M.B.; Tebaldi, C.; Mastrandrea, M.D.; Falcon, W.P.; Naylor, R.L. Prioritizing climate change adaptation needs for food security in 2030. *Science* **2008**, *319*, 607–610. [CrossRef] [PubMed]

4. Araus, J.L.; Slafer, G.A.; Reynolds, M.P.; Royo, C. Plant breeding and water relations in C3 cereals: What should we breed for? *Ann. Bot.* **2002**, *89*, 925–940. [CrossRef] [PubMed]

5. Seelan, S.K.; Laguette, S.; Casady, G.M.; Seielstad, G.A. Remote sensing applications for precision agriculture: A learning community approach. *Remote Sens. Environ.* **2003**, *88*, 157–169. [CrossRef]

6. Chapman, S.C. Use of crop models to understand genotype by environment interactions for drought in real-world and simulated plant breeding trials. *Euphytica* **2008**, *161*, 195–208. [CrossRef]

7. Araus, J.L.; Cairns, J.E. Field high-throughput phenotyping: the new crop breeding frontier. *Trends Plant Sci.* **2014**, *19*, 52–61. [CrossRef] [PubMed]

8. Glenn, E.P.; Huete, A.R.; Nagler, P.L.; Nelson, S.G. Relationship between remotely-sensed vegetation indices, canopy attributes and plant physiological processes: what vegetation indices can and cannot tell us about the landscape. *Sensors* **2008**, *8*, 2136–2160. [CrossRef] [PubMed]

9. Jones, H.G.; Serraj, R.; Loveys, B.R.; Xiong, L.Z.; Wheaton, A. Thermal infrared imaging of crop canopies for the remote diagnosis and quantification of plant responses to water stress in the field. *Funct. Plant Biol.* **2009**, *36*, 978–989. [CrossRef]

10. Sui, J.; Qin, Q.; Ren, H.; Sun, Y.; Zhang, T.; Wang, J.; Gong, S. Winter wheat production estimation based on environmental stress factors from satellite observations. *Remote Sens.* **2018**, *10*, 962. [CrossRef]

11. Zhang, C.; Pattey, E.; Liu, J.; Cai, H.; Shang, J.; Dong, T. Retrieving leaf and canopy water content of winter wheat using vegetation water indices. *IEEE J. Sel. Top. Appl. Earth Obs. Remote Sens.* **2018**, *11*, 112–126. [CrossRef]

12. Segovia-Cardozo, D.A.; Rodríguez-Sinobas, L.; Zubelzu, S. Water use efficiency of corn among the irrigation districts across the Duero river basin (Spain): Estimation of local crop coefficients by satellite images. *Agric. Water Manag.* **2019**, *212*, 241–251. [CrossRef]

13. Serrano, J.; Shahidian, S.; Marques da Silva, J. Evaluation of normalized difference water index as a tool for monitoring pasture seasonal and inter-annual variability in a Mediterranean agro-silvo-pastoral system. *Water* **2019**, *11*, 62. [CrossRef]

14. Zhang, T.; Su, J.; Liu, C.; Chen, W.H.; Liu, H.; Liu, G. Band selection in Sentinel-2 satellite for agriculture applications. In Proceedings of the 2017 23rd International Conference on Automation and Computing (ICAC), Huddersfield, UK, 7–8 September 2017; IEEE: Huddersfield, UK, 2017; pp. 1–6.

15. Varinderpal-Singh; Bijay-Singh; Yadvinder-Singh; Thind, H.S.; Gobinder-Singh; Satwinderjit-Kaur; Kumar, A.; Vashistha, M. Establishment of threshold leaf colour greenness for need-based fertilizer nitrogen management in irrigated wheat (*Triticum aestivum* L.) using leaf colour chart. *Field Crops Res.* **2012**, *130*, 109–119. [CrossRef]

16. Varinderpal-Singh; Bijay-Singh; Yadvinder-Singh; Thind, H.S.; Buttar, G.S.; Kaur, S.; Kaur, S.; Bhowmik, A. Site-specific fertilizer nitrogen management for timely sown irrigated wheat (*Triticum aestivum* L. and *Triticum turgidum* L. ssp. durum) genotypes. *Nutr. Cycl. Agroecosyst.* **2017**, *109*, 1–16. [CrossRef]

17. Bija-Singh; Varinderpal-Singh; Yadvinder-Singh; Thind, H.S.; Kumar, A.; Choudhary, O.P.; Gupta, R.K.; Vashistha, M. Site-specific fertilizer nitrogen management using optical sensor in irrigated wheat in the Northwestern India. *Agric. Res.* **2017**, *6*, 159–168. [CrossRef]

18. Van Loon, J.; Speratti, A.; Govaerts, B. Precision for smallholder farmers: a small-scale-tailored variable rate fertilizer application kit. *Agriculture* **2018**, *8*, 48. [CrossRef]

19. Araus, J.L.; Kefauver, S.C. Breeding to adapt agriculture to climate change: Affordable phenotyping solutions. *Curr. Opin. Plant Biol.* **2018**, *45*, 237–247. [CrossRef]

20. Colaço, A.F.; Bramley, R.G. Do crop sensors promote improved nitrogen management in grain crops? *Field Crops Res.* **2018**, *218*, 126–140. [CrossRef]

21. Bramley, R.G.V.; Ouzman, J. Farmer attitudes to the use of sensors and automation in fertilizer decision-making: nitrogen fertilization in the Australian grains sector. *Precis. Agric.* **2019**, *20*, 157–175. [CrossRef]

22. Fernandez-Gallego, J.A.; Kefauver, S.C.; Vatter, T.; Gutiérrez, N.A.; Nieto-Taladriz, M.T.; Araus, J.L. Low-cost assessment of grain yield in durum wheat using RGB images. *Eur. J. Agron.* **2019**, *105*, 146–156. [CrossRef]

23. Hansen, P.M.; Schjoerring, J.K. Reflectance measurement of canopy biomass and nitrogen status in wheat crops using normalized difference vegetation indices and partial least squares regression. *Remote Sens. Environ.* **2003**, *86*, 542–553. [CrossRef]

24. Babar, M.A.; Reynolds, M.P.; Van Ginkel, M.; Klatt, A.R.; Raun, W.R.; Stone, M.L. Spectral reflectance indices as a potential indirect selection criteria for wheat yield under irrigation. *Crop Sci.* **2006**, *46*, 578–588. [CrossRef]

25. Marti, J.; Bort, J.; Slafer, G.A.; Araus, J.L. Can wheat yield be assessed by early measurements of normalized difference vegetation index? *Ann. Appl. Biol.* **2007**, *150*, 253–257. [CrossRef]

26. Wright, D.L.; Rasmussen, V.P.; Ramsey, R.D. Comparing the use of remote sensing with traditional techniques to detect nitrogen stress in wheat. *Geocarto. Int.* **2005**, *20*, 63–68. [CrossRef]

27. Aparicio, N.; Villegas, N.; Casadesus, J.; Araus, J.L.; Royo, C. Spectral vegetation indices as non-destructive tools for determining durum wheat yield. *Agron. J.* **2000**, *92*, 83–91. [CrossRef]

28. Filella, I.; Serrano, L.; Serra, J.; Penuelas, J. Evaluating wheat nitrogen status with canopy reflectance indices and discriminant analysis. *Crop Sci.* **1995**, *35*, 1400–1405. [CrossRef]

29. Royo, C.; Aparicio, N.; Villegas, D.; Casadesus, J.; Monneveux, P.; Araus, J.L. Usefulness of spectral reflectance indices as durum wheat yield predictors under contrasting Mediterranean conditions. *Int. J. Remote Sens.* **2003**, *24*, 4403–4419. [CrossRef]

30. Araus, J.L.; Amaro, T.; Zuhair, Y.; Nachit, M.M. Effect of leaf structure and water status on carbon isotope discrimination in field-grown durum wheat. *Plant Cell Environ.* **1997**, *20*, 1484–1494. [CrossRef]

31. Casadesus, J.; Kaya, Y.; Bort, J.; Nachit, M.M.; Araus, J.L.; Amor, S.; Ferrazzano, G.; Maalouf, F.; Maccaferri, M.; Martos, V.; et al. Using vegetation indices derived from conventional digital cameras as selection criteria for wheat breeding in water-limited environments. *Ann. Appl. Bot.* **2007**, *150*, 227–236. [CrossRef]

32. Vergara-Diaz, O.; Kefauver, S.C.; Elazab, A.; Nieto-Taladriz, M.T.; Araus, J.L. Grain yield loss assessment for winter wheat associated with the fungus *Puccinia striiformis* f.sp. *tritici* using digital and conventional parameters under field conditions. *Crop J.* **2015**, *3*, 200–210. [CrossRef]

33. Mullan, D.J.; Reynolds, M.P. Quantifying genetic effects of ground cover on soil water evaporation using digital imaging. *Funct. Plant. Biol.* **2010**, *37*, 703–712. [CrossRef]

34. Deery, D.; Jimenez-Berni, J.; Jones, H.; Sirault, X.; Furbank, R. Proximal remote sensing buggies and potential applications for field-based phenotyping. *Agronomy* **2014**, *4*, 349–379. [CrossRef]

35. Araus, J.L.; Kefauver, S.C.; Zaman-Allah, M.; Olsen, M.S.; Cairns, J.E. Translating high throughput phenotyping into genetic gain. *Trends Plant Sci.* **2018**, *23*, 451–466. [CrossRef] [PubMed]

36. Blum, A.; Mayer, J.; Gozlan, G. Infrared thermal sensing of plant canopies as a screening technique for dehydration avoidance in wheat. *Field Crops Res.* **1982**, *5*, 137–146. [CrossRef]

37. Amani, I.; Fischer, R.A.; Reynolds, M.P. Canopy temperature depression association with yield of irrigated spring wheat cultivars in hot climate. *J. Agron. Crop Sci.* **1996**, *176*, 119–129. [CrossRef]

38. Reynolds, M.P.; Singh, R.P.; Ibrahim, A.; Ageeb, O.A.A.; Larque-Saavedra, A.; Quick, J.S. Evaluating physiological traits to complement empirical selection for wheat in warm environments. *Euphytica* **1998**, *100*, 84–95. [CrossRef]

39. Reynolds, M.; Dreccer, F.; Trethowan, R. Drought-adaptive traits derived from wheat wild relatives and landraces. *J. Exp. Bot.* **2007**, *58*, 177–186. [CrossRef]

40. Araus, J.L.; Slafer, G.A.; Royo, C.; Serret, M.D. Breeding for yield potential and stress adaptation in cereals. *Crit. Rev. Plant Sci.* **2008**, *27*, 377–412. [CrossRef]

41. Inoue, Y. Remote detection of physiological depression in crop plants with infrared thermal imagery. *Jpn. J. Crop Sci.* **1990**, *59*, 762–768. [CrossRef]

42. O'Shaughnessy, S.A.; Evett, S.R.; Colaizzi, P.D.; Howel, T.A. Using radiation thermography and thermometry to evaluate crop water stress in soybean and cotton. *Agric. Water Manag.* **2011**, *98*, 1523–1535. [CrossRef]

43. Idso, S.B.; Jackson, R.D.; Pinter, P.J.; Reginato, R.J.; Hatfield, J.L. Normalizing the stress-degree-day parameter for environmental variability. *Agric. Meteorol.* **1981**, *24*, 45–55. [CrossRef]

44. Araus, J.L.; Cabrera-Bosquet, L.; Serret, M.D.; Bort, J.; Nieto-Taladriz, M.T. Comparative performance of δ^{13}C, δ^{18}O and δ^{15}N for phenotyping durum wheat adaptation to a dryland environment. *Funct. Plant Biol.* **2013**, *40*, 595–608. [CrossRef]

45. Van Leeuwen, C.; Tregoat, O.; Chone, X.; Bois, B.; Pernet, D.; Gaudillere, J.P. Vine water status is a key factor in grape ripening and vintage quality for red Bordeaux wine. How can it be assessed for vineyard management purposes? *J. Int. Sci. Vig.* **2009**, *43*, 121–134. [CrossRef]

46. Farquhar, G.D.; O'Leary, M.H.; Berry, J.A. On the relationship between carbon isotope discrimination and the intercellular carbon dioxide concentration in leaves. *Aust. J. Plant. Physiol.* **1982**, *9*, 121–137. [CrossRef]

47. Farquhar, G.D.; Richards, R.A. Isotopic composition of plant carbon correlates with water-use-efficiency of wheat genotypes. *Aust. J. Plant Physiol.* **1984**, *11*, 539–552. [CrossRef]

48. Condon, A.G.; Richards, R.A. Broad sense heritability and genotypes-environment interaction for carbon isotope discrimination in field-grown wheat. *Aust. J. Agric. Res.* **1992**, *43*, 921–934. [CrossRef]

49. Yousfi, S.; Serret, M.D.; Voltas, J.; Araus, J.L. Effect of salinity and water stress during the reproductive stage on growth, ion concentrations, $\Delta^{13}C$, and $\delta^{15}N$ of durum wheat and related amphiploids. *J. Exp. Bot.* **2010**, *61*, 3529–3542. [CrossRef] [PubMed]

50. Araus, J.L.; Amaro, T.; Casadesus, J.; Asbati, A.; Nachit, M.M. Relationships between ash content, carbon isotope discrimination and yield in durum wheat. *Aust. J. Plant Physiol.* **1998**, *25*, 835–842. [CrossRef]

51. Susstrunk, S.; Robert-Buckley, R.; Swen, S. Standard RGB color spaces. In Proceedings of the Color and Imaging Conference Final Program and Proceedings, Scottsdale, AZ, USA, 16–19 November 1999; Society for Imaging Science and Technology: Springfield, MO, USA, 1999; pp. 127–134.

52. Gracia-Romero, A.; Vergara-Díaz, O.; Thierfelder, C.; Cairns, J.E.; Kefauver, S.C.; Araus, J.L. Phenotyping Conservation Agriculture Management Effects on Ground and Aerial Remote Sensing Assessments of Maize Hybrids Performance in Zimbabwe. *Remote Sens.* **2018**, *10*, 349. [CrossRef]

53. Farquhar, G.D.; Ehleringer, J.R.; Hubick, K.T. Carbon isotope discrimination and photosynthesis. *Annu. Rev. Plant Physiol.* **1989**, *40*, 503–537. [CrossRef]

54. Yousfi, S.; Kellas, N.; Saidi, L.; Benlakhel, Z.; Chaou, L.; Siad, D. Comparative performance of remote sensing methods in assessing wheat performance under Mediterranean conditions. *Agric. Water Manag.* **2016**, *164*, 137–147. [CrossRef]

55. Arbuckle, J.L. *Amos Users' Guide, Version 3.6*; Small Waters Corporation: Chicago, IL, USA, 1997.

56. Elazab, A.; Bort, J.; Zhou, B.; Serret, M.D.; Nieto-Taladriz, M.T.; Araus, J.L. The combined use of vegetation índices and stable isotopes to predict durum wheat grain yield under contrasting water conditions. *Agric. Water Manag.* **2015**, *158*, 196–208. [CrossRef]

57. Lopes, M.S.; Reynolds, M.P. Stay-green in spring wheat can be determined by spectral reflectance measurements (normalized difference vegetation index) independently from phenology. *J. Exp. Bot.* **2012**, *63*, 3789–3798. [CrossRef] [PubMed]

58. Li, Y.; Chen, D.; Walker, C.N.; Angus, J.F. Estimating the nitrogen status of crops using a digital camera. *Field Crop Res.* **2010**, *118*, 221–227. [CrossRef]

59. Vergara-Díaz, O.; Zaman-Allah, M.; Masuka, B.; Hornero, A.; Zarco-Tejada, P.; Prasanna, B.M.; Cairns, J.E.; Araus, J.L. A novel remote sensing approach for prediction of maize yield under different conditions of nitrogen fertilization. *Front. Plant Sci.* **2016**, *7*, 666. [CrossRef] [PubMed]

60. Jackson, R.D.; Idso, S.B.; Reginato, R.J.; Pinter, P.J., Jr. Canopy temperature as a crop water stress indicator. *Water Resour. Res.* **1981**, *17*, 1133–1138. [CrossRef]

61. Jones, H.G. Irrigation scheduling: advantages and pitfalls of plant-based methods. *J. Exp. Bot.* **2004**, *55*, 2427–2436. [CrossRef]

62. Gutierrez, M.; Reynolds, M.P.; Raun, W.R.; Stone, M.L.; Klatt, A.R. Spectral water indices for assessing yield in elite bread wheat genotypes grown under well irrigated, water deficit stress, and high temperature conditions. *Crop Sci.* **2010**, *50*, 197–214. [CrossRef]

63. Balota, M.; Amani, I.; Reynolds, M.P.; Acevedo, E. *Evaluation of Membrane Thermostability and Canopy Temperature Depression as Screening Traits for Heat Tolerance in Wheat*; Wheat Special Report No.20; CIMMYT: Mexico, D.F., Mexico, November 1993.

64. Morgounov, A.; Gummadov, N.; Belen, S.; Kaya, Y.; Keser, M.; Mursalova, J. Association of digital photo parameters and NDVI with winter wheat grain yield in variable environments. *Turk. J. Agric. For.* **2014**, *38*, 624–632. [CrossRef]

65. Zhou, B.; Elazab, A.; Bort, J.; Vergara, O.; Serret, M.D.; Araus, J.L. Low-cost assessment of wheat resistance to yellow rust through conventional RGB images. *Comput. Electron. Agric.* **2015**, *116*, 20–29. [CrossRef]

66. Verhulst, N.; Govaerts, B. *The Normalized Difference Vegetation Index (NDVI) GreenSeekerTM Handheld Sensor: Toward the Integrated Evaluation of Crop Management. Part A: Concepts and Case Studies*; CIMMYT: Mexico, D.F., Mexico, 2010.

67. Fischer, R.A.; Rees, D.; Sayre, K.D.; Lu, Z.M.; Condon, A.G.; Larque-Saavedra, A. Wheat yield progress associated with higher stomatal conductance and photosynthetic rate, and cooler canopies. *Crop Sci.* **1998**, *38*, 1467–1475. [CrossRef]

68. Balota, M.; Payne, W.A.; Evett, S.R.; Lazar, M.D. Canopy temperature depression sampling to assess grain yield variation and genotypic differentiation in winter wheat. *Crop Sci.* **2007**, *47*, 1518–1529. [CrossRef]
69. Winter, S.R.; Musick, J.T.; Porter, K.B. Evaluation of screening techniques for breeding drought-resistant Winter wheat. *Crop Sci.* **1998**, *28*, 512–516. [CrossRef]
70. Royo, C.; Villegas, D.; Garcia del Moral, L.F.; Elhani, S.; Aparicio, N.; Rharrabti, Y.; Araus, J.L. Comparative performance of carbon isotope discrimination and canopy temperature depression as predictors of genotypes differences in durum wheat yield in Spain. *Aust. J. Agric. Res.* **2002**, *53*, 561–569. [CrossRef]
71. Clevers, J.G.P.W. A simplified approach for yield prediction of sugar beet on optical remote sensing data. *Remote Sens. Environ.* **1997**, *61*, 221–228. [CrossRef]
72. Cabrera-Bosquet, L.; Molero, G.; Nogués, S.; Araus, J.L. Water and nitrogen conditions affect the relationships of Δ^{13}C and Δ^{18}O to gas exchange and growth in durum wheat. *J. Exp. Bot.* **2009**, *60*, 1633–1644. [CrossRef] [PubMed]

Article

Sentinel 2-Based Nitrogen VRT Fertilization in Wheat: Comparison between Traditional and Simple Precision Practices

Marco Vizzari *[iD], **Francesco Santaga** and **Paolo Benincasa**[iD]

Department of Agricultural, Food, and Environmental Sciences, University of Perugia, 06121 Perugia, Italy; francescosaverio.santaga@unipg.it (F.S.); paolo.benincasa@unipg.it (P.B.)
* Correspondence: marco.vizzari@unipg.it; Tel.: +39-075-585-6059

Received: 8 April 2019; Accepted: 23 May 2019; Published: 30 May 2019

Abstract: This study aimed to compare standard and precision nitrogen (N) fertilization with variable rate technology (VRT) in winter wheat (*Triticum aestivum* L.) by combining data of NDVI (Normalized Difference Vegetation Index) from the Sentinel 2 satellite, grain yield mapping, and protein content. Precision N rates were calculated using simple linear models that can be easily used by non-specialists of precision agriculture, starting from widely available Sentinel 2 NDVI data. To remove the effects of not measured or unknown factors, the study area of about 14 hectares, located in Central Italy, was divided into 168 experimental units laid down in a randomized design. The first fertilization rate was the same for all experimental units (30 kg N ha^{-1}). The second one was varied according to three different treatments: 1) a standard rate of 120 kg N ha^{-1} calculated by a common N balance; 2) a variable rate (60–120 kg N ha^{-1}) calculated from NDVI using a linear model where the maximum rate was equal to the standard rate (Var-N-low); 3) a variable rate (90–150 kg N ha^{-1}) calculated from NDVI using a linear model where the mean rate was equal to the standard rate (Var-N-high). Results indicate that differences between treatments in crop vegetation index, grain yield, and protein content were negligible and generally not significant. This evidence suggests that a low-N management approach, based on simple linear NDVI models and VRT, may considerably reduce the economic and environmental impact of N fertilization in winter wheat.

Keywords: NDVI; remote sensing; GIS; precision farming; variable rate technology; yield mapping; protein content

---

1. Introduction

In recent decades, population growth and the expanding demand of agricultural products for multiple purposes have constantly increased the environmental pressures on land and water resources. Worldwide concern of citizens and governments on environmental issues, together with the availability of improved and cost-effective methodologies and tools for spatial data acquisition, analysis, and modeling, have promoted the development of precision agriculture (PA) techniques. PA is based on innovative system approaches that use a combination of various technologies such as Geographic Information System (GIS), Global Navigation Satellite System (GNSS), computer modeling, Remote Sensing (RS), variable rate technology (VRT), yield mapping and advanced information processing for timely in-season and between season crop management [1].

Currently, a wide range of satellite data is available that varies in terms of acquisition cost, technique (active/passive), spatial resolution, spectral range, and viewing geometry [2]. A recent step forward was made thanks to the launches of Sentinel-2A (2015) and Sentinel-2B (2017) satellites by the European Space Agency (ESA), intended for Earth observation and monitoring of land surface variability. The Sentinel-2 satellites are equipped with a multispectral sensor (MSI) including 13 spectral bands, with a spatial resolution ranging from 10 m to 60 m, which provides relevant information for supporting precision agriculture [3]. Images provided by Sentinel-2 satellites are publicly available for free through the Copernicus Open Access Hub with 5 days temporal resolution averaging (2–3 days in mid-latitudes), in L1C or L2A processing levels, which make them attractive for time-series analysis and PA applications. The Level-2A product provides Bottom Of Atmosphere (BOA) reflectance images derived from the associated Level-1C product, which provides Top Of Atmosphere (TOA) reflectance images [4].

Satellite remote sensing of vegetation is mainly based on the green (495–570 nm) and red (620–750 nm) regions of the visible spectrum, the red-edge (680–730 nm), near and mid infrared bands (850–1700 nm). These bands are often combined with a range of algebraic formulas to obtain several Vegetation Indices (VIs) for assessing vegetation status and crop management (see e.g. references [5–7]). The Normalized Difference Vegetation Index (NDVI) [8], a normalized difference between reflectance of Red and Near-Infrared (NIR) spectral bands, is still the most used index worldwide because of its ease of calculation and interpretation. It varies between 0 and 1 in cropped areas, increasing with soil cover, LAI (Leaf Area Index), chlorophyll content, and plant N-nutritional status (see e.g. references [9–11]). The NDVI has been widely tested to assess wheat N-nutritional status and yield, with encouraging results [12–14].

N is the main nutrient supplied to most crops, including wheat, and may cause environmental impacts on near-surface and deep aquifers [15]. Increasing the N rate generally increases crop yield since it increases the grain number and size [16]. However, increasing the N rates reduces the N uptake efficiency and increases the amount of residual N in the soil which is exposed to leaching risks [9,17]. Countless studies are available on wheat N nutrition and several of the recent ones have dealt with precision N fertilization, but most of them propose quite complex approaches that are not widely adoptable by non-specialists of PA. For example, Bourdin et al. [18] propose a complex model starting from LAI, estimated by remote sensing, and yields from previous years; Basso et al. [19] use a crop simulated model (SALUS) based on weather and yields from previous years. Another approach consists in assessing spatial variability to delineate homogeneous sub-field areas overlapping various thematic spatial maps (such as yield, soil properties) or applying a multivariate geostatistical approach of the factors that affect yield and grain quality [20]. In this context Song et al. [21] delineated management zones on the basis of soil, yield data and remote sensing information derived from Quickbird imagery. Besides these methods, to improve the use of N prescription maps in PA, various simplified approaches based on NDVI have also been developed through user-friendly web interfaces (e.g. CropSAT [22], Agrosat [23], OneSoil [24]) but these approaches can be applied only in specific areas or lack either quantitative analysis or validation.

While traditional flat-rate spreaders typically tend to over- or under-apply fertilizers, with VRT spreaders it is possible to modify the distribution rate according to on-the-go proximal sensors measuring, in real-time, plant properties or prescription maps derived through more or less advanced models including various spatial factors (e.g.: vegetation indices, soil variables, yield maps, crop nutrition status) [25]. However, the majority of precision fertilization approaches are generally map-based, because on-the-go sensors are too expensive, not sufficiently accurate, or not available [26]. VRT spreaders are subject to errors depending on the technology and the intrinsic characteristics of the fertilizer [27].

Grain yield mapping aims at providing or increasing knowledge about the spatial heterogeneity of yields. This technology is based on the recording of georeferenced yield data thanks to the integration of different subsystems mounted on the combines [28]: (i) harvest quantity measurement sensors (mass or volume); (ii) GNSS location sensors; (iii) reference area measurement sensors (working width, speed, time); (iv) data recording and processing units. To generate yield maps, the harvested quantities are georeferenced and associated with the corresponding reference area which is calculated by multiplying the working width by the area length (working speed * time interval). Yield mapping accuracy is influenced by many factors including flow sensor calibration, combined speed changes, grain flow variations, grain moisture, and the level of smoothing applied for mapping [29].

While many studies combine VIs (from various remote or proximal sensors) and VRT crop fertilization, only a small quantity of research integrates yield mapping sensors, and, to our knowledge, probably no study has ever combined VIs, VRT fertilization, and yield mapping in a PA case study and, in particular, to assess the effects of different variable N-rate treatments on winter wheat.

In this framework, this study was aimed at comparing two VRT N fertilization treatments (based on Sentinel 2) versus a standard flat N rate in terms of crop NDVI trend, grain yield, and protein content. In order to promote the adoption of PA techniques among farmers and non-specialists in PA, a simplified approach, based on free and open-source software, the widely used NDVI, and an easily-applicable linear model, was applied to calculate the VRT N fertilization rates.

2. Materials and Methods

2.1. Crop Management and Experimental Treatments

The experiment was carried out in the cropping season 2017-2018, on a 14 ha plain field of the middle Tiber valley, near Deruta, Umbria, Italy (170 m a.s.l., 42°95′07″ N, 12°38′18″ E), provided by the Foundation for Agricultural Education of Perugia (Figure 1). The soil was loam with increasing sand content from the west to the east side. The climate is Mediterranean, characterized by a dry season between May and September and a cold and rainy season from October-November to March-April. The cropping season 2017–2018 was unusually rainy in December and March, while temperatures were generally higher than the poly-annual trend except for a very cold end of February (Figure 2).

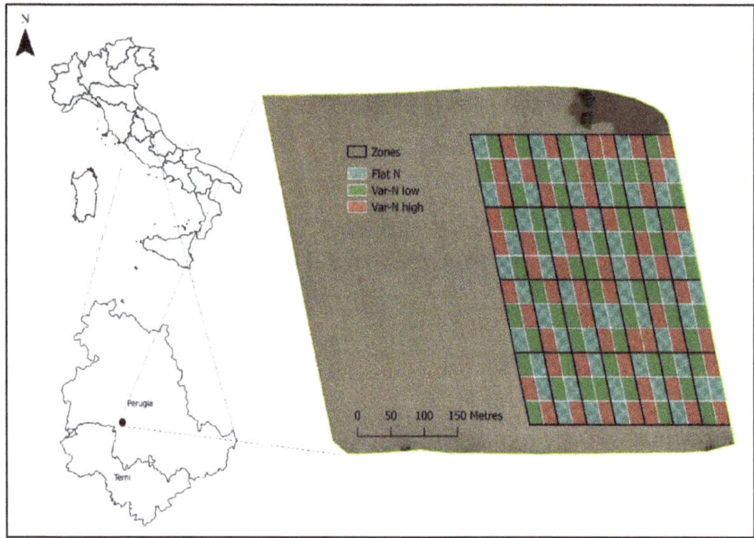

Figure 1. Geographical location of the study area and experimental layout. Flat-N (standard rate of 120 kg N ha^{-1}), Var-N-low (variable rate from 60 to 120 kg N ha^{-1}), Var-N-high (variable rate from 90 to 150 kg N ha^{-1}).

The experimental crop was a rainfed winter wheat (*Triticum aestivum* L., cv. PRR58) sown on 10 November 2017 at a nominal density of 450 viable seeds m^{-2}. The previous crop had been pea (*Pisum sativum* Asch et Gr). The crop was managed according to ordinary practices, while weeds and diseases were controlled chemically. The N rate was split in two application times in order to increase N uptake efficiency, support the formation of yield components and limit N leaching by fall-spring rainfall. The first N application (as urea) occurred on 18 January 2018 with 30 kg N ha^{-1}, while the second N fertilization (as urea), occurred on 26 March 2018 and was managed according to three experimental treatments: 1) a standard rate of 120 kg N ha^{-1} (Flat-N) derived by an N balance approach (the relatively high rate is justified by the very rainy winter); 2) a variable rate of 60 to 120 kg N ha^{-1}, based on NDVI, where the maximum rate was equal to the standard rate (Var-N-low); 3) a variable rate of 90 to 150 kg N ha^{-1}, based on NDVI, where the medium rate was equal to the standard rate (Var-N-high). An inverse linear relationship between NDVI and VRT N-rates was adopted in Var-N-low and Var-N-high on the assumption that NDVI and other correlated VIs (e.g. NIR/Red simple ratio), before the second N fertilization (7–8 Feekes' stage), are directly related to crop N nutritional status [10,13,30]. Thus, the Var-N-high was defined with the aim of keeping the average N-rate equal to the flat N-rate while optimizing the N fertilization according to the supposed N nutritional status. The range adopted in Var-N-high represents a variation of ±25% of the flat N rate, and a ±20% of the total N supplied (including the 30 kg N ha^{-1} of the first N application), which was considered as being appropriate to see appreciable effects on wheat yield. The Var-N-low was defined with the aim of testing a sensible N reduction, the results of which were very important in those areas where a more eco-compatible approach is required. So, the flat N-rate was considered as the maximum N level and the −50% of the flat N rate (−40% of the total rate, equal to 90 Kg/ha including the first N application) was adopted as a lowest reference level based on the evidence from a previous study in the same location [16].

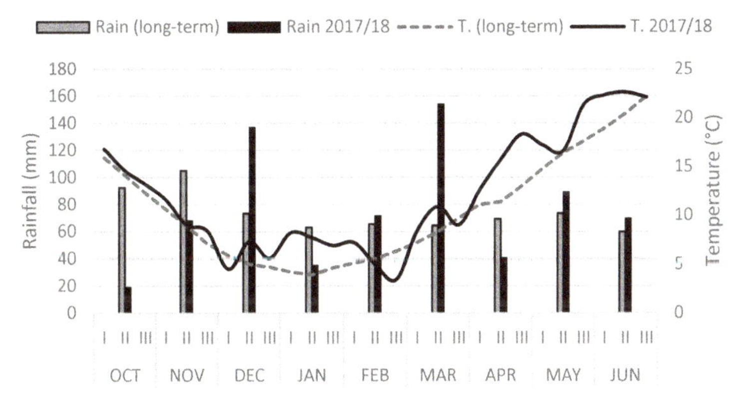

Figure 2. Monthly cumulated rainfall and average temperatures per decade during the wheat growing season 2017–2018 and in the long term (temperature 1951–2018, rainfall 1921–2018).

To account for the possible not measured or unknown factors, the 14 ha study area was divided into 168 plots of about 700 square meters each (35 m long, 21 m wide) grouped in 28 zones. The three treatments were laid down according to a randomized design with 2 replicates per treatment in each zone for a total of 56 replicates per treatment in the whole study area (Figure 1). All the precision on-field operations were performed using a tractor equipped with a Topcon GNSS automatic guide device connected to a Real Time Kinematic (RTK) network. The variable rate treatments were performed using a Sulky 40+ (ECONOV) VRT fertilizer spreader connected through ISOBUS (a widely used software protocol complaint to ISO 11783 standard) to the Topcon system console. The console, after each VRT fertilization, releases the output of each fertilization map showing the estimated distributed rates. This output was compared to the prescription map to investigate the accuracy of the N distribution.

2.2. Calculation of Rates for VRT Treatments

A Level 2A Sentinel-2 image collected on 22 March 2018 (i.e., four days before the second fertilization), georeferenced according to WGS84-UTM32, was downloaded from The Copernicus Open Access Hub. NDVI, calculated from bands 4 and 8 using QGIS Raster calculator (Figure 3a), was used for the N rates calculation of the two variable rate treatments. In both cases, a linear relationship was imposed between the average NDVI value calculated for all experimental units and fertilizer-N rates where the 5° percentile of NDVI value corresponded to the minimum fertilizer-N rates (60 or 90 kg N ha^{-1}) and the 95° percentile of NDVI value corresponded to the maximum fertilizer-N rate (120 or 150 kg N ha^{-1}) (Figure 3b). All the data processing and analysis were carried out using QGIS software, version 2.18.12 64bit [31] and MS Excel 2016. Average NDVI values were calculated using the SAGA grid statistics for the polygons algorithm included in the QGIS processing framework. To simplify this step, pixels overlapping the border of the experimental plots were not excluded from the calculation. To assess the effect of these pixels, average NDVI values were compared with those obtained while considering only non-overlapping pixels.

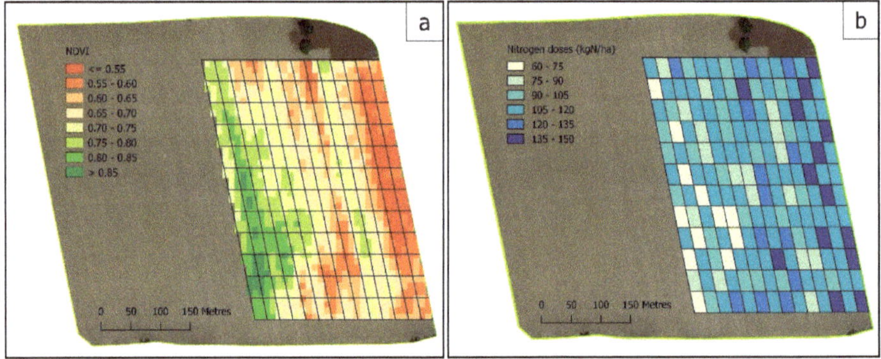

Figure 3. (**a**) NDVI calculated from Level 2A Sentinel-2 image collected on 22 March 2018; (**b**) Prescription N map developed by integrating the three different experimental treatments.

2.3. Determination of NDVI

To monitor and compare the NDVI of experimental treatments, all relevant level 2A Sentinel-2 images with no cloud cover on the study area, were collected from 22 March to the beginning of the senescence. Average NDVI values were calculated for each plot using the SAGA grid statistics for the polygons algorithm included in the QGIS processing framework. Six to nine S2 pixels fell within each experimental unit. To highlight possible differences between the effects of the experimental treatments while considering different crop vigor status before the second N fertilization, the NDVI time series analysis was performed by classifying plots according to three NDVI classes (NDVI ≤ 0.68; 0.68 < NDVI ≤ 0.79; NDVI > 0.79), as revealed by the S2 image from 22 March.

2.4. Determination of Grain Yield and Protein Content

Harvest was carried out on 26 June 2018 using a combined harvester Claas Lexion 780–740 equipped with a Topcon YieldTrakk system (processing data from the optical sensor measuring grain mass flow and moisture sensors), which produced a georeferenced yield map as an ESRI polygon shapefile. The combine harvester had a cutting width of 7.50 m and was equipped with a tilt sensor to correct the effect of slope on the sensor readings. An initial on-field calibration was performed on the combine to adjust for the actual working width and measure the unit weight of grain, which was used by the system to convert the measured mass flow (l/s) to Mg.

The output shapefile containing the georeferenced yield data was intersected with the experimental units to calculate the average yield expressed in Mg ha^{-1} which was used for the subsequent analysis. To produce an easy-to-read yield map, the shapefile was converted to a 5 m resolution raster. Subsequent gaussian kernel smoothing, based on a 50 m search radius, produced the final yield map.

Protein content was measured on 20 June 2018 (i.e., six days before final harvest). To reduce the number of samples, the sampling scheme was defined according to the three classes of NDVI previously defined for a total of 36 field samples, then merged two by two to obtain 18 lab samples, which were analyzed according to the official Kjeldahl method, a widely used chemical procedure for the quantitative determination of protein content in food, feed, feed ingredients and beverages [32].

3. Results

The NDVI values on 22 March show within-field differences mainly associated with the west-east textural gradient (Figure 3a). Average NDVI values including overlapping pixels with plot borders differ only slightly from those obtained while excluding these pixels (0.015 in average absolute value, 0.011 St. Dev.). According to the linear model used for N-rates calculation, these differences correspond only to 2.6 kg/ha (1.8 St. Dev.) for Var-N-low treatment and 2.6 kg/ha (2.1 St. Dev.) for the Var-N-high treatment.

A comparison between the prescription map and the Sulky output (Figure 4a) shows that the correlation between the prescribed rate and the rate actually supplied by the VRT machine within each plot was very good (R^2 = 0.91) (Figure 4b), with only a little bias which resulted in smoothing the extremes, raising the lower rates and lowering the higher ones (Table 1).

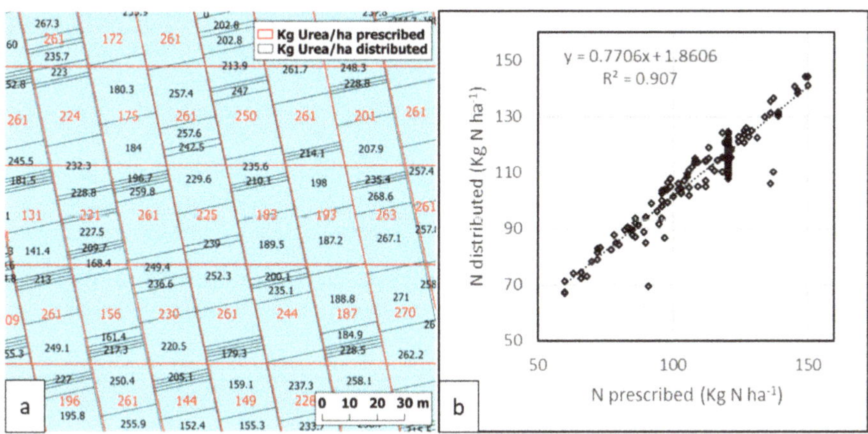

Figure 4. (**a**) Comparison between prescription map and the distributed Sulky output. (**b**) Scatter plot between the prescribed and distributed rate within each plot.

Table 1. Amounts of N prescribed and supplied with the second N application in the three treatments: Flat-N (standard rate of 120 kg N ha^{-1}), Var-N-low (variable rate from 60 to 120 kg N ha^{-1}), Var-N-high (variable rate from 90 to 150 kg N ha^{-1}).

Treatments	Average N prescribed (Kg N ha^{-1})	Average N supply (Kg N ha^{-1})	Average Δ (%)	Δ St. Dev. (Kg N ha^{-1})
Flat-N	120.0	116.5	−1.3%	0.03
Var-N-low	90.2	95.4	5.4%	0.06
Var-N-high	120.5	117.0	−3.0%	0.05

After the second fertilization, the trend of NDVI, considering the three vigor classes previously defined, was not affected by fertilization treatments (Figure 5). In the portion of the field where the NDVI was already high on 22 March (class III—about 0.85), all treatments showed a further moderate increase up to over 0.9, while in the portion of the field where NDVI was low (class I—just over 0.6), all treatments showed an increase up to 0.85 in about one month, and nearly 0.9 later on. As expected, an intermediate trend can be observed for treatments with NDVI falling within class II. Thus, whatever the crop vigor status was on March 22nd, all treatments reached a high NDVI one month after the second N application (Figure 5).

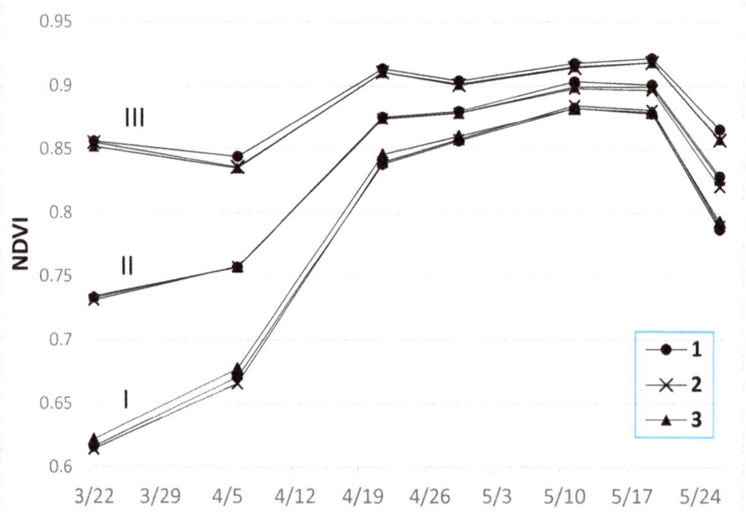

Figure 5. NDVI time series analysis, considering the three vigor classes as identified in the S2 image of March 22nd (I: NDVI ≤ 0.68; II: 0.68 < NDVI ≤ 0.79; III: NDVI > 0.79), for the three N fertilization treatments: 1) Flat-N (standard rate of 120 kg N ha^{-1}); 2) Var-N-low (variable rate from 60 to 120 kg N ha^{-1}); 3) Var-N-high (variable rate from 90 to 150 kg N ha^{-1}).

Grain Yield and Quality

The yield map used for the quantitative analysis is shown in Figure 6a, while the generalized version used only for visual analysis purposes is shown in Figure 6b. The three treatments did not differ significantly for total yield (Table 2) and no relationship was found between the N rate and yield (R^2 = 0.087). Similarly, no correlation was found between N treatments and protein content (R^2 = 0.001). However, the slight difference in protein content observed between Flat-N and Var-N-low (Table 2) was statistically significant (p = 0.02) even though it was not agronomically relevant. Finally, grain yield was weakly correlated to NDVI at any time of NDVI measurements (Table 3).

Table 2. Grain yield and protein content for the three N fertilization treatments: Flat-N (standard rate of 120 kg N ha^{-1}), Var-N-low (variable rate from 60 to 120 kg N ha^{-1}), Var-N-high (variable rate from 90 to 150 kg N ha^{-1}).

Treatment	Average Yield (Mg ha^{-1})	St. Dev. Yield (Mg ha^{-1})	Average Protein Content (%)	St. Dev. Protein Content (%)
Flat-N	6.74	0.36	9.4	0.31
Var-N-low	6.73	0.39	8.9	0.37
Var-N-high	6.76	0.38	9.2	0.48

Table 3. Correlation between yield and NDVI at any time of NDVI measurement plotted over all of the 168 plots.

Date	R^2
22 Mar	0.29
6 Apr	0.26
21 Apr	0.24
29 Apr	0.19
11 May	0.19
10 May	0.18
26 May	0.27
31 May	0.26

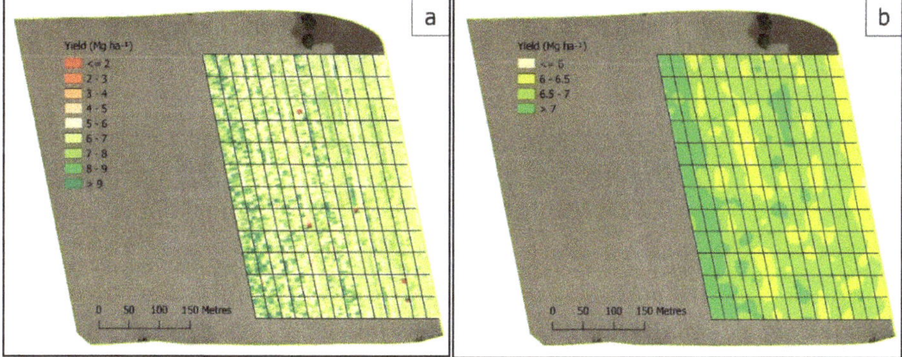

Figure 6. (**a**) yield map used for quantitative analysis; (**b**) generalized version of yield map.

4. Discussion

Our study investigated the differences between a flat N fertilization rate and two variable rates defined through a simplified method based on Sentinel-2 NDVI collected a few days before the second fertilization. One of the two variable rates was aimed at drastically reducing the overall N input (Var-N-low), the other one was aimed at maintaining the same overall input (Var-N-high) as in the flat rate (Flat-N), i.e., 120 kg N ha^{-1} while optimizing the N fertilization according to the supposed N nutritional status. Our results show that a VRT approach with a lower overall N rate may be more efficient, giving same grain yield and quality (Table 2) with a lower N input (Table 1). This result is consistent with the evidence of Raun et al. [33] which reports that the VRT method improves the NUE (Nitrogen Use Efficiency) by 15% compared to a flat rate. In vulnerable contexts, as in the study area, the N-rate reduction results environmentally and economically very relevant since it could reduce water pollution (still a critical issue in Umbria and all over the world) and decrease expenditures on N. However, differently from our case study, a Var-N-low treatment could determine a lower protein grain content. In this regard, the economic trade-off between lower N costs and reduced yield value due to lower grain protein content could be calculated by the farmers using their local costs and prices. Indeed, the treatment Var-N-low differed from the other two treatments by only about 20 kg N ha^{-1} on average, which was supposed to be not so important compared to the total amount available (i.e., about 160 vs 180 kg ha^{-1}), accounting for the residual N left by the previous pea crop (likely around 30 kg N ha^{-1}) and the amount of N supplied with the first mineral N application (30 kg N ha^{-1}). It is worth noting that about 180 kg ha^{-1} of available N was in line with the usual practice adopted by farmers in the Tiber valley of Umbria and adequate, if not limiting, in view of the high yielding potential of this area

(6–8 tons grains ha^{-1}). Nonetheless, the rainy fall-spring of 2017–2018 caused a relevant N loss from the soil, thus, in this case, a difference of 20 kg N ha^{-1} was expected to make a difference. Several reasons can be addressed to explain the lack of difference among treatments. The main one is that crop development was not associated to N availability but to the soil textural gradient, with NDVI values tendentially as lower as higher appeared the sand content in the soil. In such a condition, and considering the rainy season, forcing the N rate in plots with low NDVI (i.e., with higher sand content) resulted in losing N (mainly by leaching) without affecting yield. As reported by Asseng & Turner [34], NUE is mainly influenced by soil and rain. Thus, in our case, the N rate increase did not translate in a NUE increase as this was probably limited by the other factors. This would justify that, in some cases, the N rate should be directly related to the NDVI [35], i.e., increased where NDVI is higher and decreased where NDVI is lower, on the assumption that lower NDVI values could prove that other soil characteristics, besides N availability, are not suitable for allowing high yields. For example, a soil with low water retention is not suitable in case of dry grain filling periods (as it is in Mediterranean areas) due to water shortage and thus lower yields whatever the N availability [36–38]. The similar time course of NDVI observed in all the three treatments, distinguishing three different vigor status before the second fertilization (Figure 5), also proves that N was not the limiting factor, or at least, the 20 kg N ha^{-1} of difference among treatments was not relevant for crop NDVI.

The weak correlation between grain yield and NDVIs at any date of monitoring (R^2 always below 0.3) appears in contrast with the evidence of Sultana et al. [12] and Liu et al. [39] which show high correlation between NDVI and yield, especially for NDVI recorded during the milky-grain vegetative stage. However, the NDVI recorded in our experiment from 21 April onwards was quite high for all treatments, whereas it is known that the correlation with yield stands only when a wide range of NDVI is considered [16].

As discussed above, the textural gradient likely affected yield through N and water availability, more than the N fertilization treatments. This assumption is supported by the case study of Basso et al. [19] who observed a high correlation between yield and pedological conditions, in particular soil water capacity. Also, Zhao et al. [40] reported that fertilization, as well as the availability of water, significantly affected the content of wheat proteins, even though the latter was more influent, especially for those cultivars characterized by an intermediate protein content. Probably, further splitting the rate, with a third application at the end of shooting, would have prevented some N loss and increased the grain protein content [41], which was quite low compared to the standard of our cultivar. Again, the low protein content can be ascribed to the rainy season, as it was widespread for the harvest of 2018 in Umbria and all-over Central Italy.

Concerning the prescription map used in the study, including the overlapping pixels with plot borders in average NDVI calculation simplifies the GIS procedures without generating agronomically meaningful N-rates differences. Because of the high variability within field zones and among plots, the map could be considered as a kind of stress test for the VRT device. The very little differences between prescribed and distributed rates (Table 2) suggest that this technology is suitable for precision fertilization even with highly detailed prescription maps. The highlighted differences are clearly due to small plot size and random allotment of treatments. As a result, the VRT machine was subjected to abrupt rate variations due to the short time available to adapt the opening of the distribution valve.

The simple linear approach to calculate a N prescription map proposed in this research, even though requiring average GIS skills could be effective for VRT adoption by non-specialist farmers. Coherently with other similar experiences [22–24], to improve the overall usability of the method by non-specialists in PA and allow further validations and tuning, all the RS data management and GIS calculation could be implemented in an user-friendly web-GIS application where the user could upload (or digitize on aerial data) his own fields, choose the S2 NDVI reference date, and decide the most suitable approach for defining the N rate. NDVI S2 data could be conveniently accessed through the Sentinel Hub platform API [42].

Concerning the integration of NDVI from Sentinel 2, VRT fertilization, and yield mapping in PA, our experience confirms that these technologies, thanks also to both the web-based applications for calculating satellite-based prescription maps [22–24] and the user-friendly interfaces of system consoles, could be ordinarily used even in small or medium-sized farms. However, a preliminary cost-benefit analysis and a start-up training and support by a PA expert is often necessary.

5. Conclusions

Our results regarding crop vegetation index, grain yield, and protein content indicate that the adoption of a low-N management approach, based on simple linear models and VRT, may considerably reduce the economic and environmental impact of nitrogen fertilization in winter wheat. However, the rainfed nature of winter wheat in Mediterranean environments may cause unpredictable yield and quality variations depending on climatic trends and soil properties due to their effects on both soil nitrogen and water availability. In this view, VRT nitrogen fertilization can only partially mitigate the heterogeneity of production determined by such environmental factors. The alternative approach of providing a nitrogen supply proportional to the crop NDVI deserves to be considered when factors other than N fertilizer rate come into play, as it is with sandy soils where NDVI and yield may be limited by low N and water retention. Despite the known limitations of predictions and prescriptions based on remoted sensed vegetation indices, their use provides relevant information about within-field and between-fields variability. This information can support the implementation of crop fertilization management approaches based on GNSS/RTK and VRT technologies to replace the traditional flat-N rate, which provides results that are neither economically nor environmentally sustainable.

Author Contributions: All the Authors conceived the research and designed the methodology; M.V. and F.S. applied the methodology and analyzed the data; P.B. supervised the research; all the Authors wrote the paper.

Funding: This research was developed within the framework of the project "RTK 2.0—Prototipizzazione di una rete RTK e di applicazioni tecnologiche innovative per l'automazione dei processi colturali e la gestione delle informazioni per l'agricoltura di precisione"—RDP 2014–2020 of Umbria—Meas. 16.1—App. 84250020256.

Acknowledgments: The authors wish to thank Mauro Brunetti of the Foundation for Agricultural Education of Perugia for his very helpful and valuable support.

Conflicts of Interest: The authors declare no conflict of interest.

References

1. Liaghat, S.; Balasundram, S.K. A review: The role of remote sensing in precision agriculture. *Am. J. Agric. Biol. Sci.* **2010**, *5*, 50–55. [CrossRef]
2. Oza, S.R.; Panigrahy, S.; Parihar, J.S. Concurrent use of active and passive microwave remote sensing data for monitoring of rice crop. *Int. J. Appl. Earth Obs. Geoinf.* **2008**, *10*, 296–304. [CrossRef]
3. Zheng, Q.; Huang, W.; Cui, X.; Shi, Y.; Liu, L. New Spectral Index for Detecting Wheat Yellow Rust Using Sentinel-2 Multispectral Imagery. *Sensors* **2018**, *18*, 868. [CrossRef]
4. Copernicus Open Access Hub. Available online: https://scihub.copernicus.eu/ (accessed on 4 April 2019).
5. Silleos, N.G.; Alexandridis, T.K.; Gitas, I.Z.; Perakis, K. Vegetation indices: Advances made in biomass estimation and vegetation monitoring in the last 30 years. *Geocarto Int.* **2006**, *21*, 21–28. [CrossRef]
6. Xue, J.; Su, B. Significant remote sensing vegetation indices: A review of developments and applications. *J. Sens.* **2017**, *2017*. [CrossRef]
7. Mulla, D.J. Twenty five years of remote sensing in precision agriculture: Key advances and remaining knowledge gaps. *Biosyst. Eng.* **2013**, *4*, 358–371. [CrossRef]
8. Rouse, J.W.; Hass, R.H.; Schell, J.A.; Deering, D.W. Monitoring vegetation systems in the great plains with ERTS. *Third ERTS Symp. NASA* **1973**, *1*, 309–317.
9. Muñoz-Huerta, R.F.; Guevara-Gonzalez, R.G.; Contreras-Medina, L.M.; Torres-Pacheco, I.; Prado-Olivarez, J.; Ocampo-Velazquez, R.V. A review of methods for sensing the nitrogen status in plants: advantages, disadvantages and recent advances. *Sensors* **2013**, *13*, 10823–10843. [CrossRef]

10. Cabrera-Bosquet, L.; Molero, G.; Stellacci, A.; Bort, J.; Nogués, S.; Araus, J. NDVI as a potential tool for predicting biomass, plant nitrogen content and growth in wheat genotypes subjected to different water and nitrogen conditions. *Cereal Res. Commun.* **2011**, *39*, 147–159. [CrossRef]

11. Carlson, T.N.; Ripley, D.A. On the relation between NDVI, fractional vegetation cover, and leaf area index. *Remote Sens. Environ.* **1997**, *62*, 241–252. [CrossRef]

12. Sultana, S.R.; Ali, A.; Ahmad, A.; Mubeen, M.; Zia-Ul-Haq, M.; Ahmad, S.; Ercisli, S.; Jaafar, H.Z.E. Normalized Difference Vegetation Index as a Tool for Wheat Yield Estimation: A Case Study from Faisalabad, Pakistan. *Sci. World J.* **2014**, *2014*, 1–8. [CrossRef]

13. Zhu, Y.; Yao, X.; Tian, Y.C.; Liu, X.J.; Cao, W.X. Analysis of common canopy vegetation indices for indicating leaf nitrogen accumulations in wheat and rice. *Int. J. Appl. Earth Obs. Geoinf.* **2008**, *10*, 1–10. [CrossRef]

14. Cao, Q.; Miao, Y.; Feng, G.; Gao, X.; Li, F.; Liu, B.; Yue, S.; Cheng, S.; Ustin, S.L.; Khosla, R. Active canopy sensing of winter wheat nitrogen status: An evaluation of two sensor systems. *Comput. Electron. Agric.* **2015**, *112*, 54–67. [CrossRef]

15. Vizzari, M.; Modica, G. Environmental effectiveness of swine sewage management: a multicriteria AHP-based model for a reliable quick assessment. *Environ. Manage.* **2013**, *52*, 1023–1039. [CrossRef]

16. Benincasa, P.; Antognelli, S.; Brunetti, L.; Fabbri, C.A.; Natale, A.; Sartoretti, V.; Modeo, G.; Guiducci, M.; Tei, F.; Vizzari, M. Reliability of NDVI derived by high resolution satellite and UAV compared to in-field methods for the evaluation of early crop N status and grain yield in wheat. *Exp. Agric.* **2017**, *54*, 1–19. [CrossRef]

17. Spiertz, J.H.J. Nitrogen, sustainable agriculture and food security. A review. In *Sustainable Agriculture*; Springer: Dordrecht, The Netherlands, 2009.

18. Bourdin, F.; Morell, F.J.; Combemale, D.; Clastre, P.; Guérif, M.; Chanzy, A. A tool based on remotely sensed LAI, yield maps and a crop model to recommend variable rate nitrogen fertilization for wheat. *Adv. Anim. Biosci.* **2017**, *8*, 672–677. [CrossRef]

19. Basso, B.; Ritchie, J.T.; Cammarano, D.; Sartori, L. A strategic and tactical management approach to select optimal N fertilizer rates for wheat in a spatially variable field. *Eur. J. Agron.* **2011**, *35*, 215–222. [CrossRef]

20. Diacono, M.; Rubino, P.; Montemurro, F. Precision nitrogen management of wheat. A review. *Agron. Sustain. Dev.* **2013**, *33*, 219–241. [CrossRef]

21. Song, X.; Wang, J.; Huang, W.; Liu, L.; Yan, G.; Pu, R. The delineation of agricultural management zones with high resolution remotely sensed data. *Precis. Agricul.* **2009**, *10*, 471–487. [CrossRef]

22. CropSAT. Available online: https://cropsat.com (accessed on 4 April 2019).

23. Agrosat. Available online: https://www.agrosat.it (accessed on 4 April 2019).

24. OneSoil. Available online: https://onesoil.ai (accessed on 4 April 2019).

25. Nawar, S.; Corstanje, R.; Halcro, G.; Mulla, D.; Mouazen, A.M. Delineation of Soil Management Zones for Variable-Rate Fertilization: A Review. *Adv. Agron.* **2017**, *143*, 175–245.

26. Zhang, N.; Wang, M.; Wang, N. Precision agriculture—A worldwide overview. In Proceedings of the Computers and Electronics in Agriculture, Chicago, IL, USA, 1–3 May 2002; Volume 36, pp. 113–132.

27. Fulton, J.P.; Shearer, S.A.; Higgins, S.F.; Hancock, D.W.; Stombaugh, T.S. Distribution pattern variability of granular VRT applicators. *Trans. ASAE* **2013**, *48*, 2053–2064. [CrossRef]

28. Ross, K.W.; Morris, D.K.; Johannsen, C.J. A review of intra-field yield estimation from yield monitor data. *Appl. Eng. Agric.* **2008**, *24*, 309–317. [CrossRef]

29. Arslan, S.; Colvin, T.S. Grain yield mapping: Yield sensing, yield reconstruction, and errors. *Precis. Agric.* **2002**, *3*, 135–154. [CrossRef]

30. Vian, A.L.; Bredemeier, C.; Turra, M.A.; Giordano, C.P.d.S.; Fochesatto, E.; Silva, J.A.d.; Drum, M.A. Nitrogen management in wheat based on the normalized difference vegetation index (NDVI). *Ciência Rural* **2018**, *48*, 1–9. [CrossRef]

31. *QGIS Software, Version 2.18.12*; Quantum GIS Development Team: Berne, Switzerland, 2017.

32. *A guide to Kjeldahl Nitrogen Determination Methods and Apparatus*; ExpotechUSA: Houston, TX, USA, 1998; ISBN 9788578110796.

33. Raun, W.R.; Solie, J.B.; Johnson, G.V.; Stone, M.L.; Mullen, R.W.; Freeman, K.W.; Thomason, W.E.; Lukina, E.V. Improving nitrogen use efficiency in cereal grain production with optical sensing and variable rate application. *Agron. J.* **2002**, *94*, 815–820. [CrossRef]

34. Asseng, S.; Turner, N.C. Analysis of water- and nitrogen-use efficiency of wheat in a Mediterranean climate. *Plant Soil* **2001**, *233*, 127–143. [CrossRef]
35. Raun, W.R.; Solie, J.B.; Stone, M.L.; Martin, K.L.; Freeman, K.W.; Mullen, R.W.; Zhang, H.; Schepers, J.S.; Johnson, G.V. Optical sensor-based algorithm for crop nitrogen fertilization. *Commun. Soil Sci. Plant Anal.* **2005**, *36*, 2759–2781. [CrossRef]
36. Diacono, M.; Castrignanò, A.; Troccoli, A.; De Benedetto, D.; Basso, B.; Rubino, P. Spatial and temporal variability of wheat grain yield and quality in a Mediterranean environment: A multivariate geostatistical approach. *F. Crop. Res.* **2012**, *131*, 49–62. [CrossRef]
37. Bonciarelli, U.; Onofri, A.; Benincasa, P.; Farneselli, M.; Guiducci, M.; Pannacci, E.; Tosti, G.; Tei, F. Long-term evaluation of productivity, stability and sustainability for cropping systems in Mediterranean rainfed conditions. *Eur. J. Agron.* **2016**, *77*, 146–155. [CrossRef]
38. Miralles, D.J.; Slafer, G.A. Paper Presented at International Workshop on Increasing Wheat Yield Potential, CIMMYT, Obregon, Mexico, 20–24 March 2006. Sink limitations to yield in wheat: How could it be reduced? *J. Agric. Sci.* **2007**, *145*, 139–149. [CrossRef]
39. Liu, L.; Wang, J.; Bao, Y.; Huang, W.; Ma, Z.; Zhao, C. Predicting winter wheat condition, grain yield and protein content using multi-temporal EnviSat-ASAR and Landsat TM satellite images. *Int. J. Remote Sens.* **2006**, *27*, 737–753. [CrossRef]
40. Zhao, C.; Liu, L.; Wang, J.; Huang, W.; Song, X.; Li, C. Predicting grain protein content of winter wheat using remote sensing data based on nitrogen status and water stress. *Int. J. Appl. Earth Obs. Geoinf.* **2005**, *7*, 1–9. [CrossRef]
41. López-Bellido, L.; López-Bellido, R.J.; Redondo, R. Nitrogen efficiency in wheat under rainfed Mediterranean conditions as affected by split nitrogen application. *For. Crop. Res.* **2005**, *94*, 86–97. [CrossRef]
42. Sentinel Hub. Available online: https://www.sentinel-hub.com (accessed on 15 April 2019).

Article

Assimilation of Sentinel-2 Leaf Area Index Data into a Physically-Based Crop Growth Model for Yield Estimation

Francesco Novelli [1], Heide Spiegel [2], Taru Sandén [2] and Francesco Vuolo [1,*]

[1] Institute of Surveying, Remote Sensing & Land Information (IVFL), University of Natural Resources and Life Sciences, Vienna (BOKU), Peter Jordan Str. 82, 1190 Vienna, Austria; francesco.novelli@boku.ac.at

[2] Department for Soil Health and Plant Nutrition, Austrian Agency for Health and Food Safety (AGES), Spargelfeldstrasse 191, 1220 Vienna, Austria; adelheid.spiegel@ages.at (H.S.); taru.sanden@ages.at (T.S.)

* Correspondence: francesco.vuolo@boku.ac.at; Tel.: +43-1-47654-85735

Received: 18 April 2019; Accepted: 16 May 2019; Published: 21 May 2019

Abstract: Remote sensing data, crop growth models, and optimization routines constitute a toolset that can be used together to map crop yield over large areas when access to field data is limited. In this study, Leaf Area Index (LAI) data from the Copernicus Sentinel-2 satellite were combined with the Environmental Policy Integrated Climate (EPIC) model to estimate crop yield using a re-calibration data assimilation approach. The experiment was implemented for a winter wheat crop during two growing seasons (2016 and 2017) under four different fertilization management strategies. A number of field measurements were conducted spanning from LAI to biomass and crop yields. LAI showed a good correlation between the Sentinel-2 estimates and the ground measurements using non-destructive method. A correlating fit between satellite LAI curves and EPIC modelled LAI curves was also observed. The assimilation of LAI in EPIC provided an improvement in yield estimation in both years even though in 2017 strong underestimations were observed. The diverging results obtained in the two years indicated that the assimilation framework has to be tested under different environmental conditions before being applied on a larger scale with limited field data.

Keywords: crop growth model; data assimilation; Leaf Area Index; Sentinel-2; EPIC model; yield estimation

---

1. Introduction

Agriculture is facing great challenges to increase food production while decreasing global impact [1]. In this context, better crop management strategies can help to guarantee sustainable food production and security [2]. It is of paramount importance to set up accurate and timely monitoring tools that ensure spatially detailed and accurate information on crop yields and on the required production inputs (e.g., water, fertilizer). Remotely-sensed data can provide spatially and temporally consistent information over large agricultural areas [3] to monitor the seasonal patterns related to the biological lifecycle of the crop [4]. Through the use of some variables estimated by remotely-sensed data, it is possible to obtain crop yield predictions. Thus, different estimation approaches are described, based on remote sensing data. Statistical regression methods are the most commonly adopted for quantifying the expected yield [5]. For instance, the Normalized Difference Vegetation Index (NDVI) is commonly used as an independent variable in regression models to estimate the crop yield. The main limitations of this technique is the need for sufficiently long and consistent time series of both remote sensing data and agricultural statistics to calibrate the empirical models [6]. In order to improve the predictive power of remotely-sensed data, it is possible to add independent meteorological (or bio-climatic) variables into the regression models as shown in several studies [7,8]. The interaction of these two different sources of data provides different types of information even if the bio-climatic indicators, especially if derived from satellites as well, are not totally independent from the vegetation indices [6]. An alternative approach to obtain crop yield is to consider the crop physiological processes using semi-empirical approaches such as the Monteith's efficiency model [9]. The latter computes the daily production of dry matter (DM) integrating incident global solar radiation, usually available from the meteorological stations, and the diagnostic estimation of the status and the activity of the plant provided by remote sensing data [10]. The approach uses only the amount of photosynthetically active solar radiation (PAR) adsorbed by the plant, although it is known that the crop yield depends also on the interaction of other variables, such as fertilizer, soil, cultivar diversity, weather conditions, water availability, pests, and diseases.

A more sophisticated approach to account for the interaction of environmental factors and field management strategies is based on dynamic crop growth models (CGMs) [11]. The main advantage of CGMs is their capacity to take into account such limiting factors (e.g., soil, weather, water, nitrogen) dynamically [12]. CGMs are able to describe the behavior of the real crop by predicting the physiological mechanisms of the crop growth (e.g., phenological development, photosynthesis, dry matter, portioning, and organogenesis) using mechanistic equations [13]. Their applicability is, however, complex at field or regional scales because they are designed to describe the plant growth at the point scale, assuming that the field conditions are uniform [14]. In this context, the use of satellite data to provide spatially integrated information over large areas is widely recognized to support CGMs [11,12,15,16]. The crop phenology, which can be assimilated into the crop growth models, becomes fundamental to control the crop dry matter distribution during the growth process inside the model itself [17].

The integration of remote sensing data and CGM can be achieved in different ways [18,19]:

(1) Forcing strategy, where the remote sensing data are used to replace the crop model simulation data. For each time step of the model simulation (e.g., daily), direct ingestion of the variables into the model is done. It requires continuous data and it is subject to frequent instabilities in the overall model setting.

(2) Updating strategy, which consists of continuously updating the model state variables whenever new observations from the satellite are available. This method assumes that better-simulated state variables at day t will also improve the accuracy of the simulated state variables on succeeding days.

(3) Calibration method, which considers adjusting either the model or initial state variables to obtain an optimal agreement between simulated and observed state variables. Key model's state variables are changed according to the driving variable obtained from the satellite. Generally, a minimization algorithm is used to minimize the difference between the model state variable and the driving variable obtained from the satellite [20,21].

Various implementations exist in the literature. For example, the commonly used EPIC model (Environmental Policy Integrated Climate) [14] shows good accuracy in the estimation of yield by assimilating variables derived from remote sensing data with a calibration approach [22]. The model is suitable for simulating crop yield over a large range of soils, crops, and management scenarios [23]. Satellite-based Leaf Area Index (LAI) is one of the most commonly used variables in model assimilation for yield estimation [21,24,25]. LAI is one of the key crop state biophysical variables required by many CGMs and is able to describe the relationship between soil, plant, and atmosphere system [26]. LAI can be obtained using freely available data from operational satellites such the USGS Landsat-8 or the Copernicus Sentinel-2 missions. The latter is composed of two identical satellites (Sentinel-2 A and B) providing 5 days of revisit time [27] and this increases the possibility to have cloud-free observations in key phenological periods during the crop growing cycle. Sentinel-2 mission provides improved possibilities for retrieving biophysical parameters such as LAI, thanks to the higher spectral, spatial, and temporal resolutions [28]. Recent studies show good correlation between LAI ground measurements and LAI estimates starting from Sentinel-2 reflectance data and using machine learning approaches [29,30]. Pre-calculated LAI maps with an accuracy of around 12% at 10 m spatial resolution can be obtained from the service platform for Sentinel-2 implemented by the University of Natural Resources and Life Science, BOKU [29].

Building on the BOKU Sentinel-2 LAI data, in this study we apply a data assimilation framework using the EPIC model over a winter wheat cereal crop for two growing seasons (2016 and 2017). A re-calibration of the model parameters was done using a minimization function in order to find the minimum difference between the remote sensing LAI data and the simulated LAI data. The main goal is the evaluation of (1) the data assimilation approach and (2) of the capability of the model to estimate crop yield for possible future application at regional scale. For this reason, the yields estimated by EPIC were validated using measured yield data. The validation of the LAI values obtained from Sentinel-2 was also done through the comparison with LAI ground measurements collected during both the growing seasons.

2. Materials and Methods

2.1. Study Area

The winter wheat experimental fields used in this study are located in the Marchfeld region, an agricultural area in the east of Vienna in Lower Austria (Lat. 48.20° N, Long. 16.72° E) (Figure 1). This area was designated as a pilot area for the H2020 FATIMA (Farming Tools for external nutrient Inputs and Water Management) project [31] and field data were collected in this framework. The Marchfeld region covers about 75,000 ha and the most important crop types are: winter wheat, vegetables (e.g., carrots, onions, green peas, asparagus, spinach, green salad), sugar beet, grain maize, and potatoes [31]. The climate is semi-arid with cold winters characterized by frequently strong frosts and limited snow cover, and hot and intermittently dry summers [32]. The average annual temperature is between 9 and 10 °C and the annual precipitation ranges from 500–550 mm with a maximum in early summer, from May to July. During the vegetation period from April to September, the precipitation is between 200 and 440 mm [32]. The presence of wind, blowing at an average speed of 3.5 m/s, encourages the dry conditions in the summer and, additionally, leads to high evapotranspirative demand. Different types of soils with high spatial variability can be found [33] and the general soil conditions are characterized by a humus-rich A horizon and a sandy C horizon [34].

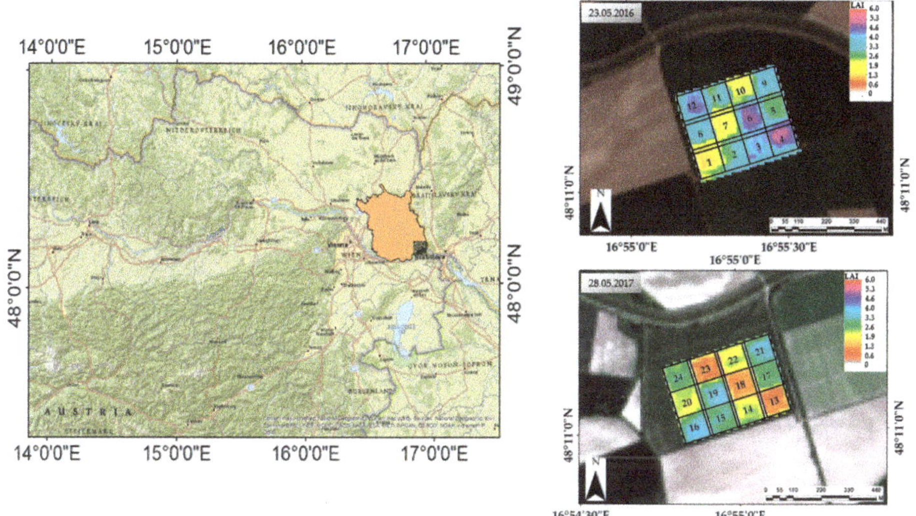

Figure 1. Study area of Marchfeld (left), Lower Austria, and the experimental plots in 2016 (top right) and 2017 (bottom right). It is possible to notice the variability in LAI values as a result of different fertilization rates given during the growing cycle.

2.2. Experimental Plots

Field measurements were conducted in 2016 and 2017 in two experimental fields of winter wheat (cultivar Capo) with total field area of 12 hectares (ha), divided in 12 plots, each of 1 ha (100 m × 100 m). Each plot received different fertilization rates with 3 repetitions of four different levels of fertilization (variant) in a range between 0 and 180 kg N/ha (0, 60, 120, 180 kg N/ha) (Figure 1).

The soil characteristics were measured before, during and after the crop growing cycle for each of the 12 plots in both years, the information was provided by the Austrian Agency for Health and Food Safety (AGES) together with the soil management practice (including date of sowing and harvest). The soil texture in the first 30 cm is silty-clay-loam or loam with small difference amongst the plots. For the experimental plots in the year 2016, the range of sand particles was between 10.9% and 33.7% with a median value of 19.65%, while presence of silt was between 40.5% and 57.4% with a median value of 48.65%. In 2017, the presence of sand was between 15.4% and 28.9% with a median value of 22.8%, while the presence of silt was between 43.9% and 50.9% with a median value of 46.3%. In both the experimental fields, the soil was slightly alkaline, indeed the pH in the first 30 cm was around 7.5–7.6 [35]. The before seeding management was characterized by a disk harrow soil processing at twelve centimeters depth and a seedbed preparation at 5 cm in depth. After sowing, the mineral fertilization was applied using calcium ammonium nitrate (N = 27%) at three times (March, April, and May). The same procedure was applied in both years. At harvest, the final yield of dry matter in kg/ha was measured and provided by AGES.

2.3. Leaf Area Index Measurements

Ground measurements of LAI were undertaken once a week during both years alternatively one week in the experimental plots and the following week in other randomly selected fields in the region. The 2016 field data collection began in the middle of April and finished in the middle of May, while for 2017 the data collection started at the end of April and ended in the middle of June.

In situ LAI measurements were collected with a non-destructive and indirect method using a Li-Cor LAI-2200C Plant Canopy Analyzer (Li-Cor Biosciences, Lincoln, NB, USA) [36]. Li-Cor

Biosciences is a privately held company based in Lincoln, Nebraska that was founded in 1971 and it is a leader in the design, production, and marketing instruments for biological and environmental research [37]. The LAI-2200C has been largely used for LAI measurement; it estimates LAI from the values of canopy transmittance by identifying the attenuation of the radiation as it passes through the canopy [36]. Therefore, measurements were taken above- (A) and below-canopy (B). The measures were done in elementary sampling units (ESU) within a radius of about 10 m of a geo-referenced point (accuracy ± 3–5 m using GPS signals); each ESU represents a homogenous area (in terms of crop development conditions). In the experimental fields, the ESUs were located in the center of each of the 12 plots. In the random fields that were additionally sampled, the centers of the ESU were placed in the corner of a square area of about 60 × 60 m, measured from an accessible border of the field. This ensures a representative sampling for the whole field and makes sure that the point also represents the Sentinel-2 pixel at 10 m and 20 m ground sampling distances. In both cases, in order to obtain a representative LAI sample within a radius of about 10 m, each ESU was ideally divided into four sectors; for each sector one measurement was done above the canopy and six below it. The latter were performed at random in each of the four sectors to well represent the ESU. A total of 24 readings were recorded for each ESU and were used to calculate the average representative LAI value used for comparison with the satellite data (Figure 2).

The LAI sampling was always carried out early in the morning or later in the afternoon under diffuse light conditions. Previous studies showed that the best results for LAI were measured in diffuse light conditions to avoid direct sunlight and, therefore, minimizing the underestimation of LAI [38,39].

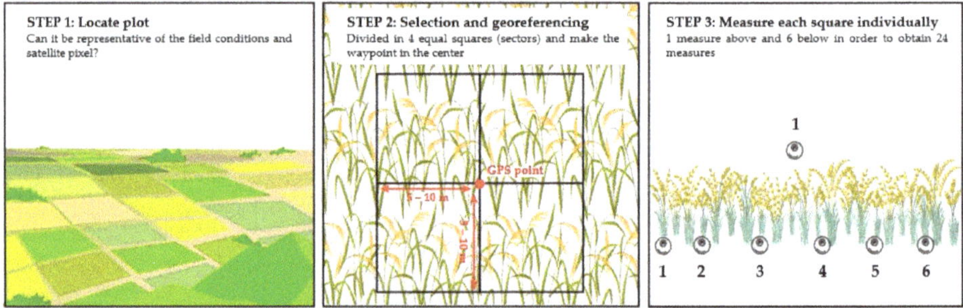

Figure 2. Workflow for collecting in situ LAI measurements using the Li-Cor LAI 2200C Plant Canopy Analyzer. Step 1 is the selection of the field; in Step 2 the GPS point, representative of the center of the ESU, is taken; and in Step 3 the measurements are performed obtaining an LAI value as a mean value of the 24 measurements below the canopy (six for each square).

2.4. Satellite Data

Leaf Area Index

Satellite-based estimates of LAI were obtained using an artificial neural network (ANN) developed at INRA. An ANN consist of a large number of simple processors, called "neurons", linked by weighted connections where each output depends only on the information that is locally available [40]. The ANN is a "black box" approach which has great capacity in predictive modelling but all the characters describing the unknown situation must be presented to the trained ANN and the identification (prediction) is then given [41]. The network was trained using PROSPECT and SAIL radiative transfer models and it uses eight Sentinel-2 spectral bands (B3, B4, B5, B6, B7, B8A, B11, and B12) [42]. On the basis of this approach, LAI was calculated and available as LAI maps at 10 m resolution at the BOKU data service platform for Sentinel-2 [29]. The LAI image data used for the 2016 and 2017 are listed in Table 1. In both years the maximum timespan between satellite acquisitions and ground measurements of LAI was five days. To guarantee that the ground reference points were not covered by clouds or shadows in the image data, the scene classification layer (SCL) was used [43]. For the validation of the LAI retrieval, LAI values were extracted for the central points of each LAI ESU measurements with a buffer of 10 m (an equal expansion in each direction from our GPS point) and used for the comparison with the field data as a mean value inside of the buffer for each ESU. This was done to compensate for the location of the center points, as some did not cover exactly one pixel or were located on a pixel border.

Table 1. List of Sentinel-2 (S2) acquisitions and dates of LAI ground measurements (three in 2016 and four in 2017) used for the LAI validation. The cloud cover indicates the percentage of the land surface covered by clouds in the image data.

	S2 Acquisition Date	Cloud Cover	LAI Ground Measurements Date
2016	April 13	0.0%	April 13–18 [1]
	April 26	18.4%	April 25
	May 06	3.8%	May 02–09
2017	April 21	10.3%	April 25
	May 18	51.9%	May 18
	May 28	0.0%	May 23–29
	June 20	0.0%	June 19

[1] Considering the timespan is possible to use LAI ground measurements from both the days.

For the assimilation of LAI in the EPIC model, LAI values were extracted from the image data using a crop mask of the fields of interest (divided in different plots). The latter involved restricting the analysis to a subset of pixels, rather than using all the pixel of the image [4]. For each polygon representing each plot, considered as an independent field, an inner buffer of 20 m (equal reduction in every direction of the space of our spatial analysis) was created to extract the median value of LAI. The inner buffer is useful in this case to exclude possible outliers in the boundaries of the field. This operation was carried out for each LAI image where the fields of interest were cloud-free and available in the Sentinel-2 data service platform. The date of acquisition and cloud cover of the image data are listed in Table 2.

Table 2. List of Sentinel-2 (S2) LAI acquisitions and cloud cover during both the growing seasons of winter wheat in the area of interest (nine in 2016 and 12 in 2017).

	S2 Acquisition Date	Cloud Cover		S2 Acquisition Date	Cloud Cover
2016	March 14	1.3%	2017	March 29	11.7%
	March 17	3.7%		April 1	0.0%
	April 13	0.0%		April 21	10.3%
	May 6	3.8%		May 11	79.3%
	May 23	35.6%		May 18	52%
	June 2	31.8%		May 28	0.0%
	June 22	4.9%		June 10	43%
	July 2	0.0%		June 20	0.0%
	July 12	4.0%		June 30	29.6%
				July 5	7.7%
				July 12	14.8%
				July 17	12%

2.5. EPIC Input and Forcing

2.5.1. EPIC Model

EPIC is a physically-based crop growth model that operates on a daily time-step. One of the most important processes simulated by EPIC is the crop growth, and a wide range of crop types can be simulated using a generic growth routine. The crop growth process includes the interception of solar radiation, conversion of intercepted light to the biomass, partition of biomass into roots, above ground biomass and economic yield, and also simulated root growth [44]. The annual crop growth from planting to harvest depends on the potential heat unit for the crop. The growing period starts when the average daily temperature exceeds the base temperature of the plant and the phenological development of the crop is based on the daily heat unit accumulation (growing degree day approach) [44]. The heat units are used to calculate the heat unit index (ranging between 0 and 1) as a fraction of the potential heat units (PHU) required for the maturation of the crop under study [45].

The potential daily increase of the biomass is estimated using Monteith's approach [9] as a function of intercepted photosynthetic active radiation. The interception of solar radiation depends on the LAI [23], which is described by five model parameters: DMLA, DLAP1, DLAP2, DLAI, and RLAD. The model LAI curve is built considering two simulation stages: from emergence to leaf decline, LAI depends on three input factors DLAP1 and DLAP2, which are two points used for the definition of the optimal leaf development curve and DMLA, namely the maximum potential LAI. The second simulation stage starts from the leaf decline (which depends on DLAI parameter, the fraction of the growing season when leaf area starts to decline) to the end of the growing season. The rate of the LAI is influenced by LAI decline parameter (RLAD). Both simulation stages are also influenced by the heat units and, theoretically, by the estimated stresses caused by water, nutrients, temperature, aeration, and radiation. The potential biomass is also adjusted daily, taking into consideration the minimum of the EPIC plant growth constraints (nutrients, water, temperature, aeration, and radiation) [23]. The crop yield is estimated using the harvest index (HI) approach as a nonlinear function of heat units from zero (at the planting stage) to the optimal value (at maturity) [44]. The HI could be altered by high temperature, low solar radiation, or water stress during critical crop stages [23]. In particular, under water stress conditions, the default parameterization of the model would adjust the HI from the second half of the growing season considering a lower limit of HI. The latter is called water stress yield factor (WSYF) [46], a fraction between 0 and the HI value that represents the lowest HI expected due to the water stress [45]. The water stress effect on HI is controlled by PARM (3), namely the fraction of the growing season where water stress starts reducing HI, which also controls the accumulation of actual and potential evapotranspiration until harvest. At harvest, the ratio between potential and actual evapotranspiration (ET) is used with an S-shaped curve to estimate the final HI [45].

2.5.2. EPIC Input Data

EPIC is designed to simulate field, farm or small watershed, homogenous with respect to climate, soil, land use, and topographic management system parameters [45]. In this study each plot was considered as an independent field. The model enables the preparation of different parameters, operations, and run files to execute the model. The soil characteristics, such as soil hydrologic group, number of the soil layers and layer depth, soil texture, pH, organic carbon concentration, calcium carbonate content of soil, the sum of bases, cation exchange capacity, and bulk density were used to build the soil parameters file. In the latter file, three soil layers were considered with 30 cm of layer's depth. The soil hydraulic parameters were also calculated by default empirical equations provided by EPIC. These equations give as a result the wilting point, field capacity, saturated soil water content, and saturated hydraulic conductivity based on measured values of bulk density, sand content, and silt content [47]. Site files were prepared for every plot while the operation schedule files were prepared for each variant. The climate data (solar radiation, maximum and minimum temperature, precipitation, relative humidity, and wind velocity) were taken from the ZAMG (Zentralanstalt für Meteorologie und Geodynamik) weather station of Gross-Enzersdorf (48°19′ N 16°55′ E) located 30 km away from the experimental fields. Weather files were built using the monthly weather statistic generator WXPM p3020 available from EPIC model website [48], using historical data from 1997.

2.5.3. Model Calibration

Four plots (Figure 1; site 1,2,3,4) with different fertilization variants from the 2016 dataset were used to calibrate two unknown parameters of the EPIC model namely the potential heat unit (PHU) and the potential evapotranspiration (PET), which was estimated using the Penman–Monteith approach. For normal operation, knowing the date of sowing and harvest, it is possible to check the output file of the model HUSC index. If this index is higher than 1.2 or smaller than 1, the PHU must be adjusted in order to obtain a value around 1. In this way it is possible to arrive to the right value of PHU from planting to maturity. The other unknown parameters inside of the model, except those involved in the temporal definition of the LAI curve, were left as the default setting.

2.5.4. Data Assimilation

The temporal evolution of the optimal LAI curve is defined by the five parameters described in Section 2.5.1. First we identify the DMLA (maximum potential LAI) as the maximum value (98th percentile) observed in the image data during the full growing season for each field. Subsequently, the re-calibration of the other parameters (DLAP1, DLAP2, DLAI, and RLAD) that define the LAI curve in EPIC was performed on the basis of the minimization of the root mean square error (RMSE) (Equation (1)) between the LAI simulated by the model and the LAI observed by satellite at the day of the satellite observation. Default values were used for the parameters that were not measured or not available for our specific experiment.

$$RMSE = \sqrt{\frac{1}{n} \sum_{i=1}^{n} (LAI_{sim} - LAI_{obs})^2} \tag{1}$$

The minimization of the cost function (Equation (1)) was done using a MATLAB optimization toolbox called Fmincon that allows to find the minimum of a constrained nonlinear multivariable function. It uses the interior-point approach for solving nonlinearity constrained optimization problems (Equation (2)) where the objective is to find the local minimum of the function x subject to constraints [49]:

$$min \ f(x) \ subject \ to \ h(x) = 0 \ g(x) \le 0 \tag{2}$$

Starting from the original problem definition in Equation (2), the crucial steps of the optimization algorithm are the formulation and solution of the equality constrained barrier sub-problems, with i steps [50] (Equation (3)). Each sub-problem is called barrier sub-problem and could be approximated at the form (for each $\mu > 0$):

$$\min_{x,s} \ f(x) - \mu \sum_{i=1}^{m} lns_i \ subject \ to \ h(x) = 0 \ g(x) + s = 0 \tag{3}$$

where μ is the barrier parameter and s_i is the slack variables that are restricted to be positive by adding ln (s_i) bounded [50]. As μ decrease close to zero, the minimum of f_μ should approach the minimum of f. The added logarithmic term is called a barrier function [50] and this is easier to solve than the original inequality-constrained problem (Equation (2)). For the optimization, we identified the bound constraints (minimum and maximum values for DLAP1, DLAP2, DLAI, and RLAD) by studying the model LAI outputs in correspondence of the four calibration plots (Section 2.5.3, Figure 1; site 1, 2, 3, 4). The bound constraints values found respectively for DLAP1, DLAP2, DLAI, and RLAD are: 05.01, 25.70, 0.50, and 0.50 for the lower bound and 20.10, 60.99, 1.00, and 3.00 for the upper bound constraints.

2.6. Accuracy Assessment

The model performance was estimated using the root mean square error (RMSE), the coefficient of determination (R^2) and relative root mean square error (RRMSE). The R^2 (Equation (4)) defines the proportion of the dependent variable variance that is predictable by the independent one [51]. It represents a normalized measure between 0 and 1 and it is given by the ratio between the "error sum of the square", that quantifies how much the data points Y_i vary around the estimated regression line $\hat{Y}_i$, and the "total sum of square", that quantifies how much the data points Y_i vary around their mean, $\overline{Y}$. In environmental research, a good value of R^2 is above 0.6 [52] even if it depends on the type of estimated variable. For instance, if we consider the range of LAI values, R^2 value larger than 0.9 is considered as excellent performance and between 0.5 and 0.8 a good one [53]. The RMSE is a performance indicator of a model and gives a quantity estimates of the deviation of the model's output from measurements [54]. The RMSE produces results in the same units as that used for measurements [55] and it is calculated as the root square of the mean of the difference between predicted values (V_p) and observed values (V_{obs}). If the RMSE is 0, there is no difference between predicted and measured values and the perfect fit is achieved. The relative RMSE (Equation (5)) is dimensionless and it is obtained by dividing the RMSE by the mean of the observed variables and thus is less sensitive to the magnitude of the variables [53]. The desired result of the RRMSE for an excellent performance of the model is less than 10% and, for a good one, is between 10% and 20% [53]:

$$R^2 = 1 - \frac{\sum_{i=1}^{n}\left(Y_i - \hat{Y}_i\right)^2}{\sum_{i=1}^{n}\left(Y_i - \overline{Y}\right)^2} \tag{4}$$

$$RMSE = \sqrt{\frac{\sum_{i=1}^{n}\left(V_p - V_{obs}\right)^2}{n}} \tag{5}$$

$$RMSE = 100\frac{RMSE}{Mean\ (obs)} \tag{6}$$

3. Results

3.1. LAI Validation

Good agreement and low error between Sentinel-2 LAI for the winter wheat and LAI reference measurements was found in both years (Figure 3). Better RMSE and RRMSE values were obtained in 2017 compared to 2016 (RMSE = 0.46 vs. 0.44) (RRMSE = 19% vs. 17%). In the 2016 year, a slightly lower R^2 value was found compared to 2017 (R^2 = 0.72 vs 0.89). Although the comparison between ground and satellite LAI estimations showed good results in both years, there are some cases in which we can find a small disagreement. In 2016, a general small overestimation by the LAI-2200 compared to satellite observations was observed. In 2017, there are three cases for which Sentinel-2 overestimated LAI remarkably.

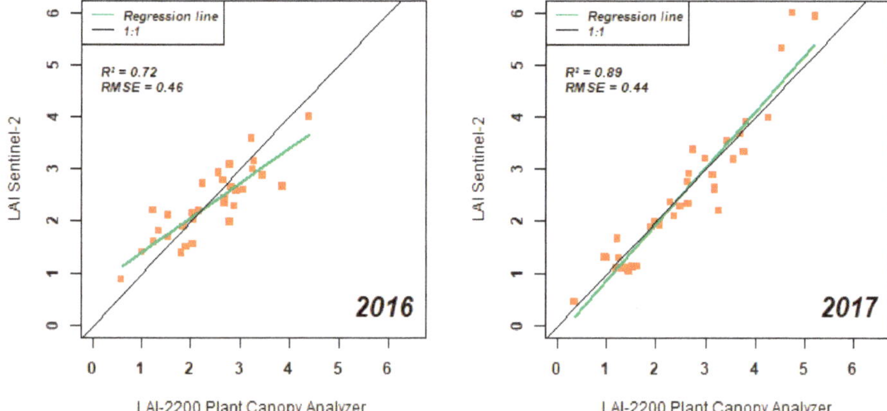

Figure 3. Scatterplot between ground and satellite LAI measurement for Marchfeld in 2016 (n = 31, Feb.–May) and in 2017 (n = 35, Feb.–June) for winter cereals.

3.2. EPIC Calibration and Assessment

Figure 4 shows the EPIC-LAI curves achieved after the assimilation of satellite data and the LAI curves built using the Sentinel-2 observations for the calibration plots in 2016. In this case, the shape of the LAI curves is similar and the only use of the bound constraints on the parameters to optimize seems sufficient to ensure a good fit between the simulated (EPIC) and the observed (Sentinel-2) LAI curves. However, Figure 4D (180 kg N/ha) shows a temporal delay in the crop development trend simulated by EPIC compared to the satellite observations.

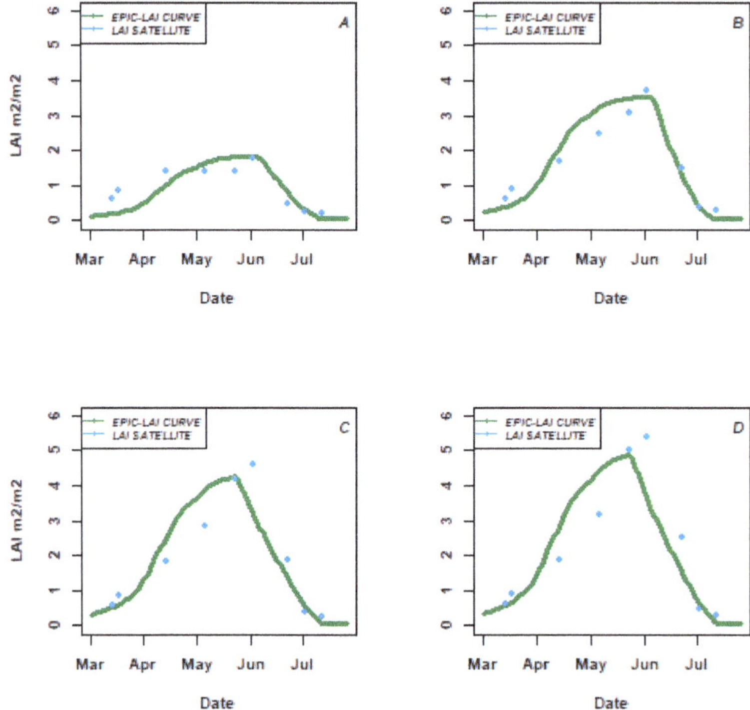

Figure 4. Calibration: the graphs show the different LAI values in four plots for 2016 for different levels of fertilization: 0 kg N/ha (**A**, id: 1-1), 60 kg N/ha (**B**, id: 2-1), 120 kg N/ha (**C**, id: 3-1), and 180 kg N/ha (**D**, id: 4-1). The green line is the daily LAI curve obtained by EPIC model, while the light blue points are the LAI measurements obtained using Sentinel-2 data.

This temporal displacement between the two LAI curves is larger in 2017 compared to 2016 as shown in Figure 5 for the highest fertilization treatment plots (Figure 5C,D), for which an incorrect parameterization appears clear especially for DLAP1, DLAP2, and DLAI. Probably, only use of the bound constraints is not always sufficient to reach a good calibration. It could also be that the model is not able to perfectly represent the satellite curve in all conditions.

3.3. Yield Estimation in 2016

Figure 6A shows a relevant good agreement (R^2 = 0.95) and low error (RMSE = 317 kg/ha and RRMSE = 6%) between the yield estimated by the EPIC model (with assimilation of LAI data from Sentinel-2) and the observed yield for the year 2016. The yield estimation notably improves with LAI assimilation compared to the one obtained using the default EPIC parameterization as shown in Figure 6B (R^2 = 0.93, RMSE = 572 kg/ha and RRMSE = 11%). Using the latter calibration, a model overestimation was found in the lower range of yield values (0 and 60 kg/ha of nitrogen during the growing season). Figure 6B also shows that the high yield values are approximately the same in both the plot with 120 and 180 kg N/ha, while using LAI assimilation from satellite, the yield values are better distributed along the regression line. Thus, the assimilation of LAI data seems to improve the sensitivity to the different amount of nitrogen provided during the growing season.

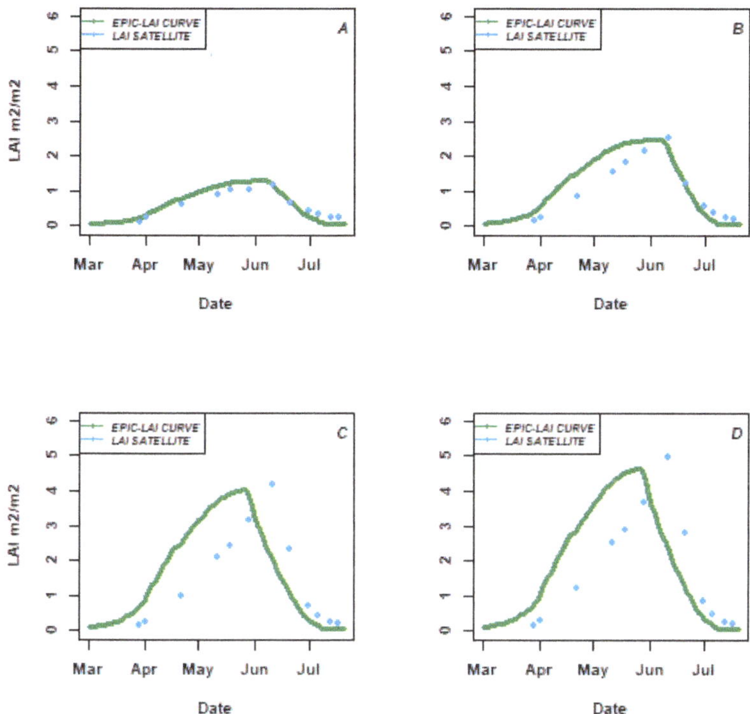

Figure 5. Assessment showing the different LAI values in four 2017 plots for different levels of fertilization: 0 kg N/ha (**A**, id: 1-1), 60 kg N/ha (**B**, id: 2-1), 120 kg N/ha (**C**, id: 3-1), and 180 kg N/ha (**D**, id: 4-1). The green line is the daily LAI curve obtained by EPIC model while the light blue points are the LAI measurements obtained using Sentinel-2 data.

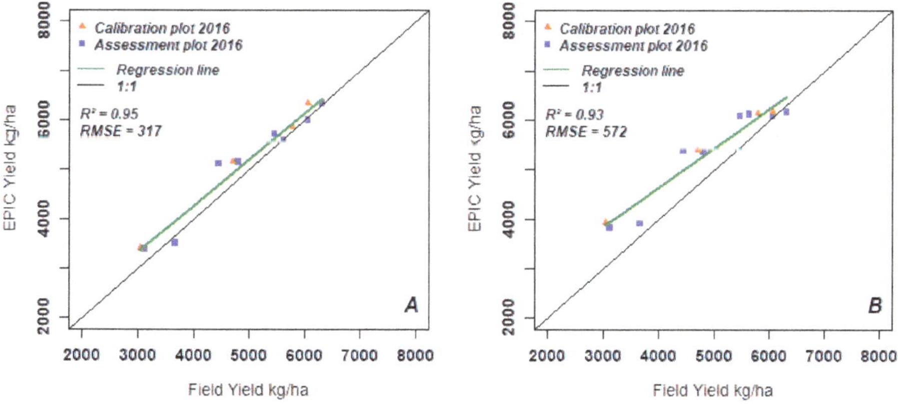

Figure 6. Yield estimation with (**A**) and without (**B**) the assimilation of LAI from Sentinel-2 data. The orange triangles are the plot used for the calibration of *fmincon* and the EPIC unknown parameters, and the blue square symbols represent plots used for the assessment only.

However, we encountered a major flaw for the assimilation of LAI data in EPIC under nitrogen stress conditions. The model lacks a correction feedback on simulated LAI. Only simulated biomass is corrected for the stress induced by limited nitrogen application. The EPIC-LAI curve is influenced mainly by the temperature and not by the variable amount of N.

Figure 7 shows the simulated LAI and biomass curves obtained using default parameters of the model for the different nitrogen fertilization applications. It appears clear that the LAI curve does not change even if the fertilization management changes. Therefore, the EPIC-LAI curve is referred to the optimal development of the LAI, without considering limiting stress conditions for the crop. With this approach, a saturation in the biomass is clearly visible for the high fertilization treatments (120–180 kg N/ha) and this causes errors in the estimation of the yield (Figure 7B).

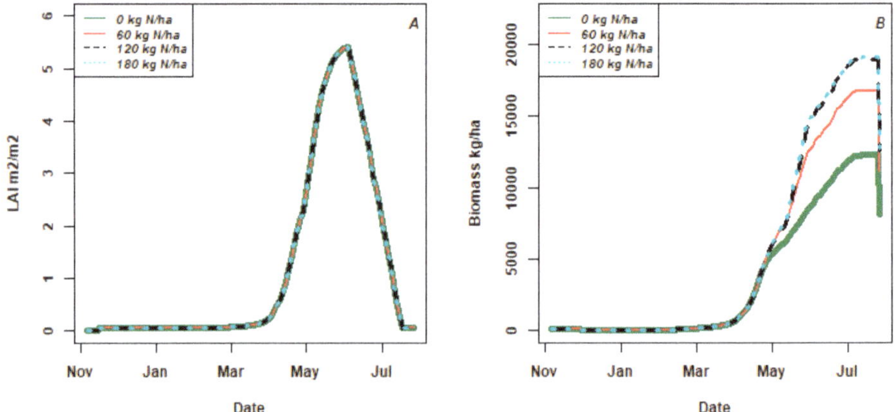

Figure 7. LAI (**A**) and biomass (**B**) curves obtained with default LAI parameters for the four different fertilization management treatments (0, 60, 120, and 180 Kg N/ha) in 2016. The graphs show a variation in the biomass levels but not in LAI.

3.4. Yield Estimation for 2017

Regarding the year 2017, we found a strong underestimation of the model compared to the yield measured in the field with $R^2 = 0.97$, RMSE = 1961, and RRMSE = 55% (Figure 8). A little improvement can be seen by assimilating LAI from the satellite (Figure 8A) compared to a parameterization using default model values (Figure 8B).

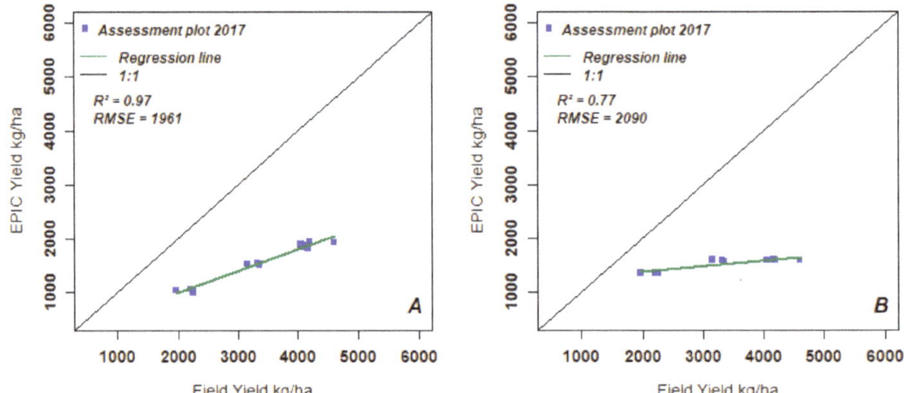

Figure 8. Comparison between (**A**) yield estimated with the assimilation of LAI from Sentinel-2 and (**B**) yield obtained with default EPIC parameters for 2017. The green line represents the regression line between the values.

The output of the model was further analyzed to understand the strong underestimation. One main reason for the underestimation was found to be due to a major decrease in the harvest index (HI) for the year 2017 (0.214–0.218) compared to 2016 (0.416–0.425). The small intra-year variations in the HI values are probably due to the different phenological development of the crop during the growing season related to the different LAI parameterization for each plot. The strong decrease in the HI in 2017 is probably related to water stress induced by limited rainfall. For comparison, Figure 9 shows the amount of precipitation during the growing season 2016 and 2017. It is known that drought stress at any growth stages decreases grain yield, especially in the stem elongation, flowering, and grain falling stages.

It appeared clear that the impact of water stress excessively reduces the HI index. Therefore, we further tuned the HI by modifying the parameter PARM(3) inside of the parameter file. For this parameter a value of 3 was set. Using the letter value, the water stress influences HI only when the heat unit accumulation is equal to three times the potential heat unit accumulation of the crop. In addition, the lower limit of HI, namely the WSYF that was set equal to the HI value. As a result of this adjustment, Figure 10 shows an improvement of the yield estimation (R^2 = 0.97, RMSE = 281, and RRMSE = 8.1%).

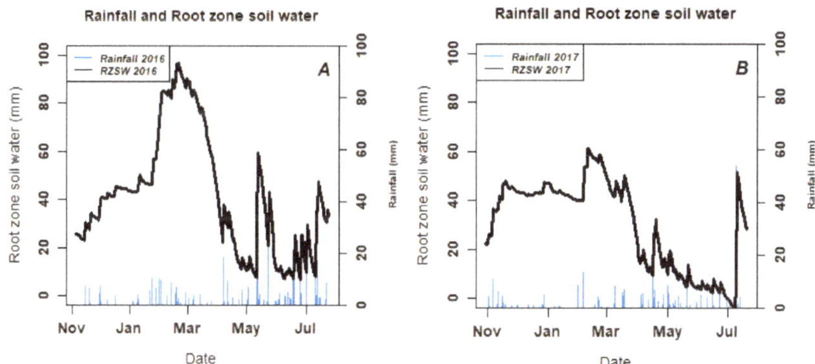

Figure 9. Precipitation (blue bars) and simulated water content in the root zone (black line) in 2016 (**A**) and in 2017 (**B**) for one of the high fertilizer plots (180 kg N /ha).

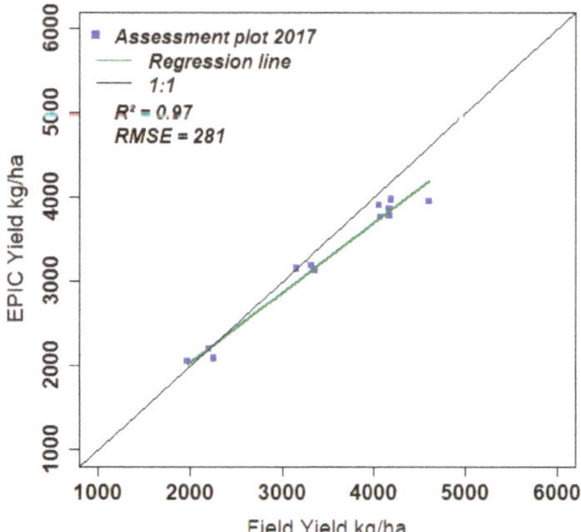

Figure 10. Yield obtained using the modified PARM(3) value that allows to reduce the impact of the water stress on the harvest index.

4. Discussion

Regarding the accuracy of LAI estimation with Sentinel-2 data, our study confirmed a previous study [29] that reported an R^2 of 0.83 and a RMSE of 0.32 m^2/m^2. When assimilated in a crop model, satellite-based LAI improves yield estimates as reported also in other studies that used a similar assimilation strategy [12,56]. The only use of the bound constraints on the model parameters, without including other restriction as nonlinear constraints, seems sufficient to find the minimum of the minimization problem but a good fit between the satellite observations and the LAI curve of the model was found only in the plot where a low amount of fertilizer was provided. However, in both the years, a yield improvement was obtained and this may be due to a forcing of the DMLA parameter using satellite observations. The use of the maximum value of LAI (98th percentile) to set DMLA can be considered as a priori knowledge of the canopy conditions. In this study, an ideal maximum LAI reachable by the crop was found for each field. In this context, the combination of Sentinel-2A and Sentinel-2B satellite data can provide timely and dense sets of observations to precisely characterize the crop growth curves and maximum LAI values.

The combination of the EPIC model and satellite-based LAI was found to be not ideal because there is no nitrogen stress feedback on the LAI in the model. In real field conditions, LAI is significantly influenced by nitrogen levels [57], while, in EPIC, the LAI curve is modeled without taking into consideration the nitrogen stress. Only simulated biomass is affected by the nitrogen constraints.

However, under high level of nitrogen application (above 120 kg/ha in our study), the simulated biomass saturates and, in this case, forcing LAI can help the model to achieve an increased precision. Therefore, the coupling of satellite-based LAI observations and potential EPIC LAI curves might improve the description of the spatial variability, for both LAI and biomass estimation, as confirmed by an improvement in yield estimation obtained in both years under study. Regarding the year specific results: in 2016, a good yield estimation was found assimilating remotely-sensed data even if only a slight improvement in R^2, RMSE, and RRMSE was obtained if compared to the yield estimation without data assimilation. This could be partially attributed to the site-specific parameterization of the model using field observations and, thus, an accurate description of the environment where the crop has been growing up. In 2017, a slight improvement in the yield results was also found especially regarding the correlation (improvement of R^2). However, both a data assimilation approach and a standard EPIC parameterization provided a strong underestimation in yield due to water stress. This is probably caused by three reasons: (1) the rainfall data may not be representative of the field condition and they can affect the model estimation, (2) the default parameters of the model are non-representative for the specific year of simulation; or (3) the impact of water stress conditions are erroneously modelled.

One of the problems that we should consider is the capability of the model to estimate yields under different scenarios and in the presence of high variability, especially in the regional scale application. Previous studies have shown promising results considering the EPIC yield estimation at the regional scale in the presence of high variability [58]. More examples should be obtained using remote sensing data that allows the improvement in the information about the crop and the environment of interest even if, due to a large amount of parameters required by the model and, therefore, the large rate of uncertainty, the regional scale application of EPIC seems difficult. Furthermore, the definition of the model parameter-bound constraints in the minimization function without a site-specific calibration is not always representative of the different geographical areas, and a multi-location calibration appears fundamental, for the application in different sub-areas (such as at the field scale) or at the regional scale. It seems necessary to add additional information in the minimization function. The application of EPIC model at the sub-field scale seems more suitable than the regional application, but specific field and crop data are needed.

5. Conclusions

The use of satellite data together with a crop growth model gives a wide range of possibilities to obtain yield estimates. However, the availability of satellite data and the ability of the model to represent both satellite observations and the characteristics of the crop system has to be evaluated. In this study, the fit between the modelled LAI and the observed LAI curves was reached for both years under investigation (2016 and 2017). However, yield was not always well estimated. This problem could be attributed to the non-capability of the model to represent the environment of study or to issues related to the data used for the model calibration. EPIC was chosen after a literature review showing the potential of the model to simulate grain yield under different environmental and management conditions. In the evaluation of our results, the adoption of a specific calibration of the model parameters, taking into consideration four plots from one growing season only (2016), has to be taken into consideration. Therefore, the cost function minimization approach, together with the EPIC capability to assimilate satellite data, need to be tested under different environmental and management conditions, at different spatial scales, and using general constraints, not specific for the study site. Limitations of the re-calibration approach used in this study are the fact that it does not consider uncertainties in the satellite observations, in the model parameterization and in the model variables' estimations. This approach is also expensive in terms of computing time. For future applications, the updating methods could be considered in order to alleviate the shortcoming of the re-calibration methods. Two classes exist: on the one hand, stochastic methods that are generally implemented using ensemble data assimilation (e.g., the ensemble Kalman filter or particle filter). On the other hand, the variational methods used in other disciplines, as for example the 4D-Var approach, seem to be promising also in the crop growth model assimilation. The 4D-Var methods can simultaneously assimilate the observational data at multiple times in an assimilation window starting with the assumption that the model is perfect. The choice of the best possible model according to the available data and the amount of the data required by the model is one of the key points, together with the quality of the satellite observations and applicability of the data assimilation method that should be considered in the data assimilation framework.

Author Contributions: Conceptualization, F.V.; formal analysis, F.N.; methodology, F.N.; field data collection and pre-processing of site data H.S., T.S.; supervision, F.V.; writing—original draft, F.N.; writing—review and editing, F.V.

Funding: This research was funded by the European Union's Horizon 2020 research and innovation programme, grant numbers 773903, 774234, and 633945. The APC was funded by the European Union's Horizon 2020 research and innovation programme, grant number 773903.

Acknowledgments: This research has received funding from the European Union's Horizon 2020 research and innovation programme under grant agreements No 773903, No 774234 and No 633945.

Conflicts of Interest: The authors declare no conflict of interest.

References

1. Mueller, N.D.; Gerber, J.S.; Johnston, M.; Ray, D.K.; Ramankutty, N.; Foley, J.A. Closing yield gaps through nutrient and water management. *Nature* **2012**, *490*, 254–257. [CrossRef]
2. Battude, M.; Al Bitar, A.; Morin, D.; Cros, J.; Huc, M.; Marais Sicre, C.; Le Dantec, V.; Demarez, V. Estimating maize biomass and yield over large areas using high spatial and temporal resolution Sentinel-2 like remote sensing data. *Remote Sens. Environ.* **2016**, *184*, 668–681. [CrossRef]
3. Yuan, W.; Chen, Y.; Xia, J.; Dong, W.; Magliulo, V.; Moors, E.; Olesen, J.E.; Zhang, H. Estimating crop yield using a satellite-based light use efficiency model. *Ecol. Indic.* **2015**, *60*, 702–709. [CrossRef]
4. Atzberger, C. Advances in remote sensing of agriculture: Context description, existing operational monitoring systems and major information needs. *Remote Sens.* **2013**, *5*, 949–981. [CrossRef]
5. Wall, L.; Larocque, D.; Léger, P.M. The early explanatory power of NDVI in crop yield modelling. *Int. J. Remote Sens.* **2008**, *29*, 2211–2225. [CrossRef]
6. Rembold, F.; Atzberger, C.; Savin, I.; Rojas, O. Using low resolution satellite imagery for yield prediction and yield anomaly detection. *Remote Sens.* **2013**, *5*, 1704–1733. [CrossRef]
7. Manjunath, K.R.; Potdar, M.B.; Purohit, N.L. Large area operational wheat yield model development and validation based on spectral and meteorological data. *Int. J. Remote Sens.* **2002**, *23*, 3023–3038. [CrossRef]

8.   Balaghi, R.; Tychon, B.; Eerens, H.; Jlibene, M. Empirical regression models using NDVI, rainfall and temperature data for the early prediction of wheat grain yields in Morocco. *Int. J. Appl. Earth Obs. Geoinf.* **2008**, *10*, 438–452. [CrossRef]

9.   Monteith, J.L.; Moss, C.J. Climate and the Efficiency of Crop Production in Britain [and Discussion]. *Philos. Trans. R. Soc. B Biol. Sci.* **2006**, *281*, 277–294. [CrossRef]

10.   Moulin, S.; Bondeau, A.; Delecolle, R. Combining agricultural crop models and satellite observations: From field to regional scales. *Int. J. Remote Sens.* **1998**, *19*, 1021–1036. [CrossRef]

11.   Huang, Y.; Zhu, Y.; Li, W.; Cao, W.; Tian, Y. Assimilating Remotely Sensed Information with the WheatGrow Model Based on the Ensemble Square Root Filter For improving Regional Wheat Yield Forecasts. *Plant Prod. Sci.* **2013**, *16*, 352–364. [CrossRef]

12.   Launay, M.; Guerif, M. Assimilating remote sensing data into a crop model to improve predictive performance for spatial applications. *Agric. Ecosyst. Environ.* **2005**, *111*, 321–339. [CrossRef]

13.   Gowda, P.T.; Satyareddi, S.A.; Manjunath, S. Crop Growth Modeling: A Review. *Res. Rev. J. Agric. Allied Sci. Crop.* **2013**, *2*, 1–11.

14.   Kasampalis, D.; Alexandridis, T.; Deva, C.; Challinor, A.; Moshou, D.; Zalidis, G. Contribution of Remote Sensing on Crop Models: A Review. *J. Imaging* **2018**, *4*, 52. [CrossRef]

15.   Prevot, L.; Chauki, H.; Troufleau, D.; Weiss, M.; Baret, F.; Brisson, N. Assimilating optical and radar data into the STICS crop model for wheat. *Agronomie* **2003**, *23*, 297–303. [CrossRef]

16.   de Wit, A.J.W.; van Diepen, C.A. Crop model data assimilation with the Ensemble Kalman filter for improving regional crop yield forecasts. *Agric. For. Meteorol.* **2007**, *146*, 38–56. [CrossRef]

17.   Jin, X.; Kumar, L.; Li, Z.; Yang, G.; Wang, J. Review article A review of data assimilation of remote sensing and crop models. *Eur. J. Agron.* **2018**, *92*, 141–152. [CrossRef]

18.   Dorigo, W.A.; Zurita-milla, R.; De Wit, A.J.W.; Brazile, J. A review on reflective remote sensing and data assimilation techniques for enhanced agroecosystem modeling. *Int. J. Appl. Earth Obs. Geoinf.* **2007**, *9*, 165–193. [CrossRef]

19.   Machwitz, M.; Giustarini, L.; Bossung, C.; Frantz, D.; Schlerf, M.; Lilienthal, H.; Wandera, L.; Matgen, P.; Hoffmann, L.; Udelhoven, T. Enhanced biomass prediction by assimilating satellite data into a crop growth model. *Environ. Model. Softw.* **2014**, *62*, 437–453. [CrossRef]

20.   Fang, H.; Liang, S.; Hoogenboom, G.; Teasdale, J.; Cavigelli, M. Corn-yield estimation through assimilation of remotely sensed data into the CSM-CERES-Maize model. *Int. J. Remote Sens.* **2008**, *29*, 3011–3032. [CrossRef]

21.   Ma, G.; Huang, J.; Wu, W.; Fan, J.; Zou, J.; Wu, S. Assimilation of MODIS-LAI into the WOFOST model for forecasting regional winter wheat yield. *Math. Comput. Model.* **2013**, *58*, 634–643. [CrossRef]

22.   Ren, J.; Yu, F.; Qin, J.; Chen, Z.; Tang, H. Integrating remotely sensed LAI with EPIC model based on global optimization algorithm for regional crop yield assessment. In Proceedings of the IEEE International Geoscience and Remote Sensing Symposium, Honolulu, HI, USA, 25–30 July 2010.

23.   Tan, G.; Shibasaki, R. Global estimation of crop productivity and the impacts of global warming by GIS and EPIC integration. *Ecol. Model.* **2003**, *168*, 357–370. [CrossRef]

24.   Liu, J.; Chen, Z.; Sun, L.; Ren, J.; Jiang, Z.; Chen, J.; Li, H.; Li, Z. Application of Crop Model Data Assimilation with a Particle Filter for Estimating Regional Winter Wheat Yields. *IEEE J. Sel. Top. Appl. Earth Obs. Remote Sens.* **2014**, *7*, 4422–4431.

25.   Dente, L.; Satalino, G.; Mattia, F.; Rinaldi, M. Assimilation of leaf area index derived from ASAR and MERIS data into CERES-Wheat model to map wheat yield. *Remote Sens. Environ.* **2008**, *112*, 1395–1407. [CrossRef]

26.   Baret, F.; Buis, S. Estimating Canopy Characteristics from Remote Sensing Observations: Review of Methods and Associated Problems. In *Advances in Land Remote Sensing*; Springer: Dordrecht, The Netherlands, 2008; pp. 173–201.

27.   European Space Agency (ESA). *SENTINEL-2 User Handbook*; ESA: Paris, France, 2015; pp. 1–64.

28.   Richter, K.; Atzberger, C.; Vuolo, F.; Weihs, P.; Urso, G.D. Experimental assessment of the Sentinel-2 band setting for RTM-based LAI retrieval of sugar beet and maize. *Can. J. Remote Sens.* **2009**, *35*, 230–247. [CrossRef]

29.   Vuolo, F.; Zóltak, M.; Pipitone, C.; Zappa, L.; Wenng, H.; Immitzer, M.; Weiss, M.; Baret, F.; Atzberger, C. Data service platform for Sentinel-2 surface reflectance and value-added products: System use and examples. *Remote Sens.* **2016**, *8*, 938. [CrossRef]

30. Upreti, D.; Huang, W.; Kong, W.; Pascucci, S.; Pignatti, S.; Zhou, X.; Ye, H.; Casa, R. A Comparison of Hybrid Machine Learning Algorithms for the Retrieval of Wheat Biophysical Variables from Sentinel-2. *Remote Sens.* **2019**, *11*, 481. [CrossRef]

31. Vuolo, F.; Essl, L.; Zappa, L.; Sandén, T.; Spiegel, H. Water and nutrient management: The Austria case study of the FATIMA H2020 project. *Adv. Anim. Biosci.* **2017**, *8*, 400–405. [CrossRef]

32. Fatima-H2020. Marchfeld Pilot Area in Austria. Available online: http://fatima-h2020.eu/pilots/austria-marchfeld/ (accessed on 14 January 2019).

33. Thaler, S.; Eitzinger, J.; Trnka, M.; Dubrovsky, M. Impacts of climate change and alternative adaptation options on winter wheat yield and water productivity in a dry climate in Central Europe. *J. Agric. Sci.* **2012**, *150*, 537–555. [CrossRef]

34. Vuolo, F.; Neugebauer, N.; Bolognesi, S.F.; Atzberger, C.; D'Urso, G. Estimation of leaf area index using DEIMOS-1 data: Application and transferability of a semi-empirical relationship between two agricultural areas. *Remote Sens.* **2013**, *5*, 1274–1291. [CrossRef]

35. USDA. Natural Resources Conservation Service Soil. Available online: https://www.nrcs.usda.gov/wps/portal/nrcs/detail/soils/ref/?cid=nrcs142p2_054253 (accessed on 7 May 2019).

36. LI_COR. *LAI-2200 Plant Canopy Analyzer Instruction Manual*; LI-COR: Lincoln, Nebraska, 2017; p. 262.

37. LI-COR Biosciences—Impacting Lives Through Science. Available online: https://www.licor.com/ (accessed on 7 May 2019).

38. Vuolo, F.; Dash, J.; Curran, P.J.; Lajas, D.; Kwiatkowska, E. Methodologies and uncertainties in the use of the terrestrial chlorophyll index for the sentinel-3 mission. *Remote Sens.* **2012**, *4*, 1112–1133. [CrossRef]

39. González-Sanpedro, M.C.; Le Toan, T.; Moreno, J.; Kergoat, L.; Rubio, E. Seasonal variations of leaf area index of agricultural fields retrieved from Landsat data. *Remote Sens. Environ.* **2008**, *112*, 810–824. [CrossRef]

40. Nelson, D.; Wang, J. Introduction to artificial neural systems. *Neurocomputing* **2003**, *4*, 328–330. [CrossRef]

41. Lek, S.; Guégan, J.F. Artificial neural networks as a tool in ecological modelling, an introduction. *Ecol. Model.* **1999**, *120*, 1–9. [CrossRef]

42. Weiss, M.; Baret, F. *Sentinel2 ToolBox Level2 Products S2ToolBox Level 2 Products: LAI, FAPAR, FCOVER Version 1.1*; INRA-CSE: Avignon, France, 2016; p. 53.

43. Müller-Wilm, U. *Sen2Cor Configuration and User Manual—Ref. S2-PDGS-MPC-L2A-SUM-V2.5.5*; Telespazio VEGA: Luton, Bedfordshire, UK, 2018; p. 54.

44. Williams, J.R.; Jones, C.A.; Kiniry, J.R.; Spanel, D.A. The EPIC Crop Growth Model. *Trans. ASAE* **1989**, *32*, 0497–0511. [CrossRef]

45. Williams, J.R.; Dagitz, S.; Magre, M.; Meinardus, A.; Staglich, E.; Taylor, R. *Environmental Policy Integrated Climate Model*; Model User Manual Version 0810; Blackland Research and Extension Center A&M AgriLife: College Station, TX, USA, 2015.

46. Kiniry, J.R.; Williams, J.R.; Major, D.J.; Izaurralde, R.C.; Gassman, P.W.; Morrison, M.; Bergentine, R.; Zentner, R.P. EPIC model parameters for cereal, oilseed, and forage crops in the northern Great Plains region. *Can. J. Plant Sci.* **2011**, *75*, 679–688. [CrossRef]

47. Huang, M.; Gallichand, J.; Dang, T.; Shao, M. An evaluation of EPIC soil water and yield components in the gully region of Loess Plateau, China. *J. Agric. Sci.* **2006**, *144*, 339–348. [CrossRef]

48. EPIC & APEX Model. Model Executables. Available online: https://epicapex.tamu.edu/model-executables/ (accessed on 9 October 2018).

49. Constrained Nonlinear Optimization Algorithms. MATLAB & Simulink. Available online: https://ch.mathworks.com/help/optim/ug/constrained-nonlinear-optimization-algorithms.html#brnpd5f (accessed on 9 November 2018).

50. Byrd, R.H.; Hribar, M.E.; Nocedal, J. An interior point method for large scale nonlinear programming. *SIAM J. Optim.* **1999**, *9*, 877–900. [CrossRef]

51. Backhaus, K.; Erichson, B.; Weiber, R. *Fortgeschrittene Multivariate Analysemethoden eine Anwendungsorientierte Einführung*; Springer-Lehrbuch: Basel, Switzerland, 2013.

52. Soliani, L. Fondamenti di Statistica Applicata All'analisi e Alla Gestione Dell'ambiente. 2001. Available online: https://dokumen.tips/documents/fondamenti-di-statistica-applicata-allanalisi-ambientalepdf25-distribuzioni.html (accessed on 7 May 2019).

53. Richter, K.; Atzberger, C.; Hank, T.B.; Mauser, W. Derivation of biophysical variables from Earth observation data: Validation and statistical measures. *J. Appl. Remote Sens.* **2012**, *6*, 063557. [CrossRef]

54. Bellocchi, G.; Rivington, M.; Donatelli, M.; Matthews, K. Validation of biophysical models: Issues and methodologies. *Agron. Sustain. Dev.* **2010**, *30*, 109–130. [CrossRef]

55. Vanino, S.; Nino, P.; De Michele, C.; Falanga Bolognesi, S.; D'Urso, G.; Di Bene, C.; Pennelli, B.; Vuolo, F.; Farina, R.; Pulighe, G. Capability of Sentinel-2 data for estimating maximum evapotranspiration and irrigation requirements for tomato crop in Central Italy. *Remote Sens. Environ.* **2018**, *215*, 452–470. [CrossRef]

56. Wang, H.; Zhu, Y.; Li, W.; Cao, W.; Tian, Y. Integrating remotely sensed leaf area index and leaf nitrogen accumulation with RiceGrow model based on particle swarm optimization algorithm for rice grain yield assessment. *J. Appl. Remote Sens.* **2014**, *8*, 083674. [CrossRef]

57. Lazauskas, S.; Povilaitis, V.; Antanaitis, Š.; Sakalauskaite, J.; Sakalauskiene, S.; Pšibišauskiene, G.; Auškalniene, O.; Raudonius, S.; Duchovskis, P. Winter wheat leaf area index under low and moderate input management and climate change. *J. Food Agric. Environ.* **2012**, *10*, 588–593.

58. Balkovič, J.; van der Velde, M.; Schmid, E.; Skalský, R.; Khabarov, N.; Obersteiner, M.; Stürmer, B.; Xiong, W. Pan-European crop modelling with EPIC: Implementation, up-scaling and regional crop yield validation. *Agric. Syst.* **2013**, *120*, 61–75. [CrossRef]

Article

Detection of Spatial and Temporal Variability of Wheat Cultivars by High-Resolution Vegetation Indices

Stefano Marino * and Arturo Alvino

Department of Agricultural, Environmental and Food Sciences (DAEFS), University of Molise, Via De Sanctis, 86100 Campobasso, Italy; alvino@unimol.it
* Correspondence: stefanomarino@unimol.it; Tel.: +390874404709; Fax: +390874404713

Received: 18 March 2019; Accepted: 3 May 2019; Published: 5 May 2019

Abstract: An on-farm research study was carried out on two small-plots cultivated with two cultivars of durum wheat (Odisseo and Ariosto). The paper presents a theoretical approach for investigating frequency vegetation indices (VIs) in different areas of the experimental plot for early detection of agronomic spatial variability. Four flights were carried out with an unmanned aerial vehicle (UAV) to calculate high-resolution normalized difference vegetation index (NDVI) and optimized soil-adjusted vegetation index (OSAVI) images. Ground agronomic data (biomass, leaf area index (LAI), spikes, plant height, and yield) have been linked to the vegetation indices (VIs) at different growth stages. Regression coefficients of all samplings data were highly significant for both the cultivars and VIs at anthesis and tillering stage. At harvest, the whole plot (W) data were analyzed and compared with two sub-areas characterized by high agronomic performance (H) yield 20% higher than the whole plot, and low performances (L), about 20% lower of yield related to the whole plot). The whole plot and two sub-areas were analyzed backward in time comparing the VIs frequency curves. At anthesis, more than 75% of the surface of H sub-areas showed a VIs value higher than the L sub-plot. The differences were evident also at the tillering and seedling stages, when the 75% (third percentile) of VIs H data was over the 50% (second percentile) of the W curve and over the 25% (first percentile) of L sub-plot. The use of high-resolution images for analyzing the frequency value of VIs in different areas can be a useful approach for the detection of agronomic constraints for precision agriculture purposes.

Keywords: UAV; vegetation indices; relative frequencies; yield; precision agriculture; cultivars

---

1. Introduction

Monitoring the spatial and temporal variability of wheat within a season is crucial to decision-making in precision farming. Precision agriculture is a modern farming management concept using digital techniques to monitor and optimize agricultural production processes. Precision agriculture can play an important role in enhancing crop yield and ensuring sustainability [1]. Among the tools used to acquire information, unmanned aerial vehicles (UAVs) equipped with visible and near-infrared cameras, provide, in a fast and easy way, field data for precision agriculture applications [2,3]. The resolution of information from satellite data typically ranges from 5 to 30 m pixels and is unsuitable in agronomy trials given the limitations of real-time monitoring and accuracy [4]. In contrast to satellite imagery and aircraft-based remote sensing, UAVs can be used frequently during the entire growth period [5]. Furthermore, vegetation indices (VIs) of UAV imagery have the same ability as ground-based recordings to quantify crop responses to experimental treatments [6]. UAVs are a useful technology for crop monitoring at different scales and can be used for agronomic experiments where space, resource, and time constraints limit manual sampling [7]. Furthermore, UAVs provide images at small pixel sizes (3–5 cm^2 pixels), with a higher resolution than an aerial or satellite platform [8–10]. Moreover, a field can be frequently surveyed to study ongoing different phenological development phenomena [3]. Unmanned aerial vehicles equipped with near-infrared (NIR) and multispectral sensors have been useful in the research environment for determining principal spectral patterns and wavebands that relate to plant stress, through the estimation of vegetation indices (VIs), which are based on formulations fitted with the canopy light reflected at different wavelengths [11,12]. Starting from wavebands and spectral patterns, different VIs have been developed and were related to vegetation canopies including plant nutrient status, plant growth rate, physiological conditions, and crop yields [13–16]. Among VIs, the normalized difference vegetation index (NDVI) and optimized soil-adjusted vegetation index (OSAVI) are generally employed as the typical quantitative data for the estimation indicator of crop growth. Crop phenotyping involves the measurement and evaluation of physical characteristics such as biomass, leaf area index (LAI), height [17,18] and that represents a bottleneck for the fine-tuning management of field crops [19]. Crop phenology is characterized by a set of growth stages; it is well known that a single crop trait in a single growth stage plays an important role in establishing the final grain yield [20,21]. These phenological events affect the vegetation index (VI) value and provide essential information for the detection of the agronomic practices and the identification of sub-field areas having the same yield-limiting factors or similar attributes that significantly affect crop yield [22].

Indeed, there are still many aspects to be analyzed in order to discriminate homogeneous agronomic areas, starting from spectroradiometric data, because many agronomic factors affect spatial and temporal crop variability. For precision agriculture purposes it is important to define, within each field, sub-areas having similar biotic or abiotic factors that are expected to affect yield significantly [23] and point out an automatic technique to detect homogeneous sub-areas. The most appropriate time to start crop monitoring for defining sub-plot areas remains unsolved [22]. Several cluster methods [24], among vegetation indices, and crop and soil parameters for delineating management zones maps have been used. Candiago et al., [3] proposed a new approach for analyzing the high-resolution contents of the VI images based on relative frequency distributions of VIs on vineyard and tomato crop, identifying best and worst vegetative areas, stating however that the main limitation of the study was the lack of ground measurements.

In the present paper, a new approach is proposed based on the combination of cluster analysis, relative frequency distribution of VIs and ground agronomic data of winter wheat cultivars.

Since each VI has its own peculiar suitability for specific purposes and practical applications, ground agronomic data of Odisseo (O) and Ariosto (A) at seedling, tillering, and anthesis have been linked to NDVI and OSAVI indices to verify the ability of VIs to detect yield and yield components.

Once the ability to detect crop parameters has been verified, within each cultivar, the relative frequency value of high-resolution VI data of the whole plot and of two homogeneous sub-areas identified by cluster analysis [22], one with high yield (HO and HA) and the second one with low yield (LO and LA) were investigated. The whole plot and the sub-areas were analyzed backward in time to assess the power of high-resolution VI data to detect, explain, and quantify the agronomic spatial and temporal variability.

2. Materials and Methods

2.1. Study Area and Field Measurements

An on-farm research study was carried out in Central Italy (411081.29 E, 4730618.48 N; UTM-WGS84 zone 33N Italy), in a flat area at 75 m above sea level, in the 2015 crop season. The experimental field (2 hectares) was cultivated with different plots of durum and winter wheat. Two cultivars of durum wheat, Odisseo and Ariosto, were cultivated on small plots 24 m × 30 m (720 m^2), previously cultivated with soybean. Sowing was performed in December 2014 (200 kg of seeds ha^{-1}) and the crop was harvested in July 2015. The soil was prepared with the help of minimum-tillage equipment and basic fertilization was performed with 200 kg ha^{-1} of P_2O_5 (27%). Nitrogen was applied at the end of March (tillering) with a slow release nitrogen fertilizer, 35% N; 23% SO_3, and at April (booting/stem elongation) with 150 kg ha^{-1} of fertilizer with 30% N; 23% SO_3 and with two foliar nitrogen application. The weed and pest control scouting was carried out with chemical pesticides during the crop cycle.

A standard agro-meteorological station was placed in the experimental fields and temperatures, rainfall, wind, humidity, and radiation were recorded. The minimum temperature (-1 °C) was recorded in December 2014, while the maximum (36.2 °C) was recorded in July 2015. Through the whole cropping season (December 2014–July 2015), precipitation amounted to 680.2 mm; a good half of the total precipitation was recorded in 6 weeks, from the final third of January to the first third of March. From tillering to anthesis, rainfall was scarce, while from anthesis to ripening, rainfall was considerable. Air temperatures were mild from emergence to anthesis and afterwards, maximum air temperatures were around 30 °C, as expected.

The phenological stages of the wheat crop were periodically recorded according to the Zadoks Scale (ZS), which is a standardized reference scale used to evaluate and measure plant growth stage in cereals [25]. During the crop growth, sampling was carried out for each cultivar: 10 at seedling growth 13—ZS (February 23), 12 at tillering 25—ZS (March 30), 11 at anthesis 65—ZS (May 14), and 11 at harvest—99 ZS (July 7). Plants from 0.5 m^2 were georeferenced and hand cut for the calculation of yield-related traits (biomass, green leaf area index—LAI, number of spikes, plant height, yield, and thousand kernels weight). For each cultivar, 44 georeferenced samplings during crop growth were collected.

Whole plant dry mass was determined after oven drying the fresh plant material at 75 °C until constant weight for 48 h. The sampling points based on visual-spatial crop variability and identified by low-resolution orthoimages of flight were processed in real time in the field. Ground truth coordinates of target locations were recorded with GPS Leica Viva GS15 (Leica Geosystems AG, Heerbrugg, Switzerland) at each sampling and flight.

Furthermore, at harvest, within each cultivar, yield data and yield components of the whole plot (720 m^2) were measured, as well as for four sub-areas, 10 m^2 each was characterized by the different yield levels in Figure 1 (H = high yield and L = low yield). The sub-areas were selected within the homogeneous areas detected using a hierarchical clustering method, Ward's minimum variance approach, as reported in a previous paper for Odisseo [25] and calculated for Ariosto. NDVI and OSAVI high-resolution data of the four areas were analyzed backward in time to deliver relative frequency histograms of VI values.

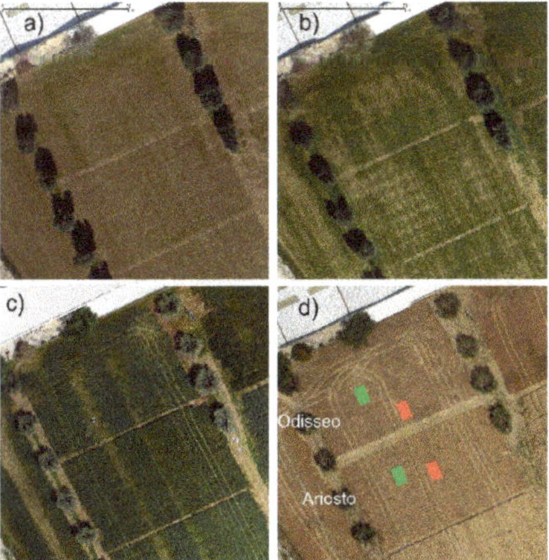

Figure 1. RGB images and samplings H (red) and L (green) areas of Odisseo (above) and Ariosto (below) at seedling (**a**), tillering (**b**), anthesis (**c**), and harvest (**d**) stages.

2.2. UAV System and Flight Missions

Four flight missions were carried out in the same dates to measure the destructive plant samples: for seedling (February 23), tillering (March 30), anthesis (May 14), harvest (July 7), with eBee UAV (senseFly, Cheseaux-sur-Lausanne, Switzerland). The eBee, designed as a fixed-wing UAV for application in precision agriculture (payload of 150 g). All flights were carried out in stable ambient light conditions from 11:30 a.m. to 1:00 p.m., with excellent visibility and wind below 5 m s^{-1}, at a flight altitude of 97 m (A.G.L.). The imaged area of the experimental field, including the surroundings, is about 15 hectares so that 96 camera stations (single flight) and 15 min flight were needed. The autopilot analyzes (continuously) the inertial measurement unit (IMU) and onboard GPS data to control every aspect of the eBee's flight. The integration of the UAV and sensors with GPS and IMU enables obtaining direct georeferencing imaging after image processing. At any acquisition date, two flights were carried out, the first one by a Canon Powershot S110 photo camera (visible spectrum, RGB-red/green/blue) for visible RGB image (orthophoto) to run a rapid analysis for visual crop variability. The second flight used a Canon Powershot S110 NIR camera (near infra-red, NIRGB—near infra-red/green/blue) which provides maximum absorption peaks at 550 nm (green), 625 nm (red), and 850 nm (NIR) wavelengths, respectively, allowing the computation of VIs.

The S110 RGB and S110 NIR camera characteristics were: weight 0.7 kg, resolution 12 million pixels, sensor size 5.58 × 7.44 mm^2, pixel pitch 1.33 µm, and format images RAW JPEG. The S110 RGB camera acquired the true-color, while the S110 NIR acquired the false-color image data at 0.55 µm (green), 0.625 µm (red), and 0.85 µm (NIR) bands.

To avoid geometric distortion due to low altitude, 96 overlapping pictures from each camera and flight were used for mosaicking to produce an ortho-image. An 80% frontal overlap and an 80% side overlap were used.

To relate and orient UAV imagery to the ground, ten ground control points (GCPs) were distributed across the field at the beginning of the season, to obtain photogrammetric imagery with uniform vertical and horizontal accuracy. The GCPs were 25 cm × 25 cm square, with a specific albedo for camera calibration (atmospheric corrections). The GCP coordinates were ensured with a Leica Viva GS15 (Leica Geosystems AG, Heerbrugg, Switzerland) GPS (horizontal accuracy of 0.025 m-vertical accuracy of 0.035 m). Fixed targets were used for a more accurate geo-referencing of UAV aerial imagery and for overlaying the measurements from multiple dates.

2.3. Data Processing

The acquired images were processed by the eMotion software (senseFly, Cheseaux-sur-Lausanne, Switzerland) to generate low-resolution visible (RGB) images of the crops. The eBee's supplied software to build a project using the drone's geo-tagged images. The project is loaded on a laptop in Pix4Dmapper Ag (senseFly, Cheseaux-sur-Lausanne, Switzerland) to run a rapid analysis for visual crop variability.

The multiple overlapped images of the whole plot were stitched and ortho-rectified to create the geo-referenced ortho-mosaicked image. In the laboratory, data processing (orthomosaicking) of acquired images were performed with Postflight Terra 3D software package, a customized version of the Pix4D digital photogrammetric solution specifically optimized for the eBee, to generate ortho-images. Postflight Terra 3D incorporates a scale-invariant feature transform (SIFT) algorithm to match key points across multiple images [26] and process data in three key steps: (1) initial processing, (2) point cloud densification, and (3) orthomosaic generation.

Ortho-rectification by aero-triangulation and mosaicking were elaborated for processing, starting from the exterior position and orientation parameters provided by the UAV internal system (roll, pitch, and yaw angles). Orthoimages were produced from the flights, with a pixel resolution of 5 cm. Ten GCPs (Leica Viva GS15 (Leica Geosystems AG, Heerbrugg, Switzerland) were used to complete the external camera orientation. The orthomosaic was georeferenced to UTM-WGS84 zone 33N Italy. The final outputs were an RGB (Visible) GeoTIFF with a resolution of 5 cm^2 pixels.

The normalized difference vegetation index (NDVI) and the soil-adjusted vegetation index (SAVI) layers were generated in a raster calculator from extracted red (R) and near infra-red (NIR) channels. The index calculator function of Postflight Terra 3D was used for generating VI maps. A certificate of calibration of Canon S110 NIR was uploaded in the software to optimize internal camera parameters, such as focal length, principal points, and lens distortions. The ten GCPs with a known albedo for Red, Green, and NIR channel (reflectance panel) were used to calibrate the camera to achieve uniform quality of image (exposure and brightness) and for atmospheric correction in the software section "processing options," point 3 DSM, orthomosaic, index and for creating VIs map. The resolution of the reflectance map has been set at 5 cm^2 pixels, the whole map was elaborated for each VI and date, and the NDVI and OSAVI formula were selected (index map function). The final outputs were an NDVI GeoTIFF and an OSAVI GeoTIFF, both with a resolution of 5 cm^2 pixels. GeoTIFF images and georeferenced sampling data were processed for agronomic purpose with QGIS 2.8.1.

2.4. Vegetation Indices

The normalized difference vegetation index (NDVI) was calculated according to Equation (1):

$$NDVI = (NIR - RED)/(NIR + RED) \tag{1}$$

The NDVI ranges from −1.0 to 1.0, where positive values indicate increasing greenness and negative values indicate non-vegetated features. It has some disadvantages though such as saturation in later growth stages [27,28].

The optimized soil-adjusted vegetation index (OSAVI) was calculated according to Equation (2). The OSAVI index was proposed by Rondeaux et al. [29] using reflectance in the red and near-infrared bands; through the following formulation:

$$OSAVI = (NIR - RED)/(NIR + RED + 0.16) \tag{2}$$

where 0.16 is the soil adjustment coefficient, selected according to the optimal value to minimize soil background variations.

2.5. Statistical Analysis

Regression analysis, coefficients of determination, significance levels, and RMSE were computed on two sets of geo-referred data: ground crop samplings (LAI, Biomass, and spikes for square meter) and remote UAV data using the statistical package Origin PRO 8 (Origin Lab Corporation, Northampton, MA, USA). H and L sub-area were selected starting from homogeneous areas identified in a previous paper [22] as follow: the yield-related traits and VIs data were analyzed using the hierarchical clustering Ward's minimum variance approach [30] to classify observations into groups, in which the group members have common properties. Statistical procedures were computed using OriginPRO 8. The difference between H and L sub-areas were recorded by the analysis of variance.

3. Results and Discussion

Ground agronomic data (biomass, green leaf area index, spikes, and plant height) of each cultivar have been correlated to high-resolution multispectral images (NDVI and OSAVI). The correlation of the whole crop cycle data was performed in accordance with van Ittersum et al. [31], who states that it is essential to study the relationship among VIs and crop traits, to evaluate and estimate the yield potential and the yield gap (Table 1). In the present study, polynomial regressions were found to be the simplest adjustment to report the time-course variability along the growth stage, starting from seedling to harvest. Moreover, significant relationships with different R^2 values were found for both the indices and crop growth traits. The spectro-radiometric response of the wheat canopy was not linear during the growing period in accordance with Aparicio et al. [32] and Dang et al. [33], since the high sensitivity of the VIs when LAI is lower (early season) and less sensitive after the canopy closes.

The relationships among VIs and crop parameters were highly significant: the R^2 for NDVI vs. biomass reached 0.953 for Odisseo and 0.859 for Ariosto, while the R^2 of OSAVI vs. biomass was 0.943 for Odisseo and 0.857 for Ariosto. The relationships among each VI and LAI showed an R^2 of 0.910 for NDVI Odisseo and R^2 of 0.777 for Ariosto, and it was 0.879 for OSAVI Odisseo and 0.763 for Ariosto. The relationships among each VI and height showed an R^2 of 0.828 for NDVI Odisseo and 0.817 for OSAVI Odisseo and R^2 of 0.518 for both NDVI and OSAVI for Ariosto. Ground truth georeferenced data of the number of spikes per square meter at harvest stages were also related to NDVI and OSAVI data; significant relationships were found with an R^2 of 0.788 for Odisseo and 0.769 for Ariosto recorded for NDVI index and an R^2 of 0.790 for Odisseo and of 0.597 for Ariosto recorded for OSAVI index.

Briefly, all R^2 of yield components were highly correlated (p-value < 0.001) with VIs, and are higher for Odisseo than the Ariosto.

Table 1. Polynomial regression analysis of vegetation index (VI) data and crop parameters each based on 44 data points. LAI = leaf area index, RMSE = root mean square error. Significance level: p-value < 0.001.

| | Odisseo | | | | Ariosto | | | |
| | NDVI | | OSAVI | | NDVI | | OSAVI | |
	R^2	RMSE	R^2	RMSE	R^2	RMSE	R^2	RMSE
Biomass	0.953	0.06	0.943	0.062	0.859	0.075	0.857	0.082
LAI	0.910	0.059	0.879	0.076	0.777	0.095	0.763	0.106
Spikes	0.788	0.067	0.790	0.075	0.769	0.042	0.597	0.065
Height	0.828	0.083	0.817	0.102	0.518	0.105	0.519	0.127

3.1. Agronomic Data and VIs at Different Crop Growth Stages

In the whole plot, the biomass at seedling stage ranged from 68 to 156 g m^{-2} for both cultivars and that of LAI (0.12 to 0.47) (Table 2), while VIs ranged from 0.26 to 0.57 for NDVI and from 0.38 to 0.65 for OSAVI, for both cultivars. The light differences recorded within yield components as well as VIs value gave no significant relationships among VIs and yield components (Table 3). This was due to a low amount of accumulated biomass and LAI values, as well as the difficulties of simple indices to discriminate between soil and vegetation as found also by many authors, such as Aparicio et al. [34] and Chlingaryan et al. [24].

At the tillering stage, the spatial variability of crop traits was highly evident; thus, the most productive data was shown by the biomass and LAI which was four times higher than the less favored zones (Table 2).

In the most productive zones, VI values ranged from 0.85 to 0.99 (for both cultivars), which revealed 2.5 times more than the lower value (Figure 2).

The relationships between VIs and crop parameters were highly significant for both cultivars (Table 3). For biomass, the R^2 reached 0.88 for Odisseo and 0.66 for Ariosto for LAI, the R^2 reached 0.85 for Odisseo and 0.69 for Ariosto. The plant height and VI regression were not significant for Ariosto. These results are in accordance with findings of several authors at tillering, for example Dalla Marta et al. [35] found a highly significant (positive) relationship with biomass weight, while Reyniers and Vrindts [36] and Magney et al. [37] found respectively an R^2 of 0.76 and $R^2 = 0.62$ between NDVI and biomass.

At anthesis, the high spatial variability of the crop traits value and VI values were still evident for both cultivars (Table 2, Figure 2). VIs were highly correlated to crop traits as reported in Table 3; the values of the R^2 for NDVI vs. biomass reached 0.89 for Odisseo and 0.82 for Ariosto, while the R^2 of OSAVI vs. biomass was 0.80 for Odisseo and 0.78 for Ariosto (p < 0.001).

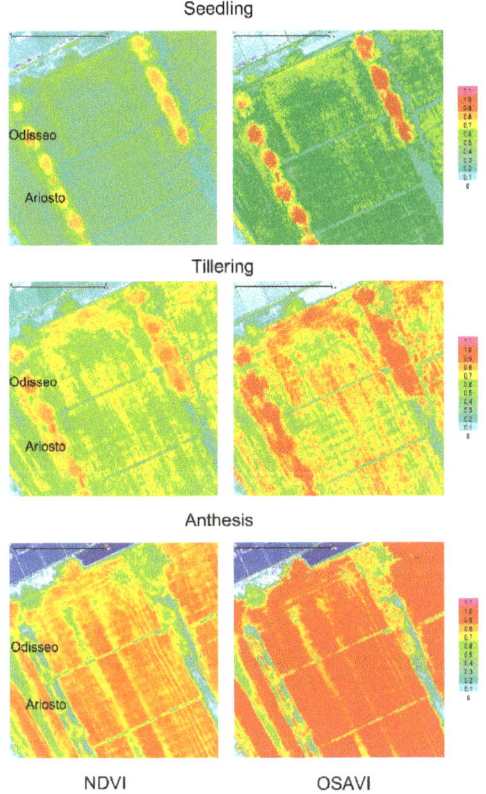

Figure 2. Normalized difference vegetation index (NDVI) and optimized soil-adjusted vegetation index (OSAVI) images for Odisseo (upper plot) and Ariosto (lower plot), at seedling, tillering, and anthesis growth stages.

Table 2. Range value (min and max), mean and standard deviation (SD) of yield components (biomass, LAI, plant height, and spike) and vegetation indices (VIs) (NDVI and OSAVI) measured at seedling, tillering, and anthesis stages.

Growth Stages	Yield Components and VIs	Odisseo				Ariosto			
		Min	Max	Mean	SD	Min	Max	Mean	SD
Seedling	Biomass (g m^{-2})	72.2	156.1	102.1	30.0	67.5	145.0	97.8	26.6
	LAI	0.167	0.472	0.267	0.086	0.123	0.424	0.269	0.089
	NDVI	0.266	0.467	0.383	0.083	0.263	0.569	0.414	0.098
	OSAVI	0.386	0.596	0.499	0.087	0.381	0.650	0.532	0.088
Tillering	Biomass (g m^{-2})	116	558	332	149	185.2	740.8	363.05	185.9
	LAI	0.328	1.610	0.922	0.409	0.500	2.259	1.106	0.566
	Plant height (cm)	34.5	54.5	45.8	5.43	28.3	45.6	39.1	5.2
	NDVI	0.338	0.828	0.639	0.165	0.434	0.857	0.637	0.145
	OSAVI	0.400	0.978	0.755	0.195	0.500	0.991	0.743	0.166
Anthesis	Biomass (g m^{-2})	282	1116	761	258	587.5	1453.9	911.5	251.6
	LAI	0.732	2.56	1.722	0.546	1.32	2.69	1.94	0.459
	Spikes (n° m^{-2})	217	527	411	91	258	651	415	119
	Plant height (cm)	39.0	73.0	63.2	10.7	60.0	73.0	69.5	3.9
	NDVI	0.527	0.838	0.753	0.092	0.654	0.876	0.814	0.073
	OSAVI	0.631	1.004	0.892	0.112	0.725	1.067	0.955	0.105

Table 3. Polynomial regression analysis of VIs data and crop parameters (whole plots) during crop cycle (seedling, tillering, and anthesis). LAI = leaf area index, RMSE = root mean square error, n.s. = not significant.

Growth Stages	VIs	Yield Components	Odisseo			Ariosto		
			R^2	RMSE	*p*-Value	R^2	RMSE	*p*-Value
Seedling	NDVI	Biomass	n.s.	0.0588	0.039	n.s.	0.082	0.012
		LAI	n.s.	0.0577	0.034	n.s.	0.079	0.097
	OSAVI	Biomass	n.s.	0.5175	0.032	n.s.	0.071	0.092
		LAI	n.s.	0.0684	0.079	n.s.	0.076	0.144
Tillering	NDVI	Biomass	0.882	0.063	0.001	0.665	0.077	0.007
		LAI	0.852	0.070	0.001	0.689	0.089	0.005
		Plant height	0.866	0.091	0.002	n.s.	0.141	0.320
	OSAVI	Biomass	0.937	0.851	0.001	0.672	0.105	0.007
		LAI	0.844	0.085	0.001	0.687	0.103	0.005
		Plant height	0.859	0.110	0.002	n.s.	0.16	0.250
Anthesis	NDVI	Biomass	0.893	0.033	0.001	0.818	0.007	0.001
		LAI	0.843	0.040	0.001	0.694	0.073	0.422
		Spikes	0.725	0.060	0.001	0.768	0.044	0.001
		Plant height	0.736	0.070	0.030	0.820	0.046	0.010
	OSAVI	Biomass	0.804	0.756	0.001	0.777	0.055	0.002
		LAI	0.745	0.063	0.002	0.672	0.106	0.470
		Spikes	0.680	0.074	0.008	0.697	0.060	0.001
		Plant height	0.680	0.091	0.061	0.803	0.070	0.016

These results are in accordance with the Marti et al. [38], who found the highest correlation values between growth and NDVI measurements when performed around anthesis, such as with Villegas et al. [39] that reported the highest values of R^2 for the relationship between NDVI and crop dry weight at anthesis. Cabrera-Bosquet et al. [40] have demonstrated a strong linear regression between durum wheat and VIs. On the contrary, Dalla Marta et al. [35] found completely different results, with no correlations observed between the crop parameters and the indices, due to VIs saturation.

3.2. Backward Analysis

Table 4 reports yield and yield components at harvest for the complete plot (W), as well as data of the two sub-areas, H_O and H_A with high yield and L_O and L_A with low yield.

Table 4. Yield (t ha^{-1}), biomass (g m^{-2}), spike number (n° m^{-2}), and plant height (cm) of Odisseo and Ariosto at harvest stage. Mean data are related to the whole plot (W) and to the sub-areas; H_o (high yield of Odisseo) and H_A (high yield of Ariosto), L_O (low yield Odisseo), and L_A (low yield Ariosto). The *p*-level values are related to H and L significant differences within each cultivar.

	Odisseo				Ariosto			
	W	H_O (SD)	L_O (SD)	*p*-level	W	H_A (SD)	L_A (SD)	*p*-level
Yield (t ha^{-1})	5.73	6.77 (0.83)	4.66 (0.61)	<0.0064	5.05	6.17 (0.44)	3.94 (0.25)	1.26×10^{-4}
Biomass (g m^{-2})	760	1029 (106)	500 (208)	<0.0398	911	1171 (200)	682 (77)	<0.0039
Spikes (n° m^{-2})	412	509 (74.3)	320 (27.3)	<0.0034	360	465 (25)	230 (31)	2.39×10^{-5}
Plant height (cm)	63.2	72.2 (2.95)	50.5 (8.5)	<0.0236	69.5	72.5 (5.57)	66.25 (4.78)	n.s.

The yield data of both cultivars were in accordance with those reported by Visioli et al. [41] in Italy. The crop yield of the whole plots was 5.73 t ha^{-1} for Odisseo (W_O) and 5.05 t ha^{-1} for Ariosto (W_A). The yield of H_O and H_A were about 20% higher than the whole plot yield, which in turn were 20% higher than L_O and L_A. Yield components were ranked according to the yield levels: biomass and spikes per square meter of H were about 25–35% higher than the W and 37–50% higher than L sub-plot areas. Differences in plant heights were less marked.

The subplot detected at harvest was analyzed backward in time from anthesis to seedling.

At anthesis, the analysis of VI value frequencies showed the maximum evidence of different crop status of H and L sub-areas and W plots. More than the 75% of the surface of the sub-areas showed a value higher (H sub-plot) or lower (L sub-plot) than W plot (Figure 3). The H and L areas curves showed a low overlapping. The H curve appeared narrow and tall. Data were confirmed by all the yield components that were significantly different from each other in the two sub-areas and W plot.

At tillering, the histogram of VIs frequency (Figure 4) shows differences among the whole plot data, both H and L areas. These differences were lower than those detected at the anthesis stage.

For both the cultivars and indices, the H zones showed values of the frequency which were higher than the 75% of H_O and H_A, and L_A, L_O VI values were higher and lower respectively than the median value of the whole plot.

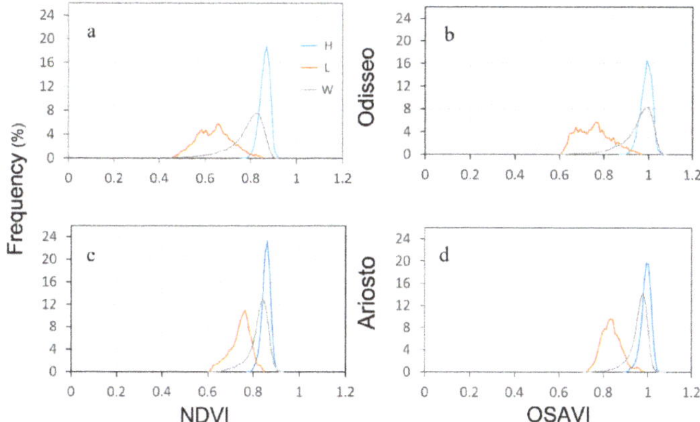

Figure 3. Relative frequency curves of the two VIs, NDVI (**a–c**) and OSAVI (**b–d**) within each cultivar, Odisseo (**a–b**) and Ariosto (**c–d**). In grey, the whole plot (W) curves have been shown, in red the low-yield sub-areas (L), and in blue the high-yield sub-areas (H) at the anthesis stage.

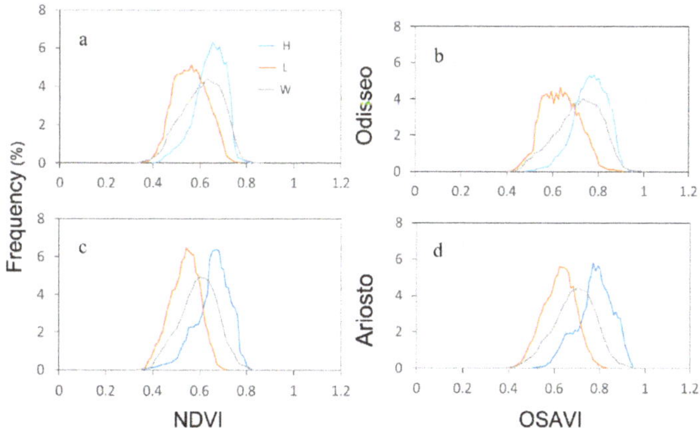

Figure 4. Relative frequency curves of the two VIs, NDVI (**a–c**) and OSAVI (**b–d**) within each cultivar, Odisseo (**a–b**) and Ariosto (**c–d**). In grey, the whole plot (W) curves have been shown, in red the low-yield sub-areas (L), and in blue the high-yield sub-areas (H) at the tillering stage.

At the seedling stage, the relative frequency curves of the two VIs are presented in Figure 5. In all the graphs, the curves appeared to be very close, except for the OSAVI of Ariosto (d) at the seedling stage.

In all graphs, about 75% of the H data (third percentile) were higher than the second percentile of the W curve as well as the first percentile of L data. The highest differences between H and L curves were detected for the OSAVI index, Ariosto (d), because the index that considers soil brightness are perform slightly better than the NDVI, as reported in different studies [42,43].

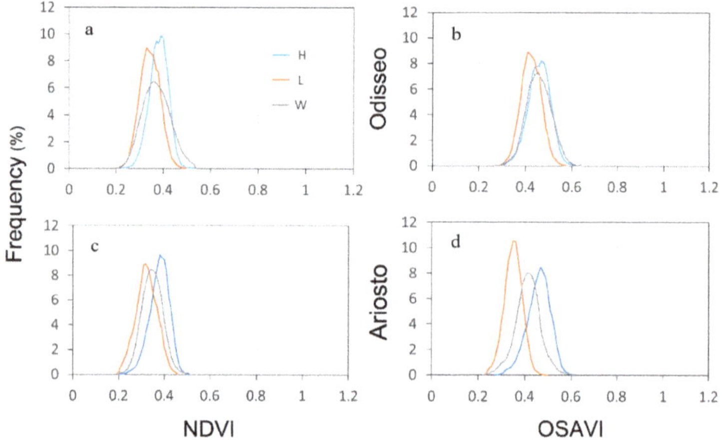

Figure 5. Relative frequency curves of the two VIs, NDVI (**a–c**) and OSAVI (**b–d**) within each cultivar, Odisseo (**a–b**) and Ariosto (**c–d**). In grey, the whole plot (W) curves have been shown, in red the low-yield sub-areas (L), and in blue the high-yield sub-areas (H) at the seedling stage.

4. Conclusions

In the last few years, UAVs have been used as tools that provide a high number of information per square meter for precision agriculture management and its development. The potentials of the information provided by the drone images have not yet been fully explored. In this paper, we have analyzed the frequency distribution curves of high-resolution VIs images of the two cultivars of durum wheat.

Within each cultivar, two yield sub-areas were detected by cluster analysis at the harvest stage and analyzed backward at different growth stages, in comparison with the whole plot data. The crop yield of the whole plots revealed 5.73 t ha^{-1} for Odisseo (W$_O$) and 5.05 t ha^{-1} for Ariosto (W$_A$). The yield for H$_O$ and H$_A$ recorded 20% higher than the whole plot yield, which in turn were 20% higher than L$_O$ and L$_A$, moreover the biomass and the spikes per square meter of H was noted 25–35% higher than the W and 37–50% higher than L sub-plot areas, confirming the ability of cluster analysis to detect groups with the same variability.

The analysis of frequencies of high-resolution VI images showed that the 75% of H data (third percentile) were recorded higher than the second percentile of the W curve as well as for the first percentile of L data. The maximum evidence of different crop status of H and L sub-areas and W plots were detected at anthesis by both indices and cultivars.

The analysis of frequency curves provided significant differences at the seedling stage where agronomic parameters showed only slight differences which were revealed as not significant following the traditional approach (regression between agronomic traits and VIs). For Ariosto, the seedling OSAVI index showed an interesting differentiation among areas, confirming the ability of OSAVI to consider the soil brightness.

The analysis of frequencies of high-resolution VI images provided detailed information on crop spatial and temporal variability and can allow developing a new approach for automatic detection of suitable vegetation indices, and spatial and temporal crop variability for precision agriculture practices.

Author Contributions: The authors, S.M. and A.A. contributed equally to this work.

Acknowledgments: The authors are grateful to Giancarmine Montalto and Emiliano Camarda (Molise Geodetica) who were in charge of the UAV flights.

Conflicts of Interest: The authors declare no conflict of interest.

References

1. Cavallo, G.; De Benedetto, D.; Castrignanò, A.; Quarto, R.; Vonella, A.V.; Buttafuoco, G. Use of geophysical data for assessing 3D soil variation in a durum wheat field and their association with crop yield. *Biosyst. Eng.* **2016**, *152*, 28–40. [CrossRef]

2. Zhang, C.; Kovacs, J.M. The application of small unmanned aerial systems for precision agriculture: A review. *Precis. Agric.* **2012**, *13*, 693–712. [CrossRef]

3. Candiago, S.; Remondino, F.; De Giglio, M.; Dubbini, M.; Gattelli, M. Evaluating multispectral images and vegetation indices for precision farming applications from UAV images. *Remote Sens.* **2015**, *7*, 4026–4047. [CrossRef]

4. Guo, L.; An, N.; Wang, K. Reconciling the discrepancy in ground- and satellite-observed trends in the spring phenology of winter wheat in China from 1993 to 2008. *J. Geophys. Res. Atmos.* **2016**, *121*, 1027–1042. [CrossRef]

5. Schirrmann, M.; Giebel, A.; Gleiniger, F.; Pflanz, M.; Lentschke, J.; Dammer, K.H. Monitoring Agronomic Parameters of Winter Wheat Crops with Low-Cost UAV Imagery. *Remote Sens.* **2016**, *8*, 706. [CrossRef]

6. Rasmussen, J.; Ntakos, G.; Nielsen, J.; Svensgaard, J.; Poulsen, R.N.; Christensen, S. Are vegetation indices derived from consumer-grade cameras mounted on UAVs sufficiently reliable for assessing experimental plots? *Eur. J. Agron.* **2016**, *74*, 75–92. [CrossRef]

7. Duan, T.; Chapman, S.C.; Guo, Y.; Zheng, B. Dynamic monitoring of NDVI in wheat agronomy and breeding trials using an unmanned aerial vehicle. *F. Crop. Res.* **2017**, *210*, 71–80. [CrossRef]

8. Abdollahnejad, A.; Panagiotidis, D.; Surový, P.; Ulbrichová, I. UAV capability to detect and interpret solar radiation as a potential replacement method to hemispherical photography. *Remote Sens.* **2018**, *10*, 423. [CrossRef]

9. Alvino, A.; Marino, S.; Alvino, A.; Marino, S. Remote Sensing for Irrigation of Horticultural Crops. *Horticulturae* **2017**, *3*, 40. [CrossRef]

10. Mulla, D.J. Twenty five years of remote sensing in precision agriculture: Key advances and remaining knowledge gaps. *Biosyst. Eng.* **2013**, *114*, 358–371. [CrossRef]

11. Kyratzis, A.C.; Skarlatos, D.P.; Menexes, G.C.; Vamvakousis, V.F.; Katsiotis, A. Assessment of Vegetation Indices Derived by UAV Imagery for Durum Wheat Phenotyping under a Water Limited and Heat Stressed Mediterranean Environment. *Front. Plant Sci.* **2017**, *8*, 1–14. [CrossRef] [PubMed]

12. Liebisch, F.; Kirchgessner, N.; Schneider, D.; Walter, A.; Hund, A. Remote, aerial phenotyping of maize traits with a mobile multi-sensor approach. *Plant Methods* **2015**, *11*. [CrossRef] [PubMed]

13. Basso, B.; Cammarano, D.; Cafiero, G.; Marino, S.; Alvino, A. Cultivar discrimination at different site elevations with remotely sensed vegetation indices. *Ital. J. Agron.* **2011**, *6*, 1–5. [CrossRef]

14. Marino, S.; Alvino, A. Proximal sensing and vegetation indices for site-specific evaluation on an irrigated crop tomato. *Eur. J. Remote Sens.* **2014**, *47*, 271–283. [CrossRef]

15. Marino, S.; Aria, M.; Basso, B.; Leone, A.P.; Alvino, A. Use of soil and vegetation spectroradiometry to investigate crop water use efficiency of a drip irrigated tomato. *Eur. J. Agron.* **2014**, *59*, 67–77. [CrossRef]

16. Marino, S.; Cocozza, C.; Tognetti, R.; Alvino, A. Use of proximal sensing and vegetation indexes to detect the inefficient spatial allocation of drip irrigation in a spot area of tomato field crop. *Precision Agric.* **2015**, *16*, 613–629. [CrossRef]

17. Yue, J.; Feng, H.; Jin, X.; Yuan, H.; Li, Z.; Zhou, C.; Yang, G.; Tian, Q. A comparison of crop parameters estimation using images from UAV-Mounted Snapshot Hyperspectral Sensor and High-Definition Digital Camera. *Remote Sens.* **2018**, *10*, 1138. [CrossRef]

18. Holman, F.H.; Riche, A.B.; Michalski, A.; Castle, M.; Wooster, M.J.; Hawkesford, M.J. High throughput field phenotyping of wheat plant height and growth rate in field plot trials using UAV based remote sensing. *Remote Sens.* **2016**, *8*. [CrossRef]

19. Araus, J.L.; Cairns, J.E. Field high-throughput phenotyping: The new crop breeding frontier. *Trends Plant Sci.* **2014**, *19*, 52–61. [CrossRef] [PubMed]

20. Erdle, K.; Mistele, B.; Schmidhalter, U. Spectral high-throughput assessments of phenotypic differences in biomass and nitrogen partitioning during grain filling of wheat under high yielding Western European conditions. *F. Crop. Res.* **2013**, *141*, 16–26. [CrossRef]

21. Fischer, R.A. Wheat physiology: A review of recent developments. *Crop Pasture Sci.* **2011**, *62*, 95–114. [CrossRef]

22. Marino, S.; Alvino, A. Detection of homogeneous wheat areas using multi-temporal UAS images and ground truth data analyzed by cluster analysis. *Eur. J. Remote Sens.* **2018**, *51*, 266–275. [CrossRef]

23. Damian, J.M.; De Castro Pias, O.H.; Santi, A.L.; Di Virgilio, N.; Berghetti, J.; Barbanti, L.; Martelli, R. Delineating management zones for precision agriculture applications: A case study on wheat in sub-tropical Brazil. *Ital. J. Agron.* **2016**, *11*, 171–179. [CrossRef]

24. Chlingaryan, A.; Sukkarieh, S.; Whelan, B. Machine learning approaches for crop yield prediction and nitrogen status estimation in precision agriculture: A review. *Comput. Electron. Agric.* **2018**, *151*, 61–69. [CrossRef]

25. Zadoks, J.C.; Chang, T.T.; Konzak, C.F. A decimal code for the growth stages of cereals. *Weed Res.* **1974**, *14*, 415–421. [CrossRef]

26. Küng, O.; Strecha, C.; Beyeler, A.; Zufferey, J.-C.; Floreano, D.; Fua, P.; Gervaix, F. The accuracy of automatic photogrammetric technique on ultra-light UAV imagery. *Int. Arch. Photogramm. Remote Sens. Spat. Inf. Sci.* **2012**, *XXXVIII-1/C22*, 125–130. [CrossRef]

27. Lopresti, M.F.; Di Bella, C.M.; Degioanni, A.J. Relationship between MODIS-NDVI data and wheatyield: A case study in Northern Buenos Airesprovince, Argentina. *Inf. Process. Agric.* **2015**, *2*, 73–84. [CrossRef]

28. Colomina, I.; Molina, P. Unmanned aerial systems for photogrammetry and remote sensing: A review. *ISPRS J. Photogramm. Remote Sens.* **2014**, *92*, 79–97. [CrossRef]

29. Rondeaux, G.; Steven, M.; Baret, F. Optimization of soil-adjusted vegetation indices. *Remote Sens. Environ.* **1996**, *55*, 95–107. [CrossRef]

30. Ward, J.H. Hierarchical Grouping to Optimize an Objective Function. *J. Am. Stat. Assoc.* **1963**, *58*, 236–244. [CrossRef]

31. Van Ittersum, M.K.; Cassman, K.G.; Grassini, P.; Wolf, J.; Tittonell, P.; Hochman, Z. Yield gap analysis with local to global relevance—A review. *Field Crop. Res.* **2013**, *143*, 4–17. [CrossRef]

32. Aparicio, N.; Villegas, D.; Casadesus, J.; Araus, J.L.; Royo, C. Spectral Vegetation Indices as Nondestructive Tools for Determining Durum Wheat Yield. *Agron. J.* **2000**, *92*, 83–91. [CrossRef]

33. Dang, Y.P.; Pringle, M.J.; Schmidt, M.; Dalal, R.C.; Apan, A. Identifying the spatial variability of soil constraints using multi-year remote sensing. *F. Crop. Res.* **2011**, *123*, 248–258. [CrossRef]

34. Aparicio, N.; Villegas, D.; Royo, C.; Casadesus, J.; Araus, J.L. Effect of sensor view angle on the assessment of agronomic traits by ground level hyper-spectral reflectance measurements in durum wheat under contrasting Mediterranean conditions. *Int. J. Remote Sens.* **2004**, *25*, 1131–1152. [CrossRef]

35. Dalla Marta, A.; Grifoni, D.; Mancini, M.; Orlando, F.; Guasconi, F.; Orlandini, S. Durum wheat in-field monitoring and early-yield prediction: Assessment of potential use of high resolution satellite imagery in a hilly area of Tuscany, Central Italy. *J. Agric. Sci.* **2015**, *153*, 68–77. [CrossRef]

36. Reyniers, M.; Vrindts, E. Measuring wheat nitrogen status from space and ground-based platform. *Int. J. Remote Sens.* **2006**, *27*, 549–567. [CrossRef]

37. Magney, T.S.; Eitel, J.U.H.; Huggins, D.R.; Vierling, L.A. Proximal NDVI derived phenology improves in-season predictions of wheat quantity and quality. *Agric. For. Meteorol.* **2016**, *217*, 46–60. [CrossRef]

38. Marti, J.; Bort, J.; Slafer, G.A.; Araus, J.L. Can wheat yield be assessed by early measurements of Normalized Difference Vegetation Index? *Ann. Appl. Biol.* **2007**, *150*, 253–257. [CrossRef]

39. Villegas, D.; Aparicio, N.; Blanco, R.; Royo, C. Biomass accumulation and main stem elongation of Durum wheat grown under Mediterranean conditions. *Ann. Bot.* **2001**, *88*, 617–627. [CrossRef]

40. Cabrera-Bosquet, L.; Molero, G.; Stellacci, A.; Bort, J.; Nogués, S.; Araus, J. NDVI as a potential tool for predicting biomass, plant nitrogen content and growth in wheat genotypes subjected to different water and nitrogen conditions. *Cereal Res. Commun.* **2011**, *39*, 147–159. [CrossRef]
41. Visioli, G.; Bonas, U.; Dal Cortivo, C.; Pasini, G.; Marmiroli, N.; Mosca, G.; Vamerali, T. Variations in yield and gluten proteins in durum wheat varieties under late-season foliar versus soil application of nitrogen fertilizer in a northern Mediterranean environment. *J. Sci. Food Agric.* **2018**, *98*, 2360–2369. [CrossRef] [PubMed]
42. Marino, S.; Alvino, A. Hyperspectral vegetation indices for predicting onion (*Allium cepa* L.) yield spatial variability. *Comput. Electron. Agric.* **2015**, *116*. [CrossRef]
43. Marino, S.; Basso, B.; Leone, A.P.; Alvino, A. Agronomic traits and vegetation indices of two onion hybrids. *Sci. Hortic.* **2013**, *155*, 56–64. [CrossRef]

 *agronomy*

Article

Towards Predictive Modeling of Sorghum Biomass Yields Using Fraction of Absorbed Photosynthetically Active Radiation Derived from Sentinel-2 Satellite Imagery and Supervised Machine Learning Techniques

Ephrem Habyarimana [1,*] , Isabelle Piccard [2], Marcello Catellani [1,3], Paolo De Franceschi [1] and Michela Dall'Agata [1]

1   CREA Research Center for Cereal and Industrial Crops, Bologna 40128, Italy; marcello.catellani@enea.it (M.C.); paolo.defranceschi@crea.gov.it (P.D.F.); michela.dallagata@crea.gov.it (M.D.)
2   Vlaamse instelling voor technologisch onderzoek N.V., MOL 2400, Belgium; isabelle.piccard@vito.be
3   Department for Sustainability, Italian National Agency for New Technologies, Energy and Sustainable Economic Development (ENEA), Rotondella (MT) 75026, Italy
*   Correspondence: ephrem.habyarimana@crea.gov.it; Tel.: +39-334-120-8984

Received: 23 March 2019; Accepted: 18 April 2019; Published: 20 April 2019

Abstract: Sorghum crop is grown under tropical and temperate latitudes for several purposes including production of health promoting food from the kernel and forage and biofuels from aboveground biomass. One of the concerns of policy-makers and sorghum growers is to cost-effectively predict biomass yields early during the cropping season to improve biomass and biofuel management. The objective of this study was to investigate if Sentinel-2 satellite images could be used to predict within-season biomass sorghum yields in the Mediterranean region. Thirteen machine learning algorithms were tested on fortnightly Sentinel-2A and Sentinel-2B estimates of the fraction of Absorbed Photosynthetically Active Radiation (fAPAR) in combination with in situ aboveground biomass yields from demonstrative fields in Italy. A gradient boosting algorithm implementing the xgbtree method was the best predictive model as it was satisfactorily implemented anywhere from May to July. The best prediction time was the month of May followed by May–June and May–July. To the best of our knowledge, this work represents the first time Sentinel-2-derived fAPAR is used in sorghum biomass predictive modeling. The results from this study will help farmers improve their sorghum biomass business operations and policy-makers and extension services improve energy planning and avoid energy-related crises.

Keywords: sorghum biomass; prediction modeling; machine learning; fAPAR; Sentinel-2 satellite imagery; big data technology; remote sensing

---

1. Introduction

Sorghum (*Sorghum bicolor* (L.) Moench) is a cereal with a C4 carbon fixation (the Hatch—Slack pathway) cultivated mainly for food, feed, forage, and fuel [1]. Sorghum grain was historically used for human consumption in developing countries but, because it is gluten-free, with low glycemic index and high contents of macronutrients and antioxidants, its utilization as food extended worldwide.

There are several types of sorghum. Grain sorghums are generally shorter (usually having recessive alleles at three of the four Dw genes) than biomass sorghums (having recessive alleles at two Dw genes at most), and have been selected to have the grain as the primary sink for photosynthates.

Biomass sorghum of biofuel production interest was used in this work and includes dual purpose (showing high grain and biomass yields), forage, sweet, and biomass sorghum types [2–5]. The sweet sorghum type translocates photosynthates to the seeds and stem; their stems are juicy (d recessive to D) instead of dry and sweet (x recessive to X) instead of nonsweet [6]. Sweet sorghums are high biomass and sugar yielding crops, and were traditionally bred for syrup or molasses production. Forage sorghum's main characteristics include digestible fiber and low lignin content, while biomass and dual purpose sorghums include high biomass yielding genotypes with high contents of structural carbohydrates, which are being developed for feedstock production [5]. Sorghum can therefore supply several products including starch, soluble sugars, structural carbohydrates, and organic matter for energy production purposes. Several countries worldwide, including in higher latitudes [7], are increasingly developing dedicated biomass sorghums in response to the pressing issue for nations to get independent of foreign energy sources and to cut carbon emissions into the atmosphere [7,8]. As a biofuel-dedicated biomass business, growing sorghum will have to meet critical requirements of high and cost-effective productivity of biobased commodities.

Crop yield forecasting is one of the most important strategies in agriculture, which enables sustainable development and helps avoid famines and shortages in several commodities [9–12]. In industrialized countries, crop yield forecasting provides data to governmental structures, companies, and farmers, which results in strategic advantages such as the rationalization of policy adjustments, price predictions and stabilization, efficient agricultural trade, and simplification of business operations particularly through planning harvest and delivery of the produce, better deployments of machineries and logistics, and a better management at the end user level (e.g., bioreactor owner).

Conventionally, and particularly in developed countries, the information on crop production is collected and disseminated through field surveys and censuses, but this system is rather costly and associated with significant uncertainties [13]. Using satellite imagery resulted in a superior solution [14–16]. Remote sensing data has been used for many years to build operational crop yield forecasting systems like the FAO's Global Information and Early Warning System (GIEWS). The use of remote sensing satellite data for crop yield forecasting is further motivated by wide coverage, near-real time delivery of data and products, and the ability to provide vegetation indicators at low cost. Many studies have shown that forecasting models based on remote sensing data can give similar or better performance comparing to the more sophisticated physiological crop growth models [13,17–19].

The use of remote sensing parameters as proxies for biomass yields was documented in previous works. Normalized difference vegetation index (NDVI), leaf area index (LAI), and fAPAR (fraction of absorbed photosynthetically active radiation) are among the most frequently used parameters [14,16,20]. Recently, the use of biophysical parameters, such as fAPAR, gained more attention relative to using vegetation indices [21,22]. Biophysical parameters reflect the state of the crops more adequately and thus could be better suited for predicting crop yield and production [15,23,24]. fAPAR is defined as the fraction of radiation absorbed by the green vegetation elements in the 400 to 700 nm spectral domain under specified illumination conditions [25]. It is directly linked to photosynthesis, and therefore expresses a canopy's energy absorption capacity [26]. fAPAR values range from 0 to 1, indicating, respectively, bare soil and fully crop covered soil.

Most of the aforementioned studies were focused on common field crops like corn, wheat, barley or soybeans [16,20,27–30]. A few studies dealt with remote sensing-based yield monitoring and prediction [16] in sorghum. For instance, Shafian et al. [31] described the use of unmanned aerial vehicle-based remote sensing to investigate sorghum crop physiological properties. Yang et al. [32] combined airborne digital videography with ground sampling, regression analysis, and image processing to map spatial sorghum grain yield variability within fields and across the cropping season. Johnson [16] presented a comprehensive assessment of the correlations between commonly used Moderate Resolution Imaging Spectroradiometer (MODIS) products and field crop yields, including sorghum, and used these correlations for biomass yield estimation and prediction.

In most studies, however, remote sensing-based biomass yield estimation or prediction makes use of low- or medium-resolution satellite images from sensors such as SPOT-VEGETATION [14,15,23,24] or MODIS [16]. These satellite products have a coarser spatial resolution (250 to 1000 m) compared to the data collected from the two Sentinel-2 satellites in this work (10-m spatial resolution). With the launch of the Sentinel-2 constellation of satellites the overpass frequency (five days and locally even two to three days) the temporal resolution is nearly as good as for SPOT-VEGETATION and MODIS satellites (one to two days). The high spatial resolution of the Sentinel-2 images is an important asset when monitoring crops in agricultural regions characterized by many small fields. To our knowledge no previous studies assessed the efficiency of high resolution Sentinel-2-derived fAPAR data in predicting within-season biomass sorghum yields, and this paper is therefore aimed at addressing this gap.

Deriving yield information from satellite imagery has shown promising results but this technology is not extensively applied across farmers and crop species worldwide [27–29]. In this work, we developed models for within-season prediction of annual and perennial sorghum biomass yields in Emilia-Romagna, Italy, based on fAPAR measurements from Sentinel 2A and Sentinel 2B satellite images on 42 mostly full-fledged commercial sorghum fields. We used machine learning algorithms to create yield prediction equations. These equations can be implemented in decision support systems to allow farmers and/or farming stakeholders to predict biomass yields from sorghum fields of interest early on in the cropping season. This information is very helpful to efficiently schedule fleets of harvesting machinery, transport vehicles, and storage facilities. The fAPAR-derived predictive models for biomass yields can also be implemented by extension services and policy-makers for several purposes, including the possibility to anticipate potential biomass availability and plan ahead, to avoid specific crises such as fuel shortage.

2. Materials and Methods

2.1. Trial Set-Up

Forty-two demonstration trials were run in this work, 23 and 19 of which were evaluated in 2017 and 2018, respectively. In 2017, the experimental sites were located in Conselice, Nonantola, Mirandola, and Anzola dell'Emilia, in the Italian region of Emilia Romagna (Table 1, Figure 1), while in 2018 the sites were established in Anzola, Mirandola, and Conselice (Figure 1, Table 1). The experimental sites were strategically selected to maximize extension impact by conducting most of the trials in the farmers' fields. The experimental fields in Mirandola and Conselice belonged to respective two big farming cooperatives with more than 2000 members each. In Nonantola, the fields belonged to individual farmers, while in Anzola the fields were established in the experimental station of the Council for Agricultural Research and Economics (CREA). The fields were generally of big size relative to plot sizes commonly used under standard experimental settings [1] in order to serve the purpose of demonstrative pilots with the objective of transferring into the production environment the technology of sorghum crop monitoring using satellite imagery. The fields areas ranged from 0.06 ha to 50.00 ha, with a mean and median of 5.70 ha and 1.10 ha, respectively. All the fields were planted with biomass sorghums including biomass per se (high tonnage), sweet, forage, and dual purpose types. One-grain sorghum trials were established in Anzola in 2017, but it was not included in this work in virtue of a different kind of experiment management and a diverse market of the grain sorghum produce relative to biomass sorghum. Thirty-five out of the 42 trials were sown with a single genotype of *Sorghum bicolor* (annual), while the 17IT_mat was sown with a diversity panel of 228 biomass *Sorghum bicolor* genotypes, and six (15R17, 16R17, 16R18, 15R18, 17R18, and 17US_mat) of the trials installed in Anzola were made up of a diversity panel consisting of advanced perennial interploid biomass hybrids deriving from *S. bicolor* × *S. halepense* (SB × SH) crosses. The original SB × SH materials originated from The Land Institute (Salina, KS, United States of America). *S. bicolor* × *S. halepense* breeding strategy was amply detailed in Piper and Kulakow [33] and in Habyarimana et al. [1]. The 15R, 16R, 17R, and 17US_mat trials were sown in 2015, 2016, and 2017, respectively, meaning that regrowth-derived biomass was evaluated for the 15R, 16R, and 17R trials, while for the 17US_mat trial, the biomass evaluated in this study was produced from direct sowing.

Table 1. Pilots descriptors: name, location, variety, season, and productivity.

Serial Number	Field/Pilot Name	Variety Name	Variety Type	Area (ha)	Dry Biomass Yield (t ha^{-1})	Cropping Season	Location Name
1	Botte 1	Harmattan	Dual purpose	9.00	14.13	2018	Conselice
2	Saracca 5	Harmattan	Dual purpose	6.50	10.52	2018	Conselice
3	*V. serrata*	Harmattan	Dual purpose	44.87	9.69	2018	Conselice
4	Magnana	P845F	Forage	32.05	11.43	2018	Conselice
5	Cà bianca	P845F	Forage	3.72	11.11	2018	Conselice
6	Gamberina 3	Aralba	Dual purpose	7.86	9.67	2018	Conselice
7	Sagrate	Harmattan	Dual purpose	50.00	8.90	2017	Conselice
8	Prato_Mensa	Harmattan	Dual purpose	3.29	19.10	2017	Conselice
9	Comuna	P845F	Forage	27.86	24.50	2017	Conselice
10	Gamberina_1	Aralba	Dual purpose	7.60	12.80	2017	Conselice
11	Botte	Harmattan	Dual purpose	5.33	23.50	2017	Conselice
12	Carafolo_G	Bulldozer	Biomass	2.00	17.10	2017	Nonantola
13	Cavriani_S	Merlin	Biomass	2.00	19.50	2017	Nonantola
14	Ferrari_R	Bulldozer	Biomass	1.00	3.40	2017	Nonantola
15	Mattioli_R	Bulldozer	Biomass	1.00	4.90	2017	Nonantola
16	Serafini_G	Bulldozer	Biomass	2.00	15.30	2017	Nonantola
17	Zavatti_E	Bulldozer	Biomass	0.89	7.60	2017	Mirandola
18	Grandi_Magonza	Bulldozer	Biomass	0.80	12.80	2017	Mirandola
19	Grandi_Ponte	Bulldozer	Biomass	1.20	9.50	2017	Mirandola
20	Zini_L	Palo Alto	Biomass	2.50	8.00	2017	Mirandola
21	Villa_verdetta	Bulldozer	Biomass	1.00	6.25	2018	Mirandola
22	Cama_grande	Bulldozer	Biomass	4.40	14.94	2018	Mirandola
23	Cama_piccolo	Bulldozer	Biomass	4.00	15.35	2018	Mirandola
24	Golinelli_Raimondo	Bulldozer	Biomass	2.02	10.48	2018	Mirandola
25	Barozzi_Lidia	Bulldozer	Biomass	3.00	7.19	2018	Mirandola
26	Molon_A	Palo Alto	Biomass	5.00	8.30	2017	Mirandola
27	T1_Anzola	Sole	Biomass	0.74	13.00	2017	Anzola
28	T2_Anzola	Trudan	Forage	0.71	19.00	2017	Anzola
29	T3_Anzola	Hannibal	Sweet	0.71	17.00	2017	Anzola
30	T4_Anzola	Harmattan	Dual purpose	0.70	14.00	2017	Anzola
31	T1_Anzola	Bulldozer	Biomass	0.74	17.00	2018	Anzola
32	T2_Anzola	Hannibal	Sweet	0.71	19.00	2018	Anzola
33	T3_Anzola	Tarzan	Biomass	0.71	21.00	2018	Anzola
34	T4_Anzola	Trudan	Forage	0.70	13.00	2018	Anzola
35	T5_Anzola	Harmattan	Dual purpose	0.70	15.00	2018	Anzola
36	15R17	Perennial	Biomass	0.06	2.39	2017	Anzola
37	16R17	Perennial	Biomass	0.15	8.27	2017	Anzola
38	17IT_mat	Bicolor	Biomass	0.17	18.30	2017	Anzola
39	17US_mat	Bicolor	Biomass	0.15	21.00	2017	Anzola
40	16R18	Perennial	Biomass	0.15	14.35	2018	Anzola
41	15R18	Perennial	Biomass	0.06	10.73	2018	Anzola
42	17R18	Perennial	Biomass	0.15	17.47	2018	Anzola

Crop management followed local extension services guidelines and was well described in Habyarimana et al. [1]. Planting density was 26 (0.75 m spacing between rows; 0.052 m spacing of hills within row) plants per square meter for most (35) trials, and 13 (0.75 m spacing between rows; 0.10 m spacing of hills within row) plants per square meter for 15R, 16R, 17R, 17IT_mat, and 17US_mat trials. In terms of weather (Figures S1–S4), summer was generally dry across years and locations as expected. In 2017, all sites had relatively wet spring except in Anzola, while in 2018 spring was relatively wet in Anzola and Conselice, but dry in Mirandola.

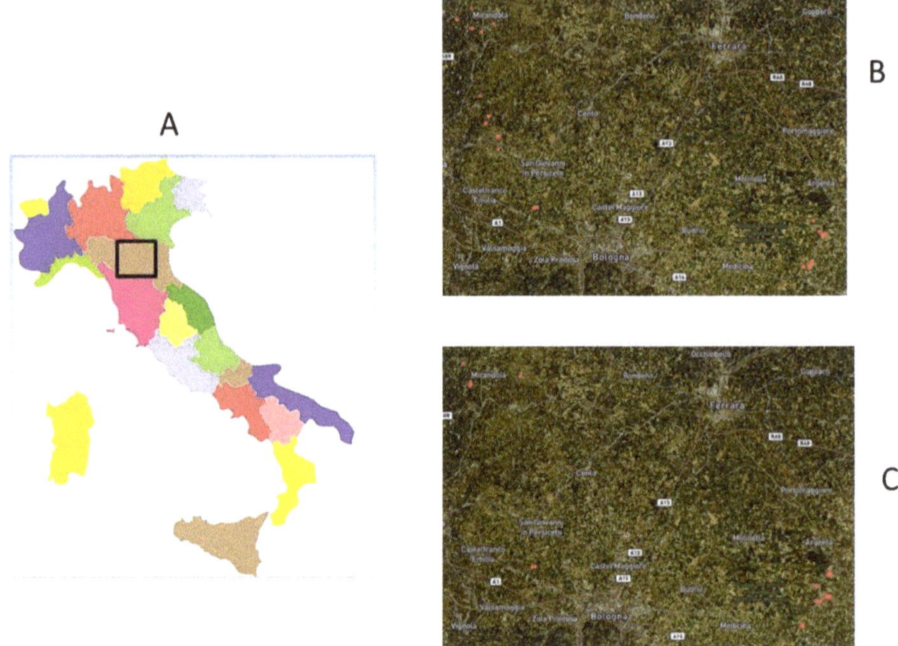

Figure 1. Map of Italy (**A**) with a rectangle inset indicating the geographical location of the experimental sites (red dots) for pilots established in 2017 (**B**) and 2018 (**C**).

2.2. Biomass Data Collection

Trials in Nonantola, Mirandola, and Conselice were harvested at industrial scale from end of August to late November, while all trials in Anzola were harvested end of November using a single-row chopper harvester. Our experience showed that postponing harvest to later times may increase the likelihood for lodging, which may lead to the crop touching the ground, adding grit to the biomass material and possibly reducing biomass quality; delayed harvest also leads to kernel loss and kernel quality deterioration particularly due to molds, insect, and bird damages. The trials were harvested according to two machinery options: forage chopper or swathing the material into windrows and then baling it in large square bales or large round bales. Chopped biomass was weighed immediately at harvest, while baled biomass was weighed when bales were transported to the bioreactor. Chopped and baled biomasses were supplied to private biogas and combustion bioreactors. From each field, a 1kg-composite sample was taken from the sold biomass at the time of shipment to the end user in order to determine the dry mass content in the commercialized produce and calculate the dry biomass yield for each entire field that will be used in modeling. For the diversity panels, samples were taken from each genotype. Fresh samples were weighed and dried at 80 °C to constant weight in a forced air oven. The fresh and dry weights of the samples, and the fresh weight of the entire field's harvest, were used to derive dry mass fraction of the fresh material and dry biomass yield of the entire field. For the diversity panel fields, the final yields integrated the contributions of the component genotypes.

2.3. Satellite Data Acquisition

For this study we used Sentinel-2 optical satellite imagery. The Sentinel-2 mission is based on a constellation of two satellites—Sentinel-2A and Sentinel-2B—both orbiting Earth at an altitude of 786 km, but 180 ° apart to optimize coverage and global revisit times. Swath width, i.e., the image width across the satellite path when scanning the Earth, is 290 km. As a constellation, the revisit time is 5 days. This means that the same spot over the equator is revisited every five days, and even faster at higher latitudes. Sentinel-2 data are acquired on 13 spectral bands in the VNIR (visible and near-infrared) and SWIR (short-wave infrared) range, of which four bands with a spatial resolution of 10 meters (blue, green, red, and near-infrared (NIR)), six bands at 20 meters (three red edge bands, a narrow NIR, and two SWIR bands), and three bands at 60 meters (a coastal aerosol, water vapor, and cirrus band). Spatial resolution refers to the surface area measured on the ground and represented by an individual pixel. Once the Sentinel data are acquired on-board, they are sent to ground and processed by a network of Processing and Archiving Centers. Next, all data products are united, archived, and disseminated online to the users by ESA's Copernicus Space Component (CSC) Ground Segment via the CSC Data Access Coordinated System. To facilitate image transfer and use, the projected Sentinel-2 images are converted to tiles with a fixed size of 100 square kilometers, each of which is approximately 500 MB.

For this study, Sentinel-2A and Sentinel-2B images from tile 32TQQ (including pilots from Conselice) and 32TPQ (including pilots from Anzola, Mirandola, and Nonantola) were downloaded from ESA and processed by Vlaamse Instelling voor Technologisch Onderzoek N.V. (VITO). Processing included atmospheric correction with iCOR [34] and cloud and shadow detection using Sen2COR v2.5.5 (ESA-STEP, ESA, Paris, France). Biophysical parameters fAPAR, fCover, and leaf area index (LAI) were calculated from the top of canopy normalized reflectances following the BV-NET (tool for mapping surface and vegetation variables) method described by Weiss and Baret [35]. The BV-NET methodology is based on neural networks which are trained on a synthetic dataset of ~50,000 simulations using the PROSAIL (PROSPECT and SAIL radiative transfer models) model [36]. The BV-NET version used in this study was calibrated with green, red and near infrared bands, all having a spatial resolution of 10 meters. Sen2Cor and BV-NET are publicly available through ESA's SNAP (Sentinel Application Platform, ESA, Paris, France) toolbox.

Previous studies such as Duveiller et al. [15], López-Lozano et al. [24], and Johnson et al. [16] illustrated the good performance of satellite derived fAPAR for estimating and predicting biomass yields of large field crops, including corn and sugarcane, which, together with sorghum, make-up the world's three economically important C4 crops of the Poaceae family with similar growth habits [37]. We therefore decided to use fAPAR for this study as well. The fAPAR estimates generated with BV-NET from Sentinel-2A and 2B top of canopy reflectances over selected tiles in Emilia-Romagna had a spatial resolution of 10 meters and a temporal resolution of 5 days up to 2–3 days in those areas where the different satellite overpasses overlapped.

For monitoring the sorghum fields in this study "WatchITgrow" (Vlaamse Instelling voor Technologisch Onderzoek N.V., MOL, Belgium) was used. WatchITgrow is a web-based application for crop monitoring developed by VITO. It provides information on crop growth and development as well as possible anomalies derived from Sentinel-2 satellite images and weather data, and it allows the user to store all kinds of collected field data, such as planting and harvest dates and development stages, but also information on crop treatments such as fertilization, spraying, or irrigation. Prior to monitoring, the fields used in this study were geolocalized (Figure 1) using Field GPS (global positioning syatem) application for iPhone with a final field boundary correction using Google Earth. The field polygons were saved as kml files and then imported into WatchITgrow for monitoring. For each field, fAPAR or "greenness" maps were created (see example in Figure 2), and a growth curve was built, showing the evolution of the fAPAR values throughout the cropping season (see example in Figure 3). To build the growth curve the fAPAR values of all pixels within the field were averaged, thereby accounting for an inside buffer of ten meters (one pixel) in order to avoid capturing signals from neighboring fields or other objects. To correct for artifacts in the resulting fAPAR curve such as abnormally low fAPAR values due to undetected clouds, shadows or haze and to interpolate fAPAR values between subsequent acquisition dates, a Whittaker smoothing filter was applied on the curve [38,39].

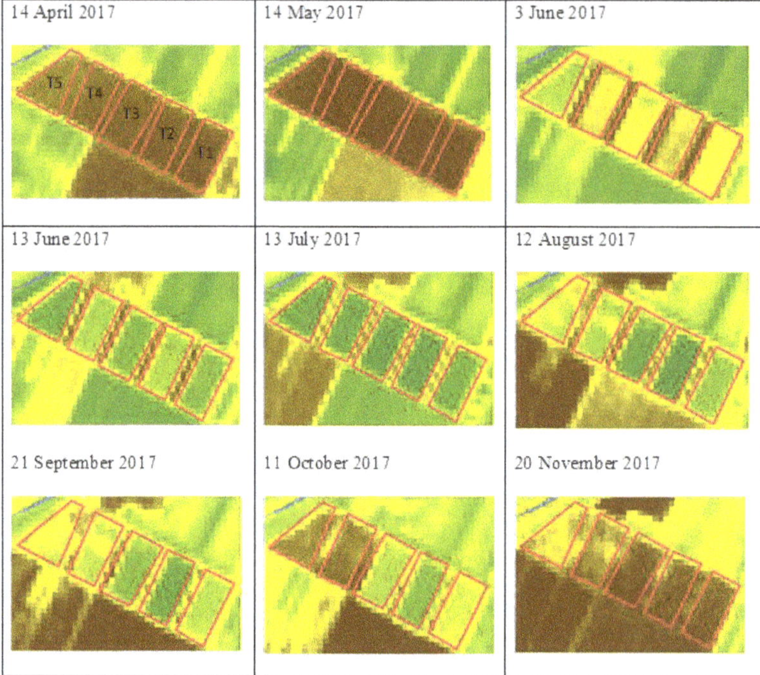

Figure 2. Greenness (fAPAR) maps derived from Sentinel-2 satellite imagery for five sorghum fields in Anzola (from left to right: T5-grain sorghum, T4-dual purpose sorghum, T3-sweet sorghum, T2-forage sorghum, T1-biomass sorghum) for a selected number of dates in 2017, as available via WatchITgrow. T5-grain sorghum was not included in this study (refer to Section 2.1 for detail).

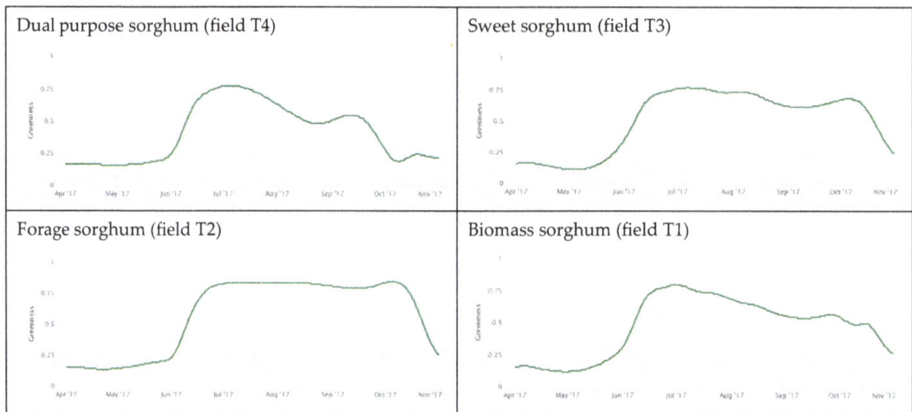

Figure 3. *Cont.*

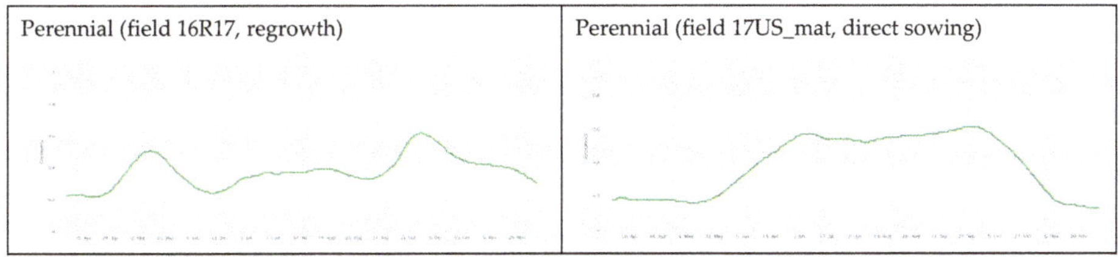

Figure 3. Greenness (fAPAR) graphs derived from Sentinel-2 satellite imagery from 2017 for six sorghum fields in Anzola (T4-dual purpose sorghum, T3-sweet sorghum, T2-forage sorghum, T1-biomass sorghum, and 16R17-perennial, 17US_mat perennial), available via WatchITgrow.

2.4. Modeling Total Aboveground Biomass Yields

Thirteen models were assessed in this study to predict sorghum biomass yields. The models included partial least square discriminant analysis (PLS-DA), principal component analysis discriminant analysis (PCA-DA), neural network (NN), random forest (RF), support vector machine (SVM) with linear classifier (SVML), nonlinear kernel (SVML_G), radial basis kernel (SVM_R), radial basis kernel with polynomial basis kernel (SVM_P), neural network (NNET), eXtreme Gradient Boosting-xgbtree method (GBT), eXtreme Gradient Boosting- xgbDART method (GBD), eXtreme Gradient Boosting-xgbLinear method (GBL), simple linear model (LM), and Neural Network-neuralnet method (NLNET). The simple linear model was used as a benchmark to gauge the performance of the models implemented. The models evaluated in this work were selected based on their robustness as reported in previous studies [40].

The field-based daily interpolated fAPAR estimates extracted from WatchITgrow were converted to fortnightly fAPAR averages. In this study, preference was given to the use of fortnightly fAPAR data as major morphophysiological changes in crops also occur fortnightly [41]. In addition pilots established in experimental stations and in farmers' fields were fortnightly visited at or close to the times the fAPAR images used in this work were acquired. This time management was also favorable and accommodated the busy schedules of farmers and scientists.

Six fortnightly fAPAR values registered from May to July—here referred to as six "days of year" (DOY), that is, DOY 135 and 150 in May, 165 and 180 in June, and 195 and 210 in July—were used as regressor variables in successive predictive modeling of sorghum biomass yields. May, June, and July are important months concerning the predictive modeling, mimicking the 1 to 2 months required to release yield predictions before harvest [42]; taking into account that biomass sorghum in the Mediterranean region is harvested from August to November.

The research questions addressed in this work are (1) how accurately can we predict the yield of a biomass sorghum field based of Sentinel-2-derived fAPAR profile early in the cropping season? (2) Which months and/or days-of-year best contribute useful information for predicting biomass yields in commercial sorghum fields? The solutions to the above problems were evaluated by solving the below linear model for n trials or experimental locations ($i = 1, \ldots, n$) and p prediction times or days of year ($j = 1, \ldots, p$). This model is represented by

$$y_i = \mu + \sum_{j=1}^{p} x_{ij}\beta_j + e_i \tag{1}$$

where μ is the overall mean, y_i is the phenotypic observation (biomass yields) from field i, e_i is the residual comprising all other nongenetic and environmental factors, x_{ij} is the days of year covariates, and β_j is the effect of the *jth* day of year covariate on y_i [43]. Note that it is beyond the scope being presented here to identify and/or predict within-field yield variability for any potential applications. In addition, different sorghum types were combined in this study as they qualified for commercial aboveground biomass production and to mimic farming practices in the region of the study. We also assumed that the test region was homogeneous with respect to climatic conditions.

All statistical analyses were carried out using R software [44]. The predictive models were fitted using the caret R package. In this work, the "one standard error" rule of Breiman et al. [45] was implemented to avoid overfitting, and the caret built-in features were invoked to automatically choose the tuning parameters associated with the best performance of the regression routines. During data preparation, zero-variance regressors were removed and those remaining were centered and scaled in order to avoid regressors with zero or near-zero variance, which often constitute a problem as they behave as second intercepts in predictive models [40]. The dataset was randomly partitioned into training (80% of the entire dataset; 34 observations) and testing set (20% of the entire dataset; eight observations). The training set was used to run a cross-validation experiment to train and assess the models using a 10× repeated 5-random fold cross-validation (CV) iterations, rendering a total of 50 estimates of accuracy and prediction error; a large number of repetitions is expected to compensate for the high variance stemming from a reduced number of folds. Models were validated on the testing set which was an external test (validation) sample set needed so that the model performance can be characterized on data that were not used in the model training. The models were evaluated based on the prediction accuracy, the mean absolute error (MAE), and the mean absolute percentage error (MAPE). The MAE built within the repeated cross validation procedure (model calibration) was used to assess the variability (dependability) of the model performance. On the other hand, the MAE, MAPE, and accuracy obtained on the testing set were used to assess the model predictive ability. The MAPE allows us to compare the prediction of different dependent variables in different scales. The MAE measured the average magnitude of the errors in the set of predictions of biophysical variable values produced in this work, without considering their direction. It represented the average over the test sample of the absolute differences between prediction and actual observation where all individual differences had equal weight. The MAE was chosen for the model verification because it provides an unambiguous measure of the magnitude of the average error and is therefore more appropriate than the Root Mean Square Error (RMSE) for dimensioned evaluations of average model performance error [46]. The distribution of the 50 MAE estimates from the optimal cross-validated models was characterized using boxplot, while the comparison of mean accuracies across models and across prediction times was performed using Duncan's test [47]. The importance of the regressor variables (useful prediction times) was determined using a 0 to 100 index, with 0 corresponding to no effect and 100 corresponding to the highest magnitude of the regressor's importance. The accuracy was defined as the Pearson correlation coefficient between the predicted and the observed biomass yield values in the testing set [5]. From the computed accuracy, r-squared values can be derived in order to better compare, for each model, the proportion of the variance in the dependent variable that is predictable from the regressors.

3. Results

3.1. fAPAR Index Pattern Across Sorghum Types

Three fAPAR curve and map patterns were consistently observed as illustrated in the above Figures 2 and 3 using data from the 2017 cropping season. In dual purpose and biomass sorghums, a major peak was observed earlier in July followed by a drop and then a weak increase at the beginning of the second half of September. For the sweet, forage, and the perennial sorghum (SB × SH) grown from seeds, the fAPAR increased significantly in early July to reach a plateau from then up to late September/early October, whereas, in October, the curve decreases sharply to reach the minimum value in early November. On the other hand, in perennial sorghum regrown from rhizomes, two fAPAR peaks (smaller peak in mid-May, bigger peak in late September/early October) were observed that were separated by a deep drop extending from June to August.

3.2. Assessment and Validation of the Predictive Models, and Importance of Regressors in Total Biomass Prediction

Thirteen models implemented in this work were assessed using the prediction accuracy (Table 2) in the validation set, the mean absolute error (MAE) and mean absolute percentage error (MAPE) metrics (Table 2, Figure 4) produced during the repeated cross-validation, and during the validation stage in the independent sample set. A repeated cross-validation iteration was run for each model, resulting in MAE resample vectors, each containing 50 elements. Over the months evaluated, the range and mean accuracy (Table 2) for NNET, RF, SVM-R, PCA-DA, PLS-DA, SVM-P, SVML, SVML-G, GBT, GBD, GBL, LM, and NLNET, were 0.16–0.78 and 0.56, 0.39–0.82 and 0.63, −0.36–0.88 and 0.16, −0.13–0.76 and 0.43, −0.02–0.77 and 0.47, 0.09–0.81 and 0.50, 0.49–0.80 and 0.64, 0.46–0.80 and 0.64, 0.56–0.81 and 0.66, −0.01–0.84 and 0.49, 0.03–0.93 and 0.47, 0.46–0.78 and 0.61, and 0.01–0.79 and 0.47, respectively.

Table 2. Predictive models × prediction time accuracies and May MAE for the training and validation sets.

(1) Model	(2) Accuracy							(3) May_MAE.T t ha^{-1}	May_MAE.V	
	May	June	July	May–June	June–July	May–July	Mean		t ha^{-1}	%
PLS-DA	0.77	0.49	−0.02	0.69	0.32	0.56	0.47 ab	5.01 bcd	3.62	26.81
PCA-DA	0.76	0.49	−0.13	0.67	0.29	0.52	0.43 ab	5.00 bcd	2.91	21.56
RF	0.82	0.39	0.64	0.68	0.53	0.74	0.63 a	5.05 bcd	2.27	16.81
SVML	0.80	0.49	0.58	0.67	0.59	0.70	0.64 a	4.82 cd	3.74	27.70
SVML-G	0.80	0.49	0.58	0.66	0.61	0.72	0.64 a	4.84 cd	3.74	27.70
SVM-R	0.88	−0.36	0.51	0.02	−0.15	0.08	0.16 b	4.95 bcd	1.87	13.85
SVM-P	0.81	0.49	0.09	0.63	0.42	0.53	0.50 a	4.64 d	6.22	46.07
NNET	0.78	0.56	0.16	0.75	0.38	0.70	0.56 a	11.99 a	12.50	92.59
GBT	0.78	0.56	0.69	0.58	0.81	0.57	0.66 a	5.29 bc	2.68	19.85
GBD	0.84	0.37	−0.01	0.76	0.48	0.51	0.49 a	4.80 d	2.18	16.15
GBL	0.45	0.11	0.89	0.93	0.03	0.43	0.47 ab	5.43 b	3.40	25.19
LM	0.78	0.50	0.56	0.73	0.46	0.65	0.61 ab	4.88 cd	4.53	33.56
NLNET	0.79	0.09	0.33	0.79	0.01	0.79	0.47 ab	5.36 b	2.34	17.33
MEAN	0.77a	0.36b	0.37b	0.66a	0.37b	0.58ab	0.52	5.54	4.00	29.63

(1) PLS-DA, PCA-DA, RF, SVML, SVML-G, SVM-R, SVM-P, NNET, GBT, GBD, GBL, LM, and NLNET, respectively, partial least squares discriminant analysis, principal component analysis discriminant analysis, random forest, Support Vector Machines with Linear Kernel, Support Vector Machines with Linear Kernel grid search, Support Vector Machines with Radial Basis Function Kernel, Support Vector Machines with Polynomial Kernel, neural network, eXtreme Gradient Boosting- xgbtree method, eXtreme Gradient Boosting- xgbDART method, eXtreme Gradient Boosting-xgbLinear method, Linear model, and Neural Network neuralnet method. (2) Accuracy represents the Pearson correlation coefficient between the predicted and the observed values in the validation set. (3) May_MAE.T mean absolute error relative to the optimal prediction model in the month of May using repeated cross-validation in the training set; May MAE.V (MAE and MAPE) magnitude of the error relative to the predicted values in the validation (testing) set; means with the same letter in a same column or row, are not significantly different at the 5% probability level using Duncan's multiple range test.

The mean comparison showed that SVM-R was the least accurate model. The other models showed comparable accuracies, but RF, SVML, SVML-G, SVM-P, NNET, GBT, and GBD showed prediction ability greater than SVM-R. GBT's prediction ability was consistently greater than 0.5 across the prediction times. Apart from GBL, the prediction ability of all models was high (prediction accuracy greater than or equal to 0.76) and/or better in the month of May (Table 2). The mean accuracy across models was high and not significantly different in May, May–June, and May–July. The across-model average accuracy computed in May was significantly superior to the mean accuracy obtained in June, June–July, and July. June, June–July, and July were statistically equally worst times for predicting biomass yields in sorghum under the Mediterranean region.

The range and mean MAE values (in t ha^{-1}) produced for each model during the calibration experiments at one (May) of the best prediction times were 9.3–14.4 and 12.0, 2.4–7.4 and 5.0, 3.1–7.0 and 4.9, 2.8–7.3 and 5.0, 2.7–7.2 and 5.0, 3.1–6.7 and 4.6, 3.1–7.1 and 4.8, 3.1–7.1 and 4.8, 3.19–7.95 and 5.29, 2.64–7.34 and 4.80, 2.77–10.21 and 5.43, 2.82–7.09 and 4.88, and 2.18–7.63 and 5.36, for NNET, RF, SVM-R, PCA-DA, PLS-DA, SVM-P, SVML, SVM-G, GBT, GBD, GBL, LM, and NLNET, respectively (Figure 4, Table 2). The average MAE values during the training process were statistically higher in

NNET followed by NLNET and GBL. Prediction error (MAE) was lower in SVM-P and GBD, while it was not statistically different in PLS-DA, PCA-DA, RF, SVML, SVML-G, SVM-R, GBT, and LM (Figure 4, Table 2). The MAE values (in t ha^{-1}) calculated using the validation (testing) set and the best prediction time (May) were, in increasing order, 1.87 (13.85%), 2.18 (16.15%), 2.27 (16.81%), 2.34 (17.33%), 2.68 (19.85%), 2.91 (21.56%), 3.40 (25.19%), 3.62 (26.81%), 3.74 (27.70%), 3.74 (27.7%), 4.53 (33.56%), 6.22 (46.07%), and 12.5 (92.59%), for SVM-R, GBD, RF, NLNET, GBT, PCA-DA, GBL, PLS-DA, SVML, SVML-G, LM, SVM-P, and NNET algorithms, respectively.

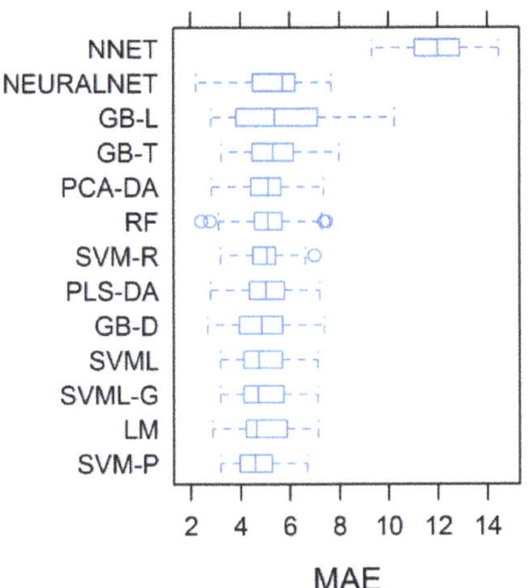

Figure 4. Visualization of models MAE (t ha^{-1}) dispersion using boxplot approach and fAPAR acquired in May. PLS-DA, PCA-DA, RF, SVML, SVML-G, SVM-R, SVM-P, NNET, GBT, GBD, GBL, LM, and NLNET, respectively, partial least squares discriminant analysis, principal component analysis discriminant analysis, random forest, Support Vector Machines with Linear Kernel, Support Vector Machines with Linear Kernel grid search, Support Vector Machines with Radial Basis Function Kernel, Support Vector Machines with Polynomial Kernel, neural network, eXtreme Gradient Boosting xgbtree method, eXtreme Gradient Boosting xgbDART method, eXtreme Gradient Boosting xgbLinear method, Linear model, and Neural Network neuralnet method.

Spearman's rank correlation coefficient (Spearman's rho) between model accuracy and MAE values (t ha^{-1} and %) corresponding to the testing set was −0.40. The Spearman's rho method assesses how well the relationship between two variables can be described using a monotonic function between ordered sets that preserves or reverses the given order [48]. The Spearman's rho approach was selected to account for the small size of the samples whose pairwise statistical dependences could not be correctly assessed with parametric approaches that have to be implemented on normally distributed data. Indeed the Shapiro–Wilk test of normality for the vectors of model accuracies and MAE values, was very highly significant ($p < 0.001$), meaning that we couldn't assume the normality.

Over the May to July prediction time interval, six days of year corresponding to fortnightly fAPAR indices, were used as regressors in this work. Among these regressors, the most important times to predict the aboveground sorghum biomass yields were investigated using the GBT algorithm as this model showed high and dependable performance that was insensitive to the prediction times across the cropping season. The model showed that the day of year 150 was the most important (index = 100) followed by DoY 165 (index = 80), DoY 135 (index = 30), DoY 195 (index = 20), and DoY 210 (index = 10) (Figure 5). The day of year 180 was associated with no importance in terms of fAPAR-based prediction of the aboveground biomass yields in sorghum under the Mediterranean environment.

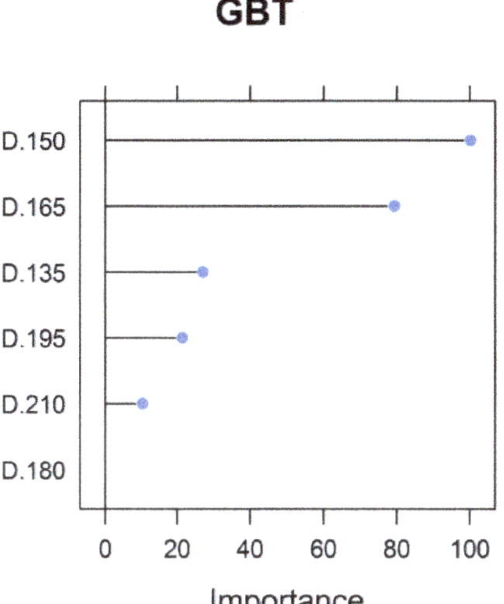

Figure 5. Relative importance of regressors (day of year, D) on sorghum biomass yields in 2017 and 2018, using eXtreme Gradient Boosting xgbtree (GBT) method.

4. Discussion

The fAPAR biophysical variable used in this work was derived from satellite imagery, which is part of Earth Observation's big data. Big data technology (BDT) is a new technological paradigm that is driving entire economies, including low-tech industries such as agriculture where it is implemented under the banner of precision farming (PF) [49]. In this work, BDT was built on geocoded maps of agricultural experiment fields and the real-time monitoring of sorghum crops on commercial farms in order to assess the possibility to monitor sorghum growth and development, with the ultimate aim of predicting the aboveground biomass yields. Early prediction of biomass production has positive implications including increased efficiency in biomass, biofuel, and farming resource management [50], and avoidance of energy crises. Forty-two sorghum pilot trials were evaluated in this work using fAPAR and different sorghum varieties belonging to four biomass producing sorghum types of dual purpose, sweet, forage, and biomass per se. Combining different types of biomass producing sorghums in this study was motivated by the need to mimic farming practice in the Mediterranean region. In this region, farmers, farming cooperatives, and third-party biomass harvesting and biodigesting companies manage the above-mentioned sorghum types indiscriminately on a regular basis. It made therefore sense not to discriminate the biomass producing sorghum types as sources of variation in the models implemented in this work. Similar investigations were reported in previous studies working on different crop species [16,51]. Important regressors of interest were identified and used in the predictive algorithms as suggested in literature [14,51].

The fAPAR index produced unique curves and maps that discriminated between the types of sorghums evaluated in this work. The fAPAR profile paralleled the evolution of leaf senescence across sorghum types [52] under the Mediterranean environment. The fAPAR curves presented in this work were purposely derived from sorghum fields established side by side in the same location in Anzola dell'Emilia. These pilots were sown and harvested on the same dates and managed identically, which allows a coherent comparison. The above-described shapes of the curves were generally similar also across locations in this study, though with slight discrepancies for some sorghum types. All sorghum trials reported in this work were conducted under a rainfed regime. Given that Mediterranean region is characterized by a semiarid climate wherein summer crops rely heavily on winter soil-stored moisture and experience postanthesis drought stress, it can be inferred that the fAPAR in dual purpose and biomass sorghum types did not rise during the reproduction growth stage probably due to a combined effect of sink demand and soil water scarcity in dual purpose, and mostly soil water scarcity in biomass sorghum. Postanthesis drought stress in sorghum under the Mediterranean environment was amply described by Habyarimana et al. [52–55]. The fAPAR profile in the sweet, forage, and SB × SH grown from seed reflects the reduced importance of the sink and the delayed leaf senescence in these types. In these sorghum types, a slow fAPAR increase toward the harvest can be explained by the precipitations registered in early fall in most locations (Figures S1 and S2), which stimulated the growth of axillary tillers in annual Sorghum bicolor [54,55] and the growth of axillary tillers and ramets in perennial SB × SH sorghum [54–57]. The above explanation holds also in the case of the SB × SH regrown from rhizomes. In these plants, the deep fAPAR drop from mid-June (anthesis) to early fall corresponds to the observed dry summers (Figures S1 and S2) and testifies to the increased susceptibility to drought stress in these plants. The conclusions drawn on fAPAR profile held particularly for fields established in the same location. Therefore, further investigations with replications in time and space are in order before any generalization is made.

In this work, high levels of model prediction accuracy ($r \geq 0.70$ or $r^2 \geq 0.50$) were obtained for 12 out of the 13 models deployed at the best prediction time (May). The models were therefore able to explain 50% of the variability that existed in the sorghum biomass yield data, while the remaining variance can be related to other factors non accounted for in this study such as the heterogeneity of external environmental and anthropogenic factors including rainfall distribution, soil types, and planting/tilling practices that could lead to different yield responses across the farms. The modeling performance metrics achieved in this work are nonetheless comparable to previous findings. For instance, Battude et al. [58], Shafian et al. [31], and Panda et al. [30] came across similar accuracy in their work on maize biomass and grain yields, and sorghum yields, respectively. On the other hand, the accuracy realized in this work was greater or equal to the values reported in Gao et al. [51], Diouf et al. [14], and the optimal values in sorghum as presented in Johnson [16]. Linear and nonlinear models performed comparably in terms of accuracy and mean absolute error implying that the relationship between fAPAR and biomass yield was mainly linear, which was expected and also supported by previous findings [14]. At the best prediction time (month of May), the correlation between the model accuracies and the MAE values was negative, denoting the expected inverse relationship between the two metrics of model prediction performance.

The simple linear model was implemented in this work to serve as a benchmark with respect to the most complex models requiring parameters optimization. Since thirteen models were implemented in this study, it is interesting to select the best algorithms. As biomass sorghum in the Mediterranean region is harvested from end of August to late November, it can be interesting to be able to predict the biomass production from May to July, allowing the farmer to know the amount to be produced one to six months ahead of harvest [42,51]. SVML, SVML-G, GBT, and LM showed good prediction accuracy ($r \geq 0.50$) across the evaluated prediction times, with MAE values (%) of 27.70, 27.70, 19.85, and 33.56, respectively. The GBT model was therefore the best algorithm as it performed consistently well ($r \geq 0.60$) across the prediction times, and was associated with low prediction error. The GBT model can therefore be recommended for sorghum biomass yield prediction using Sentinel-2-derived fAPAR as biophysical variable under the Mediterranean region. This model can be deployed anywhere from May to July without significant loss function.

In terms of biomass yields prediction times, June, July, and June–July were the worst times. May, May–June, and May–July showed comparable average accuracies, but accuracy in May was generally high ($r \geq 0.70$) across models except GBL. The month of May can therefore be recommended as the best time to predict sorghum biomass yields in the Mediterranean region. In this work, several types of sorghum were used, including high tonnage, sweet, forage, and dual purpose types. The suitability of the month of May for sorghum biomass yields prediction can be partly explained by the fact that in early sorghum growth stages, particularly in the period of time around the fast growth stage, the four sorghum types exhibit similar levels of growth and development. Furthermore, sorghum crop as currently grown in the Mediterranean region, reaches the fast growth stage generally in the month of May, meaning that predictions run in May are carried out on populations of sorghum types that are mostly at the same stage of growth and development. Overall, the days of year 150 and 165 were the most important regressors followed by days of year 135, 195, and 210 in decreasing order. The two regressors acquired in May (DoY 150 and 135) had important direct effects on the sorghum biomass, which justifies the good prediction accuracies obtained in this month. On the other hand, the two regressors corresponding to the month of July showed poor importance on biomass yields, while one of the two regressors corresponding to the month of June had meaningless effect on biomass yields, all of which explains the poor prediction accuracies obtained in June, July, and June–July (Table 2). In the Mediterranean region, sorghum is sown mid-to-late April. Therefore, being able to perform accurate sorghum biomass yields prediction in May, i.e., up to six months ahead of harvesting is a remarkable opportunity for the farmer and farming cooperatives that can use this information for several business-related purposes. They can efficiently organize the biomass business operations including the rational mobilization of the fleets of harvesting machinery, transport vehicles, and storage facilities. The predictive models developed in this work can also be used by extension services and policy-makers for strategic purposes. Obtaining the information on potential within-season biomass availability early on before actual harvest will help assess alternative means for energy supply internally, import or export, which is expected to help avoid specific crises such as fuel shortage. The findings in this work are limited in scope to one province in Italy, within the Mediterranean region. The prediction equations produced in this work can therefore be safely used in analogous modeling experiments in other Mediterranean areas. However, for these equations to be extended to modeling activity at a global level, the training populations of farms would require updates with inclusion of data accounting for sampling additional latitudes and longitudes relevant for sorghum cultivation.

5. Conclusions

The importance of sorghum as food, feed, and biofuel crop was amply described in several scientific literatures. Biomass sorghum demonstrated higher yields with better energy balance relative to major crops of agroindustrial interest. As dedicated biomass sorghum crops are steadily increasing and precision farming is driving agricultural economies worldwide, the harnessing satellite technology is well-poised to bring about agricultural advantages including cutting farming operational costs. Sentinel-2-derived fraction of absorbed photosynthetically active radiation was found to satisfactorily explain primary productivity and was used in this study as biophysical variable in the predictive modeling of aboveground biomass yields in annual and perennial sorghums. Across month combinations from May to July and the thirteen machine learning prediction algorithms used in this work, the gradient boosting machine learning algorithm implementing xgbtree was identified as the best predictive model. The best prediction time for sorghum biomass was particularly the month of May, followed by May–June and May–July using fortnightly fAPAR indices. To the best of our knowledge, the present work represents the first time Sentinel-2-derived fAPAR is used in predictive modeling of sorghum biomass yields. The outcome from this study is important and can serve several purposes including farmers being able to improve their sorghum biomass business operations. Policy-makers and extension services will also benefit from the findings in this work allowing them early on within season information on potential biomass availability, which is critical to wider energy planning and avoiding energy-related crises.

Supplementary Materials: The following are available online at http://www.mdpi.com/2073-4395/9/4/203/s1, Figure S1: Fifteen-day averaged temperatures and rainfall in Anzola dell'Emilia in 2017 and 2018 cropping seasons, Figure S2: Fifteen-day averaged temperatures and rainfall in Conselice in 2017 and 2018 cropping seasons, Figure S3: Fifteen-day averaged temperatures and rainfall in Mirandola in 2017 and 2018 cropping seasons, Figure S4: Fifteen-day averaged temperatures and rainfall in Nonantola in 2017 cropping season.

Agronomy **2019**, *9*, 203

Author Contributions: Conceptualization, E.H.; Data Curation, E.H. and I.P.; Formal Analysis, E.H.; Funding Acquisition, E.H. and I.P.; Investigation, E.H.; Methodology, E.H.; Project Administration, E.H.; Writing—Original Draft, E.H.; Writing—Review & Editing, I.P., M.C., P.D.F., and M.D.

Acknowledgments: Part of this work was supported (beneficiary: first author) by the project Data-driven Bioeconomy (www.databio.eu), GA number: 732064 (H2020-ICT-2016-1—innovation action), and the project Risorse GeneticheVegetali (RGV/FAO) 2014e2016 of the Ministero delle PoliticheAgricole, Alimentari e Forestali, Rome. The authors would like to thank the anonymous reviewers and editor for their informative remarks that contributed to improving the quality of this paper.

References

1. Habyarimana, E.; Lorenzoni, C.; Redaelli, R.; Alfieri, M.; Amaducci, S.; Cox, S. Towards a perennial biomass sorghum crop: A comparative investigation of biomass yields and overwintering of *Sorghum bicolor* x *S. halepense* lines relative to long term *S. bicolor* trials in northern Italy. *Biomass Bioenergy* **2018**, *111*, 187–195. [CrossRef]

2. Damasceno, C.M.B.; Schaffert, R.E.; Duweikat, I. *Mining Genetic Diversity of Sorghum as a Bioenergy Feedstock*; Springer: New York, NY, USA, 2014; pp. 81–106.

3. Hoffmann, L., Jr.; Rooney, W.L. Cytoplasm has no effect on the yield and quality of biomass sorghum hybrids. *JSBS* **2013**, *3*, 129–134. [CrossRef]

4. Prakasham, R.S.; Nagaiah, D.; Vinutha, K.S.; Uma, A.; Chiranjeevi, T.; Umakanth, A.V. Sorghum biomass: A novel renewable carbon source for industrial bioproducts. *Biofuels* **2014**, *5*, 159–174. [CrossRef]

5. Habyarimana, E. Genomic prediction for yield improvement and safeguarding of genetic diversity in CIMMYT spring wheat (*Triticum aestivum* L.). *Aust. J. Crop Sci.* **2016**, *10*, 127–136.

6. Rooney, W.L. Genetics and cytogenetics. In *Sorghum: Origin, History, Technology, and Production*; Smith, C.W., Frederiksen, R.A., Eds.; John Wiley & Sons: New York, NY, USA, 2000; pp. 261–307.

7. El Bassam, N. *Handbook of Bioenergy Crops: A Complete Reference to Species, Development and Applications*; Earthscan Ltd.: London, UK, 2010; pp. 45–477.

8. Stefaniak, T.R.; Dahlberg, J.A.; Bean, B.W.; Dighe, N.; Wolfrum, E.J.; Rooney, W.L. Variation in biomass composition components among forage, biomass, sorghum-sudangrass, and sweet sorghum types. *Crop Sci.* **2012**, *52*, 1949–1954. [CrossRef]

9. Kussul, N.; Sokolov, B.V.; Zyelyk, Y.I.; Zelentsov, V.A.; Skakun, S.V.; Shelestov, A.Y. Disaster risk assessment based on heterogeneous geospatial information. *J. Autom. Inform. Sci.* **2010**, *42*, 32–45. [CrossRef]

10. Kussul, N.; Shelestov, A.; Skakun, S. *Flood Monitoring from SAR Data Use of Satellite and In-Situ Data to Improve Sustainability*; Springer: Dordrecht, The Netherlands, 2011; pp. 19–29.

11. Skakun, S.; Kussul, N.; Kussul, O.; Shelestov, A. Quantitative estimation of drought risk in Ukraine using satellite data. In Proceedings of the IEEE International Geoscience and Remote Sensing Symposium (IGARSS), Quebec City, QC, Canada, 13–18 July 2014; pp. 5091–5094.

12. Skakun, S.; Kussul, N.; Shelestov, A.; Kussul, O. The use of satellite data for agriculture drought risk quantification in Ukraine. *Geomat. Nat. Hazards Risk* **2016**, *7*, 901–917. [CrossRef]

13. Gallego, J.; Kravchenko, A.N.; Kussul, N.N.; Skakun, S.V.; Shelestov, A.Y.; Grypych, Y.A. Efficiency assessment of different approaches to crop classification based on satellite and ground observations. *J. Autom. Inform. Sci.* **2012**, *44*, 67–80. [CrossRef]

14. Diouf, A.A.; Brandt, M.; Verger, A.; El Jarroudi, M.; Djaby, B.; Fensholt, R.; Ndione, J.A.; Tychon, B. Fodder Biomass Monitoring in Sahelian Rangelands Using Phenological Metrics from FAPAR time series. *Remote Sens.* **2015**, *7*, 9122–9148. [CrossRef]

15. Duveiller, G.; López-Lozano, R.; Baruth, B. Enhanced Processing of 1-km Spatial Resolution fAPAR Time Series for Sugarcane Yield Forecasting and Monitoring. *Remote Sens.* **2013**, *5*, 1091–1116. [CrossRef]

16. Johnson, D.M. A comprehensive assessment of the correlations between field crop yields and commonly used MODIS products. *Int. J. Appl. Earth Obs. Geoinf.* **2016**, *52*, 65–81. [CrossRef]

17. Kogan, F.; Kussul, N.; Adamenko, T.; Skakun, S.; Kravchenko, O.; Kryvobok, O.; Shelestov, A.; Kolotii, A.; Kussul, O.; Lavrenyuk, A. Winter wheat yield forecasting in Ukraine based on Earth observation, meteorological data and biophysical models. *Int. J. Appl. Earth Obs. Geoinf.* **2013**, *23*, 192–203. [CrossRef]

18. Kogan, F.; Kussul, N.; Adamenko, T.; Skakun, S.; Kravchenko, O.; Kryvobok, O.; Shelestov, A.; Kolotii, A.; Kussul, O.; Lavrenyuk, A. Winter wheat yield forecasting: A comparative analysis of results of regression and biophysical models. *Int. J. Autom. Inform. Sci.* **2013**, *45*, 68–81. [CrossRef]

19. Kowalik, W.; Dabrowska-Zielinska, K.; Meroni, M.; Raczka, T.U.; de Wit, A. Yield estimation using SPOTVEGETATION products: A case study of wheat in European countries. *Int. J. Autom. Inform. Sci.* **2014**, *32*, 228–239.

20. Kross, A.; McNairn, H.; Lapen, D.; Sunohara, M.; Champagne, C. Assessment of RapidEye vegetation indices for estimation of leaf areaindex and biomass in corn and soybean crops. *Int. J. Appl. Earth Obs. Geoinf.* **2015**, *34*, 235–248. [CrossRef]

21. Camacho, F.; Cernicharo, J.; Lacaze, R.; Baret, F.; Weiss, M. GEOV1: LAI, FAPAR Essential Climate Variables and FCOVER global time series capitalizing over existing products. Part 2: Validation and intercomparison with reference products. *Remote Sens. Environ.* **2013**, *137*, 310–329. [CrossRef]

22. Shelestov, A.; Kolotii, A.; Camacho, F.; Skakun, S.; Kussul, O.; Lavrenuik, M. Mapping of biophysical parameters based on high resolution EO imagery for JECAM test site in Ukraine. In Proceedings of the 2015 IEEE International Geoscience and Remote Sensing Symposium (IGARSS), Milan, Italy, 26–31 July 2015.

23. Kussul, N.; Kolotii, A.; Skakun, S.; Shelestov, A.; Kussul, O.; Oliynuk, T. Efficiency estimation of different satellite data usage for winter wheat yield forecasting in Ukraine. In Proceedings of the 2014 IEEE International Geoscience and Remote Sensing Symposium (IGARSS), Quebec City, QC, Canada, 13–18 July 2014; pp. 5080–5082.

24. López-Lozano, R.; Duveiller, G.; Seguini, L.; Meroni, M.; García-Condado, S.; Hooker, J.; Leo, O.; Baruth, B. Towards regional grain yield forecasting with 1km-resolution EO biophysical products: Strengths and limitations at pan-European level. *Agric. For. Meteorol.* **2015**, *206*, 12–32. [CrossRef]

25. Baret, F.; Weiss, M.; Lacaze, R.; Camacho, F.; Makhmara, H.; Pacholcyzk, P.; Smets, B. Geov1: LAI and FAPAR essential climate variables and FCOVER global time series capitalizing over existing products. Part1: Principles of development and production. *Remote Sens. Environ.* **2013**, *137*, 299–309. [CrossRef]

26. Fensholt, R.; Sandholt, I.; Rasmussen, M.S.; Stisen, S.; Diouf, A. Evaluation of satellite based primary production modelling in the semi-arid Sahel. *Remote Sens. Environ.* **2006**, *105*, 173–188. [CrossRef]

27. Tucker, C.J.; Holben, B.N.; Elgin, J.H.; McMurtrey, J.E. Relationship of spectral data to grain yield variation. *Photogramm. Eng. Remote Sens.* **1980**, *46*, 657–666.

28. Barnett, T.L.; Thompson, D.R. The use of large-area spectral data in wheatyield estimation. *Remote Sens. Environ.* **1982**, *12*, 509–518. [CrossRef]

29. Hatfield, J.L. Remote sensing estimators of potential and actual crop yield. *Remote Sens. Environ.* **1996**, *13*, 301–311. [CrossRef]

30. Panda, S.S.; Ames, D.P.; Panigrahi, S. Application of vegetation indices for agricultural crop yield prediction using neural network techniques. *Remote Sens.* **2010**, *2*, 673–696. [CrossRef]

31. Shafian, S.; Rajan, N.; Schnell, R.; Bagavathiannan, M.; Valasek, J.; Shi, Y.; Olsenholler, J. Unmanned aerial systems-based remote sensing for monitoring sorghum growth and development. *PLoS ONE* **2018**, *13*, e0196605. [CrossRef] [PubMed]

32. Yang, C.; Everitt, J.H.; Bradford, J.M.; Escobar, D.E. Mapping grain sorghum growth and yield variations using airborne multispectral digital imagery. *Trans. ASAE* **2000**, *43*, 1927–1938. [CrossRef]

33. Piper, J.K.; Kulakow, P.A. Seed yield and biomass allocation in Sorghum bicolor and F1 and backcross generations of *S. bicolor* x *S. halepense* hybrids. *Can. J. Bot.* **1994**, *72*, 468–474. [CrossRef]

34. De Keukelaere, L.; Sterckx, S.; Adriaensen, S.; Knaeps, E.; Reusen, I.; Giardino, C.; Bresciani, M.; Hunter, P.; Neil, C.; Van der Zande, D.; et al. Atmospheric correction of Landsat-8/OLI and Sentinel-2/MSI data using iCOR algorithm: Validation for coastal and inland waters. *Eur. J. Remote Sens.* **2018**, *51*, 525–542. [CrossRef]

35. Weiss, M.; Baret, F. ATBD S2ToolBox Level 2 Products: LAI, FAPAR, FCOVER (Version 1.1). 2016. Available online: http://step.esa.int/docs/extra/ATBD_S2ToolBox_L2B_V1.1.pdf (accessed on 7 November 2018).

36. Jacquemoud, S.; Verhoef, W.; Baret, F.; Bacour, C.; Zarco-Tejada, P.J.; Asner, G.P.; François, C.; Ustin, S.L. PROSPECT + SAIL models: A review of use for vegetation characterization. *Remote Sens. Environ.* **2009**, *113*, S56–S66. [CrossRef]

37. Sage, R.F. A portrait of the C4 photosynthetic family on the 50th anniversary of its discovery: Species number, evolutionary lineages, and Hall of Fame. *J. Exp. Bot.* **2016**, *67*, 4039–4056. [CrossRef]

38. Eilers, P.H.C. A perfect smoother. *Anal. Chem.* **2003**, *75*, 3631–3636. [CrossRef]

39. Atzberger, C.; Eilers, P.H.C. A smoothed 1-km resolution NDVI time series (1998–2008) for vegetation studies in South America. *Int. J. Digit. Earth* **2010**, *4*, 365–386. [CrossRef]

40. Kuhn, M. Building predictive models in R using the caret Package. *J. Stat. Softw.* **2008**, *28*, 1–26. [CrossRef]

41. Benor, D.; Baxter, M. *Training and Visit Extension*; The World Bank: Washington, DC, USA, 1984; pp. 21–212.

42. Hoefsloot, P.; Ines, A.V.; van Dam, J.; Duveiller, G.; Kayitakire, F.; Hansen, J. Combining crop models and remote sensing for yield prediction: Concepts, applications and challenges for heterogeneous smallholder environments. In *JRC Scientific and Policy Reports*; Report of CCFAS-JRC Workshop at Joint Research Centre; Joint Research Centre of the European Commission: Ispra, VA, Italy, 2012; pp. 7–41.

43. Meuwissen, T.H.E.; Hayes, B.J.; Goddard, M.E. Prediction of total genetic value using genome-wide dense marker maps. *Genetics* **2001**, *157*, 1819–1829. [PubMed]

44. R Core Team. *R: A Language and Environment for Statistical Computing*; R Foundation for Statistical Computing: Vienna, Austria, 2013.

45. Breiman, L.; Friedman, J.H.; Olshen, R.A.; Stone, C.J. *Classification and Regression Trees*; Wadsworth Inc.: Blelmont, CA, USA, 1984.

46. Willmott, C.J.; Matsuura, K. Advantages of the mean absolute error (MAE) over the root mean square error (RMSE) in assessing average model performance. *Clim. Res.* **2005**, *30*, 79–82. [CrossRef]

47. Duncan, D.B. Multiple range and multiple F tests. *Biometrics* **1955**, *11*, 1–42. [CrossRef]

48. Spearman, C. The proof and measurement of association between two things. *Am. J. Psychol.* **1904**, *15*, 72–101. [CrossRef]

49. Schellberg, J.; Hill, M.J.; Gerhards, R.; Rothmund, M.; Braun, M. Precision agriculture on grassland: Applications, perspectives and constraints. *Eur. J. Agron.* **2008**, *29*, 59–71. [CrossRef]

50. Segarra, E. *Precision Agriculture Initiative for Texas High Plains*; Annual Comprehensive Report; Texas A&M University Research and Extension Center: Lubbock, TX, USA, 2002.

51. Gao, F.; Anderson, M.; Daughtry, C.; Johnson, D. Assessing the Variability of Corn and Soybean Yields in Central Iowa Using High Spatiotemporal Resolution Multi-Satellite Imagery. *Remote Sens.* **2018**, *10*, 1489. [CrossRef]

52. Habyarimana, E.; Lorenzoni, C.; Busconi, M. Search for new stay-green sources in *Sorghum bicolor* (L.) Moench. *Maydica* **2010**, *55*, 187–194.

53. Habyarimana, E.; Laureti, D.; Di Fonzo, N.; Lorenzoni, C. Biomass production and drought resistance at the seedling stage and in field conditions in sorghum. *Maydica* **2002**, *47*, 303–309.

54. Habyarimana, E.; Laureti, D.; De Ninno, M.; Lorenzoni, C. Performances of biomass sorghum [*Sorghum bicolor* (L.) Moench] under different water regimes in Mediterranean region. *Ind. Crop Prod.* **2004**, *20*, 23–28. [CrossRef]

55. Habyarimana, E.; Bonardi, P.; Laureti, D.; Di Bari, V.; Cosentino, S.; Lorenzoni, C. Multilocational evaluation of biomass sorghum hybrids under two stand densities and variable water supply in Italy. *Ind. Crop Prod.* **2004**, *20*, 3–9. [CrossRef]

56. Cox, T.S.; Van Tassel, D.L.; Cox, C.M.; DeHaan, L.R. Progress in breeding perennial grains. *Crop. Pasture Sci.* **2010**, *61*, 513–521. [CrossRef]

57. Nabukalu, P.; Cox, T.S. Response to selection in the initial stages of a perennial sorghum breeding program. *Euphytica* **2016**, *209*, 103–111. [CrossRef]

58. Battude, M.; Al Bitar, A.; Morin, D.; Cros, J.; Huc, M.; Sicre, C.M.; Le Dante, V.; Demarez, V. Estimating maize biomass and yield over large areas using high spatial and temporal resolution Sentinel-2 like remote sensing data. *Remote Sens. Environ.* **2016**, *184*, 668–681. [CrossRef]

Article

Development and Evaluation of a Leaf Disease Damage Extension in Cropsim-CERES Wheat

Georg Röll [1,*], William D. Batchelor [2], Ana Carolina Castro [3,4], María Rosa Simón [3,5] and Simone Graeff-Hönninger [1]

[1] Department of Agronomy, Institute of Crop Science, University of Hohenheim, 70599 Stuttgart, Germany; Simone.graeff@uni-hohenheim.de

[2] Biosystems Engineering Department, Auburn University, Auburn, AL 36849, USA; wdb0007@auburn.edu

[3] Cerealicultura, Facultad de Ciencias Agrarias y Forestales, Universidad Nacional de La Plata, Av. 60 y 119, La Plata 1900, Argentina; ana.castro@agro.unlp.edu.ar (A.C.C.); mrsimon@agro.unlp.edu.ar (M.R.S.)

[4] Consejo Nacional de Investigaciones Científicas y Técnicas (CONICET), Centro Científico Tecnológico (CCT), La Plata 1900, Argentina

[5] Comisión de Investigaciones Científicas, Pcia. de Buenos Aires, 526, 10 y 11, La Plata 1900, Argentina

* Correspondence: georg.roell@uni-hohenheim.de; Tel.: +49-711-459-22380

Received: 10 January 2019; Accepted: 1 March 2019; Published: 2 March 2019

Abstract: Developing disease models to simulate and analyse yield losses for various pathogens is a challenge for the crop modelling community. In this study, we developed and tested a simple method to simulate septoria tritici blotch (STB) in the Cropsim-CERES Wheat model studying the impacts of damage on wheat (*Triticum aestivum* L.) yield. A model extension was developed by adding a pest damage module to the existing wheat model. The module simulates the impact of daily damage on photosynthesis and leaf area index. The approach was tested on a two-year dataset from Argentina with different wheat cultivars. The accuracy of the simulated yield and leaf area index (LAI) was improved to a great extent. The Root mean squared error (RMSE) values for yield (1144 kg ha^{-1}) and LAI (1.19 m^2 m^{-2}) were reduced by half (499 kg ha^{-1}) for yield and LAI (0.69 m^2 m^{-2}). In addition, a sensitivity analysis of different disease progress curves on leaf area index and yield was performed using a dataset from Germany. The sensitivity analysis demonstrated the ability of the model to reduce yield accurately in an exponential relationship with increasing infection levels (0–70%). The extended model is suitable for site specific simulations, coupled with for example, available remote sensing data on STB infection.

Keywords: wheat; disease; yield; septoria tritici blotch; leaf area index; crop modelling; decision support system for agrotechnology transfer (DSSAT); Cropsim-CERES Wheat

1. Introduction

Wheat (*Triticum aestivum* L.) is the second most important staple food crop for human nutrition. It is grown worldwide on approximately 220 million hectares under different climatic conditions. It is projected that wheat production must increase by 1.6% annually to meet the expected global demand by 2050 [1]. However, increasing temperatures and changing global rainfall patterns will likely influence breeding, management, fertilization and crop protection strategies for wheat [2] and also influence disease patterns [3,4]. Hence, crop protection measures will play an important role under future climate change, as rising temperatures and changes in rainfall pattern, will cause more favourable conditions for pests and diseases, especially in the warming north, where wheat production is predominant [2].

On a global scale, there are approximately 50 diseases and pests, which have the potential to damage wheat and reduce farmer's income [5–7]. On a global level, the most widely adapted wheat fungal diseases are leaf rust caused by *"Puccinia triticina* E.," stripe rust caused by *"Puccinia striiformis* W.," stem or black rust caused by *"Puccinia graminis* E.," powdery mildew caused by *"Blumeria graminis* P."* and septoria tritici blotch (STB) caused by *"Zymoseptoria tritici* D." [1]. The infection by *"Zymoseptoria tritici* D." is the most economically damaging wheat disease worldwide [8]. It can cause yield reductions of 50% to 60% [9] by creating leaf lesions resulting in defoliation and reduced photosynthesis. It has been estimated that 70% of the annual usage of fungicides in Europe is related to the treatment of this disease [10].

During the past decade, there has been an increasing resistance of STB to azole and strobilurin fungicides in Europe [9–11]. Breeding for STB disease resistance is complicated, due to the variability of the pathogen reproduction cycle [12,13]. Researchers have studied different strategies including tillage, crop rotation, delayed sowing, fungicide application and a proper level of fertilizer application to reduce or control the infection of STB [14]. It appears that moderate fungicide application coupled with the right amount of fertilizer is a strategy that holds promise for environmentally friendly wheat production, while reducing at the same time STB infection.

Crop models are suitable for decision support and contribute to a better understanding in the development of new wheat production strategies. They can play a vital role in understanding plant growth processes, the impact of different weather scenarios as well as management strategies on disease outbreak, final yield and grain quality. Hence, crop models might help to spread the production of wheat in more economic and sustainable ways.

Crop models can also provide an insight into yield losses due to pests and diseases, including STB. Several mechanistic wheat crop growth models have been developed over the last several decades, including APSIM [15], WheatGrow [16], STICS [17], Sirius [18] and DSSAT [19]. These models were developed to study crop-environment interaction and to evaluate optimum management strategies.

The Cropsim-CERES-Wheat (CCW) model [20–22] included in the DSSAT version 4.6 [23] was developed to study the impact of genetics, management, weather and climate change on wheat growth and yield. The model simulates daily plant development based on daily maximum and minimum temperature, daylength and vernalisation requirements. Growth is computed on a daily basis using a radiation use efficiency approach. Carbon is allocated daily to different plant parts based on the development stage. The CCW model has been linked with remote sensing data [24] and was successfully tested with different cultivars, soil characteristics as well as in different climatic conditions including Canada [25], Argentina [26], Southern Italy [27] and the United Kingdom [28]. Currently the CCW model does not account for damage due to weeds, pests or diseases [29]. As a consequence, inaccurate simulations of crop growth and yield result when simulating datasets that include pests and diseases [30,31].

Developing and incorporating a disease damage extension would expand the use of the CCW model to simulate and study the impact of disease damage on crop growth and yield. Batchelor et al. [32] incorporated a pest damage into the CROPGRO [33] family of models distributed with the DSSAT [34]. In their approach, they defined pest coupling points as daily rate and state variable modifiers to simulate the impact of daily pest damage on leaf, stem, seed, shell and root state variables and daily photosynthesis rate based on daily inputs of pest damage. They tested this approach for different pest damage types for peanut and soybean crops. They evaluated this approach using a dataset to simulate the impact of velvetbean caterpillar on soybean.

Using the same approach, the purpose of this work was to: (i) develop a disease model extension for the simulation of STB in wheat, to (ii) evaluate the model performance using a dataset from Argentina; and (iii) to conduct a sensitivity analysis for the impact of different disease progress curves on leaf area index and yield using a dataset from Germany.

2. Materials and Methods

2.1. Model Development

Currently the CCW model does not account for competition with weeds, pests or diseases. To solve this problem, modifications of the current CCW version are necessary to include the impact of leaf diseases on final crop yield.

Plant dry matter accumulation and yield can be expressed as a function of leaf area index (LAI), radiation use efficiency and the loss of assimilates due to respiration. Pathogens can modify both leaf area index and daily photosynthesis [35].

The primary damage resulting from STB is defoliation, which reduces both leaf area and leaf photosynthetic rate [36,37].

To apply the damage theory, it was necessary to integrate the pest damage module [32] structure into the current CCW wheat model (Figure 1). These changes included the adding and linkage of the following subroutines to the original version: PEST.for, LINDM.for, PESTCP.for, VEGDM.for and OPPEST.for. A pest damage definition file was created to define the coupling point "leaf area" for the leaf disease STB (Figure 1), where daily damage could be applied to state and rate variables in the model. Percent cumulative leaf area destroyed (PCLA) was chosen as major coupling point in the model.

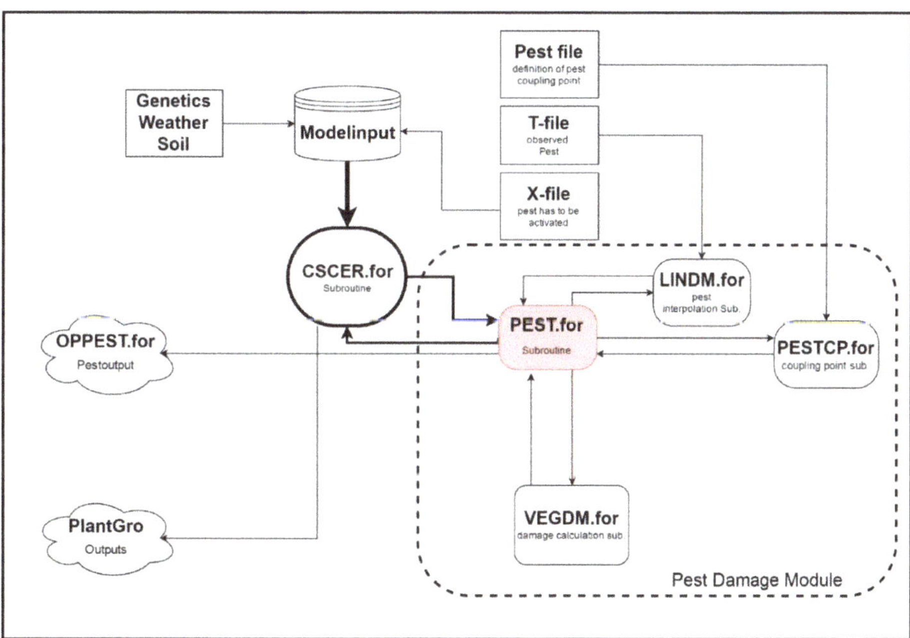

Figure 1. A simplified diagram of the Cropsim-CERES-Wheat (CCW) model with the pest damage module incorporated.

Field observed damage levels were entered in the time series file, referred to as File T in the DSSAT family of models. In this file, the year and day of year (DOY) are entered, along with STB infection in percent. Observed disease infection was linked to the percent cumulative leaf area destroyed (PCLA) coupling point. This damage type (i.e., PCLA) is defined in the pest damage definition model input file, which links field observed damage type and levels to the internal model pest damage coupling point. The model uses a linear interpolation to compute daily damage from periodic field observations. In this work, it was assumed that STB infection began ten days before the first infection symptoms were observed in the field. This assumption was made, based on Sánchez-Vallet et al. [38] who reported a latent period for STB between day 8 and 14 after infection depending on the environmental conditions. The daily percentage of damage (N_{pt}) was calculated between field observations using

$$N_{pt} = P_{rt*} + \frac{(P_{rt} - P_{rt*})}{(D_{pt} - D_{pt*})} \times (D_s - D_{pt*}) \tag{1}$$

N_{pt} = daily reported damage for damage type p at time t; D_{pt} = DOY of next field observation of damage (P_{rt}); D_{pt*} = DOY of previous field observation of damage (P_{rt*}); D_s = day of current simulation; P_{rt} = damage level reported in the next field observation; P_{rt*} = damage level reported in the previous field observation.

The daily damage calculation, which was applied to the leaf area coupling point (P_{it}) is calculated in the PESTCP. for subroutine (Figure 1) by Equation (2):

$$P_{it} = (N_{pt})(C_{ip}) \tag{2}$$

The pest coefficient (C_{ip}) allows the model to convert units of damage into units used for the model state or rate variable that is being damaged.

After calculating the daily damage to be applied to the diseased leaf area based on interpolations from field observations, the daily damage (D_{ipt}) to be applied to the leaf area state variable (defoliated leaf area) is calculated by the following equation:

$$D_{ipt} = X_{it}{}^* - (Xt_{it} - Xs_{it}) \times \left(1 - \frac{P_{it}}{100}\right) \tag{3}$$

$X_{it}{}^*$ = state or model variable i on day t, before application of damage; X_{it} = state or model variable i on day t, after application of damage; Xt_{it} = cumulative amount of coupling point I; Xs_{it} = cumulative senescence of coupling point I; D_{ipt} = amount of damage applied on state or model variable i on day t; P_{it} = coupling point leaf area.

Finally, the model state or rate variable is adjusted by subtracting the computed defoliation from the leaf area state variable by Equation (4):

$$X_{it} = X_{it}{}^* - D_{ipt} \tag{4}$$

X_{it} = state or model variable i on day t, after application of damage; $X_{it}{}^*$ = state or model variable i on day t, before application of damage; D_{ipt} = amount of damage applied on state or model variable i on day t.

2.2. Field Trials

In this study, datasets from two different locations were used for model development. The first dataset was recorded on the Experimental Station Julio Hirschhorn in La Plata (34°56′ S, 57°57′ W, 15 m above sea level, 16.3 °C average temperature; 946 mm mean annual precipitation) National University of La Plata in Argentina. The second experiment was carried out at the Experimental Station Ihinger Hof (48°44′ N, 8°55′ E; 480 m above sea level, mean annual temperature 9.1 °C and 714 mm mean annual precipitation) University of Hohenheim in Germany.

The trial in Argentina was conducted in two consecutive years (2010 and 2011) and published by Castro and Simón [39]. The objective of this trial was to test the tolerance of ten different Argentinean wheat cultivars (*Triticum aestivum* L.) for STB and to evaluate the disease impact on grain yield and grain quality. The sowing dates were on 15th of July in 2010 and 16th of June in 2011. The soil type was a silty loam. Nitrogen was applied as urea at 100 kg N ha^{-1} at sowing and 80 kg N ha^{-1} at the end of tillering. Three different inoculation levels with *Zymoseptoria tritici* D. were performed. The first level was the control treatment, the second was considered to be a low inoculation level (with 5×10^5

spores mL^{-1} suspension), while the third treatment was considered as high inoculation treatment (5×10^6 spores ml^{-1} suspension). All inoculations were performed at growth stage 22 (beginning of tillering) [40] and at growth stage 39 (flag leaf emergence). For model development weather data (daily temperature, rainfall, solar radiation) from the weather station La Plata (34°56′ S, 57°57′ W), disease severity ratings (%) from three growth stages (GS 39, 60, 82), leaf area index (LAI) which was calculated from the green leaf area (GLAI) plus non-green leaf area (NGLAI), yield, soil properties and management information were collected.

The second trial in Germany was conducted in 2006 using the cultivar Monopol with three inoculation levels (control treatment; low inoculation 50%; high inoculation 100%) of *Zymoseptoria tritici*. Inoculation was imposed by spraying 50% or 100% of a spore suspension (1×10^6 spores per mL, strain CBS 292.69) onto the plots at growth stage 32 [41]. The sowing date was 22nd of October 2005 on a silty clay soil. Nitrogen in form of ammonium nitrate was applied at three growth stages: 100 kg N ha^{-1} at GS 30, 80 kg N ha^{-1} at GS 32 and 40 kg N ha^{-1} at GS 49. The objective of this field trial was to use different vegetation indices to determine the occurrence of plant diseases in winter wheat (*Triticum aestivum* L.). For model sensitivity analysis, data including temperature, rainfall and solar radiation from the weather station Ihinger Hof, as well as growth stages, yield monitoring data, disease severity ratings and the LAI at growth stages GS 31, 34 and 49 were collected. Further information on the trial layout can be found in Gröll [41].

2.3. Model Calibration and Evaluation

The modified CCW model extension was incorporated into the DSSAT 4.6 software. Model inputs were created for both datasets from Argentina and Germany. The dataset from La Plata of 2010, which included phenological, yield, soil data (Table 1) and weather data, was used for calibration to test the ability of the model to simulate the impact of STB on wheat growth and yield.

Table 1. Soil properties for experiments in La Plata and Ihinger Hof used in the simulation.

Location La Plata	Clay Content %	Sand Content %	Silt Content %	LLL *	DUL **	SAT ***
0–30 cm	20.7	28.9	50.4	0.226	0.457	0.561
30–60 cm	20.7	28.9	50.4	0.226	0.457	0.561
60–90 cm	20.7	28.9	50.4	0.226	0.457	0.561
Location Ihinger Hof						
0–30 cm	43.3	9.9	46.8	0.247	0.412	0.467
30–60 cm	43.3	9.9	46.8	0.247	0.412	0.467
60–90 cm	25.0	18.8	56.2	0.142	0.313	0.503

* Lower limit ≘ permanent wilting point (pF 4.2). ** Drained upper limit ≘ field capacity (pF 1.8). *** Saturated ≘ saturated water content (pF 0).

Genetic coefficients for growth and development were calibrated using the 2010 dataset and the control treatment for each cultivar. Calibration was performed by sequentially adjusting the genetic coefficients (Table 2) to minimize the error between measured and simulated values [19]. The existing species file was set as default and the existing ecotype UKWH01 as well as the cultivar file were modified. Coefficients for phenological development (P1V, P1D, P1–P5 and PHINT) were calibrated in the first step, followed by crop growth coefficients (G1, G2 and G3). The RMSE, index of agreement (d-Index) and modelling efficiency (EF) statistics were used to assess the quality of the calibration (see section statistical evaluation). After calibration of individual cultivars, the percentage infection with STB of the low and high inoculation was applied to test the model response. The dataset from La Plata 2011 was used for model validation.

Table 2. Cultivar coefficients in Cropsim-CERES-Wheat (CCW) model used to simulate crop development, crop growth and crop yield for different cultivars.

Parameter	Definition	Ihinger Hof		La Plata					
		Monopol	ACA801	B75 Aniversario	Buck Brasil	Buck Guapo * Baguette 10 * Klein Zorro * Relmó Centinela *	Klein Escorpion	Klein Flecha	Klein Chaja
					Cultivar file				
P1D	Percentage reduction in development rate in a photoperiod 10 h shorter than the threshold relative to that at the threshold	110	100	100	100	100	100	100	100
P1V	Days at optimum vernalizing temperature required to complete vernalisation	90	20	20	20	20	20	20	20
G1	Kernel number per unit canopy weight at anthesis (kernels g^{-1})	15	25	25	20	22	22	22	25
G2	Standard kernel size under optimum conditions (mg)	48	25	30	30	30	35	30	39
G3	Standard, non-stressed dry weight of a single tiller at maturity (g)	2.0	4.0	4.0	4.0	4.0	4.0	4.0	4.0
P5	Duration of the grain filling phase (°C d)	600	420	430	420	420	420	420	420
PHINT	Interval between successive leaf tip appearances (°C d)	90	92	92	92	92	80	93	92
					Ecotype file				
P1	Duration of phase end juvenile to terminal spikelet	350	350	350	350	350	350	350	350
P2	Duration of phase from double ridges to the end of leaf growth (°C d)	200	250	250	250	250	250	250	250
P3	Duration of phase from the end of leaf growth to the end of spike growth (°C d)	300	220	220	220	220	220	220	220
P4	Duration of phase from the end of spike growth to the end of the grain fill lag (°C d)	380	300	300	300	300	300	300	300
PARUE	Conversion rate from photosynthetically active radiation to dry matter before the last leaf stage (g MJ^{-1})	2.3	2.7	2.7	2.7	2.7	2.7	2.7	2.7
SLAS	Specific leaf area, standard first leaf (cm^2/g)	400	450	450	450	450	450	450	450

* various wheat cultivars with the same cultivar coefficients.

The second dataset from Ihinger Hof was used for sensitivity analysis to test the model on a different location to proof the concept and to test the responsiveness of the model of different STB infection levels. The calibration was performed in the same way as in La Plata with the control treatment by modifying the necessary cultivar coefficients (Table 2). We applied 0%, 10%, 20%, 30%, 50% and 70% damage rates at maximum LAI (GS 39) and started the damage application at growth stage GS 31 to estimate the corresponding yield loss. The different damage rates were used to test the model responsiveness on a broad disease range, which typically starts to impact yield after growth stage 31 [42].

Statistical Evaluation

The statistical model evaluation was conducted by comparing the simulated and observed LAI and yield of the different inoculation treatments (dataset from La Plata).

For statistical analysis, the root mean squared error (RMSE. Equation (5)), the index of agreement (d-Index, Equation (6)) [43] and the modelling efficiency (EF, Equation (7)) were used. The RMSE was used to quantify the amount of variation between simulated and measured values on a metric scale. The d-Index shows if the model is under -or over-estimating the measurements. The EF parameter compares simulated values with the average of the measurements. For a perfect fit between simulated and observed data, the RMSE should be at 0 and the d-Index and EF parameter should have a value of 1.0.

The statistical evaluation was done for simulation runs of the original CCW version and the modified CCW for LAI and yield from both years (2010; 2011) on the location La Plata over all cultivars and inoculation treatments.

Root mean square error (RMSE):

$$RMSE = \left[\frac{1}{n} \sum_{i=1}^{n} (S_i - O_i)^2 \right]^{0.5} \tag{5}$$

Index of agreement (d):

$$d = 1 - \left[\frac{\sum_{i=1}^{n} (S_i - O_i)^2}{\sum_{i=1}^{n} \left(|S_i - \overline{O}| + |O_i - \overline{O}| \right)^2} \right] \tag{6}$$

Modelling efficiency (EF):

$$EF = 1 - \left[\frac{\sum_{i=1}^{n} (S_i - O_i)^2}{\sum_{i=1}^{n} (O_i - \overline{O})^2} \right] \tag{7}$$

where: O_i = observed values; S_i = simulated values; n = numbers of samples; $\overline{O}$ = mean of observed data

3. Results and Discussion

3.1. Model Calibration for La Plata

3.1.1. Leaf Area Index

The calibration was performed on the 2010 dataset by fitting the relevant genetic coefficients (Table 2) for phenology and growth. One essential prerequisite for model development is an accurate simulation of growth stages. In this study, growth stages (GS 39; GS 60; GS 82) were predicted by the model conclusively: For all ten cultivars the flowering date (GS 65) was documented approximately 110 days after sowing (DAS), the model simulated this growth stage 115 DAS. A similar result was obtained by comparing observed and simulated DAS of the early dough stage (GS 82) (observed approximately at 131 DAS, simulated 132 DAS).

The main focus of this model calibration was on the adjustment of leaf area as a major coupling point for disease damage. Figures 2 and 3 illustrate the simulated and the observed values for leaf area index and grain yield across different inoculation treatments along with the statistics (Table 3).

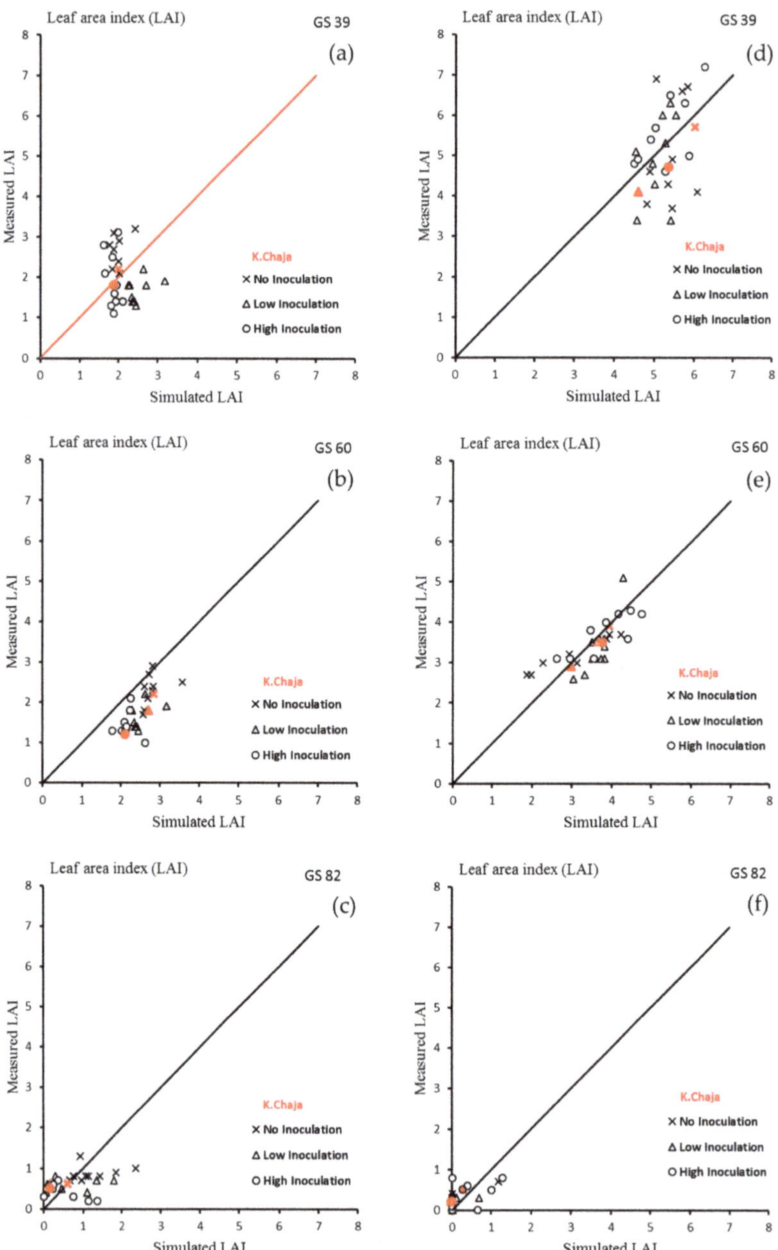

Figure 2. Simulated vs. measured leaf area index (LAI) for calibration (year 2010 **a–c**) and validation (year 2011 **d–f**) for all ten cultivars on the location La Plata. Different symbols represent the different inoculation treatments where × = No Inoculation; Δ = Low Inoculation and O = High Inoculation.

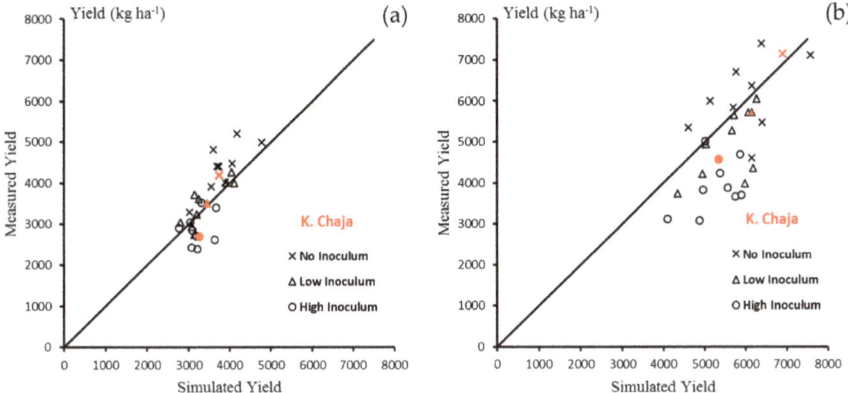

Figure 3. Simulated vs. measured yield (kg DM ha^{-1}) for calibration and validation for all ten cultivars on the location La Plata 2010 (**a**) and La Plata 2011 (**b**).

Table 3. Statistical evaluation of the simulation of leaf area index and grain yield of the original CCW model and the developed CCW model extension for diseases using root mean square error (RMSE), Willmott's d statistic (d-Index) and modelling efficiency (EF).

Variable	Experiment	Original CCW			Modified CCW		
		RMSE	d-Index	EF	RMSE	d-Index	EF
Leaf area	La Plata 2010	1.19	0.33	−2.69	0.69	0.51	−1.07
index	La Plata 2011	2.88	0.24	−0.98	1.11	0.70	0.68
Yield	La Plata 2010	1144	0.47	−1.19	499	0.81	0.58
	La Plata 2011	1755	0.50	−1.19	1285	0.66	−0.18

For demonstration of the overall model behaviour in regard to LAI changes induced by three different STB inoculation treatments over time, the wheat cultivar K. Chaja was selected. This cultivar was considered to be highly susceptible to STB infection [39]. Figure 4a shows the impact of disease infestation on LAI according to different inoculation treatments 90 days after sowing. All three simulation runs reached the maximum LAI at day 100. For the control treatment a maximum LAI of 3.5 was simulated. A difference of 0.5 LAI was found between the control and the high-inoculation treatment. Comparing simulated and observed LAI values, the model predicted the LAI over the vegetation period in an accurate manner (RMSE 0.47, d-Index 0.9).

Similar results are displayed in Figure 2a–c, which illustrates the simulated versus observed LAI across three different inoculation treatments for different cultivars. Regardless of susceptibility, in GS 39 all ten cultivars showed a homogenous distribution of all data points around the 1:1 line with no strong outliers. A slight tendency for an overestimation of LAI was given at GS 39 in the low and high inoculation treatments, whereas for the control treatment a slight underestimation was shown over all cultivars. In GS 60 and GS 82, a slight overestimation of LAI was found for all inoculation treatments.

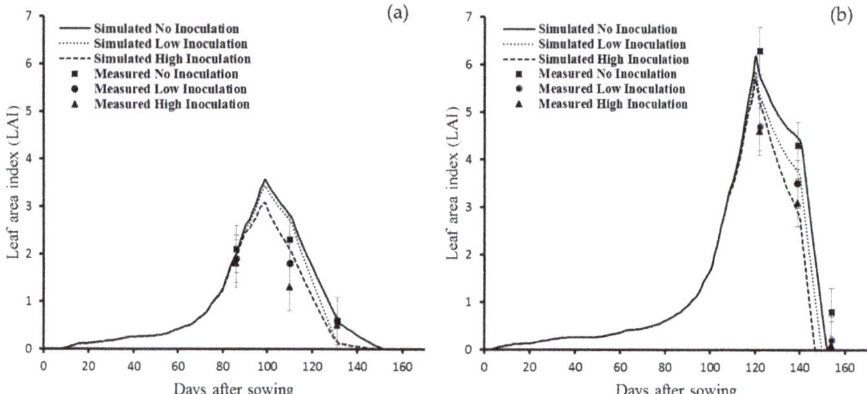

Figure 4. Simulated and measured leaf area index values for cultivar K. Chaja, year 2010 (**a**) and 2011 (**b**) including different inoculation treatments with septoria tritici blotch (STB). The error bars demonstrate the LSD of the leaf area index.

Finally, the modified CCW model (RSME 0.69; d-Index 0.51) performed better compared to the original CCW model (RSME 1.19; d-Index 0.33), as indicated by the corresponding statistics. Outliers in Figure 2a–c can be explained by the LSD ranging from 0.4 to 1.0 depending on the sampling date, reported from Castro and Simón [39].

Nevertheless, the model was able to account for all ten cultivars representing different tolerance levels to STB at different growth stages and disease severities accurately.

3.1.2. Yield

A reduction in LAI after infection with STB also leads to a reduction in yield (Figure 5a). Yield formation started for the cultivar K. Chaja (Figure 5a) on day 122 and was negatively correlated with the inoculation treatment. Yield of the control treatment (3800 kg ha^{-1}) was slightly underestimated and the high inoculation treatment (3400 kg ha^{-1}) showed a slight overestimation.

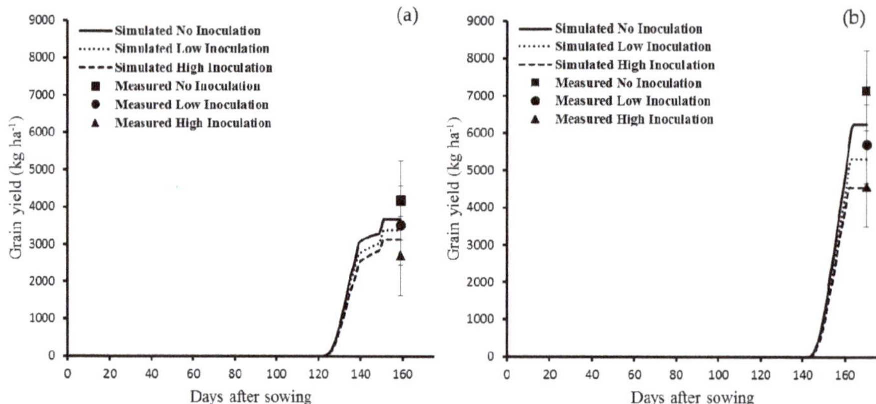

Figure 5. Simulated and measured grain yield values for cultivar K. Chaja year 2010 (**a**) and 2011 (**b**) including different inoculation treatments with STB. The error bars demonstrate the LSD of the yield.

Figure 3a represents the observed versus the simulated yield for all ten cultivars and showed a dense clustering of the different inoculation treatments around the 1:1 line. Overall, it indicated the highest simulated yield for the control and the lowest yield for the high inoculation treatment. The results illustrated the capability of the modified CCW model to account for disease damage. This is expressed in the statistical evaluation (Table 3), where a reduction of the RMSE from 1144 (original version) to 499 (modified version) was observed. The d-Index also underlined these findings, which increased from 0.47 (original version) to 0.81 (modified version).

The modification of the existing CCW showed very good results in LAI and yield simulation (Table 3). It indicated a clear improvement for all statistical parameters compared with the existing CCW included in the current DSSAT version. The calibration successfully minimized the error between measured and simulated data for both, LAI and yield.

3.2. Model Validation for La Plata

3.2.1. LAI

Illustrating LAI (Figure 2d–f) and yield (Figure 3b) for the cultivar K. Chaja and all cultivars in the validation year 2011.

In 2011, a maximum LAI of 6.3 was observed in the control treatment for K. Chaja (Figure 4b) at day 120. The model simulated a maximum LAI of 6.2 for the control treatment on the same day. For the low-inoculation treatment, a maximum LAI of 4.7 was observed, whereas the model simulated a maximum LAI of 5.8. Regarding the highest inoculation treatment, a LAI of 4.6 was observed in the field experiment. The model simulated for the same treatment a maximum LAI of 5.7. The model was capable to simulate the maximum LAI for the control treatment exactly but it slightly overestimated the maximum LAI both for the lowest and highest inoculation treatment.

In general LAI was higher in 2011 than in 2010 independent of cultivars, growth stages and inoculation treatments (Figure 2). For 2011 and GS 39, the 1:1 plot showed no strong outliers and a slight overestimation for the control treatment and a slight underestimation for the highest inoculation treatment. This can be caused by an earlier onset of disease in the inoculated treatments which was not reported and cause a slightly underestimation in the model. For GS 60 and GS 82 the model predicted the observed LAI values accurately.

3.2.2. Yield

Yield formation started 140 days after sowing for K. Chaja (Figure 5b), while full maturity was reached on day 165. A maximum yield of 6000 kg ha^{-1} was reached in the control treatment compared with the lowest yield of 5400 kg ha^{-1} in the high inoculation treatment. The corresponding error bars of the measured values were met by the simulated curves, which indicated a high accuracy of the simulation. Under consideration of all cultivars and inoculation treatments (Figure 3b) data points scattered around the 1:1 line on a broader range compared to the calibration (Figure 3a). An inverse relationship between inoculation level and yield was shown (Figure 5).

Overall, the developed model extension was able to account for STB disease damage. This is also shown by the statistics (Table 3), where a 30% improvement of the RMSE in the modified CCW version was achieved compared with the original model. This improvement was also shown by the d-Index and EF values. Further, the calibration showed a higher model accuracy when compared with the validation. Jing et al. [44] and Attia et al. [45] also reported a slightly weaker simulation accuracy regarding the validation dataset.

For 2010, a 20 days shorter growing period due to a 30 days later sowing date and a 130 mm lower precipitation compared to 2011 [39] was reported. Both factors resulted in a reduction of LAI and yield in 2010. Despite these differences the model performed very well for each inoculation treatment and showed its robustness when growing conditions differ between years. Measured yields in the inoculation treatments were simulated quite accurately, while the measured mean value of the control showed a 5% off-set. An explanation for this offset might be given in the way the disease ratings were performed and represented in the model. The model used the mean values from the disease ratings of all repetitions and did not represent each individual plot. It is also possible that the trials had a slight infection of other diseases, which were not measured and caused a slight model offset. Another conceivable reason is the defined onset of disease ten days before disease rating was reported. This assumption was made because of the reported latent period for STB between day 8 and 14 after infection from Sánchez-Vallet et al. [38].

For 2010 and 2011, the d-Index values of the original CCW version are in a similar range for both LAI and yield. In the developed model extension, the d-Index, which represents the model accuracy, increased strongly even though in different intensities for each year. This may point to one possible shortcoming of the current model extension, as it does not account for spore disposal [46]. Spore disposal model use different leaf layers, rain intensity thresholds, droplets, sporulation and concentrations of starting spore pools and can therefore extend the simulation accuracy further.

Nevertheless, other STB models show a strong performance, if a minimum dataset is provided, in which more inputs like leaf wetness are included [47] or in which the initial state of infection of the first leaves is known [46]. Magarey et al. [47] also reported the necessity of hourly weather data for many disease models. In previous datasets this information is not given [48] and they cannot be used for disease modelling, which is literally a loss of information for agricultural decision making. This clearly shows the advantage of the developed CCW extension, in which only the percentage of disease rating, weather- and soil data is needed as a minimum dataset. This leads to a more accurate simulation as shown in Table 3 and makes the CCW applicable for a broader use.

3.3. Sensitivity Analysis

In order to test the general responsiveness of the developed model extension, a sensitivity analysis was carried out by comparing different inoculation treatments with the corresponding disease infections [41]. The model was calibrated by using an independent dataset with disease infections from Germany [41]. The disease infections varied between 4%, 13% and 15%. Disease infection started in GS 31 (DAS 200). Figure 6a depicts the simulated curves for the three different inoculation treatments. A maximum LAI of 7.3 was reached 40 days after GS 31. Simulated curves illustrate a clear separation between the 4%, 13% and the 15% disease infection. A maximum LAI of 7.2, 6.3, 6.1 was reached at day 240 at 4%, 13% and the 15% disease infection rating. Comparing simulated infection scenarios with measured values, the model simulated the LAIs of the three different disease infection levels accurately.

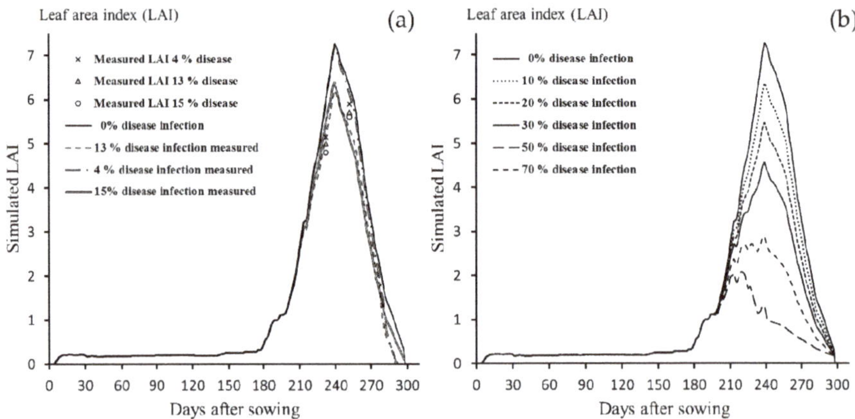

Figure 6. Sensitivity analysis of the CCW disease extension for measured disease infections (4%; 13%; 15%) (**a**) and for five infection scenarios (0%; 10%; 20%; 30%; 50%; 70% disease infection) (**b**) with STB at the location Ihinger Hof.

In the next step an artificial disease infection level of up to 70% was applied (Figure 6b) to test the general responsiveness and the boundaries of the developed model and to test the leaf damage theory on the leaf area coupling point (PCLA).

In Figure 6a a maximum LAI of 7.3 was reached 40 days after GS 31 in the control treatment. LAI increased almost linearly from day 200 to day 240 before the onset of senescence led to a constant decrease in LAI up to final harvest date. For the depicted disease infection scenarios of 10%, 20%, 30% and 50 % a maximum LAI of 7.3, 6.2, 5.4, 4.6, 2.9 and 1.3 was reached at day 240 (Figure 6b). The 70% disease infection scenario showed that a maximum LAI of 2.0 was reached earlier at 220 DAS (Figure 6b). Due to the massive destruction of leaf area, a shortage of assimilate production occurred, which affected in a next step the growth of new leaves. Simulated LAI reduction for maximum LAI in the different disease levels followed the magnitude of 12.5 % (10% diseased LAI), 24.8% (20% diseased LAI), 37.1% (30% diseased LAI), 60.3% (50% diseased LAI) and 82.4% (70% diseased LAI) (Table 4).

Table 4. Yield evaluation of the sensitivity analysis from the Ihinger Hof dataset, by comparing the percentage disease infection with STB and the corresponding simulated percentage yield reduction in kg DM ha^{-1} for the cultivar Monopol.

% Disease Infection	Simulated Yield kg ha^{-1}	Measured Yield kg ha^{-1}	% Yield Reduction	% LAI Reduction at Maximum LAI
0	4384		0	0
4 *	4332	4409	1.2	0.8
10	4190		4.4	12.5
13 *	4159	3934	5.1	14.5
15 *	4124	3965	5.9	12.0
20	3990		9.0	24.8
30	3737		14.8	37.1
50	3242		26.0	60.3
70	2380		45.7	82.4

* measured disease infection.

Table 4 shows the corresponding yields of the applied and measured disease infection levels. The maximum observed yield obtained with the disease infection level of 4% was 4409 kg ha^{-1}. Higher disease infection levels (13%; 15%) resulted in lower yield (3934 kg ha^{-1}; 3965 kg ha^{-1}). Simulated yield decreased gradually with higher infection levels. The control treatment resulted in a maximum yield of 4384 kg ha^{-1}, while the 70% disease infection level resulted in a total grain yield of 2380 kg ha^{-1} which corresponded to a 45.7% yield reduction. Over all tested disease infection levels, the simulated yield reduction followed an exponential shape, indicating that yield reductions became more severe and are more than doubled at higher disease infection levels. An exponential relationship between yield loss and disease infection was also shown by King et al. [49].

Comparing simulated and measured yield, the model showed a slightly underestimation for the 4% level and a slight overestimation for the 13% and 15% disease infection level. These variations are in an acceptable range.

Regarding the accuracy of the current simulation, similar results for yield reduction based on the occurrence of leaf diseases are reported by Ziv and Eyal [50]. Ziv and Eyal [50] tested different inoculation treatments in different spring wheat cultivars and reported yield losses of up to 53% at a disease infection of 73%. The developed CCW model extension gave comparable results to a previous study of Bhathal et al. [51] also at lower infection level scenarios. Bhathal et al. [51] tested different inoculation treatments in wheat to evaluate the relationship between disease infection and yield. Notably, they showed an onset of the disease as it was used in the sensitivity analysis of this study, at GS 31 and demonstrated a 10% yield loss due to a natural disease infection of 26%. King et al. [49], also confirmed this model theory on an independent dataset from the United Kingdom carried out at four different locations. Similar observations and an exponential yield loss curve due to STB disease were obtained. In addition, a yield loss of 30% by a disease infection of 55.1% as well as a yield reduction of 8% by a disease infection of 14.5% occurred. This confirmed the model theory and clearly showed the capability to simulate leaf disease infection with STB by using the coupling point leaf area (PCLA).

3.4. Future Model Applications

Despite the good simulation results, the developed model concept can currently only be used for STB. The concept was not tested on other wheat diseases like stripe rust caused by *"Puccinia striiformis* W.," stem or black rust caused by *"Puccinia graminis* E." or powdery mildew caused by *"Blumeria graminis* P."* It can be assumed that this concept will also work for other diseases, by changing the pest coefficient in the pest file to account for different damage types. Bastiaans [52] showed a ß–value for STB, which represents the correlation between visible affected leaf area and the affected photosynthetic rate. A ß–value > 1 indicates a stronger effect on photosynthetic rate as it visually appears. For STB this value is close to 1 wherefore the pest coefficient in the CCW model extension was set to 1. Bastiaans [52] reported a ß–value of 8.7 for *"Erisyhe graminis"* or 1.3 for *"Puccinia recondita"* in winter wheat. It is assumed, that the pest coefficient has to be increased in a similar manner but it has to be proven by real data. However, the structure of the model extension is set up in a flexible way and has the possibility to be transferred to other leaf diseases.

Further, the disease extension routine can suite as a gateway between crop models and remote sensing data, like it was published by Thorp et al. [53]. Thorp et al. [53] showed an improvement of simulation results by updating the plant leaf area state variable with green LAI generated by remote sensing. This offers the opportunity to simulate a given field on a site-specific scale, which means the CCW model extension can be updated by the percentage diseased leaf area detected by for example, remote sensing. In this way, the model could serve as decision support tool to give farmers an economic advise on a field level as Ficke et al. [54] proposed.

4. Conclusions

In this study a disease extension for the CCW model was developed to simulate the damage effect of STB disease on LAI and yield in wheat. The model was tested successfully in a sensitivity analysis on a German dataset and on a dataset obtained from La Plata, Argentina. Results of the study clearly showed the effect of the implementation of the coupling point "PCLA" and on the corresponding LAI and yield for different locations. For the location La Plata, the obtained simulation results of the modified CCW model indicated a higher model accuracy which almost doubled and clearly showed an improved model behaviour. Especially for the cultivar K. Chaja, the CCW model extension showed a high modelling accuracy. The LAI and yield were simulated very accurate in both years. Furthermore the sensitivity analysis also displayed the flexibility of the CCW model extension to account for disease damage over a broad range between 0 and 70% of STB disease infection.

Nevertheless, further research is needed to test the developed model on other leaf diseases like leaf rust, powdery mildew or stripe rust in wheat. The model extension could be used in future studies as decision support system for example, coupled with remote sensing technologies to obtain the necessary disease ratings for the model input files.

Author Contributions: Conceptualization, G.R., W.D.B. and S.G.-H.; methodology, W.D.B., G.R. and S.G.-H.; software, G.R. and W.D.B.; validation, G.R.; formal analysis, G.R.; investigation, G.R., A.C.C. and M.R.S.; resources, G.R., A.C.C. and M.R.S.; data curation, G.R.; writing—original draft preparation, G.R.; writing—review and editing, G.R., W.D.B. and S.G.-H.; visualization, G.R.; supervision, W.D.B. and S.G.-H.; project administration, S.G.-H.; funding acquisition, S.G.-H.

Funding: This research was funded by German Federal Environmental Foundation (DBU) (ProjectNr. 33143/01) and by the National Institute of Food and Agriculture, U.S. Department of Agriculture, Hatch project (ALA-14-1-16016).

Acknowledgments: The authors would like to thank the Universitätsbund Hohenheim e.V. for supporting the journey to Auburn University, Alabama.

Conflicts of Interest: The authors declare no conflict of interest. The funders had no role in the design of the study; in the collection, analyses or interpretation of data; in the writing of the manuscript or in the decision to publish the results.

References

1. Singh, R.P.; Singh, P.K.; Rutkoski, J.; Hodson, D.P.; He, X.; Jørgensen, L.N.; Hovmøller, M.S.; Huerta-Espino, J. Disease Impact on Wheat Yield Potential and Prospects of Genetic Control. *Annu. Rev. Phytopathol.* **2016**, *54*, 303–322. [CrossRef] [PubMed]
2. Olesen, J.E.; Trnka, M.; Kersebaum, K.C.; Skjelvåg, A.O.; Seguin, B.; Peltonen-Sainio, P.; Rossi, F.; Kozyra, J.; Micale, F. Impacts and adaptation of European crop production systems to climate change. *Eur. J. Agron.* **2011**, *34*, 96–112. [CrossRef]
3. Eigenbrode, S.D.; Binns, W.P.; Huggins, D.R. Confronting Climate Change Challenges to Dryland Cereal Production: A Call for Collaborative. *Prod. Engagem.* **2018**, *5*. [CrossRef]
4. Ahanger, R.A.; Bhat, H.A.; Bhat, T.A.; Ganie, S.A.; Lone, A.A.; Wani, I.A.; Ganai, S.A.; Haq, S.; Khan, O.A.; Junaid, M.J.; Bhat, T.A. Impact of Climate Change on Plant Diseases. *Int. J. Modern Plant & Anim. Sci. USA* **2013**, *3*, 105–115.
5. Weiss, M.V. *Compendium of Wheat Diseases*, 2nd ed.; APS Press: St. Paul, MN, USA, 1987.
6. Forrer, H.R.; Zadoks, J.C. Yield reduction in wheat in relation to leaf necrosis caused by Septoria tritici. *Neth. J. Plant Pathol.* **1983**, *89*, 87–98. [CrossRef]
7. Eyal, Z. The septoria tritici and stagonospora nodorum blotch diseases of wheat. *Eur. J. Plant Pathol.* **1999**, *105*, 629–641. [CrossRef]
8. Bearchell, S.J.; Fraaije, B.A.; Shaw, M.W.; Fitt, B.D.L.; Cowling, E.B. Wheat Archive Links Long-Term Fungal Pathogen Population Dynamics to Air Pollution. *Proc. Natl. Acad. Sci. USA* **2005**, *102*, 5438–5442. [CrossRef] [PubMed]
9. Eyal, Z.; Amiri, Z.; Wahl, I. Physiological Specialization of Septoria tritici. *Phytopathology* **1973**, *63*, 1087–1091. [CrossRef]
10. Fones, H.; Gurr, S. The impact of Septoria tritici Blotch disease on wheat: An EU perspective. *Fungal Genet. Biol.* **2015**, *79*, 3–7. [CrossRef] [PubMed]
11. Fraaije, B.; Cools, H.J.; Fountaine, J.; Lovell, D.J.; Motteram, J.; West, J.S.; Lucas, J. A Role of Ascospores in Further Spread of QoI-Resistant Cytochrome b Alleles (G143A) in Field Populations of Mycosphaerella graminicola. *Phytopathology* **2005**, *95*, 933–941. [CrossRef] [PubMed]
12. Estep, L.K.; Torriani, S.F.F.; Zala, M.; Anderson, N.P.; Flowers, M.D.; Mcdonald, B.A.; Mundt, C.C.; Brunner, P.C. Emergence and early evolution of fungicide resistance in North American populations of Zymoseptoria tritici. *Plant Pathol.* **2015**, *64*, 961–971. [CrossRef]
13. Simón, M.R.; Cordo, C.A.; Castillo, N.S.; Struik, P.C.; Börner, A. Population Structure of Mycosphaerella graminicola and Location of Genes for Resistance to the Pathogen: Recent Advances in Argentina. *Int. J. Agron.* **2012**, *2012*, 680275. [CrossRef]
14. Rodrigo, S.; Cuello-Hormigo, B.; Gomes, C.; Santamaria, O.; Costa, R.; Poblaciones, M.J. Influence of fungicide treatments on disease severity caused by Zymoseptoria tritici, and on grain yield and quality parameters of bread-making wheat under Mediterranean conditions. *Eur. J. Plant Pathol.* **2014**, *141*, 99–109. [CrossRef]
15. Keating, B.A.; Carberry, P.S.; Hammerb, G.L.; Probert, M.E.; Robertson, M.J.; Holzworth, D.; Huth, N.I.; Hargreaves, J.N.G.; Meinke, H.; Hochman, Z.; et al. An overview of the crop model APSIM. *Eur. J. Agron.* **2003**, *18*, 267–288. [CrossRef]
16. Lv, Z.; Liu, X.; Cao, W.; Zhu, Y. Agricultural and Forest Meteorology Climate change impacts on regional winter wheat production in main wheat production regions of China. *Agric. For. Meteorol.* **2013**, *171–172*, 234–248. [CrossRef]
17. Brisson, N.; Gary, C.; Justes, E.; Roche, R.; Mary, B.; Ripoche, D.; Zimmer, D.; Sierra, J.; Bertuzzi, P.; Burger, P.; et al. An overview of the crop model Stics. *Eur. J. Agron.* **2003**, *18*, 309–332. [CrossRef]
18. Jamieson, P.D.; Semenov, M.A.; Brooking, I.R.; Francis, G.S. Sirius a mechanistic model of wheat respinse to environmental variation. *Eur. J. Agron.* **1998**, *8*, 161–179. [CrossRef]

19. Jones, J.W.; Hoogenboom, G.; Porter, C.H.; Boote, K.J.; Batchelor, W.D.; Hunt, L.A.; Wilkens, P.W.; Singh, U.; Gijsman, A.J.; Ritchie, J.T. The Dssat Cropping System Model. *Eur. J. Agron.* **2003**, *18*, 235–265. [CrossRef]

20. Hunt, L.A.; Pararajasingham , S. CROPSIM—WHEAT: A model describing the growth and development of wheat. *Can. J. Plant Sci.* **1995**, 619–632.

21. Ritchie, J.T.; Singh, U.; Godwin, D.C.; Bowen, W.T. Cereal growth, development and yield. *Underst. Opt. Agric. Prod.* **1998**, 79–98. [CrossRef]

22. Ritchie, J.T.; Otter, S. Description and performance of CERES-Wheat: A user-orientes wheat yield model. *ARS Wheat Yield Proj.* **1985**, *38*, 159–175.

23. Hoogenboom, G.; Jones, J.W.; Wilkens, P.W.; Porter, C.H.; Boote, K.J.; Hunt, U.S.; Lizaso, J.I.; White, J.W.; Uryasev, O.; Ogoshi, R.; et al. *Decision Support System for Agrotechnology Transfer (DSSAT) [CD-ROM]*; University of Hawaii: Honululu, HI, USA, 2015.

24. Thorp, K.R.; Hunsaker, D.J.; French, A.N.; White, J.W.; Clarke, T.R. Evaluation of the CSM-CROPSIM-CERES-Wheat Model as a Tool for Crop Water Management. *Trans. ASABE* **2010**, *53*, 1–17. [CrossRef]

25. Chipanshi, A.C.; Ripley, E.A.; Lawford, R.G. Large-scale simulation of wheat yields in a semi-arid environment using a crop-growth model. *Agric. Syst.* **1999**, *59*, 57–66. [CrossRef]

26. Savin, R.; Satorre, E.H.; Hall, A.J.; Slafer, G.A. Assessing strategies for wheat cropping in the monsoonal climate of the Pampas using the CERES-Wheat simulation model. *Field Crops Res.* **1995**, *42*, 81–91. [CrossRef]

27. Sardinia, S.; Dettori, M.; Cesaraccio, C.; Motroni, A.; Spano, D.; Duce, P. Field Crops Research Using CERES-Wheat to simulate durum wheat production and phenology. *Field Crops Res.* **2011**, *120*, 179–188. [CrossRef]

28. Bannayan, M.; Crout, N.M.J.; Hoogenboom, G. Application of the CERES-Wheat model for within-season prediction of winter wheat yield in the United Kingdom. *Agron. J.* **2003**, *95*, 114–125. [CrossRef]

29. Gbegbelegbe, S.; Cammarano, D.; Asseng, S.; Robertson, R.; Chung, U.; Adam, M.; Abdalla, O.; Payne, T.; Reynolds, M.; Sonder, K.; et al. Baseline simulation for global wheat production with CIMMYT mega-environment specific cultivars. *Field Crops Res.* **2017**, *202*, 122–135. [CrossRef]

30. Batchelor, W.D.; Jones, J.W.; Boote, K.J.; Porter, C.H. *Pest and Disease Damage Module*; University of Florida: Gainsville, FL, USA, 2004.

31. Boote, K.J.; Bennet, J.M.; Jones, J.W.; Jowers, H.E. On-farming testing of peanut and soybean models in north Florida. *Paper Am. Soc. Agric. Eng. USA* **1989**. Available online: http://agris.fao.org/agris-search/search.do?recordID=US9165910 (accessed on 10 January 2019).

32. Batchelor, W.D.; Jones, J.W.; Boote, K.J.; Pinnschmidt, H.O. Extending the use of crop models to study pest damage. *Trans. Am. Soc. Agric. Eng. Gen. Ed.* **1993**, *36*, 551–558. [CrossRef]

33. Boote, K.J.; Jones, J.W.; Hoogenboom, G.; Pickering, N.B. The CROPGRO model for grain legumes. In *Understanding Options for Agricultural Production*; Tsuji, G.Y., Hoogenboom, G., Thornton, P.K., Eds.; Springer: Dordrecht, The Netherlands, 1998; pp. 99–128. ISBN 978-94-017-3624-4.

34. Andarzian, B.; Hoogenboom, G.; Bannayan, M.; Shirali, M.; Andarzian, B. Determining optimum sowing date of wheat using CSM-CERES-Wheat model. *J. Saudi Soc. Agric. Sci.* **2015**, *14*, 189–199. [CrossRef]

35. Waggoner, P.E.; Berger, R.D. Defoliation, Disease, and Growth. *Phytopathology* **1987**, *77*, 1495–1497.

36. Robert, C. Analysis and modelling of effects of leaf rust and Septoria tritici blotch on wheat growth. *J. Exp. Bot.* **2004**, *55*, 1079–1094. [CrossRef] [PubMed]

37. Robert, C.; Bancal, M.O.; Lannou, C.; Ney, B. Quantification of the effects of Septoria tritici blotch on wheat leaf gas exchange with respect to lesion age, leaf number, and leaf nitrogen status. *J. Exp. Bot.* **2006**, *57*, 225–234. [CrossRef] [PubMed]

38. Sánchez-Vallet, A.; McDonald, M.C.; Solomon, P.S.; McDonald, B.A. Is Zymoseptoria tritici a hemibiotroph? *Fungal Genet. Biol.* **2015**, *79*, 29–32. [CrossRef] [PubMed]

39. Castro, A.C.; Simón, M.R. Effect of tolerance to Septoria tritici blotch on grain yield, yield components and grain quality in Argentinean wheat cultivars. *Crop Prot.* **2016**, *90*, 66–76. [CrossRef]

40. Zadoks, J.C.; Chang, T.T.; Konzak, C.F. A decimal code for the growth stages of cereals. *Weed Res.* **1974**, *14*, 415–421. [CrossRef]

41. Gröll, K. *Use of Sensor Technologies to Estimate and Assess the Effects of Various Plant Diseases on Crop Growth and Development*; Universty of Hohenheim: Stuttgart, Germany, 2008; Available online: http://opus.uni-hohenheim.de/volltexte/2008/296/ (accessed on 10 January 2019).

42. Thomas, M.; Cook, R.; King, J. Factors affecting development of Septoria tritici in winter wheat and its effect on yield. *Plant Pathol.* **1989**, 246–257. [CrossRef]
43. Willmott, C.J. Some Comments on the Evaluation of Model Performance. *J. Appl. Phys.* **1982**, *36*, 1309–1313. [CrossRef]
44. Jing, Q.; Qian, B.; Shang, J.; Huffman, T.; Liu, J.; Pattey, E.; Dong, T.; Tremblay, N.; Drury, C.F.; Ma, B.L.; et al. Assessing the options to improve regional wheat yield in eastern canada using the csm–ceres–wheat model. *Agron. J.* **2017**, *109*, 510–523. [CrossRef]
45. Attia, A.; Rajan, N.; Xue, Q.; Nair, S.; Ibrahim, A.; Hays, D. Application of DSSAT-CERES-Wheat model to simulate winter wheat response to irrigation management in the Texas High Plains. *Agric. Water Manag.* **2016**, *165*, 50–60. [CrossRef]
46. Robert, C.; Fournier, C.; Andrieu, B.; Ney, B. coupling a 3D virtual wheat (Triticum aestivum) plant model with a Septoria tritici epidemic model (Septo3D): A new approach to investigate plant-pathogen interactions linked to caonpy architecture. *Funct. Plant Biol.* **2008**, *35*, 997–1013. [CrossRef]
47. Magarey, R.D.; Sutton, T.B.; Thayer, C.L. A Simple Generic Infection Model for Foliar Fungal Plant Pathogens. *Phytopathology* **2005**, *95*, 92–100. [CrossRef] [PubMed]
48. Donatelli, M.; Magarey, R.D.; Bregaglio, S.; Willocquet, L.; Whish, J.P.M.; Savary, S. Modelling the impacts of pests and diseases on agricultural systems. *Agric. Syst.* **2017**, *155*, 213–224. [CrossRef] [PubMed]
49. King, J.E.; Jenkins, J.E.E.; Morgan, W.A. The estimation of yield losses in wheat from severity of infection by Septoria species. *Plant Pathol.* **1983**, *32*, 239–249. [CrossRef]
50. Ziv, O.; Eyal, Z. Assessment of Yield Component Losses Caused in Plants of Spring Wheat Cultivars by Selected Isolates of Septoria tritici. *Phytopathology* **1977**, *68*, e796. [CrossRef]
51. Bhathal, J.S.; Loughman, R.; Speijers, J. Yield reduction in wheat in relation to leaf disease from yellow (tan) spot and septoria nodorum blotch. *Eur. J. Plant Pathol.* **2003**, *109*, 435–443. [CrossRef]
52. Bastiaans, L. Ecology and Epidemiology Ratio Between Virtual and Visual Lesion Size as a Measure to Describe Reduction in Leaf Photosynthesis of Rice Due to Leaf Blast. *Phytopathology* **1991**, *81*, 611–615. [CrossRef]
53. Thorp, K.R.; Hunsaker, D.J.; French, A.N. Assimilating leaf area index estimates from remote sensing into the simulations of a cropping systems model. *Trans. ASABE* **2010**, *53*, 251–262. [CrossRef]
54. Ficke, A.; Cowger, C.; Bergstrom, G.; Brodal, G. Understanding Yield Loss and Pathogen Biology to Improve Disease Management: Septoria Nodorum Blotch—A Case Study in Wheat. *Plant Dis.* **2018**, *102*, 696–707. [CrossRef] [PubMed]

Article

Development of Chlorophyll-Meter-Index-Based Dynamic Models for Evaluation of High-Yield Japonica Rice Production in Yangtze River Reaches

Ke Zhang [1,2], Xiaojun Liu [1,2], Syed Tahir Ata-Ul-Karim [1,2,3], Jingshan Lu [1,2], Brian Krienke [4], Songyang Li [1,2], Qiang Cao [1,2], Yan Zhu [1,2], Weixing Cao [1,2] and Yongchao Tian [1,2,*]

[1]   National Engineering and Technology Center for Information Agriculture, Key Laboratory for Crop System Analysis and Decision Making, Ministry of Agriculture, 1 Weigang Road, Nanjing 210095, China; 2017201080@njau.edu.cn (K.Z.); liuxj@njau.edu.cn (X.L.); ataulkarim@issas.ac.cn (S.T.A.-U.-K.); 2016201019@njau.edu.cn (J.L.); 2016101006@njau.edu.cn (S.L.); qiangcao@njau.edu.cn (Q.C.); yanzhu@njau.edu.cn (Y.Z.); caow@njau.edu.cn (W.C.)
[2]   Key Laboratory for Information Agriculture, Jiangsu, Collaborative Innovation Center for Modern Crop Production, Nanjing Agricultural University, 1 Weigang Road, Nanjing 210095, China
[3]   Key Laboratory of Soil Environment and Pollution Remediation, Institute of Soil Science, Chinese Academy of Science, Nanjing 210008, China
[4]   Department of Agronomy and Horticulture, University of Nebraska-Lincoln, Lincoln, NE 68583, USA; krienke.brian@unl.edu
*   Correspondence: yctian@njau.edu.cn; Tel.: +86-25-84399050; Fax: +86-25-84396672

Received: 28 January 2019; Accepted: 20 February 2019; Published: 22 February 2019

Abstract: Accurate estimation of the nitrogen (N) spatial distribution of rice (*Oryza sativa* L.) is imperative when it is sought to maintain regional and global carbon balances. We systematically evaluated the normalized differences of the soil and plant analysis development (SPAD) index (the normalized difference SPAD indexes, NDSIs) between the upper (the first and second leaves from the top), and lower (the third and fourth leaves from the top) leaves of Japonica rice. Four multi-location, multi-N rate (0–390 kg ha^{-1}) field experiments were conducted using seven Japonica rice cultivars (9915, 27123, Wuxiangjing14, Wunyunjing19, Wunyunjing24, Liangyou9, and Yongyou8). Growth analyses were performed at different growth stages ranging from tillering (TI) to the ripening period (RP). We measured leaf N concentration (LNC), the N nutrition index (NNI), the NDSI, and rice grain yield at maturity. The relationships among the NDSI, LNC, and NNI at different growth stages showed that the NDSI values of the third and fourth fully expanded leaves more reliably reflected the N nutritional status than those of the first and second fully expanded leaves (LNC: NDSI$_{L3,4}$, R^2 > 0.81; NDSI$_{others}$, 0.77 > R^2 > 0.06; NNI: NDSI$_{L3,4}$, R^2 > 0.83; NDSI$_{others}$, 0.76 > R^2 > 0.07; all p < 0.01). Two new diagnostic models based on the NDSI$_{L3,4}$ (from the tillering to the ripening period) can be used for effective diagnosis of the LNC and NNI, which exhibited reasonable distributions of residuals (LNC: relative root mean square error (RRMSE) = 0.0683; NNI: RRMSE = 0.0688; p < 0.01). The relationship between grain yield, predicted yield, and NDSI$_{L3,4}$ were established during critical growth stages (from the stem elongation to the heading stages; R^2 = 0.53, p < 0.01, RRMSE = 0.106). An NDSI$_{L3,4}$ high-yield change curve was drawn to describe critical NDSI$_{L3,4}$ values for a high-yield target (10.28 t ha^{-1}). Furthermore, dynamic-critical curve models based on the NDSI$_{L3,4}$ allowed a precise description of rice N status, facilitating the timing of fertilization decisions to optimize yields in the intensive rice cropping systems of eastern China.

Keywords: SPAD; leaf nitrogen concentration; nitrogen nutrition index; grain yield; dynamic model

1. Introduction

Nitrogen (N) is one of the most important yield-limiting factors [1]. Appropriate N management is essential to achieve relatively high yields with low N input, particularly to ensure maximum rice yield and quality [2,3]. Yield-target-based N fertilization plays an important role when developing profitable and environmentally friendly rice production systems (which is good in environment protection during field production) [4,5]. Accurate and remote in-season estimation of crop N status and its site-specific applications in intensive rice cropping systems is challenging [6]. Rapid, non-destructive, and cost-effective N diagnostic tools are imperative for accurate and timely diagnosis of rice N status at critical growth stages—it is imperative to match N requirements to soil N supply [2,3].

Non-destructive diagnostic strategies use various devices to monitor crop growth and N status [7,8]. A chlorophyll meter (Soil Plant Analysis Development, SPAD-502, Minolta Camera Co., Osaka, Japan) has been widely used for simple, rapid, and non-destructive assessment of leaf chlorophyll concentrations [9], however, the readings are significantly influenced by growth stage, plant leaf position, leaf measurement location, leaf thickness, leaf weight, the cultivar, solar radiation, and environmental stress [1,10,11].

Previous studies sought to correlate specific leaf weight (SLW) with SPAD values and leaf N concentrations (LNC) [12,13], which range from 2–3.2% [4]; multiplying upper leaf SPAD values by the leaf area index (LAI) [14,15], or linking sensor data to the product of SPAD and height [16]. Many researchers have developed SPAD indices [17], including SPAD positional difference sufficiency index [18], a relative SPAD index [19], and a normalized SPAD index [20]. Chlorophyll meter readings have also been linked with digital still canopy color images to improve chlorophyll meter data [13]. However, linking SPAD values to tissue N concentrations remains challenging due to controversies in their reliability [4].

During the plant growth cycle, N and carbon (C) levels are in dynamic balance in crops; this is particularly significant when paddy rice leaves turn from green to yellow [21]. Either an N deficit or excess will retard crop growth. Cropping duration is controlled principally by genotype and N nutritional status. Various rice canopy leaves reflect N status, the color difference between 4LFT (the fourth fully expanded leaf from the top) and 3LFT (the third fully expanded leaf from the top) can be used to diagnose the critical N concentration at the booting stage; this is 27 g kg^{-1} dry matter weight (DW) for Japonica rice and 25 g kg^{-1} DW for Indica rice [22]. However, other indices, such as the relative SPAD index (RSI) [23], the difference SPAD index (DSI) [24], the relative difference SPAD index (RDSI) [22,25], and the normalized difference SPAD index [24], have been developed using the SPAD readings of 1LFT (the first fully expanded leaf from the top), 2LFT (the second fully expanded leaf from the top), and 3LFT to assess in-season crop N status. Recently, Yuan et al. reported that the 2LFT, 3LFT, and 4LFT SPAD values were related to various N indicators (e.g., the N nutrition index (NNI) and leaf N accumulation (LNA); the LNC index). The cited authors concluded that the normalized SPAD index of 4LFT (the NSI4) could be used to increase grain yield and nitrogen efficiency [10]. Therefore, SPAD indices of lower leaves are better than those of upper leaves regarding diagnosing rice N status.

Several attempts have been made to establish quantitative relationships between SPAD indicators and the NNIs of C3 and C4 crops [26], such as wheat [24], rice [27], corn [28], barley [29], etc. Given the differences among environmental conditions and genotypes, the relationships between NNI and SPAD values vary greatly. However, the relative SPAD values are not notably influenced by the cultivar, growing season, or growth stage. Calculating relative SPAD values requires the use of a non-N-limiting treatment as a control, reducing the utility of the method regarding on-farm N diagnosis [30]. SPAD value differences among varieties, and variations in production levels, can be eliminated by normalizing experimental data before modeling [31]. Therefore, some studies have used different normalized SPAD index (NDSI) values to reduce the effects of variation. LNCs of lower leaves were more sensitive to increases in N application rate, and SPAD readings of lower leaves were closely related to tissue N concentrations [28,32]. Thus, LNC of lower leaves may better reflect crop N status when N application rates vary. Although efforts have been made to link crop N status to

NDSI, no attempt has yet been made to associate NDSI values with the NNI or grain yield of rice, or to establish the relationships between the NNIs and NDSIs of the four topmost fully expanded leaves, on the one hand, and grain yield, on the other.

In recent years, many authors have sought to simulate the crop tiller number (LAI) and other dynamic indicators [31]. Dynamic models of crop growth indices should ideally be universal [33], but labor- and time-consuming due to destructive sampling required for obtaining the LAI, dry matter (DM), tissue N concentration, and other growth parameters. As the LNC correlates strongly with chlorophyll content, SPAD meters have been used as real-time, portable, non-destructive devices for estimating N levels by assessing transmittance [7]. A previous study found that SPAD readings of flag leaves correlated strongly with grain yields at different wheat growth stages, and multiple regression implied that maintenance of optimal leaf chlorophyll content over the interval of 50–75 days after sowing was essential to obtain high yields [34]. To date, few dynamic models based on spectral indices are available. In 2017, Liu et al. reported a double logistic NDVI dynamic model for high-yield production in rice, which can be used to accurately predict canopy NDVI dynamic changes during the entire growth period [35]. Further studies on SPAD index variation regarding the establishment of a SPAD index-based dynamic model are essential for monitoring and diagnosing crop nutritional status in-season.

Therefore, we defined the relationships between NDSI, LNC, and NNI during different growth periods: (1) to accurately access N nutrition status using SPAD-based index, and (2) to draw a dynamic-critical curve showing when yields were lower than required by the NDSI target to guide N fertilizer topdressing, thus aiding rice production in eastern China.

2. Materials and Methods

2.1. Sites and Experimental Design

Yangtze River Reaches is the main rice production region of China, which has a great influence on China's food security (Huang RH et al., 2002). Yangtze River Reaches is not only the major agricultural regions of China, but also the oldest niche of rice cultivation [36]. China contributes 29% of global rice production, and Yangtze River Reaches alone contributes more than 65% of the national rice production in China [37]. Thus, four field experiments using multi-N rates (0–390 kg N ha^{-1}; N0, N1, N2, N3, N4, N5, N6, and N7 were 0, 130, 150, 225, 260, 300, 375, and 390 kg N ha^{-1}, respectively) were conducted at Jiangning (E 118.98°, N 31.93°) and Wujiang (E 121.28°, N 31.33°) in eastern China from 2007 to 2009, and in 2013. Seven Japonica rice cultivars (9915, 27123, Wuxiangjing14, Wuyunjing19, Yongyou8, Wuyunjing24, Wuxiangjing19) used were the most widely cultivated cultivars in Jiangsu Province; with distinct subspecies and yield potentials. All experiments were conducted using a randomized complete block design with different N treatments and three replications. Details of the cultivars and N application rates are shown in Table 1.

Rice seedlings with three to four fully expanded leaves were transplanted on the 20 June 2007, 25 June 2008, and 26 June 2013, which were raised alone. The hill spacing was 0.30 m × 0.15 m (about 22 hills m^{-2}), with three seedlings per hill in all experiments. Each plot was of 3 m × 6 m in size. N treatment featured 30% N application at the pre-planting stage, and the remaining N was top dressed three times at the tillering (TI, 20%), booting (BT, 30%), and heading (HD, 20%) stages, as urea. In each experiment, 59 kg ha^{-1} phosphorus as P_2O_5 and 158 kg ha^{-1} potassium as K_2O were incorporated into each plot before transplantation following local standard rice production practices. Crop management practices at each site followed local recommendations to maximize grain yield (the only limiting factor was N fertilizer). Weeds, diseases, and insects were intensively controlled, as in conventional cultivation, throughout the growing period. Data from Exp. 1, Exp. 2, and Exp. 4 were used for model calibrating, while data of Exp. 3 were used to validate the models.

Table 1. Basic information about the four rice field experiments.

Experiment No.	Transplanting/Harvesting Date	Location	Cultivar	N Rate (kg N ha^{-1})	Soil Characteristic
EXP. 1 2007	20-Jun; 21-Oc	Jiangning, E 118.98°, N 31.93°	9915, 27123 (Japonica)	N0 (0) N1 (130) N4 (260) N7 (390)	Soil type = Fe-leachic -stagnic Anthrosols Soil pH = 6.5 OM = 13.5 g·kg^{-1} Total N = 1.13 g·kg^{-1} Available P = 45 mg·g^{-1} Available K = 82.6 mg·g^{-1}
EXP. 2 2008	25-Jun; 27-Oct	Jiangning, E 118.98°, N 31.93°	WXJ14, 27123 (Japonica)	N0 (0) N1 (130) N4 (260) N7 (390)	Soil type = Fe-leachic -stagnic Anthrosols Soil pH = 6.9 OM = 13.5 g·kg^{-1} Total N = 1.38 g·kg^{-1} Available P = 43 mg·g^{-1} Available K = 80 mg·g^{-1}
EXP. 3 2009	19-Jun; 20-Oct	Jiangning, E 118.98°, N 31.93°	WYJ19, YY8, WXJ14, WYJ24 (Japonica)	N0 (0) N1(130) N4 (260) N7 (390)	Soil type = Gley- stagnic Anthrosols Soil pH = 6.9 OM = 26.15 g·kg^{-1} Total N = 1.65 g·kg^{-1} Available P = 38 mg·g^{-1} Available K = 70 mg·g^{-1}
EXP. 4 2013	19-Jun; 20-Oct	Wujiang, E 121.28°, N 31°33′	WYJ19, WXJ19 (Japonica)	N0 (0) N2(150) N3 (225) N5 (300) N6 (375)	Soil type = Typic Endoaquepts Soil pH = 6.75 OM = 26.15 g·kg^{-1} Total N = 2.15 g·kg^{-1} Available P = 45.5 mg·g^{-1} Available K = 115.3 mg·g^{-1}

Note: "9915, 27123, WXJ14, WYJ19, WYJ24, YY8, and WXJ19" are rice cultivars; "WXJ14" is Wuxiangjing14; "WYJ19" is Wuyunjing19; "WYJ24" is Wuyunjing24; "YY8" is Yongyou8; and "WXJ19" is Wuxiangjing19. "N0 = 0 kg N ha^{-1}; N1 = 130 kg N ha^{-1}; N2 = 150 kg N ha^{-1}; N3 = 225 kg N ha^{-1}; N4 = 260 kg N ha^{-1}; N5 = 300 kg N ha^{-1}; N6 = 375 kg N ha^{-1}; N7 = 390 kg N ha^{-1}".

2.2. Plant Sampling and Measurement; Shoot Biomass, Nitrogen Concentration, and the NNI

Plants collected by randomly clipping 1 m^2 from the TI to the ripening period (RP) stages were separated into green leaf blades (leaves) and culm-plus-sheath (stems), oven-dried for 30 min at 105 °C to halt metabolic processes, and then dried at 80 °C in a forced-draft oven until constant weight was attained. Each component was then ground to powder, passed through a 1-mm-diameter sieve in a Wiley mill, and stored at room temperature. Samples (0.2 g) were digested with H_2O_2 and H_2SO_4. N concentrations were determined using a continuous-flow auto-analyzer AA3 (Bran + Luebbe; Norderstedt, Germany). Grain yield was measured at maturity by harvesting 1 m^2 of crop and drying to a moisture content of 14%. Leaf dry matter levels were measured using this material.

2.3. SPAD Measurements

The chlorophyll meter is a spectral instrument, it measures the difference between the transmittance of a red (650 nm) and an infrared (940 nm) light through the leaf, generating a 3-digit SPAD value, which was used to take SPAD readings from the four uppermost fully expanded leaves of 10 randomly selected plants from each plot. Three SPAD values per leaf, including one value around the midpoint of the leaf blade and two values 3 cm apart from the midpoint, were averaged to give the mean SPAD value of the leaf avoiding the midribs. These measurements were taken at each growth stage and averaged [10]. Chlorophyll meter (SPAD) readings were obtained at six growth stages: TI, stem elongation (SE), panicle initiation (PI), BT, HD, flowering (FL), grain filling (GF), and RP, using a SPAD-502 meter (Minolta Camera Co., Osaka, Japan). SPAD readings were obtained from the four, uppermost fully expanded leaves of 10 randomly selected plants in each plot. The normalized difference SPAD index (NDSI) between LFT_i and LFT_j used to evaluate N nutritional status was that for wheat developed by Zhao et al. [29], and the equation for other SPAD-based indices is described in Table 2.

$$NDSI_{Li,j} = (SPAD_i - SPAD_j)/(SPAD_i + SPAD_j) \tag{1}$$

where $SPAD_i$ and $SPAD_j$ are the SPAD readings of leaf positions i and j; i and j vary from 1 to 4, and i < j.

Table 2. Equations of soil and plant analysis development (SPAD)-based indices.

Index	Description	Algorithm	Reference
$DSI_{L1\text{-}L3}$	The difference SPAD between 1LFT and 3LFT	$S_{1LFT} - S_{3LFT}$	[24]
$SPAD_{L3\text{-}L4}$	The difference SPAD between 3LFT and 4LFT	$S_{3LFT} - S_{4LFT}$	[18]
$RSI_{L1/L3}$	The relative SPAD index between 1LFT and 3LFT	S_{1LFT}/S_{3LFT}	[23]
$RDSI_{L1,3}$	The relative difference SPAD index between 1LFT and 3LFT	$S_{1LFT}/(S_{1LFT} + S_{3LFT})$	[25]
$NDSI_{L1,3}$	The normalized differences SPAD and index between 1LFT and 3LFT	$(S_{1LFT} - S_{3LFT})/(S_{1LFT} + S_{3LFT})$	[28]
NDSI	The normalized differences SPAD and index between i LFT and j LFT, the range of i, j values is from 1 to 4, i < j	$(S_{iLFT} - S_{jLFT})/(S_{iLFT} + S_{jLFT})$	[29]

Note: "S" means SPAD values; "LFT" represents the fully expanded leaf position form the top; "1-4 LFT" means the first to fourth fully expanded leaf position form the top.

2.4. Data Analysis

2.4.1. Nitrogen Nutrition Index

The NNIs of various rice cultivars at different vegetative growth stages were calculated using the critical N concentration (*Nc*) values obtained from the *Nc* dilution curve of rice developed by Ata-Ul-Karim et al. [36], using the following equation:

$$Nc = 3.53 \times W^{-0.28} \ (12.37 \geq W \geq 1.55\,\text{t ha}^{-1}; W < 1.55\,\text{t ha}^{-1}, Nc = 3.05\%) \tag{2}$$

$$NNI = \frac{N_a}{N_c} \tag{3}$$

where *Nc* is the critical rice N, *W* is the weight of the crop, and *Na* is the crop N concentration [36].

2.4.2. Calibration of Dynamic-Critical Curve Models

The critical $NDSI_{L3,4}$ curve of high-yield was in the shape of a sigmoid curve. Different equation models were used to fit the curve, and the Boltzmann model was selected to image the changes of $NDSI_{L3,4}$, based on the R^2 and the relative root mean square error (RRMSE):

$$y = A_2 + \frac{A_1 - A_2}{1 + e^{\frac{x - x_0}{dx}}}, \; Boltzmann\; model \tag{4}$$

where A_1 is the vegetative plateau, A_2 the reproduction plateau, x_0 the x value when $NDSI_{L3,4} = 0$, and dx is a time constant. Three high-yield $NDSI_{L3,4}$ trends were similar as a sigmoid curve. In the $NDSI_{L3,4}$-based high-yield critical curve, the A_1, A_2, x_0, and dx values differed regarding the $NDSI_{L3,4}$ trend.

2.4.3. Statistical Analysis

All SPAD data were normalized using the maximum conversion ratio method. Data from each sampling date and year were subjected to analysis of variance using SPSS ver.20.0 software (IBM, Armonk, New York, NY, USA); this software was also used to compare yields. The least significant difference (LSD) test was employed to compare differences between treatment means. GraphPad Prism 5 software (GraphPad Software, San Diego, CA, USA) was used to fit the critical NDSI curve to different grain yields and to compare the intercept and slope of the regression curve at different growth stages. The R^2 value and the relative root mean square error (RRMSE, %) were used to explore the predictive accuracy of the model:

$$RRMSE = \sqrt{\frac{\sum_{i=1}^{n}(P_i - Q_i)^2}{n}} \times \frac{100}{\overline{Q_i}} \tag{5}$$

where n is the number of samples, P_i the measured values, Q_i the predicted values, and the average of Q_i. The RRMSE was used to explore agreement between model predictions and measured values.

3. Results

3.1. SPAD Readings of Different Leaves

SPAD readings were measured from the TI to the RP stages under different N treatments. The readings of Japonica rice (27123) increased with increasing N application, and ranged from 34–42 and 40–48 when low, or sufficient N, and excess N were applied, respectively (Figure 1). SPAD reading trends of 1LFT and 2LFT differed greatly from those of 3LFT and 4LFT. The N application rates had no significant effect on the SPAD values of 1LFT and 2LFT, but did affect SPAD values of 3LFT and 4LFT, especially 4LFT. However, a higher N fertilization rate did not significantly change SPAD readings. SPAD reading of 1LFT gradually increased from the TI to the BT and RP stages, in a deformed "S" manner. SPAD reading of 2LFT fell from the TI to the RP stage. SPAD readings of 3LFT and 4LFT were variable from the TI stage to the flowering stage (FL), but then decreased toward the RP stage. Regarding SPAD readings at different N rates, the N0 stage exhibited the lowest readings at every leaf, and the N3 stage the highest. N2 readings were not always greater than those for N1 prior to the 1–3LFT PI stages. Thus, N topdressing fertilizer increased SPAD values and enhanced the stability of the photosynthetic reaction.

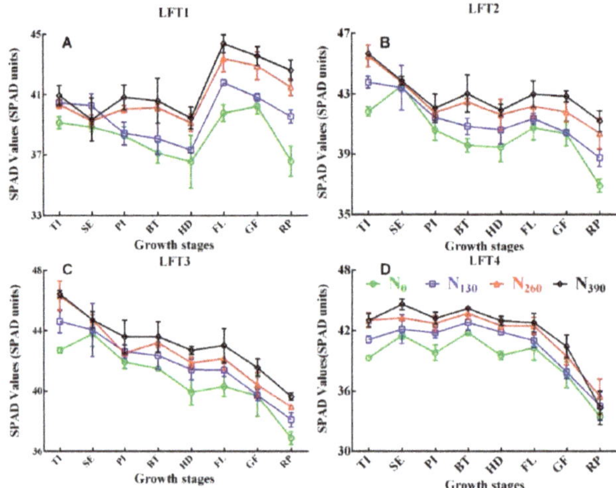

Figure 1. Changes over time in the soil and plant analysis development (SPAD) readings of different leaves ("LFT" represents the fully expanded leaf position form the top; "1-4 LFT" means one to four fully expanded leaf position form the top.; 1LFT, **A**; 2LFT, **B**; 3LFT, **C**; and 4LFT, **D**) of 27123 plants evaluated in 2007 and 2008 at N levels of 120 kg ha^{-1}. The vertical bars are standard error (TI, tillering; SE, stem elongation; PI, panicle initiation; BT, booting; HD, heading; FL, flowering; GF, grain filling; and RP, ripening).

3.2. Differences in the Normalized SPAD Indices

SPAD values were significantly affected by cultivar, growing season, and growth stage. We calculated NDSI values to evaluate leaf performance. Table 3 shows that simple linear SPAD analysis revealed that $NDSI_{L1,3}$, $NDSI_{L1,4}$, and $NDSI_{L3,4}$ did not differ significantly among the seven varieties, showing that SPAD readings were eliminated among variety differences. Furthermore, $NDSI_{L1,3}$, $NDSI_{L2,3}$, and $NDSI_{L3,4}$ values did not differ between the years. Therefore, compared to other SPAD indicators, $NDSI_{L3,4}$ better compared data from different years or varieties. Simple linear analysis of the seven N rates indicated that all the $NDSI_{L1,3}$, $NDSI_{L2,3}$, and $NDSI_{L3,4}$ differed significantly at the 0.01 probability level. For the various growth stages, all the $NDSI_{L1,3}$, $NDSI_{L1,4}$, $NDSI_{L2,3}$, and $NDSI_{L3,4}$ differed significantly at the 0.05 or 0.01 probability levels. Thus, $NDSI_{L1,3}$, $NDSI_{L2,3}$, and $NDSI_{L3,4}$ indices can be used to describe the time course of N nutritional status.

Table 3. Simple grouping linear analysis of SPAD indicator.

SPAD Indicator	Variety			Year			Treatment			Growth Stage			Residual	
	df	MS	F-Value	df	MS	F-Value	df	MS	F-Value	df	MS	F-Value	df	MS
$NDSI_{L1,2}$	6	0.00612 **	12.016	2	0.016 *	30.025	3	0.0001 **	0.064	5	0.16 ns	53.7	271	0.0001
$NDSI_{L1,3}$	6	0.013 ns	11.288	2	0.032 ns	27.303	3	0.0001 ns	0.184	5	0.04 *	81.014	271	0.001
$NDSI_{L1,4}$	6	0.034 ns	11.751	2	0.079 **	27.068	3	0.003 ns	0.863	5	0.1 **	81	271	0.002
$NDSI_{L2,3}$	6	0.002 *	1.706	2	0.003 ns	2.84	3	0.016 *	17.755	5	0.009 **	10.771	271	0.001
$NDSI_{L2,4}$	6	0.016 **	8.448	2	0.035 *	17.932	3	0.006 ns	2.62	5	0.048 ns	44.103	271	0.002
$NDSI_{L3,4}$	6	0.004 ns	6.65	2	0.009 ns	13.945	3	0.006 **	10.29	5	0.017 **	60.636	271	0.0001

"df" is degrees of freedom; "MS" is mean square; "ns" means non-significance; "*" indicates significant difference at 0.05 probability level; "**" indicates significant difference at 0.01 probability level. Variety include Japonica rice "9915, 27123, Wuxiangjing14 Wuyunjing19, 27123, Wuxiangjing14, Wuyunjing19, Wuyunjing24, Wuxiangjing19", and Indica rice "Y-Liangyou8"; years are 2007, 2008, and 2013; treatment is the N rates (N1–N6); growth stage form tillering to flowering.

3.3. The Relationship between the $NDSI_{Li,j}$ and Leaf N Concentration

Linear relationships were evident between $NDSI_{Li,j}$ and LNC, irrespective of growth stage or year (2007, 2008, and 2013; Table 4). $NDSI_{L3,4}$ exhibited less variability on regression analysis ($R^2 > 0.82$,

$p < 0.01$); lower coefficients of determination (R^2) were apparent for the other leaves ($0.77 > R^2 > 0.06$). The standard deviation (SD) ranged from 0.15–0.54; the lowest SD was that of the $NDSI_{L3,4}$ and the highest were those of $NDSI_{L1,4}$ and $NDSI_{L2,3}$. $NDSI_{L3,4}$ was thus an ideal index for diagnosis of LNC reliability. Thus, we developed an $NDSI_{L3,4}$-based model to determine LNC (Figure 2A); the variability was 84%, proving that the model afforded good retrieval accuracy (RRMSE = 0.0683; Figure 3a).

Table 4. Quantitative relationships between the SPAD readings of different rice leaves and leaf nitrogen concentrations.

Year	SPAD Index	Quantitative Relationship	R^2	SD
2007	$SPAD_{L3\text{-}L4}$	$LNC = 2.243 \times e^{-0.053SPAD}$	0.21 ns	0.37
	$RSI_{L1/L3}$	$LNC = 2.12 \times e^{-2.66RSI}$	0.56 *	0.27
	$DSI_{L1\text{-}L3}$	$LNC = 11.909 \times e^{-1.75DSI}$	0.53 *	0.33
	$RDSI_{L1,3}$	$LNC = 2.054 \times e^{-0.049RDSI}$	0.35 *	0.29
	$NDSI_{L1,2}$	$LNC = 2.193 \times e^{-11.38NDSI}$	0.77 **	0.31
	$NDSI_{L1,3}$	$LNC = 2.137 \times e^{-5.072NDSI}$	0.61 *	0.33
	$NDSI_{L1,4}$	$LNC = 2.205 \times e^{-2.229NDSI}$	0.36 *	0.46
	$NDSI_{L2,3}$	$LNC = 2.16 \times e^{-7.467NDSI}$	0.36 *	0.23
	$NDSI_{L2,4}$	$LNC = 2.257 \times e^{-2.681NDSI}$	0.19 ns	0.25
	$NDSI_{L3,4}$	$LNC = 2.249 \times e^{-10.58NDSI}$	0.83 **	0.23
2008	$SPAD_{L3\text{-}L4}$	$LNC = 2.195 \times e^{-0.042SPAD}$	0.11 ns	0.27
	$RSI_{L1/L3}$	$LNC = 1.863 \times e^{-6.747RSI}$	0.58 *	0.31
	$DSI_{L1\text{-}L3}$	$LNC = 4.131 \times e^{-0.688DSI}$	0.56 *	0.33
	$RDSI_{L1,3}$	$LNC = 1.862 \times e^{-0.163RDSI}$	0.59 *	0.24
	$NDSI_{L1,2}$	$LNC = 1.896 \times e^{-14.87NDSI}$	0.61 **	0.21
	$NDSI_{L1,3}$	$LNC = 1.898 \times e^{-14.84NDSI}$	0.61 **	0.21
	$NDSI_{L1,4}$	$LNC = 2.092 \times e^{-3.476NDSI}$	0.18 ns	0.48
	$NDSI_{L2,3}$	$LNC = 2.019 \times e^{-0.4133NDSI}$	-	0.54
	$NDSI_{L2,4}$	$LNC = 2.201 \times e^{-1.895NDSI}$	0.06 ns	0.49
	$NDSI_{L3,4}$	$LNC = 2.208 \times e^{-14.71NDSI}$	0.81 **	0.16
2013	$SPAD_{L3\text{-}L4}$	$LNC = 1.974 \times e^{-0.045SPAD}$	0.16 ns	0.41
	$RSI_{L1/L3}$	$LNC = 1.979 \times e^{-0.037RSI}$	0.49 *	0.34
	$DSI_{L1\text{-}L3}$	$LNC = 8.39 \times e^{-1.43DSI}$	0.17 ns	0.52
	$RDSI_{L1,3}$	$LNC = 1.998 \times e^{-1.434RDSI}$	0.57 *	0.36
	$NDSI_{L1,2}$	$LNC = 2.167 \times e^{-3.556NDSI}$	0.56 **	0.34
	$NDSI_{L1,3}$	$LNC = 2.219 \times e^{-2.401NDSI}$	0.57 **	0.29
	$NDSI_{L1,4}$	$LNC = 2.324 \times e^{-1.706NDSI}$	0.18 ns	0.39
	$NDSI_{L2,3}$	$LNC = 2.363 \times e^{-2.673NDSI}$	0.07 ns	0.15
	$NDSI_{L2,4}$	$LNC = 2.385 \times e^{-0.121NDSI}$	-	0.24
	$NDSI_{L3,4}$	$LNC = 2.246 \times e^{-10.56NDSI}$	0.82 **	0.18

"RSI" means relative SPAD index; "DSI" represents difference SPAD index; "RDSI" means relative difference SPAD index; "NDSI" is the normalized difference SPAD index; "SD" is standard deviation value; "LNC" means leaf nitrogen concentration; "NNI" is nitrogen nutrition index; "ns" and "-" mean non-significance; * indicates significant difference at 0.05 probability level; ** indicates significant difference at 0.01 probability level.

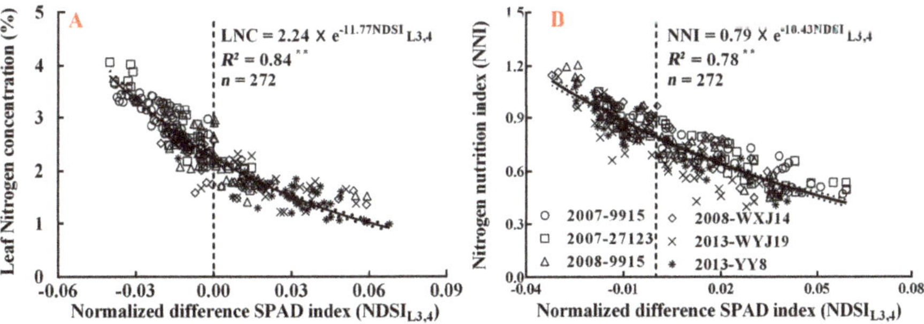

Figure 2. Regression fits between the leaf nitrogen concentration (LNC, **A**), nitrogen nutrition index (NNI, **B**), and $NDSI_{L3,4}$. The experimental years are shown as 2007, 2008, or 2013; 9915, 27123, and WXJ14 are the Wuxiangjing 14, WYJ19 means Wuyunjing19, and YY8 is Yongyou8. "**" means significant difference at 0.01 probability level.

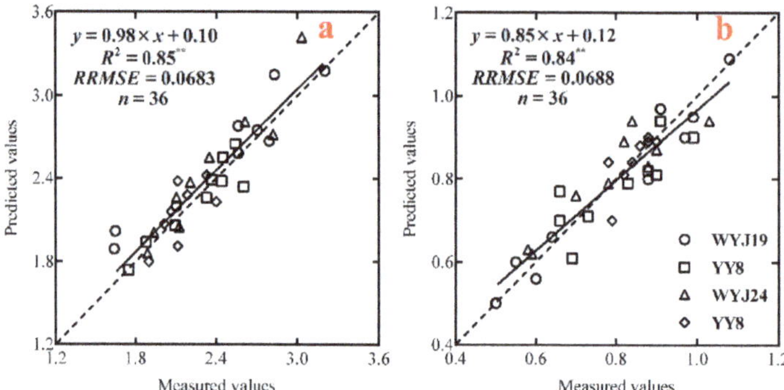

Figure 3. The relationship between measured and predicted leaf nitrogen concentration (LNC, **a**), nitrogen nutrition index (NNI, **b**) of four rice cultivars from the time of stem elongation (SE) to heading (HD) (WYJ19, Wuyungjing19; YY8, Yongyou8; WXJ14, Wuxiangjing14; and WYJ24, Wuyungjing 24; Japonica). The solid line is the linear regression line and the dotted line a line inclined at 45° to the axis. ** indicates significant difference at 0.01 probability level.

In Table 5, using the model equation of $NDSI_{Li,j}$ ($LNC = a \times e^{b \times NDSI_{Li,j}}$), the least significant difference (LSD) test was employed to measure differences in the "a" and "b" parameters among years or varieties. Regarding years, "a" differed significantly among the $NDSI_{L1,2}$ values, and "b" differed significantly among the $NDSI_{L2,4}$ values of the various varieties. In Table 5, "a" did not vary among varieties, and "b" did not change with the year. Thus, $NDSI_{L3,4}$ was an optimal indicator of rice N status.

Table 5. Simple linear regression; fitted curves between the SPAD values of different leaves and rice nitrogen indicators.

Nitrogen Indicator	Parameter	Impact Factor	Mean Square (MS)					
			$NDSI_{L1,2}$	$NDSI_{L1,3}$	$NDSI_{L1,4}$	$NDSI_{L2,3}$	$NDSI_{L2,4}$	$NDSI_{L3,4}$
LNC	a	Year	0.27 *	0.27 ns	0.15 ns	0.0355 ns	0.09 ns	0.001 ns
		variety	0.104 ns	0.035 ns	0.021 ns	0.0178 ns	0.26 ns	0.0252 ns
	b	Year	42.67 ns	42.88 ns	0.83 ns	12.97 ns	1.72 ns	5.71 ns
		variety	36.58 ns	41.24 ns	0.61 ns	8.73 ns	1.22*	3.07 ns
	Residual		0.051	0.061	0.021	0.035	0.014	0.023
NNI	a	Year	0.02 ns	0.01 ns	0.01 ns	-	-	0.02 ns
		variety	0.21 ns	0.04 ns	0.09 *	0.17 ns	0.06 ns	0.51 ns
	b	Year	5.51 ns	5.58 ns	0.54 ns	8.45 ns	5.91 ns	5.62 ns
		variety	1.39 ns	6.71 ns	0.91 ns	4.27 ns	6.53 ns	4.98 ns
	Residual		0.27	0.39	0.17	0.32	0.21	0.04

"LNC" means leaf nitrogen concentration; "NNI" is nitrogen nutrition index; "NDSI" represents normalized difference SPAD index; "ns" and "-" mean non-significance; * indicates significant difference at 0.05 probability level; ** indicates significant difference at 0.01 probability level.

3.4. Relationships between the $NDSI_{Li,j}$ and N Nutrition Index

The NNI is a widely used diagnostic indicator; when NNI = 1, N nutrition is optimal; NNI >1 and NNI <1 indicate excess and deficient N nutrition, respectively. We found a non-linear relationship between the $NDSI_{Li,j}$ and the NNI. Table 6 shows that the relationships between $NDSI_{Li,j}$ values of the differences among $LFT_{L1,2}$, $LFT_{L1,3}$, $LFT_{L3,4}$, and NNI were more stable than those of differences in the other $LFT_{Li,j}$ values across both the growth stages and the cultivars ($0.23 < R^2 < 0.84$ vs. $0.07 < R^2 < 0.16$). Regarding the SDs, these ranged from 0.15–0.54; the $NDSI_{L3,4}$ value was the lowest except in 2013; those of the $NDSI_{L1,4}$ or $NDSI_{L2,3}$ were the highest. The SDs were little affected by the LNC model chosen. Compared to the other $NDSI_{Li,j}$ values, the $NDSI_{L3,4}$ was more closely related to the NNI in both of the earlier years. We created a diagnostic model using the $NDSI_{L3,4}$ values

(Figure 2B). The $NDSI_{L3,4}$ explained 78% of the variability in the NNI; thus, the $NDSI_{L3,4}$ predicted N status (RRMSE = 0.0688; Figure 3b).

Table 6. Quantitative relationships between SPAD readings of different rice leaves and the nitrogen nutrition index.

Year	SPAD Index	Quantitative Relationship	R^2	SD
2007	$SPAD_{L3\text{-}L4}$	$NNI = 0.780 \times e^{-0.028SPAD}$	0.16 ns	0.49
	$RSI_{L1/L3}$	$NNI = 0.739 \times e^{-0.012RSI}$	0.35 *	0.38
	$DSI_{L1\text{-}L3}$	$NNI = 1.017 \times e^{-0.322DSI}$	0.26 ns	0.32
	$RDSI_{L1,3}$	$NNI = 0.746 \times e^{-0.627RDSI}$	0.56 *	0.27
	$NDSI_{L1,2}$	$NNI = 0.776 \times e^{-5.84NDSI}$	0.76 **	0.31
	$NDSI_{L1,3}$	$NNI = 0.759 \times e^{-2.35NDSI}$	0.53 *	0.33
	$NDSI_{L1,4}$	$NNI = 0.766 \times e^{-1.06NDSI}$	0.15 ns	0.46
	$NDSI_{L2,3}$	$NNI = 0.755 \times e^{-2.89NDSI}$	0.61 *	0.23
	$NDSI_{L2,4}$	$NNI = 0.773 \times e^{-1.32NDSI}$	0.14 ns	0.25
	$NDSI_{L3,4}$	$NNI = 0.809 \times e^{-8.85NDSI}$	0.83 **	0.23
2008	$SPAD_{L3\text{-}L4}$	$NNI = 0.764 \times e^{-0.36SPAD}$	0.14 ns	0.51
	$RSI_{L1/L3}$	$NNI = 0.713 \times e^{-2.91RSI}$	0.21 ns	0.34
	$DSI_{L1\text{-}L3}$	$NNI = 0.498 \times e^{0.41DSI}$	0.43 *	0.40
	$RDSI_{L1,3}$	$NNI = 0.707 \times e^{-0.07RDSI}$	0.67 *	0.26
	$NDSI_{L1,2}$	$NNI = 0.709 \times e^{-5.99NDSI}$	0.71 *	0.21
	$NDSI_{L1,3}$	$NNI = 0.710 \times e^{-6.05NDSI}$	0.62 *	0.21
	$NDSI_{L1,4}$	$NNI = 0.749 \times e^{0.50NDSI}$	-	0.48
	$NDSI_{L2,3}$	$NNI = 0.746 \times e^{-2.11NDSI}$	0.68 *	0.54
	$NDSI_{L2,4}$	$NNI = 0.742 \times e^{1.48NDSI}$	0.07 ns	0.49
	$NDSI_{L3,4}$	$NNI = 0.821 \times e^{-11.96NDSI}$	0.85 **	0.16
2013	$SPAD_{L3\text{-}L4}$	$NNI = 0.653 \times e^{-0.02SPAD}$	0.15 ns	0.39
	$RSI_{L1/L3}$	$NNI = 0.610 \times e^{-0.03RSI}$	0.32 *	0.28
	$DSI_{L1\text{-}L3}$	$NNI = 2.492 \times e^{-1.40DSI}$	0.51 *	0.36
	$RDSI_{L1,3}$	$NNI = 0.615 \times e^{-1.40RDSI}$	0.60 *	0.26
	$NDSI_{L1,2}$	$NNI = 0.708 \times e^{-1.85NDSI}$	0.57 *	0.34
	$NDSI_{L1,3}$	$NNI = 0.711 \times e^{-1.52NDSI}$	0.41 *	0.29
	$NDSI_{L1,4}$	$NNI = 0.721 \times e^{-1.96NDSI}$	0.25 *	0.39
	$NDSI_{L2,3}$	$NNI = 0.733 \times e^{-2.96NDSI}$	0.58 *	0.15
	$NDSI_{L2,4}$	$NNI = 0.764 \times e^{-3.36NDSI}$	0.16 ns	0.24
	$NDSI_{L3,4}$	$NNI = 0.734 \times e^{-13.51NDSI}$	0.84 **	0.18

"RSI" means relative SPAD index; "DSI" represents difference SPAD index; "RDSI" means relative difference SPAD index; "NDSI" is the normalized difference SPAD index; "SD" is standard deviation value; "LNC" means leaf nitrogen concentration; "NNI" is nitrogen nutrition index; "ns" and "-" mean non-significance; * indicates significant difference at 0.05 probability level; ** indicates significant difference at 0.01 probability level.

Table 5 also shows a simple exponential function regression exercise, grouping the fitted curves between SPAD values at different positions and the rice NNI. The exponential regression equation is ($NNI = a \times e^{b \times NDSI_{Li,j}}$). N either "a" nor "b" was influenced by year or variety, except for the "a" of $NDSI_{L1,4}$ (which varied by variety). The $NDSI_{L3,4}$ residual was the smallest of all leaves. Thus, Table 5 shows that $NDSI_{L3,4}$ was the optimal indicator of rice N status.

3.5. The Relationship between $NDSI_{L3,4}$ and Grain Yield

Grain yield was positively associated with the $NDSI_{L3,4}$ (Table 7), especially from the SE to HD stages. In Table 7, the linear regressions between grain yield and $NDSI_{L3,4}$ values at varying N addition rates for all growth stages are plotted; we used data from Experiments 1, 2, and 4. The coefficients of determination of the SE-to-HD stages tended to be higher than those of the TI and FL stages. Thus, the $NDSI_{L3,4}$ value was related to grain yield during the SE-to-BT stages, accounting for 53% of the variation (Figure 4). The plateau/linear relationship showed that grain yield decreased linearly with the $NDSI_{L3,4}$ value when that value was >0.001 for paddy rice; at which time the yield approached $10.28 \, t \cdot ha^{-1}$.

Table 7. Linear regressions between various normalized SPAD indices and grain yield.

Year	Variety	Growth Stage					
		TI	SE	PI	BT	HD	FL
2007	9915	0.26 *	0.73 **	0.79 **	0.68 *	0.68 *	0.32 *
	27123	0.21 *	0.48 *	0.62 *	0.75 **	0.51 *	0.43 *
2008	27123	0.20 *	0.68 *	0.73 **	0.72 **	0.65 *	0.40 *
	WXJ14	0.26 *	0.75 **	0.79 **	0.72 **	0.54 *	0.37 *
2013	WYJ19	0.22 *	0.57 *	0.63 *	0.61 *	0.52 *	0.34 *
	YY8	0.19 ns	0.74 **	0.62 *	0.47 *	0.62 *	0.42 *

"ns" and "-" mean non-significant difference; * indicates significant difference at 0.05 probability level; ** indicates significant difference at 0.01 probability level. "TI" is tillering stage; "SE" is stem elongation stage; "PI" is panicle initiation stage; "BT" is booting stage; "HD" is heading stage; "FL" is flowering stage; "9915, 27123, WXJ14, WYJ19, WYJ24, YY8, WXJ19" are different rice cultivars; "WXJ14" is Wuxiangjing14; "WYJ19" is Wuyunjing19; "WYJ24" is Wuyunjing24; "YY8" is Yongyou8; "WXJ19" is Wuxiangjing19.

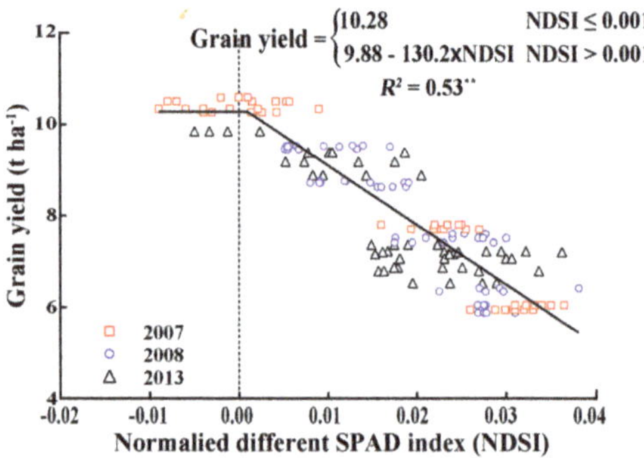

Figure 4. Correlations between $NDSI_{L3,4}$ values and grain yields at critical growth stages (from stem elongation to heading; cultivars: 9915, 27123 (2007); wuxiangjing14, 27123; 2013: wuyunjing19, wuxiangjing19 (2008)).

Table 8 shows results of the differential function models fitted. The results showed that most models had relatively high R^2 ($0.88 < R^2 < 0.99$), low RRMSE values ($16\% <$ RRMSE $< 30\%$), and high F-value ($113 < F < 266$; the higher F-value means model fitting). The equation in the shape of the sigmoid curve obtained the higher R^2 and low RRMSE values. The Boltzmann model performed the best among all regression models. Therefore, given $y = A_1 + \frac{A_1 - A_2}{1 + e^{\frac{x - x_0}{dx}}}$ (Boltzmann model) during SE-to-HD stages is critical in terms of high yield. We had many data points from SE to GF; we covered all the critical growth stages. Our studied varieties constitute distinct subspecies; all are high-yielding. Next, we created a dynamic model based on a high-yield (10.28 t·ha^{-1}) $NDSI_{L3,4}$ value (Figure 4). In the $NDSI_{L3,4}$-based high-yield critical curve, the A_1, A_2, x_0, and dx values differed in terms of the $NDSI_{L3,4}$ trend. In the high-yield critical curve, the day of the year (DOY) 205 to 220 (SE) was the vegetative plateau; vegetative and reproductive growth was in describe from DOY220 to DOY250 (thus from PI to BT), and DOY250 was the start of the reproductive stage. When stress develops, N is the pivotal factor limiting grain filling; we found that the $NDSI_{L3,4}$ reliably indicated rice N nutrition.

Table 8. Summary of fitting results of normalize difference SPAD index (NDSI) dynamic model under different fitting equation.

Regression Model	Equation	R^2	RRMSE (%)	F-Value
linear model	$y = a + b \times x$	0.899	23.5	213.8
Boltzmann model	$y = A_2 + (A_1 - A_2)/(1 + e^{((x - x0)/dx)})$	0.975	16.8	265.9
Polynomial model	$y = A + B \times x + C \times x^2 + D \times x^3 + \ldots$	0.927	25.2	113.91
Exponential model	$y = a - b \times c^x$	0.894	28.5	99.9
Bradley model	$y = a \times \ln(-b \times \ln(x))$	0.889	28.95	142.5
Power model	$y = a \times x^b$	0.453	48.3	15.3
Nelder model	$y = (x + a)/(b_0 + b_1 \times (x + a) + b_2 \times (x + a)^2)$	0.929	24.4	116.9
DoseResp model	$y = A_1 + (A_2 - A_1)/(1 + 10^{((LOGx_0 - x) \times p)})$	0.931	23.4	119.8
Hyperbola model	-	-	-	-

Note: "RRMSE" indicates relative root mean square error; "a, b, A, A_1, A_2, B, b_0, b_1, C, and D" are parameters of the equation; 'LOG' means the base-10 logarithm.

3.6. Use of the $NDSI_{L3,4}$ Curve for N Management

Estimation of the in-season N requirement (NR) is essential for the management of N topdressing during paddy rice production. However, we found that topdressing at critical growth stages (the SE–HD stages) did not support N management at every growth stage. Therefore, we developed an $NDSI_{L3,4}$ change curve for high-yield, topdressing N management at critical growth stages. We used this curve to first determine $NDSI_{L3,4}$ values (Figure 4) for the in-season N nutritional status of Japonica rice and a high-yield target, and then calculated ΔNNI values by evaluating quantitative relationships between the $NDSI_{L3,4}$ and NNI (Figure 2). Finally, the N fertilizer requirement for Japonica rice was modeled as suggested by Ata-Ul-Karim et al. [2]. This fertilization decision support method precisely estimates crop growth, grain yield, and the NR time-course (an N management strategy). However, the tool cannot be implemented in the early (active) TI stage. Moreover, our N fertilizer topdressing management strategy needs to be tested in other cultivars and rice growing regions to test its reliability for predicting grain yield and crop N status. This section may be divided by subheadings. It should provide a concise and precise description of the experimental results, their interpretation as well as the experimental conclusions that can be drawn.

4. Discussion

The present study analyzed the N nutrient status of rice leaves, determining SPAD values in four field experiments distributed in the Yangtze River Reaches. We examined N distribution by evaluating color differences between upper and lower leaves, and via differential diagnosis of rice N status. A previous study found that chlorophyll content reflected N nutritional status, but was significantly influenced by variety, site, and year of experimentation [31]. A previous study indicated that data normalization prior to modeling eliminated differences caused by variety, soil types, and management strategies, etc. [30]. Therefore, $NDSI_{Li,j}$ indicators can be used to correct traditional SPAD indicators. Most previous studies used threshold SPAD indicators during critical growth periods to diagnose N nutritional status [18]. However, most of the monitoring methods such as saturation index (SI) or NNI were mainly based on the single test of each growth stage [38]. These methods were mostly related to accessing the relationship with poor mechanism, using indicators with the complex calculation methods [39]. In addition, the existence of deviation in the identification of growth stage, environmental factors, time selection also have a greater impact on the single test. However, the single measured value was treated as each stage in the calculation, which results in a greater deviation from the actual value [40]. This minimizes field management but does not consider cropping duration. Thus, a dynamic model incorporating cropping duration, not only the critical growth stages, is better. Such models optimally display nutritional status over time, best-supporting careful agriculture.

We found a better relationship between $NDSI_{L3,4}$ and the N indicators (pooled data from five varieties in Exp. 1, 2, and 4) than between any other combination of $NDSI_{Li,j}$, and the N data, probably due to the non-linear regressions between $NDSI_{Li,j}$ and the N indicators varied over three years at the different sites. We found considerable differences between the linear regression coefficients of SE, PI, and BT stages. When the 3-year data (2007, 2008, and 2013) were pooled, R^2 decreased considerably because of significant regression slopes during different growth periods for the various varieties at different sites in different years. Such results showed that both the growth stage and leaf position significantly influenced $NDSI_{Li,j}$ values. The relationships between 4LFT and NSI4 were a linear/plateau in nature, from the SE to the BT periods. Prost and Jeuffroy indicated that the 1LFT showed poor defined relationship with N indicators, this might be due to the time difference in maturity of the first expanded leaves, due to these differences and wide variation in 1LFT [41]. Wang et al. reported that the lower leaf (3,4LFT) responded more to nitrogen supply than upper leaf (1,2LFT), and also suggested that lower leaf (4LFT) could be the ideal sample leaf for diagnosis of plant nitrogen nutrition [18]. Yuan et al. also proved that lower leaves afforded more reliable estimations of crop N status [17].

At the same target yield, different varieties require very different N topdressing. A previous study revealed that rice grain yield was positively associated with the SPAD readings of 4LFT, but the intercepts of the response curves of grain yield (as a function of SPAD value) differed markedly for the two varieties studied (Xiushui 63 and Hang 43) [27]. It has been reported that use of the NSI4 increased grain yield and N use efficiency compared to 4LFT. This may vary by cultivar and site conditions [17]. The cited authors identified the most sensitive stages at which to measure N nutrition and grain yield, and then created optimal fertilization prescriptions. However, such methods cannot diagnose N status in real time. Attempts have been made to estimate grain yield based on dynamic LAI models [42]. The LAI had the highest, positive indirect effects on grain yield, as measured by kernel number per spike, but it was difficult to describe the physiological condition of the crop directly. Wang et al. reported that SPAD measured the difference between 4LFT and 3LFT [43]. Color differences between these leaves could be used to determine N concentrations [21]. However, all models used to diagnosis N nutrition status or grain yield operate only in the critical growth period of rice [15,17]. If crop parameters are to be monitored, nutritional status should be diagnosed and regulated by simulating the dynamic changes of SPAD indicators appropriate for paddy rice. To date, few studies have sought to establish dynamic models based on spectral indices [35].

Further research on variations in such indices and establishment of index-based dynamic models is necessary to monitor the nutrition of late crops and for dynamic diagnosis [31,44]. We attempted to solve this problem by establishing $NDSI_{L3,4}$ change curves for a high-yield target, thus slightly higher than that of the high-yield rice cultivars grown Yangtze River Reaches [31]. In Figure 4 we established the relationship between $NDSI_{L3,4}$ and grain yield, and comprehensive comparisons of the simulated and observed values at critical growth stages revealed that the slope was 1.02, the $R^2 = 0.74$, and RRMSE = 0.106 (Figure 5).

We developed a dynamic $NDSI_{L3,4}$ model based on a high-yield target (Figure 6) to display the critical change trends. Using the high-yield dynamic model, it is possible to guide N fertilizer rice topdressing efficiently during the critical N growth period (from the SE to the HD). Some authors have used a spectral index or a nitrogen indicator to manage N fertilizer topdressing at critical growth stages [2,45]. N-deficient fields subject to intensive N-regulation become N-sufficient fields after application of local agricultural practices, increasing potential production. Our model allows detection of not only deficient N nutrition, but also excess N nutrition; there is no requirement for N-saturation. However, more data are needed for assessing the reliability of the model in different rice production areas.

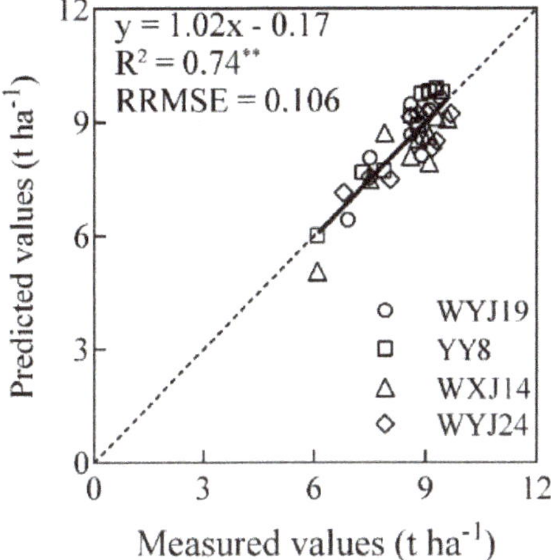

Figure 5. Relationships between measured and predicted grain yields of four rice cultivars in 2009 (WYJ19, Wuyungjing19; YY8, Yongyou8; WXJ14, Wuxiangjing14; and WYJ24, Wuyungjing24; Japonica).'*x*' represents measured values, '*y*' means predicted values. The solid line is the linear regression line and the dotted line a line inclined at 45° to the axis.

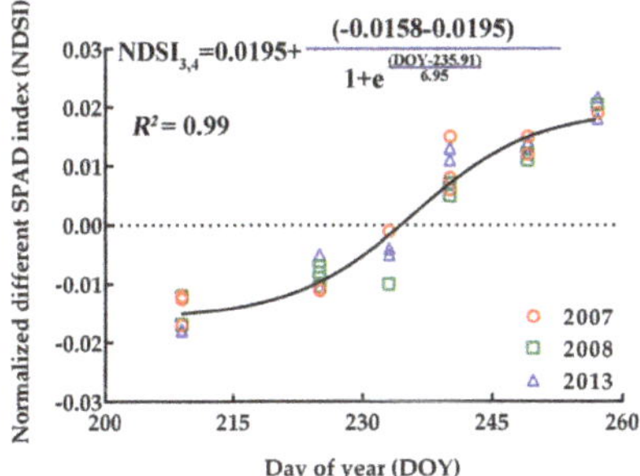

Figure 6. Critical $NDSI_{L3,4}$ data points used to define the changes in the $NDSI_{L3,4}$ curve when data of high-yield targets were pooled. The solid line is the critical $NDSI_{L3,4}$ change curve ($NDSI_{3,4} = 0.0195 + \frac{-0.0158-0.0195}{1+e^{\frac{DOY-235.91}{6.95}}}$; the A_1, A_2, x_0, and dx values differ in terms of their effect on $NDSI_{L3,4}$ trends) of high-yield target rice in the Yangtze river valley.

5. Conclusions

4LFT SPAD values measured using a chlorophyll meter correlated significantly with rice LNC and NNI values. The $NDSI_{L3,4}$ difference between the third and fourth LFT [$NDSI_{L3,4}$ = SPAD₃ − SPAD₄/(SPAD₃ + SPAD₄)] could be used to improve LNC and NNI estimations compared to those afforded by isolated SPAD readings and the differences between other leaf positions. We have developed two universal $NDSI_{L3,4}$ based diagnostic models (from the TI to the RP). Both models can be used for effective diagnosis of the LNC (R^2 > 0.81, p < 0.001) and NNI (R^2 > 0.83, p < 0.01) with a reasonable distribution of residuals (LNC: RRMSE = 0.0683; NNI: RRMSE = 0.0688; p < 0.01). To optimize N topdressing for Japonica rice, we first established the relationship between the $NDSI_{L3,4}$ and grain yield to predict the yield during the SE-to-HD stages. Secondly, a new, critically dynamic $NDSI_{L3,4}$ model was developed based on the previous experiments. Thirdly, the ΔNNI was estimated using both a dynamic model and the NNI–$NDSI_{L3,4}$ model. Finally, the N requirement was determined using the NNI model developed by Ata-Ul-Karim et al. [3]. However, parameters of the newly developed model may require adjustment under varied conditions caused by different climatic features, etc. The robustness and sensitivity of the model should be further tested using data from different rice production region.

Author Contributions: K.Z., Y.T., and X.L. conceived and designed the experiments, S.T.A.-U.-K., J.L., and S.L. performed experiments, Q.C., Y.Z., K.Z., and X.L. analyzed the data, K.Z., X.L., and Y.T. wrote the paper, W.C. and B.K. provided advice and edited the manuscript.

Funding: This work was financially supported by the National Key Research & Development Program of China (2018YFD0300805; 2016YFD0300604; 2016YFD0200602), the Science and Technology Support Program of Jiangsu [grant No.: BE2016375], the Fundamental Research Funds for the Central Universities (No.: 262201602) and the 111 project (B16026).

Acknowledgments: The authors also thank the anonymous reviews for their constructive comments and suggestions. The first author also thanks Xiaofang Duan for her support and love all the time.

Abbreviations

TI	Tillering
RP	Ripening period
BT	Booting
HD	Heading
SE	Stem elongation
PI	Panicle initiation
FL	Flowering
GF	Grain filling
SPAD	Soil and plant analysis development
NDSI	Normalized Different SPAD values
RSI	Relative SPAD index
DSI	Difference SPAD index
RDSI	Relative difference SPAD index
NNI	Nitrogen Nutrition Index
LNC	Leaf Nitrogen concentration
LAI	Leaf Area Index
DW	Dry matter weight
SSNM	Site-Specific Nitrogen Management
NUE	Nitrogen Use Efficiency
DAT	Day after Transplanting
RRMSE	Relative root mean square error
LSD	Least significant difference
LFT	The position on upper fully expanded Leaf From the rice Top

References

1. Zhao, B.; Ata-Ul-Karim, S.T.; Duan, A.; Liu, Z.; Wang, X.; Xiao, J.; Liu, Z.; Qin, A.; Ning, D.; Zhang, W. Determination of critical nitrogen concentration and dilution curve based on leaf area index for summer maize. *Field Crop. Res.* **2018**, *228*, 195–203. [CrossRef]
2. Ata-Ul-Karim, S.T.; Liu, X.; Lu, Z.; Zheng, H.; Cao, W.; Zhu, Y. Estimation of nitrogen fertilizer requirement for rice crop using critical nitrogen dilution curve. *Field Crop. Res.* **2017**, *201*, 32–40. [CrossRef]
3. Ata-Ul-Karim, S.T.; Zhu, Y.; Liu, X.; Cao, Q.; Tian, Y.; Cao, W. Comparison of different critical nitrogen dilution curves for nitrogen diagnosis in rice. *Sci. Rep.* **2017**, *7*, 42679. [CrossRef] [PubMed]
4. Ataulkarim, S.T.; Liu, X.; Lu, Z.; Yuan, Z.; Yan, Z.; Cao, W. In-season estimation of rice grain yield using critical nitrogen dilution curve. *Field Crop. Res.* **2016**, *195*, 1–8.
5. Ge, H.X.; Zhang, H.S.; Zhang, H.; Cai, X.H.; Song, Y.; Kang, L. The characteristics of methane flux from an irrigated rice farm in East China measured using the eddy covariance method. *Agric. For. Meteorol.* **2018**, *249*, 228–238. [CrossRef]

6. Yousaf, M.; Li, X.; Zhang, Z.; Ren, T.; Cong, R.; Ata-Ul-Karim, S.T.; Shah, F.; Shah, A.N.; Lu, J. Nitrogen fertilizer management for enhancing crop productivity and nitrogen use efficiency in a rice-oilseed rape rotation system in China. *Front. Plant Sci.* **2016**, *7*, 1496. [CrossRef]

7. Ata-Ul-Karim, S.T.; Cao, Q.; Zhu, Y.; Tang, L.; Rehmani, M.I.; Cao, W. Non-destructive assessment of plant nitrogen parameters using leaf chlorophyll measurements in rice. *Front. Plant Sci.* **2016**, *7*, 1829. [CrossRef]

8. Wang, W.; Yao, X.; Yao, X.F.; Tian, Y.C.; Liu, X.J.; Ni, J.; Cao, W.X.; Zhu, Y. Estimating leaf nitrogen concentration with three-band vegetation indices in rice and wheat. *Field Crop. Res.* **2012**, *129*, 90–98. [CrossRef]

9. Markwell, J.; Osterman, J.C.; Mitchell, J.L. Calibration of the Minolta SPAD-502 leaf chlorophyll meter. *Photosynth. Res.* **1995**, *46*, 467–472. [CrossRef]

10. Yuan, Z.; Qiang, C.; Ke, Z.; Ata-Ul-Karim, S.T.; Tian, Y.; Yan, Z.; Cao, W.; Liu, X. Optimal leaf positions for spad meter measurement in rice. *Front. Plant Sci.* **2016**, *7*, 719. [CrossRef]

11. Muñoz-Huerta, R.F.; Guevara-Gonzalez, R.G.; Contreras-Medina, L.M.; Torres-Pacheco, I.; Prado-Olivarez, J.; Ocampo-Velazquez, R.V. A review of methods for sensing the nitrogen status in plants: Advantages, disadvantages and recent advances. *Sensors* **2013**, *13*, 10823. [CrossRef] [PubMed]

12. Peng, S.; García, F.V.; Laza, R.C.; Cassman, K.G. Adjustment for specific leaf weight improves chlorophyll meter's estimate of rice leaf nitrogen concentration. *Agron. J.* **1993**, *85*, 987–990. [CrossRef]

13. Wang, Y.; Wang, D.; Shi, P.; Omasa, K. Estimating rice chlorophyll content and leaf nitrogen concentration with a digital still color camera under natural light. *Plant Methods* **2014**, *10*, 36. [CrossRef] [PubMed]

14. Gitelson, A.A.; Viña, A.; Ciganda, V.; Rundquist, D.C.; Arkebauer, T.J. Remote estimation of canopy chlorophyll content in crops. *Geophys. Res. Lett.* **2005**, *32*, 93–114. [CrossRef]

15. Zhang, K.; Ge, X.; Liu, X.; Zhang, Z.; Liang, Y.; Tian, Y.; Cao, Q.; Cao, W.; Zhu, Y.; Liu, X. Evaluation of the chlorophyll meter and GreenSeeker for the assessment of rice nitrogen status. In Proceedings of the 11th European Conference on Precision Agriculture (ECPA 2017), Edinburgh, UK, 16–20 July 2017; Volume 8, pp. 359–363.

16. Sudduth, K.A.; Kitchen, N.R.; Drummond, S.T. Nadir and oblique canopy reflectance sensing for n application in corn. *Liccosec* **2011**, *7*, 162–172.

17. Yuan, Z.; Ata-Ul-Karim, S.T.; Cao, Q.; Lu, Z.; Cao, W.; Zhu, Y.; Liu, X. Indicators for diagnosing nitrogen status of rice based on chlorophyll meter readings. *Field Crop. Res.* **2016**, *185*, 12–20. [CrossRef]

18. Wang, S.; Zhu, Y.; Jiang, H.; Cao, W. Positional differences in nitrogen and sugar concentrations of upper leaves relate to plant N status in rice under different N rates. *Field Crop. Res.* **2006**, *96*, 224–234. [CrossRef]

19. Ziadi, N.; Bélanger, G.; Claessens, A.; Lefebvre, L.; Tremblay, N.; Cambouris, A.N.; Nolin, M.C.; Parent, L.É. Plant-based diagnostic tools for evaluating wheat nitrogen status. *Crop Sci.* **2010**, *50*, 2580–2590. [CrossRef]

20. Hussain, F.; Bronson, K.F.; Singh, Y.; Singh, B.; Peng, S. Use of chlorophyll meter sufficiency indices for nitrogen management of irrigated rice in Asia. *Agron. J.* **2000**, *92*, 875–879.

21. Zhao, Q.Z.; Ding, Y.F.; Wang, Q.S.; Huang, P.S.; Ling, Q.H. Relationship between leaf color and nitrogen uptake of rice. *Sci. Agric. Sin.* **2006**, *39*, 916–921.

22. Wang, S.H.; Cao, W.X.; Wang, Q.S.; Ding, Y.F.; Huang, P.S.; Ling, Q.H. Positional distribution of leaf color and diagnosis of nitrogen nutrition in rice plant. *Sci. Agric. Sin.* **2002**, *192*, 45–51.

23. Shen, Z.Q.; Wang, K.; Zhu, J.Y. Preliminary study on diagnosis of the nitrogen status of two rice varieties using the chlorophyll meter. *Bull. Sci. Technol.* **2002**, *18*, 174–176.

24. Lin, F.F.; Qiu, L.F.; Deng, J.S.; Shi, Y.Y.; Chen, L.S.; Ke, W. Investigation of SPAD meter-based indices for estimating rice nitrogen status. *Comput. Electron. Agric.* **2010**, *71*, S60–S65. [CrossRef]

25. Ata-Ul-Karim, S.T.; Cao, Q.; Zhu, Y.; Rehmani, M.I.A.; Cao, W.; Tang, L. In-season assessment of rice protein and amylose content using critical nitrogen dilution curve. *Eur. J. Agron.* **2017**, *90*, 139–151. [CrossRef]

26. Greenwood, D.J.; Lemaire, G.; Gosse, G.; Cruz, P.; Draycott, A.; Neeteson, J.J. Decline in percentage n of c3 and c4 crops with increasing plant mass. *Ann. Bot.* **1990**, *66*, 425–436. [CrossRef]

27. Hu, Y.; Yang, J.P.; Lv, Y.M.; He, J.J. SPAD values and nitrogen nutrition index for the evaluation of rice nitrogen status. *Plant Prod. Sci.* **2014**, *17*, 81–92.

28. Noura, Z.; Marianne, B.; Gilles, B.; Annie, C.; Nicolas, T.; Athynan, C.; Michelc, N.; Léonétienne, P. Chlorophyll measurements and nitrogen nutrition index for the evaluation of corn nitrogen status. *Agron. J.* **2010**, *100*, 2275–2279.

29. Zhao, B.; Liu, Z.; Ata-Ul-Karim, S.T.; Xiao, J.; Liu, Z.; Qi, A.; Ning, D.; Nan, J.; Duan, A. Rapid and nondestructive estimation of the nitrogen nutrition index in winter barley using chlorophyll measurements. *Field Crop. Res.* **2016**, *185*, 59–68. [CrossRef]

30. Liu, X.; Zhang, K.; Zhang, Z.; Cao, Q.; Lv, Z.; Yuan, Z.; Tian, Y.; Cao, W.; Zhu, Y. Canopy chlorophyll density based index for estimating nitrogen status and predicting grain yield in rice. *Front. Plant Sci.* **2017**, *8*, 1829. [CrossRef]

31. Liu, X.; Ferguson, R.B.; Zheng, H.; Cao, Q.; Tian, Y.; Cao, W.; Zhu, Y. Using an active-optical sensor to develop an optimal NDVI dynamic model for high-yield rice production (Yangtze, China). *Sensors* **2017**, *17*, 672. [CrossRef]

32. Debaeke, P.; Rouet, P.; Justes, E. Relationship between the normalized spad index and the nitrogen nutrition index: Application to Durum Wheat. *J. Plant Nutr.* **2006**, *29*, 75–92. [CrossRef]

33. Hou, Y.H.; Chen, C.Y.; Guo, Z.Q.; Hou, L.B.; Dong, Z.Q.; Zhao, M. Establishment of dry matter accumulation dymamic simulation model and analysis of growth charateristc for high-yielding population of spring maize. *J. Maize Sci.* **2008**, *16*, 90–95.

34. Islam, M.R.; Haque, K.M.S.; Akter, N.; Karim, M.A. Leaf chlorophyll dynamics in wheat based on SPAD meter reading and its relationship with grain yield. *Sci. Agric.* **2014**, *4*, 13–18.

35. Zheng, H.; Cheng, T.; Yao, X.; Deng, X.; Tian, Y.; Cao, W.; Zhu, Y. Detection of rice phenology through time series analysis of ground-based spectral index data. *Field Crop. Res.* **2016**, *198*, 131–139. [CrossRef]

36. Ata-Ul-Karim, S.T.; Xia, Y.; Liu, X.; Cao, W.; Yan, Z. Development of critical nitrogen dilution curve of Japonica rice in Yangtze River Reaches. *Field Crop. Res.* **2013**, *149*, 149–158. [CrossRef]

37. Ata-Ul-Karim, S.T.; Zhu, Y.; Yao, X.; Cao, W. Determination of critical nitrogen dilution curve based on leaf area index in rice. *Field Crop. Res.* **2014**, *167*, 76–85. [CrossRef]

38. Bijay-Singh; Varinderpal-Singh; Yadvinder-Singh; Thind, H.S.; Kumar, A.; Gupta, R.K.; Kaul, A.; Vashistha, M. Fixed-time adjustable dose site-specific fertilizer nitrogen management in transplanted irrigated rice (*Oryza sativa* L.) in South Asia. *Field Crop. Res.* **2012**, *126*, 63–69. [CrossRef]

39. Liu, K.; Yazhen, L.I.; Huiwen, H.U. Estimating the effect of urease inhibitor on rice yield based on NDVI at key growth stages. *Front. Agric. Sci. Eng.* **2014**, *1*, 150–157. [CrossRef]

40. Beck, P.S.A.; Atzberger, C.; Høgda, K.A.; Johansen, B.; Skidmore, A.K. Improved monitoring of vegetation dynamics at very high latitudes: A new method using MODIS NDVI. *Remote Sens. Environ.* **2006**, *100*, 321–334. [CrossRef]

41. Prost, L.; Jeuffroy, M.H. Replacing the nitrogen nutrition index by the chlorophyll meter to assess wheat N status. *Agron. Sustain. Dev.* **2007**, *27*, 321–330. [CrossRef]

42. Dente, L.; Satalino, G.; Mattia, F.; Rinaldi, M. Assimilation of leaf area index derived from ASAR and MERIS data into CERES-Wheat model to map wheat yield. *Remote Sens. Environ.* **2008**, *112*, 1395–1407. [CrossRef]

43. Wang, S.H.; Zhi-Jun, J.I.; Liu, S.H.; Ding, Y.F.; Cao, W.X. Relationships between balance of nitrogen supply-demand and nitrogen translocation and senescence of different position leaves on rice. *J. Integr. Agric.* **2003**, *2*, 747–751.

44. Escobar-Gutiérrez, A.J.; Combe, L. Senescence in field-grown maize: From flowering to harvest. *Field Crop. Res.* **2012**, *134*, 47–58. [CrossRef]

45. Xue, L.; Yang, L. Recommendations for nitrogen fertilizer topdressing rates in rice using canopy reflectance spectra. *Biosyst. Eng.* **2008**, *100*, 524–534. [CrossRef]

Article

Effect of Large-Scale Cultivated Land Expansion on the Balance of Soil Carbon and Nitrogen in the Tarim Basin

Erqi Xu [1] , Hongqi Zhang [1,*] and Yongmei Xu [2,*]

[1] Key Laboratory of Land Surface Pattern and Simulation, Institute of Geographic Sciences and Natural Resources Research, Chinese Academy of Sciences, Beijing 100101, China; xueq@igsnrr.ac.cn
[2] Institute of Soil and Fertilizer, Xinjiang Academy of Agricultural Sciences, Urumqi 830091, China
* Correspondence: zhanghq@igsnrr.ac.cn (H.Z.); xym1973@163.com (Y.X.)

Received: 23 January 2019; Accepted: 12 February 2019; Published: 14 February 2019

Abstract: Land reclamation influences the soil carbon and nitrogen cycling, but its scale and time effects on the balance of soil carbon and nitrogen are still uncertain. Taking the Tarim Basin as the study area, the impact of land reclamation on the soil organic carbon (SOC), total nitrogen (TN), and carbon to nitrogen (C:N) ratio was explored by the multiple temporal changes of land use and soil samples. Remote sensing detected that cropland nearly doubled in area from 1978 to 2015. Spatial analysis techniques were used to identify land changes, including the prior land uses and cultivation ages. Using land reclamation history information, a specially designed soil sampling was conducted in 2015 and compared to soil properties in ca. 1978. Results found a decoupling characteristic between the C:N ratio and SOC or TN, indicating that changes in SOC and TN do not correspond directly to changes in the C:N ratio. The land reclamation history coupled with the baseline effect has opposite impacts on the temporal rates of change in SOC, TN and C:N ratios. SOC and TN decreased during the initial stage of conversion to cropland and subsequently recovered with increasing cultivation time. By contrast, the C:N ratio for soils derived from grassland increased at the initial stage but the increase declined when cultivated longer, and the C:N ratio decreased for soils derived from forest and fluctuated with the cultivation time. Lower C:N ratios than the global average and its decreasing trend with increasing reclamation age were found in newly reclaimed croplands from grasslands. Sustainable agricultural management practices are suggested to enhance the accumulation of soil carbon and nitrogen, as well as to increase the C:N ratio to match the nitrogen deposition to a larger carbon sequestration.

Keywords: soil stoichiometry; land use change; soil organic carbon; nitrogen; Tarim Basin

---

1. Introduction

Soil carbon (C) and nitrogen (N) are the main sources of plant mineral nutrition and organic nutrients, which are the two primary factors affecting soil fertility [1,2]. They also play an important role in terrestrial soil C and N pools and global C and N cycles [3,4]. The ratio of soil organic carbon (SOC) and total nitrogen (TN), labeled as the C:N ratio, reliably indicates changes in soil microbial activity, decomposition, mineralization rates of SOC, and the cycle of soil C and N [5,6]. Moreover, the soil C:N ratio can be a good predictor of key parameters in the related C and N cycles, such as the dissolved organic carbon flux [7], nitrate leaching [8,9], and nitrous oxide emissions [10]. Therefore, understanding soil C and N content and their variation can help explore the mechanisms of soil C and N cycling and their coupling effects, and aid in enhancing soil C sequestration and reducing N losses in the ecosystem.

A high soil C:N ratio correlates with low organic matter (SOM) decomposition and mineralization rates while in soils with a low C:N ratio, SOM can be decomposed and mineralized relatively quickly [1]. SOC, TN, and the C:N ratio are influenced by many factors, such as climatic conditions (temperature, moisture), soil properties (soil texture, soil pH), terrain characteristics, and anthropogenic factors (land use and management) [2,11–13]. The soil C:N ratio and its variation are determined by gains (input of SOC and TN) and losses [11]. At the global scale, it has been shown that tundra, natural wetlands, and boreal forest can have a relatively wide range of soil C:N ratios, while croplands have a narrow range [2]. Land reclamation can alter the soil C and N bio-geochemical cycling and spatial-temporal characteristics of soil C:N ratios greatly [14–17]. The conversion from native vegetation to cropland decreases the input of vegetative tissues and increases soil temperatures to accelerate the litter decomposition and soil micro-organisms in the decomposition of soil C and N [18]. Reclamation activities can further cause C and N losses and change the soil C:N ratio [3]. Meanwhile, the increasing crop yields of land under sustainable agricultural management can enhance the input crop residues and roots, which can in turn, increase the SOM content and promote the accumulation of SOC and TN [19–21].

Generally, the dynamics of soil C:N ratios are determined by the relative changes in magnitude of SOC and TN contents [22,23]. Many case studies have found that forest converted to cropland results in a decrease in soil C:N ratio because of the high uptake and storage of N by trees [1,3,24,25]. In contrast, agricultural management can promote maintenance of crop residues at the soil surface, which may have beneficial impacts on soil fertility and increase the soil C:N ratio [1]. Plantation age also shows complex effects on the accumulation and decomposition of SOC and TN. For example, soil C:N ratios of cropland showed a significant increase over time at a sub-tropical site, although little change was observed at three temperate sites [11]. The above studies illustrate that the variation in soil C:N ratios during agricultural reclamation and management is complex and is not yet fully understood.

The dynamics of SOC, N, and the C:N ratio, which are affected by land use coupled with other physical factors, present significant spatial and temporal variations. Agricultural reclamation activities mean exploiting other land uses for croplands. Outcomes of reclamation are primarily the result of changes in plant types associated with the changes in the litter input, detritus and litter decomposition, root secretion and soil mineralization [14,26,27]. These changes are related to the prior land use type before the reclamation, which directly or indirectly influence the soil C, N, and their balance after the reclamation. Meanwhile, knowledge of the agricultural duration is important [28], because it is related to the cultivation intensity and the soil mellowing process. Thus, land reclamation history, including the different prior land use types and the duration age of the reclamation, will influence the fate of soil C and N. However, the effects of reclamation history remain uncertain [29], and this leads to difficulties in decision making and agriculture management. This calls for a deeper understanding of the impact that past and future land use changes can have on soil C:N ratios.

Arid and semi-arid regions are especially susceptible to human disturbance and climate change [30], which easily lead to land degradation and soil nutrient element loss [31–33]. Oasis agriculture is a specific ecosystem surrounded by the desert in an arid area, where environmental conditions are milder than the desert, ensuring fertility and allowing desert farming [34]. Aridisols, called Yermosols in the FAO-UNESCO classification (Food and Agriculture Organization-United Nations Educational, Scientific and Cultural Organization) [35], are the major original soil types, undisturbed by humans in arid regions [36]. The expansion of oasis agriculture disturbs the soil environment and may cause the loss of soil C and N. Under long-term sustainable tillage, fertilizing and irrigation of the oasis agriculture [37,38], Aridisols are gradually developed to Anthrosols with changes in C and N storage [35]. Thus, this calls for a deep understanding of how oasis agriculture influences the soil C and N.

The Tarim Basin contains the Taklimakan Desert, the world's second largest shifting sand desert [39]. It has experienced a significant expansion in oasis agriculture from its native vegetation, and even the desert since 1949 [40]. The cropland area in this region has nearly doubled since the 1980s [41,42], which was much higher than the average increase rate in the world [43]. Due to the extreme dry and hot climate with annual limited precipitation of 50–100 mm, almost the entire oasis agriculture relies on irrigation for crop growth in harsh conditions and to resist drought [44]. For the last few decades, land reclamation and the development of high-efficiency irrigation techniques has played an important role in this region [45], making the Tarim Basin one of the main agricultural production regions in China and one of the largest areas of oasis agriculture in the world. This dramatic expansion in cropland has significantly changed the nutrients in the cultivated soils, and has also caused the overuse of water sources and land desertification [46], which in turn, influence the terrestrial C and N cycles [47]. Proper management of such changes requires a deep understanding of the effects of the land reclamation history on the variation in SOC and TN and their balance. Thus, the Tarim Basin was selected as the study area to explore the relationship between the land reclamation history and changes in the soil C, N, and C:N ratio. The objectives of this study were to (1) design a soil sampling campaign from land use change histories to quantify the changes in the SOC and TN dynamics of the cultivated topsoil, and (2) investigate the effects of the land reclamation history on these changes and the coupled relationships between SOC and TN.

2. Methods

2.1. Study Area

The Tarim Basin, located in the south of Xinjiang Province, Northwest China, was selected as our study region. It is located between 38°–43° N latitude and 76°–87° E longitude. Five typical sub-basins, the Aksu River Basin, the Kaidu-Kongque River Basin, the mainstream of Tarim River Basin, the Weigan River Basin, and the Yarkant River Basin, were selected as the case study sites with a total area of 311,765 km^2 (Figure 1). They lie within an extremely dry climate zone, with high temperatures of 20–30 °C in the summer and low temperatures of -10 to -20 °C in winter, a mean annual precipitation of 50–100 mm, and an annual evaporation of 2000–3000 mm [48]. Landscapes in the study area majorly consist of the desert, grassland, cropland, and forest. Native forests with shrubs, mainly including the *Populus euphratica* Oliv. and *Tamarix ramosissima* Ledeb. [49], are located in the riparian zone with an adequate water supply for vegetation growth. These climatic conditions have resulted in a specific oasis agriculture, with widespread use of highly efficient water-saving irrigation techniques.

2.2. Conceptual Framework for Exploring Impacts of Land Reclamation History on Soil C and N

Figure 2 showed the framework for exploring how land reclamation history influences the soil properties. Using the land use maps at different time nodes and spatial overlaying techniques, the land change histories at each location were identified from ca. 1978 to 2015. Also, the specific land reclamation history, including the prior land use before the reclamation and the duration age after the reclamation were extracted. Meanwhile, the simultaneous changes of SOC, TN, and C:N ratio were collected by purposive survey sampling in 2015 and the soil dataset in ca. 1978. Based on the statistical test, the impact of land reclamation history on SOC, TN, and the C:N ratio can be analyzed.

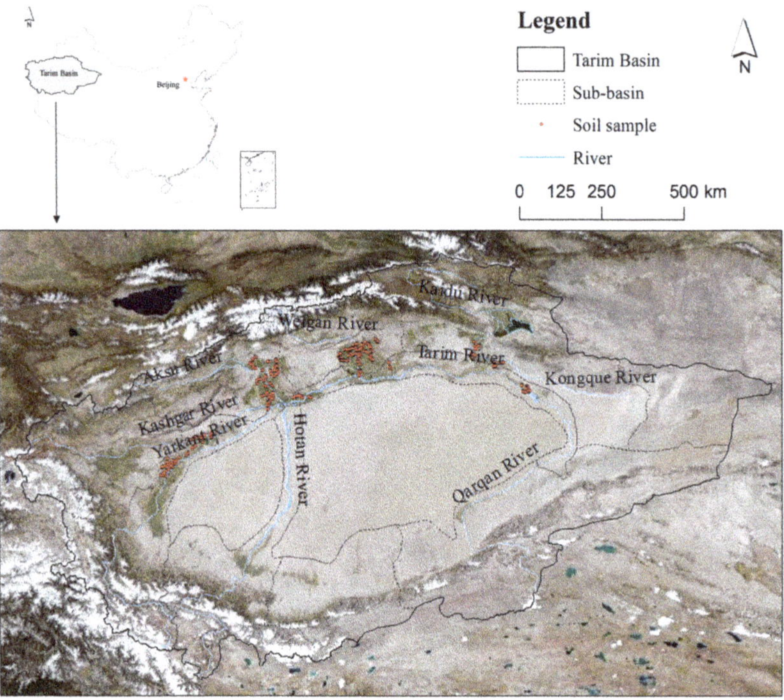

Figure 1. Location and soil sampling sites in the area of the Tarim Basin.

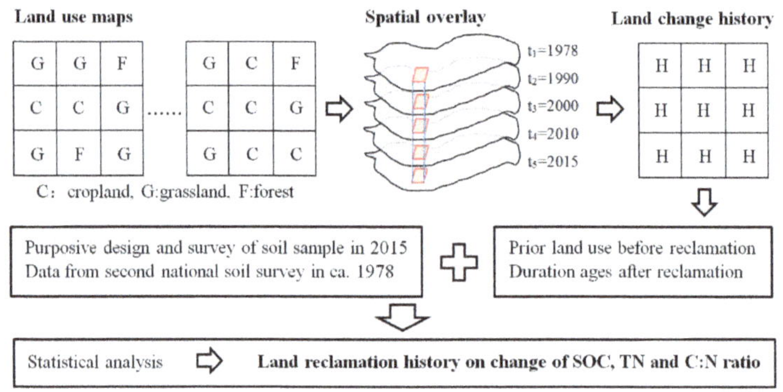

Figure 2. Conceptual framework for exploring impacts of land reclamation history on soil C and N.

2.3. Land Use Change History Detection

The soil sampling strategy was designed on the basis of the land reclamation history, which included land use prior to reclamation indicating the original soil nutrient status, while the cropland age was used to characterize the duration of management practices. To examine the land use change history on the SOC and TN dynamics from ca. 1980 (national soil surveys in 1978 to 1982) to 2015 (soil survey in this study), five time nodes, i.e., 1978, 1990, 2000, 2010, and 2015, of remote sensing images from Landsat MSS/TM and Landsat-8 Operational Land Imager (OLI) images were collected to map the land use via visual interpretation with a local expert tutorial on land use interpretation by ourselves. The paths and rows covering the study area are shown in Table S1. Preprocessing, including the application of geometric and atmospheric corrections, was conducted using ENVI 5.1 (ITT Visual

Information Solutions Inc.). Images within ± 2 y of the time node were used as a proxy when there were no effective cloud-free images from a particular node.

The land use maps were divided into seven classes: cropland, forest, grassland, water body, built-up land, bare land, and glacier. As the majority of the croplands were derived from grassland, with a minority from forest and other land use types [47], this study focused on the effects of grassland and forest conversion to croplands on the SOC and TN dynamics. Using spatial overlay techniques in ArcGIS 10.1 (ESRI Inc., Redlands, CA, USA) for the land use vector maps with seven classes at the five time nodes, the overlaying vector polygon representing the land use change trajectory from ca. 1978 to 2015 was obtained. With the overlaying land use maps, the time nodes where the reclamations occurred were obtained, and the land use prior to the reclamation was identified. Grassland, forest, and cropland were labeled as G, F, and C, respectively, and the land use types for each of the five time nodes were identified. For example, "G→G→C→C→C" indicates the reclamation of grassland to cropland in the period of 1990 to 2000. The label "C→C→C→C→C", indicates that the cropland was assarted before 1978, and land with this history of use was used as a reference for this study. Following labeling for land use, the assarted croplands were grouped according to age as follows: assarted between 2010 and 2015 (new, 1–5 years), 2000 and 2010 (young, 6–15 years), 1990 and 2000 (medium, 16–25 years), 1978 and 1990 (old, 26–37 years). In addition, in some areas land use changes occurred more than once, and cases where cropland transferred to other land use types because of abandonment or build-up and occupation were excluded in this study. Thus, cropland, grassland, and forest were the three prior land use types in this study. Finally, three land use change types, i.e., unchanged cropland, cropland from grassland, and cropland from forest over nearly four decades, with five discrete reclamation ages were identified for the design of the soil sampling campaign.

2.4. Soil Sampling Design, Collection, and Laboratory Methods

Since we were particularly interested in the effect of land reclamation history, soil sampling conducted in 2015 was only carried out for cropland with different reclamation histories. First, the range of cropland in 2015 was extracted and 280 soil plots were identified in the croplands of five sub-basins by a random sampling strategy. A minimum distance of 1000 m between plots of the same category of land use change history was set. Next, three land use change types and five discrete reclamation ages from 1978 to 2015 in the designed soil plots were identified based on the spatial overlay techniques mentioned in Section 2.1. If the area of the vector polygon for overlaying land use maps was smaller than 1 km^2, the corresponding soil plots were excluded to minimize geolocalization errors of land use interpretation and make the resolution of land use maps consistent with soil C and N data. Finally, we collected samples from a total of 270 soil sample plots, including 126, 103, and 41 plots of unchanged cropland, cropland from grassland, and cropland from forest between 1978 and 2015, respectively. At each sampling plot of 10 m $\times$ 10 m, five sites, four at the corners and one in the center of the plot, were selected. According to the Genetic Soil Classification of China [38], the main soil type of the soil samples is irrigation-silted soil (Aric Anthrols in the FAO-UNESCO classification). Irrigation-silted soil is a type of Anthrosol and formed the new anthraquic epipedon in the soil irragric process and mellowing process under cultivation, which means soil materials are silted with irrigation water [38].

The soil C and N distribution within soil profile is highly influenced by management practices, especially tillage systems [50]. As soil disturbance by agricultural tillage is a primary cause of the historical change of soil nutrients, the tillage zone or management zone, mainly including the 0–20 cm soil layer, was sampled using a soil auger of 5 cm diameter. Soil samples were transported to the laboratory and preprocessing was conducted. Soil clods were crushed, and litter and living roots, stones (>2 mm), and visible plant remains were removed. Next, the soil samples were air-dried in the shade and passed through a 2 mm sieve. SOC and TN were measured by the H_2SO_4-$K_2Cr_2O_7$ oxidation method [51] and Kjeldahl procedure [52], respectively. Soil C:N fractions were calculated as the mass ratio of SOC to TN.

To compare simultaneous changes in soil properties and land use, the SOM and TN from ca. 1978 were obtained from the soil characteristics database of China, which were measured by the same chemical analysis methods as those used in 2015. This data has a resolution of 30 × 30 arc-seconds, and was generated from soil data collected in the 1979–1982 national soil survey. As the designed soil plots were located in the overlaying land use vector polygon with an area of larger than 1 km^2 (close to 30 × 30 arc-seconds), the spatial resolution of land reclamation histories and soil properties in the plot were consistent and could be compared and analyzed. The data was provided by the Cold and Arid Regions Science Data Center at Lanzhou [53] and downloaded from their website (http://westdc.westgis.ac.cn/data). The SOC was then calculated from SOM values, using the Bemmelen index of 0.58, which was widely used in previous studies [54,55]. Since the soil data set consists of eight vertical layers with different depths, the SOC and TN at the study depth of 0–20 cm were estimated by a weighted depth method provided by Yan, et al. [19].

2.5. Statistical Analysis

Statistical analysis was carried out using the SPSS18.0 statistical package (SPSS Inc., New York, NY, USA). Differences in variation of SOC, TN, and C:N ratio for the 270 soil plots between ca. 1978 and 2015 were calculated using paired-samples *t*-tests, using 270 pairs in the same spatial coordinates. The significance of differences in the levels and variations of SOC, TN, and C:N ratio in plots with different land reclamation histories were tested with one-way analysis of variance (ANOVA) and least-significant-difference (LSD) methods, followed by Tamhane post hoc tests ($p < 0.05$).

3. Results

3.1. Land Reclamation from 1978 to 2015 in Typical Areas of the Tarim Basin

The classification accuracy of land use maps was assessed by a total of 2000 ground-reference data-points in the five sub-basins, including the field validation investigations with GPS in 2015, and points from high resolution images from Google Earth in 2010 and 2015. The ground-reference data-points were randomly collated by the stratified land use types. This assessment found a high classification accuracy, with kappa coefficients of 0.913 and 0.906 and an overall accuracy of 92.30% and 91.35% for the 2010 and 2015 data, respectively. Since the visual interpretation methods used for other year were similar, it was assumed that the classification accuracies of land use maps for 1990 and 2000 were similar to those for the 2010 and 2015 maps. Similar to previous studies using the multiple Landsat MSS/TM images for long time series of land use change detection [56,57], the land use map in 1978 interpreted by the same executor with the same satellite and similar images could be relied upon. Figure 3 shows the land use maps for 1978 and 2015.

Croplands accounted for 10.37% of the total area of the five sub-basins of the Tarim Basin in 2015. Areas of croplands were 1.72×10^4, 1.89×10^4, 2.11×10^4, 2.72×10^4, and 3.23×10^4 km^2 for the five time nodes of 1978, 1990, 2000, 2010, and 2015, respectively. The cropland area nearly doubled with a significantly increased rate of 88.37% from 1978 to 2015. Most of them mainly expanded in the vicinity of their existing areas. The assarted croplands were converted from grassland, forest, bare land, and water body, with the proportions of 70.35%, 16.89%, 9.56%, and 3.20%, respectively.

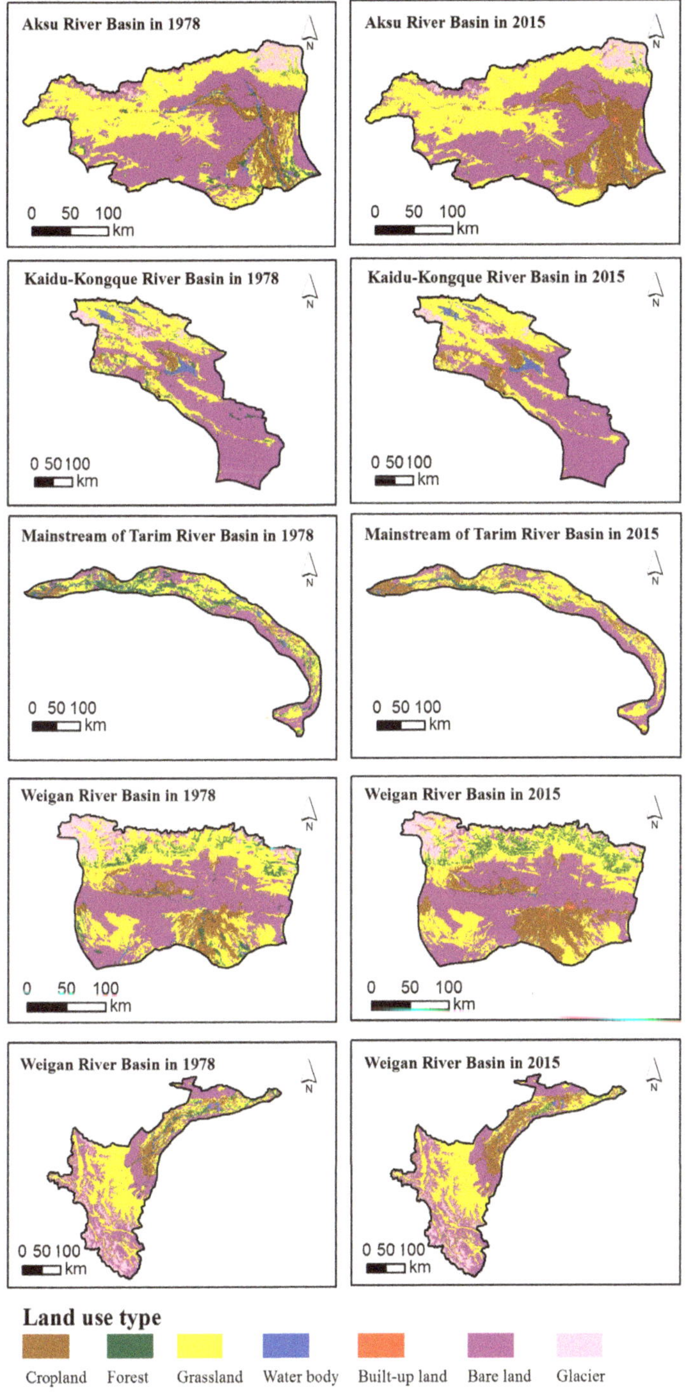

Figure 3. Land use maps of 1978 and 2015 in typical areas of the Tarim Basin.

3.2. Status and Change in Soil C, N and C:N

Table 1 shows the statistical characteristics of SOC, TN, and C:N ratios in the Tarim Basin. The mean values of these parameters increased from ca.1978 to 2015 with varying magnitudes, but do not show a significant change ($p < 0.05$). The mean SOC content is 6.61 g/kg, with an increase of 0.17 g/kg from ca.1978 to 2015. The TN contents, meanwhile, increased from 0.63 g/kg to 0.65 g/kg. It was found that the TN content is highly correlated with the SOC content in 270 soil sample plots, showing high Pearson's correlation coefficients of 0.938 and 0.925 in ca. 1978 and 2015, respectively. Although the range of contents of SOC and TN increased from ca. 1978 to 2015 (Table 1), their standard deviations decreased. With cultivation practices, the standard deviation of SOC and TN in 270 plots show increases from 1.96 g/kg and 0.22 g/kg, respectively, in ca. 1978, to 2.90 g/kg and 0.28 g/kg, respectively, in 2015.

The change in statistical characteristics for the C:N ratio is similar to those of SOC and TN content. The mean C:N ratio is 10.22 in 2015, ranging from 6.11 to 17.25. Most of these values fall in the range from 9.0 (10% quantile) to 11.5 (90% quantile). A larger increase in SOC over TN resulted in a slight increase of 0.11 in the C:N ratio from circa (ca.) 1978 to 2015, with no significant change at the 95% confidence level. Long-term cultivation has tended to homogenize the soil C:N ratio, resulting in a decrease in standard deviation, from 1.73 in ca. 1978 to 1.21 in 2015, separating the soil plots as two groups, i.e., the increased values from ca. 1978 to 2015 and decreased values at this period. Using the paired-samples *t*-test, two-third of the total observations showed an increasing trend in SOC and TN at a significant level of 99%, and other one-third showed a significant decrease. In contrast, about 59.26% of the total observations showed an increasing trend in C:N ratio at a non-significant level of 95%, with the remainder showing a decrease at a significant level of 95%. Thus, a deep exploration of change in soil C and N based on the different land reclamation histories is called for.

Table 1. Statistical characteristics of soil properties.

		Mean	Standard Deviation	Maximum	Minimum
Soil bulk density (g cm^{-3})	ca.1978	1.30	0.18	1.75	0.93
	2015	1.29	0.16	1.61	1.05
SOC content (g/kg)	ca.1978	6.44	2.90	16.85	3.42
	2015	6.61	1.96	17.65	2.26
TN content (g/kg)	ca.1978	0.63	0.28	1.67	0.29
	2015	0.65	0.22	1.76	0.19
C:N ratio	ca.1978	10.11	1.73	14.90	7.33
	2015	10.22	1.21	17.25	6.11

3.3. Effects of Land Reclamation History on SOC and TN

The selected sites presented significant differences ($p < 0.05$) in SOC and TN contents in 2015, depending on their prior land uses (Figure 4a,b). The SOC and TN contents of cropland soils for lands assarted before 1978 were 7.29 g/kg and 0.74 g/kg, respectively. These values were significantly higher than those for soils from land with prior use as grassland and forest assarted in the study period. The SOC content in previously forested land (6.21 g/kg) is slightly higher than that for land that was previously grassland (5.93 g/kg), and the TN content of soil from forest (0.59 g/kg) is close to that from grassland (0.58 g/kg). Comparing the SOC and TN contents in ca. 1978 and 2015, differences in change values according to prior land use were also significant ($p < 0.05$) (Figure 4c,d). In 2015, SOC and TN content in those sites previously used as cropland had significantly increased, as indicated by the paired-samples *t*-test ($p < 0.05$), the largest increases being 1.59 g/kg and 0.15 g/kg, respectively. Sites where prior land use was grassland showed an increase in SOC content of 0.17 g/kg and a slight decrease in TN content. In contrast, the SOC and TN content in previously forested land showed a significant decrease of 4.18 g/kg and 0.30 g/kg, respectively.

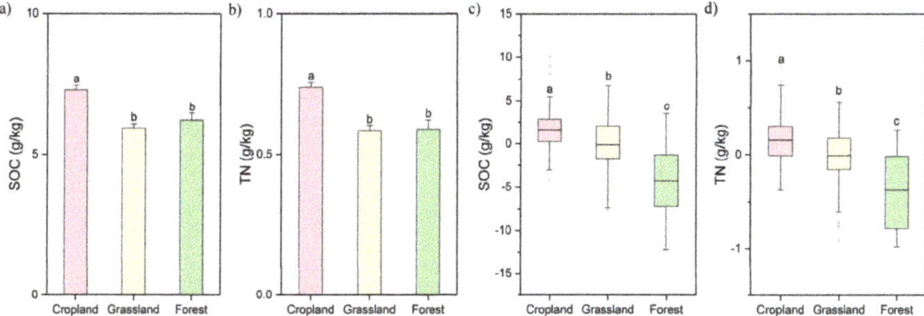

Figure 4. Soil C and N for different prior land uses: (**a**) Soil organic carbon content in 2015, (**b**) Total nitrogen content in 2015, (**c**) Change of soil organic carbon content, (**d**) Change of total nitrogen content. Note: 1. Group differences after one-way ANOVA ($p < 0.05$) was indicated by different lowercase letters. 2. Bar length gives the mean value, vertical whisker of bar for each column indicate standard errors of the mean. 3. Line within the boxes gives the median value, box means the 25th and 75th percentile, whisker of box represents 1.5 times the length of the box from either end of the box (1.5 times the interquartile range), circle represents outliers and extremes. The same as below.

With the increasing reclamation age, the accumulation of SOC and TN in soil from land derived from grassland and forest increased (Figure 5a–d). The mean SOC and TN content in the newly assarted soils were the lowest, and were significantly lower than those in the old assarted soils ($p < 0.05$). In addition, SOC and TN content for old assarted soils were lower than the reference cropland assarted before 1978, but were not significantly different. The SOC content in land that was previously grassland and forest increased significantly from 5.17 and 5.23 g/kg in the newly assarted soils to 6.60 and 6.95 g/kg in the old assarted soils, respectively (Figure 5a,c). Similarly, the TN content in land that was previously grassland and forest increased significantly from 0.51 and 0.47 g/kg in the newly assarted soils to 0.66 and 0.66 g/kg in the old assarted soils, respectively (Figure 5b,d).

An increase in the reclamation age resulted in an accumulation in SOC and TN at a significance level of 95% (Figure 5e–h). Compared to the initial SOC and TN levels prior to reclamation, changes in content for new, young, and medium assarted soils were negative. The decreases in the newly assarted soils were the largest, and increasing reclamation age resulted in increases in SOC and TN stocks. A slight decrease in SOC and TN content in the sites which were previously grassland was observed, with a larger decrease for previously forested sites. For example, the changes in SOC content for new, young, and medium croplands derived from grassland are −0.48, −0.32, and −0.07 g/kg, respectively; and those for croplands derived from forest are −6.03, −3.90, and −4.19 g/kg, respectively. With a longer cultivation age, the change in SOC and TN in old croplands derived from grassland was positive, with mean values of 0.89 and 0.09 g/kg, respectively, which were significantly higher than those of newly assarted croplands with mean changes of −0.48 and −0.06 ($p < 0.05$) (Figure 5e,f). In contrast, changes in SOC and TN in soils from old croplands derived from forested lands were still negative, with mean values of −2.91 and −0.25 g/kg, respectively (Figure 5g,h).

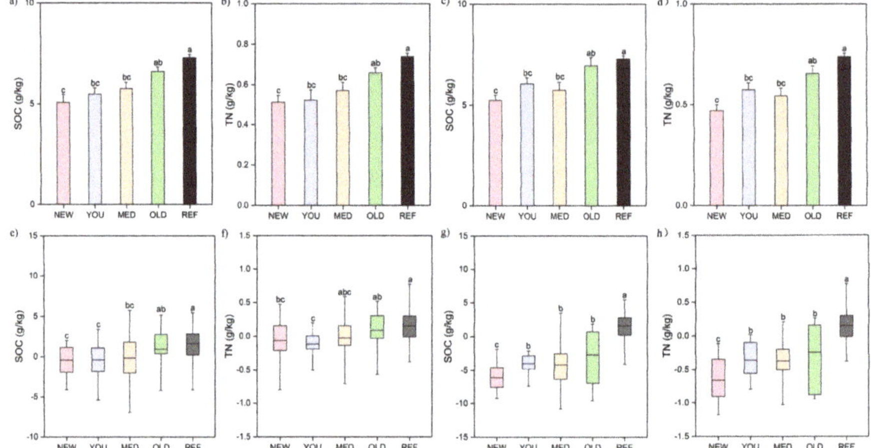

Figure 5. Impact of reclamation ages on soil C and N: (**a**) Soil organic carbon content in 2015 assarted from grassland, (**b**) Total nitrogen content in 2015 assarted from grassland, (**c**) Soil organic carbon content in 2015 assarted from forest, (**d**) Total nitrogen content in 2015 assarted from forest, (**e**) Change of soil organic carbon content assarted from grassland, (**f**) Change of total nitrogen content assarted from grassland, (**g**) Change of soil organic carbon content assarted from forest, (**h**) Change of total nitrogen content assarted from forest. NEW = newly assarted (1–5 years), YOU = young assarted (6–15 years), MED = medium assarted (16–25 years), OLD = old assarted (26–37 years), REF = referenced cropland assarted before 1978.

3.4. Effects of Land Reclamation History on Soil C:N Ratio

Compared to the difference in SOC and TN levels for different prior land uses, the C:N ratios showed opposite characteristics (Figure 6a). The C:N ratio in previously forested land (10.85) was significantly higher than where the prior land use was cropland or grassland (9.96 and 10.29, respectively). Comparing the changes with the three prior land uses in the past four decades, the C:N ratios showed non-significant variations ($p < 0.05$) (Figure 6b), of 0.05, 0.32, and −0.25 for sites with prior land use as cropland, grassland, and forest, respectively. The C:N ratio for sites with prior land use as cropland increased slightly, from 9.91 to 9.96 in two periods of ca. 1978 and 2015, with a non-significant change as indicated by the paired-samples t-test ($p < 0.05$). In contrast, the C:N ratio for soils from grassland sites increased from 9.97 to 10.29, and the ratio for forest sites decreased from 11.10 to 10.85.

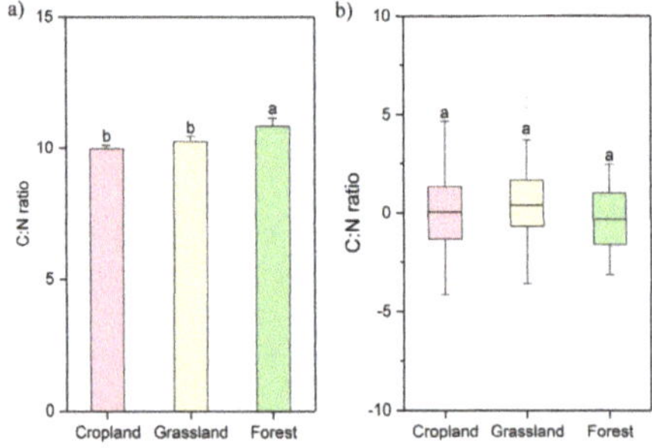

Figure 6. Soil C:N ratio for different prior land uses: (**a**) Values in 2015, (**b**) Changes from ca.1978 to 2015.

Although the SOC and TN changed significantly with increasing reclamation age, the change in soil C:N ratios shows different trends (Figure 7a,b). With an increase in the reclamation age, C:N ratios with prior land use as grassland significantly decreased from 10.79 in the newly assarted soils to 9.92 in the old assarted soils at a 95% confidence level (Figure 7a). The C:N ratio of old assarted soils was close to the C:N ratio of 9.96 in the reference cropland soils assarted prior to 1978. In contrast, relatively little variation, with a non-significant difference in C:N ratios, was found for soils from previously forested land (Figure 7b). C:N ratios from the previously forested land are 10.98, 11.01, 10.83, and 10.87 for the new, young, medium, and old assarted soils, respectively.

Changes in C:N ratios in two periods of ca. 1978 and 2015 were found to have relatively less variation (Figure 7c,d). The changes in C:N ratios for grassland tended to be positive, but that increased magnitude tended to diminish with increasing reclamation age (Figure 7c). The mean increase in C:N ratio was 0.89 for newly assarted soils, which is significantly higher than those in medium and old assarted soils, with mean values of 0.28 and 0.06 at a 95% confidence level, respectively. Moreover, for previously forested land, the change in the magnitude of the C:N ratios did not tend to vary significantly with an increase in the reclamation age, being −0.43, −0.21, −0.23, −0.15 for new, young, medium, and old assarted soils, respectively (Figure 7d).

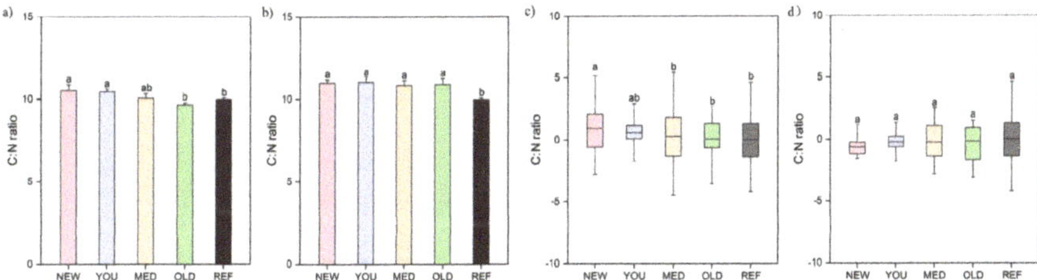

Figure 7. Impact of reclamation ages on soil C:N ratio: (**a**) Values in 2015 assarted from grassland, (**b**) values in 2015 assarted from forest, (**c**) Changes assarted from grassland, (**d**) Changes assarted from forest. NEW = newly assarted (1–5 years), YOU = young assarted (6–15 years), MED = medium assarted (16–25 years), OLD = old assarted (26–37 years), REF = referenced cropland assarted before 1978. NEW = newly assarted (1–5 years), YOU = young assarted (6–15 years), MED = medium assarted (16–25 years), OLD = old assarted (26–37 years), REF = referenced cropland assarted before 1978.

4. Discussion

Land use change history is closely related to soil C and N dynamics [1,29]. Numerous studies have explored the relationship between soil C and N, however, the balance between them and their coupling relationship remains uncertain [55]. This study investigated how large-scale cultivation has affected the SOC, TN, and C:N ratio. Using remote sensing and GIS techniques, this research identified the land use change history, which indicated that cropland has nearly doubled in typical regions of the Tarim Basin over the last four decades. According to different land reclamation histories, large-scale soil sampling was conducted to quantify the spatial variability of soil C and N. The multiple temporal changes of land use and soil proprieties using a repeated soil sampling strategy allowed the investigation of the time effects of reclamation on the soil C:N ratio [22,58]. Results found that the changes in soil C:N ratios are significantly different from the changes in SOC and TN under different land reclamation histories. This research can serve as a better reference for the soil C and N balance in ecological interactions and processes [59].

Land reclamation from grassland and forest in the study area resulted in SOC and TN loss during the initial reclamation stages, but these recovered with increasing reclamation age (Figure 5), which is in accordance with previous studies [18,60,61]. Anthropogenic disturbances frequently destroy the initial soil structure and accelerate the mineralization and decomposition process of SOM, which can intensify losses of C and N [62–64]. In the study area, the scarce water resources and low rate of bioaccumulation limited the growth of uncultivated grassland, primarily desert steppe [65], which resulted in a low level of initial SOC and TN [47,66]. After reclamation, the conditions of soil moisture and biomass growth improved in the arid regions; processes which can increase the above- and below-ground biomass, and result in soil C and N accumulation with increasing reclamation age [67]. The high-efficiency agricultural production in the Tarin Basin increases the crop yield with enhanced soil C and N input through crop residues and also enhances the anthropogenic mellowing process of soils to improve the soil structure and function, which further promotes the accumulation of soil C and N [68]. Improved agricultural practices, including applying organic manure, and conservative tillage measures are used in oasis agriculture, which enhances the soil C and N storage by providing water, organic amendments, and increased crop residue is returned to the soil [19,69,70].

C and N cycling are tightly coupled in ecosystems [71] and TN change in soils is generally assumed to follow variation in SOC, since both elements are bound into organic compounds [72,73]. The high correlation between these parameters was confirmed in this study. Land use change and management practices can alter C and N bio-geochemical cycling, and the change in soil C:N ratio can be used to determine the relative input and output of SOC and TN [23]. This study found that land reclamation history change had an opposite effect on C:N ratio, compared to the trends observed for SOC and TN. The magnitude of changes observed for C:N ratios were also smaller than those observed for SOC and TN (Figures 4–7). This implied that the change in C:N ratio is complex and does not show consistency with either C or N variations.

Different land use regulates C and N source availability, soil microbial activities, and therefore litter decomposition rates [74–76]. In general, forested soils had a higher C:N ratio, which is to be expected since trees take up more N and store it within their biomass rather than in the soil, and also, microbial decomposers of forested soils have higher C:N ratios than cultivated soils [3,25]. This resulted in a significant decrease in C:N ratios when forests were converted to croplands (Figure 6). In contrast, the conversion from grassland to cropland resulted in an increase in C:N ratios but a decrease in SOC and TN (Figures 4 and 6) in the initial reclamation process, indicating a greater loss of N compared to that of SOC. This may be partly explained by baseline effects [58]. In the Tarim Basin, the uncultivated grassland, primarily desert steppe, was limited by water resources and a lower rate of bioaccumulation [65], resulting in a lower level of initial SOC than TN [47,66]. Cultivation with adequate water supplies and fertilizer application would significantly improve soil physicochemical properties [77]. Land reclamation enhanced the C input of the litter returned to the soil and sped up nutrient cycling, resulting in an increase in the soil C:N ratio [23,78]. This, in turn, can result in relatively lower SOC losses, and more SOC accumulation compared to TN to increase the C:N ratio at the initial reclamation stage.

Moreover, as the age of cultivated soils increased, the different accumulation rates of C and N in the soils resulted in a change of C:N ratio [79]. With increasing reclamation age, it was found that the increases in C:N ratios for sites previously used as grassland tended to be less (Figure 7c). The increased N accumulation in the soil may be partly due to the overuse of N fertilizer, an effective way to increase crop biomass and yield, enhancing the N deposition in soils [80]. An experimental site receiving synthetic N fertilization over a 40–50 year period indicated a net decline in soil C [81]. Thus, the C:N ratio decreased according to the smaller change rate of SOC accumulation than N deposition. In addition, C:N ratios from previously forested land stayed relatively stable, indicating similar accumulation rates and a relative balance between SOC and TN. The variations in soil C:N ratio are not only caused by C and N inputs, but also by decomposition in soils [11]. Soil micro-organisms in the previously forested soils may enhance nitrogen-induced increases in the C uptake in the soil [82]. The underlying mechanisms of these variations and the longer time monitoring of C:N ratios require further study.

The global mean value of soil C:N ratios (0–30 cm) in cropland is 12.5 [2], which is much higher than the mean level for our study area (10.22), and the highest value for sites derived from forest (10.85). In addition, the C:N ratios for sites derived from grassland with low cropland ages tended to decrease with increasing reclamation age. A reduction in the soil C:N ratio could disrupt the balance between soil C and N cycling and cause loss of N through leaching and denitrification processes [1,22,83]. These results highlight the importance of effective agricultural management for the soils in the study area. Although the change in C:N ratios for cropland assarted before 1978 has remained relatively stable over the last four decades, it was found that over 10% of these C:N ratios were larger than 12.50, indicating that it is possible to increase the C:N ratio with the use of sustainable agricultural management. For example, no-tillage can increase soil C:N ratios as compared to the use of traditional pillow tillage [1,84,85]. In addition, using N fertilizer more efficiently and decreasing the overall amounts used would effectively decrease the input of N to soils [80,86]. Management efforts in the region should focus on how to enhance the accumulation of soil C and N, and also to better understand their coupling characteristics in order that N deposition might be matched with C sequestration [87].

5. Conclusions

To reveal the effect of land use changes on the balance of soil C and N, a purpose designed sampling strategy, along with the gathering of information on different land use change histories, was conducted in representative areas within the Tarim Basin. Based on the visual interpretation of data from remote sensing, cropland area nearly doubled in four decades in the study area, where over 70% of new cropland was assarted mainly from grasslands, a small fraction from forest, and few from other land use types. Using GIS spatial analysis techniques, three prior land uses and five types of cropland age were identified. The status and change in soil C, N, and their fractions were investigated under the different land reclamation histories. Results found that the prior land use significantly influenced the soil C and N. Meanwhile, the relative magnitudes of C:N ratios for the various land uses were,: forest > grassland > cropland. For soils assarted before 1978, SOC and TN significantly increased, but the C:N ratio increased slightly by 0.05 without a significant difference. An increase was found in the SOC, TN, and C:N ratio where prior land use was grassland, but these decreased in soils derived from forest between 1978 and 2015. The change in magnitudes of SOC and TN with a significant difference between three prior land uses is larger than that C:N ratio with a non-significant difference. SOC and TN decreased in the initial stage of the reclamation process and did not recover in the short term. After 30 years of cultivation, the decrease in initial SOC and TN values for land recovered from grassland had recovered. SOC and TN loss from land derived from forest, however, was not completely recovered even after 40 years of agricultural management. With increasing reclamation age, the increase in C:N ratios of soils from land derived from grassland was less, but remained relatively stable for land derived from forest. The findings in this study indicated near opposite effects of land reclamation history on the soil C and N and C:N ratios. Differences in the variation of the soil C:N ratio are determined by the relative input and output of SOC and TN under the different land reclamation processes and associated cultivation practices. Since the C:N ratios are much lower than the global average and show a decreasing trend in soils derived from grasslands, the application of sustainable agricultural management is suggested to increase not only SOC and TN, but also the C:N ratio, matching N deposition with the currently larger amount of carbon sequestration.

Supplementary Materials: The following are available online at http://www.mdpi.com/2073-4395/9/2/86/s1, Table S1: Landsat image description.

Author Contributions: E.X., H.Z. and Y.X. conceived and designed the experiments. E.X. and Y.X. analyzed the data. E.X. and H.Z. wrote and revised the paper.

Funding: This work was financially supported by the National Natural Science Foundations of China (Grant No. 41671097, 41601095 and 41561070).

Acknowledgments: We are thankful to the National Natural Science Foundations of China for their financial support of the present study and the comments and suggestions of the editor and reviewers.

Conflicts of Interest: The authors declare no conflict of interest.

References

1. Liu, M.; Ussiri, D.A.; Lal, R. Soil organic carbon and nitrogen fractions under different land uses and tillage practices. *Commun. Soil Sci. Plant Anal.* **2016**, *47*, 1528–1541. [CrossRef]
2. Xu, X.; Thornton, P.E.; Post, W.M. A global analysis of soil microbial biomass carbon, nitrogen and phosphorus in terrestrial ecosystems. *Glob. Ecol. Biogeogr.* **2013**, *22*, 737–749. [CrossRef]
3. Murty, D.; Tarinm, M.U.; Mcmurtrie, R.E.; Mcgilvray, H. Does conversion of forest to agricultural land change soil carbon and nitrogen? A review of the literature. *Glob. Chang. Biol.* **2002**, *8*, 105–123. [CrossRef]
4. Stiglitz, R.; Mikhailova, E.; Sharp, J.; Post, C.; Schlautman, M.; Gerard, P.; Cope, M. Predicting Soil Organic Carbon and Total Nitrogen at the Farm Scale Using Quantitative Color Sensor Measurements. *Agronomy* **2018**, *8*, 212. [CrossRef]
5. Huang, B.; Sun, W.; Zhao, Y.; Zhu, J.; Yang, R.; Zou, Z.; Ding, F.; Su, J. Temporal and spatial variability of soil organic matter and total nitrogen in an agricultural ecosystem as affected by farming practices. *Geoderma* **2007**, *139*, 336–345. [CrossRef]
6. Wiesmeier, M.; Hübner, R.; Barthold, F.; Spörlein, P.; Geuß, U.; Hangen, E.; Reischl, A.; Schilling, B.; von Lützow, M.; Kögel-Knabner, I. Amount, distribution and driving factors of soil organic carbon and nitrogen in cropland and grassland soils of southeast Germany (Bavaria). *Agric., Ecosyst. Environ.* **2013**, *176*, 39–52. [CrossRef]
7. Kindler, R.; Siemens, J.; Kaiser, K.; Walmsley, D.C.; Bernhofer, C.; Buchmann, N.; Cellier, P.; Eugster, W.; Gleixner, G.; Grünwald, T. Dissolved carbon leaching from soil is a crucial component of the net ecosystem carbon balance. *Glob. Change Biol.* **2011**, *17*, 1167–1185. [CrossRef]
8. Lovett, G.M.; Weathers, K.C.; Arthur, M.A. Control of Nitrogen Loss from Forested Watersheds by Soil Carbon: Nitrogen Ratio and Tree Species Composition. *Ecosystems* **2002**, *5*, 712–718. [CrossRef]
9. Cools, N.; Vesterdal, L.; De Vos, B.; Vanguelova, E.; Hansen, K. Tree species is the major factor explaining C:N ratios in European forest soils. *For. Ecol. Manag.* **2014**, *311*, 3–16. [CrossRef]
10. Klemedtsson, L.; Von Arnold, K.; Weslien, P.; Gundersen, P. Soil CN ratio as a scalar parameter to predict nitrous oxide emissions. *Glob. Chang. Biol.* **2005**, *11*, 1142–1147. [CrossRef]
11. Cong, R.; Wang, X.; Xu, M.; Zhang, W.; Xie, L.; Yang, X.; Huang, S.; Wang, B. Dynamics of soil carbon to nitrogen ratio changes under long-term fertilizer addition in wheat-corn double cropping systems of China. *Eur. J. Soil Sci.* **2012**, *63*, 341–350. [CrossRef]
12. Obu, J.; Lantuit, H.; Myers-Smith, I.; Heim, B.; Wolter, J.; Fritz, M. Effect of terrain characteristics on soil organic carbon and total nitrogen stocks in soils of Herschel Island, Western Canadian Arctic. *Permaf. Periglac. Processes* **2017**, *28*, 92–107. [CrossRef]
13. Lozano-García, B.; Muñoz-Rojas, M.; Parras-Alcántara, L. Climate and land use changes effects on soil organic carbon stocks in a Mediterranean semi-natural area. *Sci. Total Environ.* **2017**, *579*, 1249–1259. [CrossRef] [PubMed]
14. Yimer, F.; Ledin, S.; Abdelkadir, A. Changes in soil organic carbon and total nitrogen contents in three adjacent land use types in the Bale Mountains, south-eastern highlands of Ethiopia. *For. Ecol. Manag.* **2007**, *242*, 337–342. [CrossRef]
15. Soosaar, K.; Mander, Ü.; Maddison, M.; Kanal, A.; Kull, A.; Lõhmus, K.; Truu, J.; Augustin, J. Dynamics of gaseous nitrogen and carbon fluxes in riparian alder forests. *Ecol. Eng.* **2011**, *37*, 40–53. [CrossRef]
16. Edmondson, J.L.; Davies, Z.G.; McCormack, S.A.; Gaston, K.J.; Leake, J.R. Land-cover effects on soil organic carbon stocks in a European city. *Sci. Total Environ.* **2014**, *472*, 444–453. [CrossRef] [PubMed]
17. Mishra, U.; Ussiri, D.A.; Lal, R. Tillage effects on soil organic carbon storage and dynamics in Corn Belt of Ohio USA. *Soil Till. Res.* **2010**, *107*, 88–96. [CrossRef]
18. Kopittke, P.M.; Dalal, R.C.; Finn, D.; Menzies, N.W. Global changes in soil stocks of carbon, nitrogen, phosphorus, and sulphur as influenced by long-term agricultural production. *Glob. Chang. Boil.* **2017**, *23*, 2509–2519. [CrossRef]

19. Yan, X.; Cai, Z.; Wang, S.; Smith, P. Direct measurement of soil organic carbon content change in the croplands of China. *Glob. Chang. Biol.* **2011**, *17*, 1487–1496. [CrossRef]

20. Maaroufi, N.I.; Nordin, A.; Hasselquist, N.J.; Bach, L.H.; Palmqvist, K.; Gundale, M.J. Anthropogenic nitrogen deposition enhances carbon sequestration in boreal soils. *Glob. Chang. Boil.* **2015**, *21*, 3169–3180. [CrossRef]

21. Alvarez, R. A review of nitrogen fertilizer and conservation tillage effects on soil organic carbon storage. *Soil Use Manag.* **2005**, *21*, 38–52. [CrossRef]

22. Gao, Y.; He, N.; Yu, G.; Chen, W.; Wang, Q. Long-term effects of different land use types on C, N, and P stoichiometry and storage in subtropical ecosystems: A case study in China. *Eco. Eng.* **2014**, *67*, 171–181. [CrossRef]

23. Brady, N.; Weil, R. *The Nature and Properties of Soils*; Pearson Prentice Hall: Upper Saddle River, NJ, USA, 2008.

24. Liu, X.; Ma, J.; Ma, Z.-W.; Li, L.-H. Soil nutrient contents and stoichiometry as affected by land-use in an agro-pastoral region of Northwest China. *Catena* **2017**, *150*, 146–153. [CrossRef]

25. Livesley, S.J.; Ossola, A.; Threlfall, C.G.; Hahs, A.K.; Williams, N.S.G. Soil Carbon and Carbon/Nitrogen Ratio Change under Tree Canopy, Tall Grass, and Turf Grass Areas of Urban Green Space. *J. Environ. Qual.* **2016**, *45*, 215–223. [CrossRef] [PubMed]

26. Hobbie, S.E.; Reich, P.B.; Oleksyn, J.; Ogdahl, M.; Zytkowiak, R.; Hale, C.; Karolewski, P. Tree species effects on decomposition and forest floor dynamics in a common garden. *Ecology* **2006**, *87*, 2288–2297. [CrossRef]

27. Li, Z.; Liu, C.; Dong, Y.; Chang, X.; Nie, X.; Liu, L.; Xiao, H.; Lu, Y.; Zeng, G. Response of soil organic carbon and nitrogen stocks to soil erosion and land use types in the Loess hilly–gully region of China. *Soil Till. Res.* **2017**, *166*, 1–9. [CrossRef]

28. Post, W.M.; Kwon, K.C. Soil carbon sequestration and land-use change: Processes and potential. *Glob. Chang. Boil.* **2000**, *6*, 317–327. [CrossRef]

29. Sartori, F.; Lal, R.; Ebinger, M.H.; Parrish, D.J. Potential soil carbon sequestration and CO2 offset by dedicated energy crops in the USA. *Crit. Rev. Plant Sci.* **2006**, *25*, 441–472. [CrossRef]

30. Hulme, M. Recent climatic change in the world's drylands. *Geophys. Res. Lett.* **1996**, *23*, 61–64. [CrossRef]

31. Reynolds, J.F.; Smith, D.M.S.; Lambin, E.F.; Turner, B.; Mortimore, M.; Batterbury, S.P.; Downing, T.E.; Dowlatabadi, H.; Fernández, R.J.; Herrick, J.E. Global desertification: Building a science for dryland development. *Science* **2007**, *316*, 847–851. [CrossRef]

32. Qiu, L.; Wei, X.; Zhang, X.; Cheng, J.; Gale, W.; Guo, C.; Long, T. Soil organic carbon losses due to land use change in a semiarid grassland. *Plant Soil* **2012**, *355*, 299–309. [CrossRef]

33. Lal, R. Carbon sequestration in dryland ecosystems. *Environ. Manag.* **2004**, *33*, 528–544. [CrossRef] [PubMed]

34. Mekki, I.; Jacob, F.; Marlet, S.; Ghazouani, W. Management of groundwater resources in relation to oasis sustainability: The case of the Nefzawa region in Tunisia. *J. Environ. Manag.* **2013**, *121*, 142–151. [CrossRef] [PubMed]

35. FAO-UNESCO. *Soil Map of the World: Revised Legend (with Corrections and Updates), World Soil Resources Report 60*; FAO: Rome, Italy, 1988.

36. Eswaran, H.; Van Den Berg, E.; Reich, P. Organic carbon in soils of the world. *Soil Sci. Soc. Am. J.* **1993**, *57*, 192–194. [CrossRef]

37. Su, Y.-Z.; Wang, F.; Zhang, Z.-H.; Du, M.-W. Soil Properties and Characteristics of Soil Aggregate in Marginal Farmlands of Oasis in the Middle of Hexi Corridor Region, Northwest China. *Agric. Sci. China* **2007**, *6*, 706–714. [CrossRef]

38. Gong, Z.; Zhang, G.; Luo, G. Diversity of Anthrosols in China. *Pedosphere* **1999**, *9*, 193–204.

39. Liu, F.; Zhang, H.; Qin, Y.; Dong, J.; Xu, E.; Yang, Y.; Zhang, G.; Xiao, X. Semi-natural areas of Tarim Basin in northwest China: Linkage to desertification. *Sci. Total Environ.* **2016**, *573*, 178–188. [CrossRef] [PubMed]

40. Chen, Y. *Water Resources Research in Northwest China*; Springer Science & Business Media: Dordrecht, The Netherlands, 2014.

41. Liu, J.; Kuang, W.; Zhang, Z.; Xu, X.; Qin, Y.; Ning, J.; Zhou, W.; Zhang, S.; Li, R.; Yan, C. Spatiotemporal characteristics, patterns, and causes of land-use changes in China since the late 1980s. *J. Geogr. Sci.* **2014**, *24*, 195–210. [CrossRef]

42. Bai, L.Y.; Feng, J.Z.; Ma, Y.X.; Ran, Q.Y.; Wang, K.; Zhao, Y. Analysis of spatial pattern Change of LU/LC over the upper Tarim River region since 1990 using remote sensing data. *IOP Conference Series: Earth Environ. Sci.* **2017**, *57*, 012038. [CrossRef]

43. Song, X.-P.; Hansen, M.C.; Stehman, S.V.; Potapov, P.V.; Tyukavina, A.; Vermote, E.F.; Townshend, J.R. Global land change from 1982 to 2016. *Nature* **2018**, *560*, 639. [CrossRef]

44. Wichern, F.; Luedeling, E.; Müller, T.; Joergensen, R.G.; Buerkert, A. Field measurements of the CO_2 evolution rate under different crops during an irrigation cycle in a mountain oasis of Oman. *Appl. Soil Ecol.* **2004**, *25*, 85–91. [CrossRef]

45. Liu, Y.; Fang, F.; Li, Y. Key issues of land use in China and implications for policy making. *Land Use Policy* **2014**, *40*, 6–12. [CrossRef]

46. Wang, T.; Yan, C.; Song, X.; Xie, J. Monitoring recent trends in the area of aeolian desertified land using Landsat images in China's Xinjiang region. *ISPRS J. Photogramm. Remote Sens.* **2012**, *68*, 184–190. [CrossRef]

47. Wang, Y.; Luo, G.; Zhao, S.; Han, Q.; Li, C.; Fan, B.; Chen, Y. Effects of arable land change on regional carbon balance in Xinjiang. *Acta Geogr. Sin.* **2014**, *69*, 110–120. (In Chinese) [CrossRef]

48. Yaning, C.; Changchun, X.; Xingming, H.; Weihong, L.; Yapeng, C.; Chenggang, Z.; Zhaoxia, Y. Fifty-year climate change and its effect on annual runoff in the Tarim River Basin, China. *Quat. Int.* **2009**, *208*, 53–61. [CrossRef]

49. Zhang, Y.; Chen, Y.; Pan, B. Distribution and floristics of desert plant communities in the lower reaches of Tarim River, southern Xinjiang, People's Republic of China. *J. Arid Environ.* **2005**, *63*, 772–784. [CrossRef]

50. Olson, K.R.; Al-Kaisi, M.M. The importance of soil sampling depth for accurate account of soil organic carbon sequestration, storage, retention and loss. *CATENA* **2015**, *125*, 33–37. [CrossRef]

51. Nelson, D.W.; Sommers, L.E. Total carbon, organic carbon, and organic matter. *Methods Soil Anal. Part 3—Chem. Methods* **1996**, 961–1010.

52. Gallaher, R.; Weldon, C.; Boswell, F. A Semiautomated Procedure for Total Nitrogen in Plant and Soil Samples 1. *Soil Sci. Soc. Am. J.* **1976**, *40*, 887–889. [CrossRef]

53. Shangguan, W.; Dai, Y.; Liu, B.; Zhu, A.; Duan, Q.; Wu, L.; Ji, D.; Ye, A.; Yuan, H.; Zhang, Q.; et al. A China data set of soil properties for land surface modeling. *J. Adv. Model. Earth Syst.* **2013**, *5*, 212–224. [CrossRef]

54. Yang, Y.; Fang, J.; Smith, P.; Tang, Y.; Chen, A.; Ji, C.; Hu, H.; Rao, S.; Tan, K.; HE, J.S. Changes in topsoil carbon stock in the Tibetan grasslands between the 1980s and 2004. *Glob. Chang. Biol.* **2009**, *15*, 2723–2729. [CrossRef]

55. Zhou, Z.; Wang, C. Reviews and syntheses: Soil resources and climate jointly drive variations in microbial biomass carbon and nitrogen in China's forest ecosystems. *Biogeosciences* **2015**, *12*, 6751–6760. [CrossRef]

56. Renó, V.F.; Novo, E.M.; Suemitsu, C.; Rennó, C.D.; Silva, T.S. Assessment of deforestation in the Lower Amazon floodplain using historical Landsat MSS/TM imagery. *Remote Sens. Environ.* **2011**, *115*, 3446–3456. [CrossRef]

57. Vittek, M.; Brink, A.; Donnay, F.; Simonetti, D.; Desclée, B. Land cover change monitoring using Landsat MSS/TM satellite image data over West Africa between 1975 and 1990. *Remote Sens.* **2014**, *6*, 658–676. [CrossRef]

58. Chen, L.; Smith, P.; Yang, Y. How has soil carbon stock changed over recent decades? *Glob. Change Boil.* **2015**, *21*, 3197–3199. [CrossRef] [PubMed]

59. Sterner, R.W.; Elser, J.J. *Ecological Stoichiometry: The Biology of Elements from Molecules to the Biosphere*; Princeton University Press: Princeton, NJ, USA, 2002.

60. Houghton, R.; Skole, D.; Nobre, C.A.; Hackler, J.; Lawrence, K.; Chomentowski, W.H. Annual fluxes of carbon from deforestation and regrowth in the Brazilian Amazon. *Nature* **2000**, *403*, 301–304. [CrossRef] [PubMed]

61. Lal, R. Soil carbon dynamics in cropland and rangeland. *Environ. Pollut.* **2002**, *116*, 353–362. [CrossRef]

62. Gelaw, A.M.; Singh, B.; Lal, R. Soil organic carbon and total nitrogen stocks under different land uses in a semi-arid watershed in Tigray, Northern Ethiopia. *Agric., Ecosyst. Environ.* **2014**, *188*, 256–263. [CrossRef]

63. Berihu, T.; Girmay, G.; Sebhatleab, M.; Berhane, E.; Zenebe, A.; Sigua, G.C. Soil carbon and nitrogen losses following deforestation in Ethiopia. *Agron. Sustain. Dev.* **2017**, *37*, 1. [CrossRef]

64. Lozano-García, B.; Parras-Alcántara, L.; Cantudo-Pérez, M. Land use change effects on stratification and storage of soil carbon and nitrogen: Application to a Mediterranean nature reserve. *Agric., Ecosyst. Environ.* **2016**, *231*, 105–113. [CrossRef]

65. Shan, L.; Zhang, X.; Wang, Y.; Wang, H.; Yan, H.; Wei, J.; Xu, H. Influence of moisture on the growth and biomass allocation in Haloxylon ammodendron and Tamarix ramosissima seedlings in the shelterbelt along the Tarim Desert Highway, Xinjiang, China. *Chin. Sci. Bull.* **2008**, *53*, 93–101. [CrossRef]

66. Huang, Y.; Sun, W. Changes in topsoil organic carbon of croplands in mainland China over the last two decades. *Chin. Sci. Bull.* **2006**, *51*, 1785–1803. [CrossRef]

67. Li, X.-G.; Li, F.-M.; Rengel, Z.; Wang, Z.-F. Cultivation effects on temporal changes of organic carbon and aggregate stability in desert soils of Hexi Corridor region in China. *Soil Till. Res.* **2006**, *91*, 22–29. [CrossRef]

68. Shen, W.; Lin, X.; Gao, N.; Zhang, H.; Yin, R.; Shi, W.; Duan, Z. Land use intensification affects soil microbial populations, functional diversity and related suppressiveness of cucumber Fusarium wilt in China's Yangtze River Delta. *Plant Soil* **2008**, *306*, 117–127. [CrossRef]

69. Congreves, K.; Hooker, D.; Hayes, A.; Verhallen, E.; Van Eerd, L. Interaction of long-term nitrogen fertilizer application, crop rotation, and tillage system on soil carbon and nitrogen dynamics. *Plant Soil* **2017**, *410*, 113–127. [CrossRef]

70. Smith, P. Land use change and soil organic carbon dynamics. *Nutr. Cycl. Agroecosyst.* **2008**, *81*, 169–178. [CrossRef]

71. Luo, Y.; Hui, D.; Zhang, D. Elevated CO_2 stimulates net accumulations of carbon and nitrogen in land ecosystems: A meta-analysis. *Ecology* **2006**, *87*, 53–63. [CrossRef] [PubMed]

72. Manzoni, S.; Porporato, A. Soil carbon and nitrogen mineralization: Theory and models across scales. *Soil Biol. Biochem.* **2009**, *41*, 1355–1379. [CrossRef]

73. Yuan, J.; Ouyang, Z.; Zheng, H.; Xu, W. Effects of different grassland restoration approaches on soil properties in the southeastern Horqin sandy land, northern China. *Appl. Soil Ecol.* **2012**, *61*, 34–39. [CrossRef]

74. Fierer, N.; Craine, J.M.; McLauchlan, K.; Schimel, J.P. Litter quality and the temperature sensitivity of decomposition. *Ecology* **2005**, *86*, 320–326. [CrossRef]

75. Kemmitt, S.J.; Wright, D.; Goulding, K.W.; Jones, D.L. pH regulation of carbon and nitrogen dynamics in two agricultural soils. *Soil Biol. Biochem.* **2006**, *38*, 898–911. [CrossRef]

76. Pietri, J.A.; Brookes, P. Nitrogen mineralisation along a pH gradient of a silty loam UK soil. *Soil Biol. Biochem.* **2008**, *40*, 797–802. [CrossRef]

77. Singh Brar, B.; Singh, J.; Singh, G.; Kaur, G. Effects of Long Term Application of Inorganic and Organic Fertilizers on Soil Organic Carbon and Physical Properties in Maize–Wheat Rotation. *Agronomy* **2015**, *5*, 220–238. [CrossRef]

78. Su, Y.Z.; lin Zhao, H. Soil properties and plant species in an age sequence of Caragana microphylla plantations in the Horqin Sandy Land, north China. *Eco. Eng.* **2003**, *20*, 223–235. [CrossRef]

79. O'Brien, S.L.; Jastrow, J.D.; Grimley, D.A.; Gonzalez-Meler, M.A. Moisture and vegetation controls on decadal-scale accrual of soil organic carbon and total nitrogen in restored grasslands. *Glob. Chang. Biol.* **2010**, *16*, 2573–2588. [CrossRef]

80. Liu, X.; Zhang, Y.; Han, W.; Tang, A.; Shen, J.; Cui, Z.; Vitousek, P.; Erisman, J.W.; Goulding, K.; Christie, P. Enhanced nitrogen deposition over China. *Nature* **2013**, *494*, 459. [CrossRef] [PubMed]

81. Khan, S.A.; Mulvaney, R.L.; Ellsworth, T.R.; Boast, C.W. The Myth of Nitrogen Fertilization for Soil Carbon Sequestration. *J. Environ. Qual.* **2007**, *36*, 1821–1832. [CrossRef] [PubMed]

82. Reay, D.S.; Dentener, F.; Smith, P.; Grace, J.; Feely, R.A. Global nitrogen deposition and carbon sinks. *Nat. Geosci.* **2008**, *1*, 430. [CrossRef]

83. Li, Z.-P.; Zhang, T.-L.; Chen, B.-Y. Changes in organic carbon and nutrient contents of highly productive paddy soils in Yujiang county of Jiangxi province, China and their environmental application. *Agric. Sci. China* **2006**, *5*, 522–529. [CrossRef]

84. Mazzoncini, M.; Antichi, D.; Di Bene, C.; Risaliti, R.; Petri, M.; Bonari, E. Soil carbon and nitrogen changes after 28 years of no-tillage management under Mediterranean conditions. *Eur. J. Agron.* **2016**, *77*, 156–165. [CrossRef]

85. Mazzoncini, M.; Sapkota, T.B.; Barberi, P.; Antichi, D.; Risaliti, R. Long-term effect of tillage, nitrogen fertilization and cover crops on soil organic carbon and total nitrogen content. *Soil Till. Res.* **2011**, *114*, 165–174. [CrossRef]

86. Yang, Y.-C.; Zhang, M.; Zheng, L.; Cheng, D.-D.; Liu, M.; Geng, Y.-Q. Controlled release urea improved nitrogen use efficiency, yield, and quality of wheat. *Agron. J.* **2011**, *103*, 479–485. [CrossRef]
87. Gu, B.; Ju, X.; Wu, Y.; Erisman, J.W.; Bleeker, A.; Reis, S.; Sutton, M.A.; Lam, S.K.; Smith, P.; Oenema, O. Cleaning up nitrogen pollution may reduce future carbon sinks. *Glob. Environ. Chang.* **2018**, *48*, 56–66. [CrossRef]

 *agronomy*

Article

Mapping of Olive Trees Using Pansharpened QuickBird Images: An Evaluation of Pixel- and Object-Based Analyses

Isabel Luisa Castillejo-González 

Department of Graphic Engineering and Geomatics, University of Cordoba, Campus de Rabanales, 14071 Córdoba, Spain; ilcasti@uco.es; Tel.: +34-957-218-537

Received: 2 November 2018; Accepted: 27 November 2018; Published: 2 December 2018

Abstract: This study sought to verify whether remote sensing offers the ability to efficiently delineate olive tree canopies using QuickBird (QB) satellite imagery. This paper compares four classification algorithms performed in pixel- and object-based analyses. To increase the spectral and spatial resolution of the standard QB image, three different pansharpened images were obtained based on variations in the weight of the red and near infrared bands. The results showed slight differences between classifiers. Maximum Likelihood algorithm yielded the highest results in pixel-based classifications with an average overall accuracy (OA) of 94.2%. In object-based analyses, Maximum Likelihood and Decision Tree classifiers offered the highest precisions with average OA of 95.3% and 96.6%, respectively. Between pixel- and object-based analyses no clear difference was observed, showing an increase of average OA values of approximately 1% for all classifiers except Decision Tree, which improved up to 4.5%. The alteration of the weight of different bands in the pansharpen process exhibited satisfactory results with a general performance improvement of up to 9% and 11% in pixel- and object-based analyses, respectively. Thus, object-based analyses with the DT algorithm and the pansharpened imagery with the near-infrared band altered would be highly recommended to obtain accurate maps for site-specific management.

Keywords: Á Trous algorithm; conservation agriculture; crop inventory; remote sensing; spectral-weight variations in fused images

---

1. Introduction

Nowadays, one of the most important objectives in agriculture is to perform precision agriculture (PA) in most possible scenarios to control efficiently the input data and, consequently, reduce the production cost and the environmental pollution produced by this activity. To facilitate and assist this change in farm management, government institutions tend to regulate and encourage different techniques based on PA. The European Commission is greatly concerned about the new challenges in agriculture and promote the changes by different legal instruments and key texts [1]. Diverse action areas such as farming, protection of natural or agricultural environment, food safety, security and traceability, or even climate change mitigation are regulated. Some of the recommended or mandatory practices are supported in an accurate control of the spatial distribution of crops. To obtain economical funds from the common agricultural policy [2], PA promotes among many other actions, the application of site-specific management or integrated management systems in crops production to reduce the use of fertilisers, herbicides, or pesticides and the establishment of certain conservation agro-environmental measures such as cover crops in olive orchards [3]. To control the correct application of PA techniques, different monitoring systems were developed. The expensive, time-consuming, and imprecise system based on sample and ground visit to small percentage of fields has forced a search for new techniques that reduce costs and increase the controlled area, maintaining high accuracy in the analyses. Remote sensing data can significantly improve the deficiencies of ground visits, allowing accurate maps.

Several studies have focused on addressing diverse PA topics by remote sensing, such as obtaining accurate maps of crops [4–6]; detecting the location of weeds [7–9], pests [10–12], and diseases [13–15] to apply site-specific management, or determining the level of water stress to design optimal irrigation systems [16–18]. Nevertheless, most of the precision agriculture studies with remote sensing analysed the characteristics of herbaceous crops, as these types of crops usually cover all the field and are easier to study with digital imagery. Woody crops present very different spectral responses between tree canopies, soils, and other covers presented in the field. Thus, very high spatial resolution images are needed. Most of the studies which aim is to characterize the architecture of the trees used airborne or Unmanned Aerial Vehicle (UAV) images to obtain the canopy information [19–21]. Nevertheless, few studies used satellite imagery [22–24] and most of them were aid with LiDAR information [25–27].

High resolution satellite imagery can be useful to accurately map tree canopies. Companies that distribute Earth Observation Satellite images usually offer the user community two separate products: a high-spatial resolution panchromatic image and a low-spatial resolution multispectral one. While the multispectral image facilitates the discrimination of land covers types, the panchromatic image allows to delimit accurately each land cover [28]. To use simultaneously the advantages of both resolutions in one image, fusion techniques have been developed. The pansharpen fusion method allows the injection of spatial detail information from the panchromatic image into each band of the multispectral image [29]. These new characteristics of the pansharpened image can help to accurately delineate the tree canopies to apply PA techniques.

Supervised classification methods are extensively used in land use classification studies [30]. These procedures extrapolate the spectral characteristics obtained from the image training sites defined for classification by the user to other areas of the image. Nowadays, there are classification routines based on spectral or angular distances, probability analysis, and more advanced data mining techniques. There is no one ideal classification routine. The most appropriate method is determined by the needs and requirements of each study [31]. Many of the remote sensing classifications are based on pixels as the minimum spatial information unit. These analyses provide very good results in homogeneous land uses. Nevertheless, the increase of spatial resolution causes an increase in the intraclass spectral variability and a reduction in classification performance and accuracy when pixel-based analyses are used [32]. This is particularly true when the pixel size is significantly smaller than the average size of the objects of interest [33]. To overcome this problem, different segmentation techniques, in which adjacent pixels are grouped into spectrally and spatially homogeneous objects, have been developed. The main segmentation algorithms can be classified into two general classes: edge-based and object-merging algorithms [34]. Most of the segmentation procedure developed are object merging algorithms, which take some pixels as seeds and grow the regions around them based on certain homogeneity criteria [35]. Since Kettig and Landgrebe [36], the object-based approach has hardly been used in favour of easier pixel-based analyses. Some researchers have reported that the segmentation techniques used in classifications reduce the local variation caused by textures, shadows and shape in forestry trees [37,38] and agricultural trees [39] classifications. However, object-based classifications in typical agricultural dryland Mediterranean areas to map accurately olive tree canopies using only high spatial resolution satellite imagery are lacking.

Therefore, the main objective of this paper was to evaluate the potential of four supervised classification routines, applied to pixel- and object-based classifications, to delineate olive tree canopies using a pansharpened QuickBird image. An additional goal was to check the effect of the variation of the spectral weight in the pansharpen process, to emphasise the spectral information of different wavelengths over another.

2. Materials and Methods

2.1. Study Area and Satellite Image Acquisition

This study was focused on dryland olive (*Olea europaea* L.) orchards representative of the typical continental Mediterranean climate [40]. The analysis was conducted in five olive orchards fields named A, B, C, D and E (Figure 1) located near Montilla, province of Córdoba (Andalusia, southern Spain, centre UTM *X-Y* coordinates 355,746–4,164,520 m, datum WGS84). This agricultural region has a typical continental Mediterranean climate characterized with short mild winters and long dry summers [41]. The study sites were located in a farmer-managed area where the farmers made decisions individually. Thus, different characteristics of the studied fields such as size and morphology of olive crowns, presence or not of vegetation cover or soil tillage, were found. The total areas of the A, B, C, D and E fields were 2.16 ha, 22.80 ha, 15.66 ha, 24.55 ha and 28.44 ha, respectively.

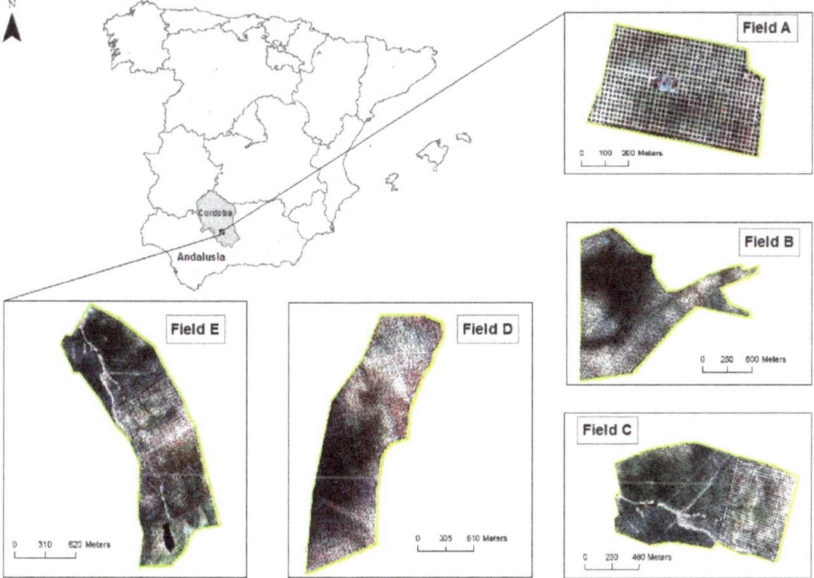

Figure 1. Location of the study area in Andalusia, southern Spain. Detailed olive orchards fields are depicted by QuickBird pansharpened images.

On 10 July 2004, a QuickBird (QB) satellite digital image was acquired for the study area. The QB satellite provided four multispectral bands (blue, B: 450–520 nm; green, G: 520–600 nm; red, R: 630–690 nm; and near-infrared, NIR: 760–900 nm) with a spatial resolution of 2.8 m, and a panchromatic band (PAN: 450–900 nm) with a spatial resolution of 0.7 m. Radiometric resolution of the QB image was 11 bit. A QB Standard image product was ordered, which included radiometric, sensor and geometric corrections previously carried out by the distributor [42]. No atmospheric corrections were needed as long as each classification was carried out in a single date image on the same relative scale [43].

2.2. Data Fusion: Pansharpening of Multispectral Images

To obtain an image of high spectral and spatial resolution, a pansharpening process was carried out with the QB bands. The pansharpen techniques allow to obtain new bands with the spectral resolution of the multispectral bands and the spatial resolution of the panchromatic band. In this study, a weighting variant of the fusion algorithm based on the wavelet transform calculated using the Á Trous algorithm was used to fuse the multispectral and panchromatic bands [44]. This fusion method consists basically of successive convolutions between the analyzed image and a low-pass filter called the scaling function, which commonly is the b3-spline function. The filters to be applied in the subsequent decomposition levels are obtained from the filter applied in the previous level, intercalating it with zeros in the rows and columns. The wavelet coefficients are obtained from the difference between two consecutive decomposition levels.

Knowing that the red and especially the near infrared bands are very important in the discrimination of vegetation, these two bands were weighted differently. Thus, to evaluate the effect of different weighted pansharpened bands in the classifications, three weight combinations were proposed: (a) 1-1-1-1, (b) 1-1-5-5 and (c) 1-1-1-10 as weight factor for B-G-R-NIR bands, respectively. As a result of this fusion or pansharpening process, three different images with four multispectral bands (B, G, R and NIR) and with a spatial resolution of 0.7 m were obtained.

Usually, the global quality of the resulting pansharpened image is estimated with the ERGAS index (Erreur Relative Globale Adimensionnelle de Synthèse) [45]. This relative dimensionless global error in synthesis offers a global picture of the spectral quality of the fused product. Nevertheless, in pansharpening techniques, high spectral quality implies low spatial quality and vice versa, which suggests the necessity to control not only the spectral quality of the process, but also the spatial. Thus, a spatial index based on ERGAS concepts and translated to the spatial domain [46] was used.

2.3. Segmentation

A segmentation procedure was performed to partition the QB pansharpened images into homogeneous objects using the Fractal Net Evolution Approach (FNEA) segmentation algorithm [47] (Figure 2). This algorithm allows the multiresolution bottom-up region-merging segmentation, a process in which individual pixels merge to objects in successive fusing iterations. The merging process continues until a threshold derived from the user-defined parameters is reached. The result is an image in which the pixels are aggregated in highly homogeneous objects at an arbitrary resolution.

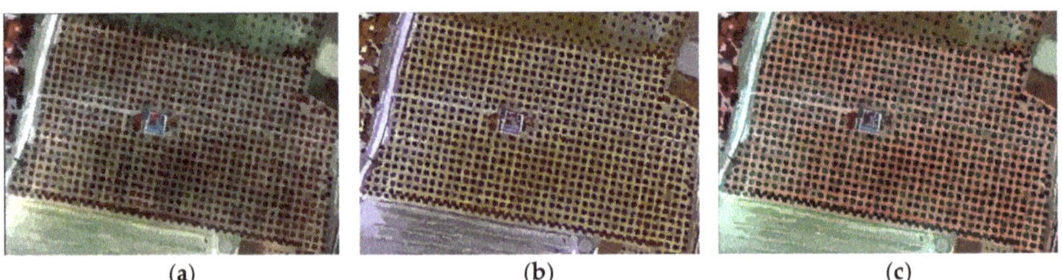

(a)   (b)   (c)

Figure 2. Multiresolution segmentation of pansharpened QuickBird (QB) imagery in field A. Pansharpen weight (B-G-R-NIR): (**a**) 1-1-1-1; (**b**) 1-1-5-5; and (**c**) 1-1-1-10.

The segmentation process can be controlled by the weighting of the input data and the definition of three parameters. The scale parameter controls the size of the objects, while the colour and shape parameters define the importance of the spectral and morphological information involved in the object generation, respectively. The setting of the segmentation parameters were determined by testing different segmentation input scenarios to evaluate their ability to delineate olive crowns. For each field, the first parameter to adjust was the scale parameter to control the size of the objects depending on the characteristics of the field. Then, with the scale parameter fixed, the spectral and morphological weight of the information was defined. The morphological information is divided into two characteristics, the compactness and the smoothness of the objects. For this study, as the trees crowns present a compact structure, these two characteristics were fixed in all scenarios tested to 0.8 and 0.2 for compactness and smoothness, respectively.

The segmentation procedure generates not only the mean spectral information of the objects, which is derived from the spectral information of the pixels that form each object, but also a large amount of data divided mainly in three categories: spectral, morphological and textural. In this study, only some spectral and morphological variables derived from the segmentation process were used to characterize the olive orchards. Therefore, to perform this analysis eight object-based features, three spectral and five morphological variables, were included (Table 1). The spectral feature *Mean* is calculated for each multispectral band independently, obtaining 4 final features: Mean (Blue), Mean (Green), Mean (Red) and Mean (Near-Infrared).

Table 1. Object-based features derived from segmentation.

Categories	Features	Brief Description
Spectral	Mean	Mean of the intensity values of all pixels forming an image object
	NDVI	Normalized Difference Vegetation Index [48]
	RDVI	Renormalized Difference Vegetation Index [49]
Shape	Area	Number of pixels forming an image object
	Asymmetry	Relative length of an image object compared to a regular ellipse polygon
	Border index	Ratio between the border lengths of the image object and the smallest enclosing rectangle
	Length	Multiplication between the number of pixels and the length-to-width ratio of an image object
	Width	Ratio between the number of pixels and the length-to-width ratio of an image object

2.4. Classification and Accuracy Assessments

Both pixel- and object-based analyses with four different supervised classifiers were conducted on the five olive orchards fields. The classification algorithms were Minimum Distance (MD), Spectral Angle Mapper (SAM), Maximum Likelihood (ML) and Decision Tree (DT). MD classifies each pixel in the category that presents the minimum spectral distance between the spectral signal of the pixel and the spectral average of the class. The spectral distance is determined by Euclidean distance in N-bands spectral space [50]. Similarly, SAM measures spectral similarity but assigns the category of each pixel to the class that presents the minimum spectral angle, instead the spectral distance. The spectral angle between two spectra is calculated by taking the arccosine of the dot product of the two spectral vector [51]. ML creates classification rules based on probabilistic algorithms considering the spectral average of each class and the variance. This algorithm assigns a pixel to the most probable class and thus minimizes the probability of error using Bayesian theory [52]. Finally, DT classifier creates models of decision based on conditional control statements. In this study, the DT classification was performed with the data mining C4.5 algorithm, a top-down inductor of decision trees that expands nodes in depth-first order for each step using the divide and conquer strategy [53]. Ground-truth land use was randomly defined to substantiate and validate the classification procedures. For each field, a sampling with distant and independent locations was digitized directly on the image. Approximately 25% of the sampled surface were used to collect the spectral signature in the training process, and the remaining 75% were used to assess the accuracy of the classifications. To avoid any subjective estimation, the training and verification procedure did not change in any of the classifications.

To determine the accuracy obtained with every classifier in each olive orchard field, the confusion matrix of the classification and the Kappa test were analysed. The confusion matrix compares the percentage of classified pixels of each class with the verified ground truth class, indicating the correct assessment and the errors among the studied classes [38]. In addition to detailed accuracies obtained in every classification category, the confusion matrix obtain the overall accuracy (OA), which indicates the overall percentage of correctly classified pixels in the classification. The Kappa test yields the Kappa coefficient (K), which determines if the results obtained in the classification are significantly better than the results obtained in a random classification. The combination of both accuracy values is more conservative than a simple percent agreement value [54,55].

In pixel-based classification it is frequent to observe isolated-misclassified pixels dispersed within the area of another class. To reduce this commonly named salt and pepper noise and increase the accuracy of the classifications, a majority filter of 3×3 was applied to improve the classified maps. In object-based classification this noise is practically eliminated when pixels group in the segmentation process.

The pansharpened QB bands were generated with the IJFUSION software (Polytechnic University of Madrid, Spain). To obtain the segmented bands used in object-based classifications, the eCognition Developer 8 software (Definiens AG) was used. The Weka 3.8 software (University of Waikato, New Zealand) determined the decision tree sequences. Finally, the software ENVI 5.1 (Harris Geospatial Solutions) was used to carry out all the pixel- and object-based classifications.

3. Results and Discussion

3.1. Data Fusion: Pansharpening of Multispectral Images

Three pansharpen procedures were performed with the entire QB image. To control the quality of the process, two ERGAS indexes were calculated. Table 2 shows the spectral and spatial ERGAS indexes obtained in the study for each pansharpen combination. The spectral ERGAS index showed values of 0.72, 1.84 and 1.83 whereas the spatial ERGAS index exhibited slightly higher values of 1.3, 1.73, and 1.86, for pansharpen B-G-R-NIR combinations 1-1-1-1; 1-1-5-5 and 1-1-1-10, respectively. Being a value of 0 of each ERGAS index the maximum quality, the lower the ERGAS value, the higher the spectral quality of fused images. ERGAS error lower than 3 are considered as good quality for fused product [56]. In this study, the ERGAS errors obtained were considerably low (lower than 2 in all combinations), which implied a high spectral and spatial quality in the images obtained [57]. This premise is important in this type of study, as an accurate isolation of olive crowns need a very high spatial resolution but without losing the spectral information, especially in complex areas where a mixture of olive tree, natural cover, and soil spectral data can be observed. The increase of the ERGAS indexes values in the pansharpened image, where the red and near-infrared bands were over-weighed, was predictable. Gonzalo and Lillo-Saavedra [44] conceived the pansharpen algorithm performed in this study to apply the exact weighted factor to every band to obtain same spatial and spectral quality of the image, it means, to obtain the "best fused image". In this study, and controlling that the quality of the fused images does not exceed the ERGAS index limit, the weight of the bands was based on the necessity of the study but not in the control of the pansharpen quality.

Table 2. Spectral and spatial indexes to control the quality of the pansharpened images.

Pansharpen Weight (B-G-R-NIR)	Spectral ERGAS	Spatial ERGAS
1-1-1-1	0.72	1.13
1-1-5-5	1.84	1.73
1-1-1-10	1.83	1.86

3.2. Segmentation

For the fifteen pansharpened fields analysed, three pansharpen combinations $\times$ five olive orchard fields, a considerable number of input parameters were tested to obtain the criteria that provided the most satisfactory inputs scenarios (Table 3). Each olive orchard field presented different characteristics, which involved different segmentation parameters. As the aim of the segmentation is to obtain objects with a similar or smaller area than an olive tree canopy, the values of the scale parameters were low, ranging from 12 in the fields A and E to 25 in the fields B and C. The pansharpen images with the weight 1-1-1-1 always required a smaller scale parameter than the other two pansharpen images. In the segmentation process, the spectral information (colour) presented more weight than the morphology of the objects (shape) in most of the scenarios evaluated, although colour never exceeded the 70% of the weight. Only in the field D, the percentage of both types of information was divided equally (50%).

As mentioned in Section 2.3, the compactness and smoothness characteristics of the morphology of the objects were fixed to 0.8 and 0.2, respectively.

Table 3. Most satisfactory segmentation parameters obtained for each pansharpened field.

Field	Pansharpen Weight (B-G-R-NIR)	Scale Parameter	Colour	Shape	Compactness	Smoothness
	1-1-1-1	12	0.6	0.4	0.8	0.2
A	1-1-5-5	20	0.7	0.3	0.8	0.2
	1-1-1-10	20	0.7	0.3	0.8	0.2
	1-1-1-1	15	0.7	0.3	0.8	0.2
B	1-1-5-5	25	0.7	0.3	0.8	0.2
	1-1-1-10	25	0.5	0.5	0.8	0.2
	1-1-1-1	15	0.6	0.4	0.8	0.2
C	1-1-5-5	25	0.7	0.3	0.8	0.2
	1-1-1-10	17	0.6	0.4	0.8	0.2
	1-1-1-1	14	0.6	0.4	0.8	0.2
D	1-1-5-5	14	0.5	0.5	0.8	0.2
	1-1-1-10	14	0.5	0.5	0.8	0.2
	1-1-1-1	12	0.6	0.4	0.8	0.2
E	1-1-5-5	19	0.7	0.3	0.8	0.2
	1-1-1-10	22	0.7	0.3	0.8	0.2

3.3. Olive Orchard Fields Classification

The accuracy assessments, OA and K, of the different pixel- and object-based classifications carried out in the different weighted pansharpened images of every field are displayed in Table 4. Figure 3 shows an example of the least and the most accurate olive crowns classifications in an individual field. Table 4 reveals slight differences between classifiers, yielding most of the scenarios evaluated very high classification accuracies and showing that olive trees canopies could be discriminated very accurately in most of the test carried out. All the classifiers achieved high comparable classification results, although some classifiers stood out above others. In pixel-based classifications, MP exhibited the highest average of OA and K with a value of 94.2% and 0.89, respectively. In object-based classification, MP obtained the highest accuracies values in eight of the 15 images analysed and DT obtained the highest precision in the remaining seven classifications. Nevertheless, the average results were slightly higher with the DT classifier offering OA and K values of 96.6% and 0.94%, while ML obtained values of 95.3% and 0.91%. ML and DT classification algorithms exhibited high reliability in all classifications performed, while SAM and MD yielded more erratic results, showing always the lowest accuracies of each analysis. As an example, in pixel-based classification of field E, MD obtained the highest OA values for the pansharpen combinations 1-1-5-5 and 1-1-1-10 with 93.4% and 95.2%, respectively, whereas this classifier showed very low OA value for the pansharpen combination 1-1-1-1, with a value of 77.3%. Nevertheless, all the olive orchard fields could be classified very accurately with at least one classifier, showing OA values greater than 90%. The results observed in this study satisfied the commonly accepted requirements to consider an accurate classification when the OA value is at least an 85% [58] and the Kappa coefficient exceeds the 0.75 [59]. Such high accuracies are essential if the olive orchard map obtained is going to be used in precision agriculture to design site specific management. To emphasise one classifier from all of them, ML classifier could be selected considering that yielded one of the highest precisions in all classifications performed and that the computational and expertise requirements involved in this classification method is lower than the other most accurate classifier, DT, a data mining algorithms which demands deeper knowledge. Additionally, ML algorithm is implemented in most of the image processing software.

Table 4. Classification accuracies of olive orchards at the five fields analysed in the three pansharpened images using different classification algorithms.

		Analyses														
		Pixel-Based								Object-Based						
Field	Image [1]	MD [2]		SAM		ML		DT		MD		SAM		ML		DT	
		OA [3]	K	OA	K	OA	K	OA	K	OA	K	OA	K	OA	K	OA	K
A	1-1-1-1	94.4	0.91	92.6	0.88	95.0	0.92	91.7	0.86	96.2	0.93	96.6	0.94	97.8	0.96	94.9	0.91
	1-1-5-5	91.7	0.86	90.8	0.84	94.4	0.91	98.7	0.97	96.4	0.94	94.5	0.91	97.9	0.96	98.9	0.98
	1-1-1-10	89.4	0.82	88.0	0.79	92.8	0.87	98.6	0.98	91.7	0.86	95.7	0.93	98.8	0.98	98.7	0.98
B	1-1-1-1	91.3	0.83	94.2	0.89	98.7	0.97	95.7	0.91	88.8	0.78	91.3	0.82	99.1	0.98	96.9	0.95
	1-1-5-5	95.2	0.91	95.4	0.91	98.7	0.98	92.1	0.89	94.6	0.89	97.3	0.94	99.3	0.99	99.3	0.98
	1-1-1-10	95.7	0.86	86.5	0.73	97.5	0.95	94.3	0.90	94.1	0.88	77.9	0.59	98.7	0.97	98.5	0.97
C	1-1-1-1	90.4	0.81	93.4	0.87	97.7	0.95	96.5	0.94	88.8	0.78	88.4	0.77	97.5	0.95	97.0	0.94
	1-1-5-5	95.8	0.92	93.1	0.86	97.7	0.95	92.4	0.89	95.6	0.91	96.5	0.93	97.9	0.96	98.7	0.98
	1-1-1-10	95.5	0.91	85.3	0.71	96.6	0.93	94.1	0.90	94.9	0.90	64.9	0.31	98.5	0.97	98.6	0.97
D	1-1-1-1	69.7	0.39	75.4	0.51	88.1	0.76	84.8	0.80	78.9	0.58	86.1	0.72	79.7	0.59	87.2	0.82
	1-1-5-5	87.5	0.75	87.9	0.76	94.2	0.88	88.5	0.84	87.1	0.74	85.6	0.71	88.4	0.77	98.3	0.96
	1-1-1-10	89.5	0.79	85.3	0.71	92.3	0.85	87.3	0.83	85.7	0.72	88.0	0.76	91.6	0.83	98.7	0.97
E	1-1-1-1	77.3	0.55	76.7	0.54	86.7	0.73	86.8	0.83	82.0	0.64	79.2	0.59	89.4	0.79	89.0	0.84
	1-1-5-5	93.4	0.87	91.5	0.83	93.0	0.86	90.7	0.89	97.1	0.94	96.8	0.94	97.5	0.95	97.9	0.96
	1-1-1-10	95.2	0.91	88.1	0.76	89.9	0.80	88.8	0.85	96.1	0.92	82.1	0.64	97.3	0.95	97.1	0.95

[1] Pansharpen weight (B-G-R-NIR); [2] Method of classification: MD, Minimum Distance; SAM, Spectral Angel Mapper; ML, Maximum Likelihood; DT, Decision Tree; [3] Accuracy values: OA, overall accuracy (%); K, Kappa coefficient.

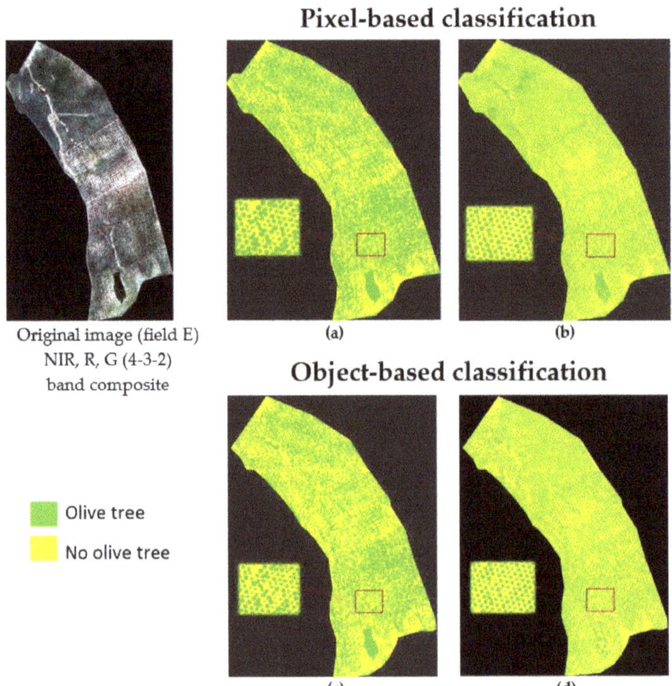

Figure 3. Result of the least (**a**,**c**) and most accurate (**b**,**d**) olive orchard classifications of field E.

Between pixel- and object-based analyses, no clear difference was observed. Despite that object-based classifications analysed a greater number of segmented variables than pixel-based, the precision of the classifications yielded were slightly higher. In object-based analyses, the increase of average OA values was approximately 1% for all classifiers except DT, which improved its general performance with an average increase of 4.5% and showed the most accurate average OA (96.6%). Twelve of the fifteen pansharpen combinations classified improved the accuracies when the object-based analysis were performed, but these increases usually were minimal, not exceeding in many cases the 1% of improvement. The most significant variations between object- and pixel-based analyses were observed in fields D and E. A 6.4% of improvement was observed in the pansharpen combination 1-1-1-10 of the field D, when the most accurate pixel-based analysis was performed with the MP classifier and yielded an OA value of 92.3%, whereas the most accurate object-based classification was obtained with DT algorithm and obtained an OA of 98.7%. Similarly, accuracy increases superior to 4% could be observed with the pansharpen combination 1-1-5-5 of both, field D and E, obtaining the greatest OA values of 98.3% and 97.9% with DT algorithm. Little advantage when applying object-based techniques can be observed in other studies focused on precision agriculture when pixel-based analyses offer reasonable performance. Pérez-Ortiz et al. [60] tested different scenarios of pixel- and object-based classifications to detect weeds in sunflower crops and obtained similar results, with improvements of up to 6%. Similarly, Castillejo González et al. [61] evaluated pixel- and object based classifications to distinguish late-season wild oat weed patches in wheat fields and suggested that the small sizes of the objects and the excellent behaviour of the classification algorithms in the pixel-based classifications did not produce a significant improvement over the precision obtained in the object-based classifications.

In object-based analyses eight different segmented variables were classified, which implies more information that can enhance or worsen the capacity to distinguish among categories. MD, SAM, and ML algorithms give equal weight to all the variables involved in the classification process, and use all these variables in the process, independently of the level of improvement or deterioration that the classification can suffer. Nevertheless, the DT algorithm analyses all the variables and selects only the information that really help to distinguish among categories, increasing its efficiency. Figure 4 shows the percentage of use for each variable in DT classifications. From the eight segmented variables managed in this study, DT algorithm only used six different variables in the total set of analyses, and only two or three were necessary in most of the DT classifications. All spectral variables were used in the DT analysis, but the bands that were selected more frequently were the NIR mean layer and the NDVI index. Whereas the NIR mean layer showed the highest level of intervention in DT analysis with a 44.4% for pansharpen combination 1-1-1-10 and a 37.5% for combination 1-1-5-5, the NDVI index band was used in the three pansharpen combinations with a 27.3%, 22.2%, and 12.5% of intervention for combinations 1-1-1-1, 1-1-1-10, and 1-1-5-5, respectively. From the five geometrical segmented variables, only Width, Length, and Border index were used in those classifications. With feature was the most useful, with a 45.5% of intervention for combination 1-1-1-1 and a 12.5% for combination 1-1-1-5. The scarce use of morphological variables can be explained because the olive is a tree that presents different canopy architecture depending on characteristics such as age, farm management, variety, pruning, etc. (Figure 5). The very high accuracies obtained with limited variables in DT classifications agrees with the idea exposed in [62], when they concluded that DT tends to be very efficient and robust when a large volume of predicted variables are introduced in a model, generally performing fast and being insensitive to noise in input data. This behaviour explain the rise of the accuracies that DT classifications exhibited in object-based analyses compared to ML algorithm.

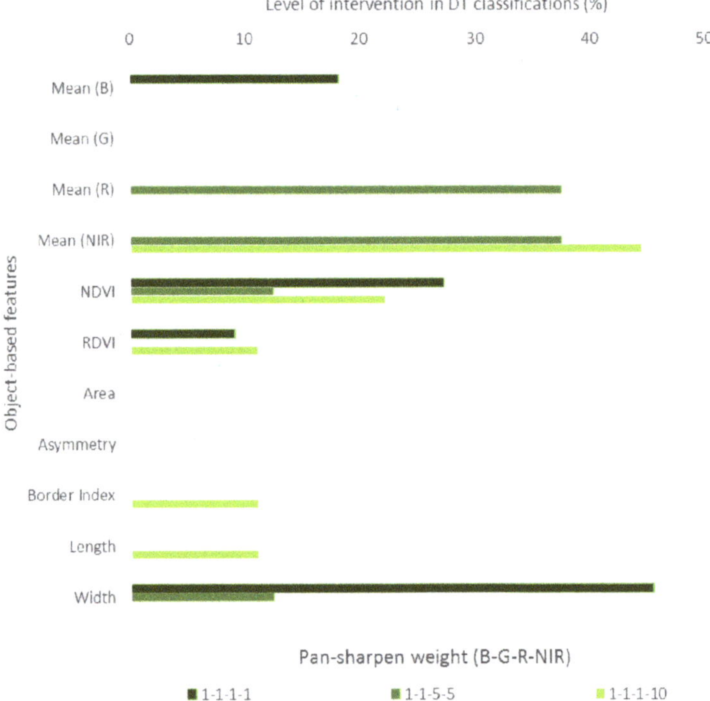

Figure 4. Relative contribution of object-based variables for DT classifications.

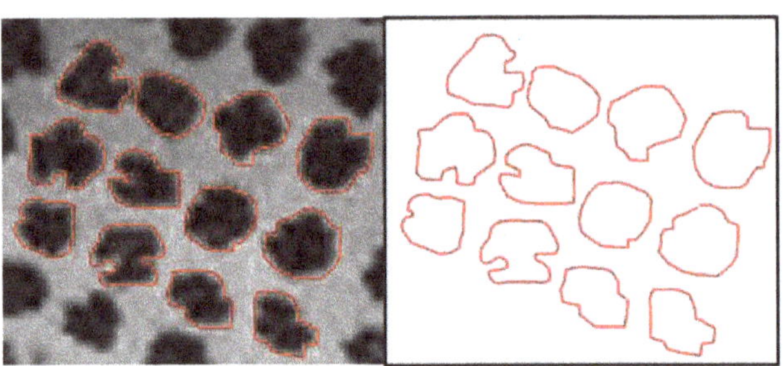

Figure 5. Example of differences in the morphology of the olive crowns.

The pansharpen process was necessary to obtain enough spatial resolution in the images to accurately distinguish the olive trees canopies. The original idea of alter the spectral information of the pansharpened images to emphasize the most useful spectral bands to distinguish vegetation from other land uses exhibited satisfactory results, offering increases of accuracy in three of the five fields studied. This improvement is especially significant in fields D and E, spectral complex fields which showed more difficult to isolate the olive trees. Field D showed the greatest improvement with the spectrally altered images. In pixel-based classifications, increases of approximately 6% and 4% were observed among the combination 1-1-1-1 (OA of 88.1%), and the combinations 1-1-5-5 (94.2%) and 1-1-1-10 (92.3%). More prominent were the increases observed in object-based classifications, where the combination 1-1-1-1 obtained an OA of 87.2% whereas the pansharpen combinations 1-1-5-5 and 1-1-1-10 showed OA values of 98.3% and 98.7%, respectively, which implied accuracy increases higher than 11%. Similarly, field E showed important accuracy increases of 6% and 9% in pixel-based analyses and 8% and 7% in object-based classifications, for 1-1-5-5 and 1-1-1-10 pansharpened combinations, respectively. Although studies based on pansharpened images performed with different spectral weights were not found, the use of equal weight of the multispectral bands in the pansharpen process were evaluated in agronomic scenarios. When homogeneous land uses such as herbaceous crops were analysed with pansharpen imagery, the improvements in classification accuracy due to the use of more spectral and spatially detailed imagery could not really be considered as remarkable regarding the multispectral imagery [63]. Nevertheless, in land uses with high intraclass variability such as woody crops, where the individual trees must be isolated from soil and other covers presented in the field to design a site-specific crop management, the spatial detail is needed. García-Torres et al. [64] analyzed the capacity of different spatial resolutions to isolate olive trees and suggested that imagery with spatial resolutions from 0.25 to 1.5 m were generally suitable for olive grove characterization with olive trees of over 3–4 m^2. Scarce studies used pansharpen images to isolate trees, being more frequent to obtain the information of the tree based on the fusion of different types of information such as multispectral, hyperspectral or LiDAR data or with the analysis of very high spatial resolution UAV imagery. Johnson et al. [65] tested different pansharpened processes to map residential area trees and damaged oak trees. Since a hybrid approach including two pansharpening methods produced the best results in this study, they recommend users planning to process the pansharpen themselves rather than purchase pansharpened imagery directly from the image vendor, so that they can incorporate variations in the methodology for their analysis.

An accurate map of olive orchards fields enable the design of site-specific management treatments and can contribute to the follow up and assessment of agri-environmental regulations. Further studies will focus on establishing a hierarchical classification system that aims at discriminating all olive orchard fields present in the entire QB scene while evaluating the potential of image sharpening in classifying different land uses. In the first level of the hierarchical analysis, the olive orchard fields will be identified and isolated from the other land uses included in the whole studied region to, subsequently, characterize the olive trees of each field. Additionally, with each tree individually discriminated, different agri-environmental indicators of olive orchards related with the number and area of the trees (e.g., potential productivity of each tree and potential production of each plot) and related with bare soil and other vegetation covers (e.g., risk of erosion and run-off) can be predicted.

4. Conclusions

The olive production sector is characterized by a large number of small operators which directly affect production. The results of the present study show that pansharpened multi-spectral QuickBird imagery can be successfully used to map delineation of olive tree canopies. The knowledge of the accurate location and delimitation of the olive trees can be used as a basis for the precision management of fertilizers, pesticides and watering, since there is an obvious relationship between tree size and potential productivity based on the requirements of nutrients, watering doses and plant protection products such as fungicides. Although all classification algorithms tested offered accurate results, ML and DT were the two most precise classifiers. Knowing that the red and especially the near infrared bands are very important in the discrimination of vegetation, the alteration of the weight of these spectral bands in the pansharpen process enhanced significantly the accuracy of the olive trees delineation, with an average increase improvement of up to 9% and 11% in pixel- and object-based analyses, respectively.

With regard to recommending one methodology to define the olive trees crowns, two considerations should be made: the improvement in accuracy obtained and the computational or expertise requirements involved in the process. The decision of whether or not to carry out more complex analyses will depend on the importance of achieving maximum accuracy and the ratio of cost/efficiency wished to obtain in the objectives. If one aimed to create a crop inventory, then the performance of pixel-based classification with ML classifier and standard pansharpened images would be the best choice. However, if it desires to produce a map that is ready to be used for precision agriculture decision-making procedures, object-based analyses with the DT algorithm and the pansharpened imagery with the near-infrared band altered would be highly recommended.

Funding: This research received no external funding.

Acknowledgments: I extend thanks to Javier Carrasco for his assistance with data processing.

Conflicts of Interest: The authors declare no conflict of interest.

References

1. Precision Agriculture in Europe. Legal, Social and Ethical Considerations. European Parliament (European Parliamentary Research Service). Available online: http://www.europarl.europa.eu/RegData/etudes/STUD/2017/603207/EPRS_STU(2017)603207_EN.pdf (accessed on 24 October 2018).
2. The European Parliament and the Council of the European Union. Regulation 1306/2013 of the European Parliament and of the Council of 17 December 2013 on the financing, management and monitoring of the common agricultural policy and repealing Council Regulations (EEC) No 352/78, (EC) No 165/94, (EC) No 2799/98, (EC) No 814/2000, (EC) No 1290/2005 and (EC) No 485/2008 OJ L 347, 20.12.2013. *Off. J. Eur. Union* **2013**, *347*, 549–607.
3. The European Parliament and the Council of the European Union. Regulation 1107/2009 of the European Parliament and of the Council of 21 October 2009 concerning the placing of plant protection products on the market and repealing Council Directives 79/117/EEC and 91/414/EEC. *Off. J. Eur. Union* **2009**, *309*, 1–50.
4. de Castro, A.I.; Jiménez-Brenes, F.M.; Torres-Sánchez, J.; Peña, J.M.; Borra-Serrano, I.; López-Granados, F. 3-D characterization of vineyards using a novel UAV imagery-based OBIA procedure for precision viticulture applications. *Remote Sens.* **2018**, *10*, 584. [CrossRef]
5. Fitzgerald, G.J.; Maas, S.J.; Detar, W.R. Spider mite detection and canopy component mapping in cotton using hyperspectral imagery and spectral mixture analysis. *Precis. Agric.* **2004**, *5*, 275–289. [CrossRef]
6. Yang, C.; Everitt, J.H.; Bradford, J.M. Comparison of QuickBird satellite imagery and airborne imagery for mapping grain sorghum yield patterns. *Precis. Agric.* **2006**, *7*, 33–44. [CrossRef]
7. López-Granados, F.; Torres-Sánchez, J.; Serrano-Pérez, A.; de Castro, A.I.; Mesas-Carrascosa, F.; Peña, J. Early season weed mapping in sunflower using UAV technology: Variability of herbicide treatment maps against weed thresholds. *Precis. Agric.* **2016**, *17*, 183–199. [CrossRef]
8. Huang, Y.; Reddy, K.N.; Fletcher, R.S.; Pennington, D. UAV low-altitude remote sensing for precision weed management. *Weed Technol.* **2018**, *32*, 2–6. [CrossRef]
9. López-Granados, F.; Torres-Sánchez, J.; De Castro, A.; Serrano-Pérez, A.; Mesas-Carrascosa, F.; Peña, J. Object-based early monitoring of a grass weed in a grass crop using high resolution UAV imagery. *Agron. Sustain. Dev.* **2016**, *36*, 67. [CrossRef]
10. Du, Q.; Chang, N.; Yang, C.; Srilakshmi, K.R. Combination of multispectral remote sensing, variable rate technology and environmental modeling for citrus pest management. *J. Environ. Manag.* **2008**, *86*, 14–26. [CrossRef]
11. Percival, D.C.; Gallant, D.; Harrington, T.; Brown, G. Potential for commercial unmanned aerial vehicle use in wild blueberry production. *Acta Hortic.* **2017**, *1180*, 233–240. [CrossRef]
12. Rai, M.; Ingle, A. Role of nanotechnology in agriculture with special reference to management of insect pests. *Appl. Microbiol. Biotechnol.* **2012**, *94*, 287–293. [CrossRef] [PubMed]
13. Herrmann, I.; Vosberg, S.K.; Ravindran, P.; Singh, A.; Chang, H.; Chilvers, M.I.; Conley, S.P.; Townsend, P.A. Leaf and canopy level detection of fusarium virguliforme (sudden death syndrome) in soybean. *Remote Sens.* **2018**, *10*, 426. [CrossRef]
14. Hillnhütter, C.; Mahlein, A.; Sikora, R.A.; Oerke, E. Remote sensing to detect plant stress induced by heterodera schachtii and rhizoctonia solani in sugar beet fields. *Field Crops Res.* **2011**, *122*, 70–77. [CrossRef]
15. Santoso, H.; Gunawan, T.; Jatmiko, R.H.; Darmosarkoro, W.; Minasny, B. Mapping and identifying basal stem rot disease in oil palms in north sumatra with QuickBird imagery. *Precis. Agric.* **2011**, *12*, 233–248. [CrossRef]

16. Fisher, D.K.; Hinton, J.; Masters, M.H.; Aasheim, C.; Butler, E.S.; Reichgelt, H. Improving irrigation efficiency through remote sensing technology and precision agriculture in se Georgia. In Proceedings of the ASAE Annual International Meeting 2004, Otawa, ON, Canada, 1–4 August 2004; pp. 3461–3470.

17. Gago, J.; Douthe, C.; Coopman, R.E.; Gallego, P.P.; Ribas-Carbo, M.; Flexas, J.; Escalona, J.; Medrano, H. UAVs challenge to assess water stress for sustainable agriculture. *Agric. Water Manag.* **2015**, *153*, 9–19. [CrossRef]

18. Rossini, M.; Fava, F.; Cogliati, S.; Meroni, M.; Marchesi, A.; Panigada, C.; Giardino, C.; Busetto, L.; Migliavacca, M.; Amaducci, S.; et al. Assessing canopy PRI from airborne imagery to map water stress in maize. *ISPRS J. Photogramm. Remote Sens.* **2013**, *86*, 168–177. [CrossRef]

19. Gonzalez-Dugo, V.; Zarco-Tejada, P.; Nicolás, E.; Nortes, P.A.; Alarcón, J.J.; Intrigliolo, D.S.; Fereres, E. Using high resolution UAV thermal imagery to assess the variability in the water status of five fruit tree species within a commercial orchard. *Precis. Agric.* **2013**, *14*, 660–678. [CrossRef]

20. Launeau, P.; Kassouk, Z.; Debaine, F.; Roy, R.; Mestayer, P.G.; Boulet, C.; Rouaud, J.; Giraud, M. Airborne hyperspectral mapping of trees in an urban area. *Int. J. Remote Sens.* **2017**, *38*, 1277–1311. [CrossRef]

21. Torres-Sánchez, J.; López-Granados, F.; Serrano, N.; Arquero, O.; Peña, JM. High-throughput 3-D monitoring of agricultural-tree plantations with unmanned aerial vehicle (UAV) technology. *PLoS ONE* **2015**, *10*, e0130479. [CrossRef]

22. Kim, C.; Hong, S. Identification of tree species from high resolution satellite imagery by using crown parameters. Presented at the SPIE—The International Society for Optical Engineering 2008, Cardiff, Wales, UK, 15–18 September 2008. [CrossRef]

23. Molinier, M.; Astola, H. Feature selection for tree species identification in very high resolution satellite images. In Proceedings of the 2011 IEEE International Geoscience and Remote Sensing Symposium (IGARSS), Vancouver, BC, Canada, 24–29 July 2011; pp. 4461–4464.

24. Arockiaraj, S.; Kumar, A.; Hoda, N.; Jeyaseelan, A.T. Identification and quantification of tree species in open mixed forests using high resolution QuickBird satellite imagery. *J. Trop. For. Environ.* **2015**, *5*. [CrossRef]

25. Caughlin, T.T.; Rifai, S.W.; Graves, S.J.; Asner, G.P.; Bohlman, S.A. Integrating LiDAR-derived tree height and Landsat satellite reflectance to estimate forest regrowth in a tropical agricultural landscape. *Remote Sens Ecol. Conserv.* **2016**, *2*, 190–203. [CrossRef]

26. Hawryło, P.; Wezyk, P. Predicting growing stock volume of scots pine stands using Sentinel-2 satellite imagery and airborne image-derived point clouds. *Forests* **2018**, *9*, 274. [CrossRef]

27. Vierling, L.A.; Vierling, K.T.; Adam, P.; Hudak, A.T. Using satellite and airborne LiDAR to model woodpecker habitat occupancy at the landscape scale. *PLoS ONE* **2013**, *8*, e80988. [CrossRef] [PubMed]

28. González-Audícana, M.; Saleta, J.L.; Catalán, R.G.; García, R. Fusion of multispectral and panchromatic images using improved IHS and PCA mergers based on wavelet decomposition. *IEEE Trans. Geosci. Remote Sens.* **2004**, *42*, 1291–1299. [CrossRef]

29. Otazu, X.; González-Audícana, M.; Fors, O.; Núñez, J. Introduction of sensor spectral response into image fusion methods. Application to wavelet-based methods. *IEEE Trans. Geosci. Remote Sens.* **2005**, *43*, 2376–2385. [CrossRef]

30. Ma, L.; Li, M.; Ma, X.; Cheng, L.; Du, P.; Liu, Y. A review of supervised object-based land-cover image classification. *ISPRS J. Photogramm. Remote Sens.* **2007**, *130*, 277–293. [CrossRef]

31. Smits, P.C.; Dellepiane, S.G.; Schowengerdt, R.A. Quality assessment of image classification algorithms for land-cover mapping: A review and a proposal for a cost-based approach. *Int. J. Remote Sens.* **1999**, *20*, 1461–1486. [CrossRef]

32. Immitzer, M.; Atzberger, C.; Koukal, T. Tree species classification with random forest using very high spatial resolution 8-band WorldView-2 satellite data. *Remote Sens.* **2012**, *4*, 2661–2693. [CrossRef]

33. Whiteside, T.G.; Boggs, G.S.; Maier, S.W. Comparing object-based and pixel-based classifications for mapping savannas. *Int. J. Appl. Earth Obs. Geoinf.* **2011**, *13*, 884–893. [CrossRef]

34. Funck, J.W.; Zhong, Y.; Butler, D.A.; Brunner, C.C.; Forrer, J.B. Image segmentation algorithms applied to wood defect detection. *Comput. Electron. Agric.* **2003**, *41*, 157–179. [CrossRef]

35. Yu, Q.; Gong, P.; Clinton, N.; Biging, G.; Kelly, M.; Schirokauer, D. Object-based detailed vegetation classification with airborne high spatial resolution remote sensing imagery. *Photogramm. Eng. Remote Sens.* **2006**, *72*, 799–811. [CrossRef]

36. Kettig, R.L.; Landgrebe, D.A. Classification of multispectral image data by extraction and classification of homogeneous objects. *IEEE Trans. Geosci. Electron.* **1976**, *14*, 19–26. [CrossRef]

37. Hulet, A.; Roundy, B.A.; Petersen, S.L.; Jensen, R.R.; Bunting, S.C. An object-based image analysis of pinyon and juniper woodlands treated to reduce fuels. *Environ. Manag.* **2014**, *53*, 660–671. [CrossRef] [PubMed]

38. MacLean, M.G.; Congalton, R.G. Using object-oriented classification to map forest community types. In Proceedings of the American Society for Photogrammetry and Remote Sensing Annual Conference, Milwaukee, WI, USA, 1–4 May 2011.

39. Jiménez-Brenes, F.M.; López-Granados, F.; Castro, A.I.; Torres-Sánchez, J.; Serrano, N.; Peña, J.M. Quantifying pruning impacts on olive tree architecture and annual canopy growth by using UAV-based 3D modelling. *Plant Methods* **2017**, *13*, 55. [CrossRef] [PubMed]

40. International Olive Council (IOC). Olive Growing and Nursery Production. Available online: http://www.internationaloliveoil.org/projects/paginas/Section-a.htm (accessed on 26 November 2018).

41. Ayerza, R.; Steven Sibbett, G. Thermal adaptability of olive (*Olea europaea* L.) to the arid Chaco of Argentina. *Agric. Ecosyst. Environ.* **2001**, *84*, 277–285. [CrossRef]

42. DigitalGlobe. Information Products: Standard Imagery. Available online: https://dg-cms-uploads-production.s3.amazonaws.com/uploads/document/file/21/Standard_Imagery_DS_10-7-16.pdf (accessed on 20 November 2018).

43. Song, C.; Woodcock, C.E.; Seto, K.C.; Lenney, M.P.; Macomber, S.A. Classification and change detection using Landsat TM data: When and how to correct atmospheric effects? *Remote Sens. Environ.* **2001**, *75*, 230–244. [CrossRef]

44. Gonzalo, C.; Lillo-Saavedra, M. A directed search algorithm for setting the spectral–spatial quality trade-off of fused images by the wavelet Á Trous method. *Can. J. Remote Sens.* **2008**, *34*, 367–375. [CrossRef]

45. Ranchin, T.; Wald, L. Fusion of high spatial and spectral resolution images: The ARSIS concept and its implementation. *Photogramm. Eng. Remote Sens.* **2000**, *66*, 49–61.

46. Lillo-Saavedra, M.; Gonzalo, C. Spectral or spatial quality for fused satellite imagery? A trade-off solution using the wavelet Á Trous algorithm. *Int. J. Remote Sens.* **2006**, *27*, 1453–1464. [CrossRef]

47. Baatz, M.; Schäpe, A. Multiresolution segmentation: An optimization approach for high quality multi-scale image segmentation. In Proceedings of the 12th Symposium for Applied Geographic Information Processing (Angewandte Geographische Informationsverarbeitung XII) AGIT 2000, Salzburg, Austria, 5–7 July 2000.

48. Rouse, J.W., Jr.; Haas, R.; Schell, J.; Deering, D. Monitoring vegetation systems in the Great Plains with ERTS. *NASA Spec. Publ.* **1974**, *351*, 309–317.

49. Roujean, J.L.; Breon, F.M. Estimating PAR absorbed by vegetation from bidirectional reflectance measurements. *Remote Sens. Environ.* **1995**, *51*, 375–384. [CrossRef]

50. Hodgson, M.E. Reducing the computational requirements of the minimum-distance classifier. *Remote Sens. Environ.* **1998**, *25*, 117–128. [CrossRef]

51. Hecker, C.; Van Der Meijde, M.; Van Der Werff, H.; Van Der Meer, F.D. Assessing the influence of reference spectra on synthetic SAM classification results. *IEEE Trans. Geosci. Remote Sens.* **2008**, *46*, 4162–4172. [CrossRef]

52. Strahler, A.H. The use of prior probabilities in maximum likelihood classification of remotely sensed data. *Remote Sens. Environ.* **1980**, *10*, 135–163. [CrossRef]

53. Quinlan, R. *C4-5: Programs for Machine Learning*; Morgan Kaufmann: San Mateo, CA, USA, 1993.

54. Congalton, R.G. A review of assessing the accuracy of classifications of remotely sensed data. *Remote Sens. Environ.* **1991**, *37*, 35–46. [CrossRef]

55. Rogan, J.; Franklin, J.; Roberts, D.A. A comparison of methods for monitoring multitemporal vegetation change using thematic mapper imagery. *Remote Sens. Environ.* **2002**, *80*, 143–156. [CrossRef]

56. Wald, L. *Data Fusion. Definition and Architectures- Fusion of Images of Different Spatial Resolutions*; Presses de l'Ecole, Ecole des Mines de Paris: Paris, France, 2002; p. 200, ISBN 2-911762-38-X.

57. Gonzalo-Martín, C.; Lillo-Saavedra, M. Quickbird image fusion by a multirresolution-multidirectional joint image representation. *IEEE Lat. Am. Trans.* **2007**, *5*, 32–37. [CrossRef]

58. Foody, G.M. Status of land cover classification accuracy assessment. *Remote Sens. Environ.* **2002**, *80*, 185–201. [CrossRef]

59. Monserud, R.A.; Leemans, R. Comparing global vegetation maps with the kappa statistic. *Ecol. Model.* **1992**, *62*, 275–293. [CrossRef]

60. Pérez-Ortiz, M.; Gutiérrez, P.A.; Peña, J.M.; Torres-Sánchez, J.; Hervás-Martínez, C.; López-Granados, F. An experimental comparison for the identification of weeds in sunflower crops via unmanned aerial vehicles and object-based analysis. *Lect. Notes Comput. Sci.* **2015**, *9094*, 252–262. [CrossRef]

61. Castillejo-González, I.L.; Peña-Barragán, J.M.; Jurado-Expósito, M.; Mesas-Carrascosa, F.J.; López-Granados, F. Evaluation of pixel- and object-based approaches for mapping wild oat (*Avena sterilis*) weed patches in wheat fields using QuickBird imagery for site-specific management. *Eur. J. Agron.* **2014**, *59*, 57–66. [CrossRef]

62. Brodley, C.E.; Friedl, M.A. Decision tree classification of land cover from remotely sensed data. *Remote Sens. Environ.* **1997**, *61*, 399–409. [CrossRef]

63. Castillejo-González, I.L.; López-Granados, F.; García-Ferrer, A.; Peña-Barragán, J.M.; Jurado-Expósito, M.; de la Orden, M.S.; González-Audicana, M. Object- and pixel-based analysis for mapping crops and their agro-environmental associated measures using QuickBird imagery. *Comput. Electron. Agric.* **2009**, *68*, 207–215. [CrossRef]

64. García Torres, L.; Peña-Barragán, J.M.; López-Granados, F.; Jurado-Expósito, M.; Fernández-Escobar, R. Automatic assessment of agro-environmental indicators from remotely sensed images of tree orchards and its evaluation using olive plantations. *Comput. Electron. Agric.* **2008**, *61*, 179–191. [CrossRef]

65. Johnson, B.A.; Tateishi, R.; Hoan, N.T. Satellite image pansharpening using a hybrid approach for object-based image analysis. *ISPRS Int. J. Geo-Inf.* **2012**, *1*, 228–241. [CrossRef]

Article

Workflow to Establish Time-Specific Zones in Precision Agriculture by Spatiotemporal Integration of Plant and Soil Sensing Data

Elia Scudiero [1,2,*], Pietro Teatini [3], Gabriele Manoli [4], Federica Braga [5], Todd H. Skaggs [2] and Francesco Morari [6]

1 Department of Environmental Sciences, University of California Riverside, 900 University Ave., Riverside, CA 92521, USA
2 USDA-ARS, United States Salinity Laboratory, 450 West Big Springs Rd., Riverside, CA 92507, USA; todd.skaggs@ars.usda.gov
3 Department of Civil, Environmental, and Architectural Engineering (ICEA), University of Padua, Via Trieste 63, 35121 Padua, Italy; teatini@dmsa.unipd.it
4 ETH Zurich, Department of Civil, Environmental and Geomatic Engineering, Stefano-Franscini-Platz 5, 8093 Zürich, Switzerland; manoli@ifu.baug.ethz.ch
5 Institute of Marine Sciences (ISMAR), National Research Council, Arsenale Tesa 104, Castello 2737/F, 30122 Venezia, Italy; federica.braga@ve.ismar.cnr.it
6 Department of Agronomy, Food, Natural resources, Animals, and Environment (DAFNAE), University of Padua, Viale dell'Università 16, 35020 Legnaro, Italy; francesco.morari@unipd.it
* Correspondence: elia.scudiero@ars.usda.gov or elia.scudiero@ucr.edu; Tel.: +1-951-369-4847

Received: 20 September 2018; Accepted: 1 November 2018; Published: 7 November 2018

Abstract: Management zones (MZs) are used in precision agriculture to diversify agronomic management across a field. According to current common practices, MZs are often spatially static: they are developed once and used thereafter. However, the soil–plant relationship often varies over time and space, decreasing the efficiency of static MZ designs. Therefore, we propose a novel workflow for time-specific MZ delineation based on integration of plant and soil sensing data. The workflow includes four steps: (1) geospatial sensor measurements are used to describe soil spatial variability and in-season plant growth status; (2) moving-window regression modelling is used to characterize the sub-field changes of the soil–plant relationship; (3) soil information and sub-field indicator(s) of the soil–plant relationship (i.e., the local regression slope coefficient[s]) are used to delineate time-specific MZs using fuzzy cluster analysis; and (4) MZ delineation is evaluated and interpreted. We illustrate the workflow with an idealized, yet realistic, example using synthetic data and with an experimental example from a 21-ha maize field in Italy using two years of maize growth, soil apparent electrical conductivity and normalized difference vegetation index (NDVI) data. In both examples, the MZs were characterized by unique combinations of soil properties and soil–plant relationships. The proposed approach provides an opportunity to address the spatiotemporal nature of changes in crop genetics × environment × management interactions.

Keywords: remote sensing; proximal sensing; crop modeling; soil; plant; management zone; spatial variability; temporal variability; precision agriculture

---

1. Introduction

Crop yields and resource use efficiency (e.g., nutrients and water) have a strong spatial component, which can be observed over a wide range of scales, from regional to subfield [1]. At the field scale, yield variability in uniformly managed fields is often related to the spatial variability of soil properties and their impact on plant growth [2–6]. This variability can be addressed using precision agriculture practices [7], such as variable rate management (VRM) [8]. According to VRM principles, efficiency or crop production can be increased by varying agronomic inputs over a field according to varying soil and crop conditions [9]. Within a field, areas with similar soil properties (e.g., texture, topography, water holding capacity) that can be managed uniformly are commonly called management zones (MZs) [10]. To justify the subdivision of a field into MZs, there should be sizeable difference in soil properties between MZs [9,11]. The use of MZs for VRM has been shown to increase productivity, decrease costs, and/or reduce environmental impacts of agronomic practices [12,13]. Several authors have shown that within-MZ management should change over time [14–20], such an approach is often referred to as "dynamic VRM" [14].

Shanahan et al. [21], Long et al. [22], and Quebrajo et al. [23] indicated that coupling soil and plant data should increase the efficiency of MZ-based precision agriculture. Several sensor measurements can be used as indicators of soil spatial variability, including: apparent electrical conductivity, gamma-ray spectrometry, visible and near-infrared reflectance, and penetrometry [24–26]. Throughout the growing season, field-scale crop canopy measurements from near-ground (e.g., tractor mounted sensors) and remote sensing (e.g., unmanned aerial vehicles—UAV, satellites) can be used to provide information about plant health [27]. In particular, visible, near-infrared, and thermal sensing can be used to infer crop water status, nitrogen deficiency, and the effect of other biotic/abiotic stressors [28–31] and as an indicator of potential yield [16,32–35].

According to current practices, MZ-designs derived from soil maps, crop sensing, and/or historical yield maps are generally designed once and then implemented year after year. Nevertheless, several authors have reported inconsistent benefits for this precision management strategy [21,22]. Spatial patterns in yield tend to change from year-to-year, mostly because of changes in meteorological conditions [12,36–38]. In other words, the spatial patterns of most soil properties are fairly stable in time, but at different times plants may be limited in different ways at the same location because of the influence of transient factors affecting the soil–plant relationship, such as meteorological factors and agronomical management [12,38,39].

Sadler et al. [40] indicated that there is a need for accurate and inexpensive systems to delineate dynamic management zones, obtained by sensing within-field variability in real time, so that agricultural management can be controlled adaptively. Recent research strongly suggested that MZ designs should change over time, both intra- and inter-seasonally [19,20,38,41–44]. Myers [45] formally justified the need for time-specific and spatially dynamic VRM through the "fundamental theorem of precision agriculture production" where *Crop Yield* is a function of *Genetics* × *Environment* × *Management* × *Space* × *Time* interactions. According to this theorem, the spatiotemporal variability of crop performance should be addressed by adjusting the agronomic prescriptions over time and space. Several examples of protocols and data analysis workflow for static soil and/or plant-based MZ delineation are present in the scientific literature [46–49]. To our best knowledge, time-specific MZ delineation based on soil and in-season plant information has not been commonly discussed in the literature. Particularly, there is a lack of protocols and analytical workflow that farm managers, agricultural consultants, and scientists can use to take advantage of free/inexpensive in-season crop information (e.g., from UAV, Sentinel 2 satellite) and high-resolution soil maps.

We aim to present a novel workflow for the selection of time-specific MZs according to in-season spatial measurements of crop growth status and its relationship with high-resolution soil spatial information. The MZ should identify areas with homogeneous and unique (within a single field) soil–plant relationships. The MZ-delineation workflow will be described in detail. We will also provide two examples on how to implement the workflow. The first example is based on synthetic data. The second example uses data from a maize (*Zea mays* L.) field in northeastern Italy.

2. Materials and Methods

2.1. Time-Specific MZ-Delineation Workflow

The time-specific MZ delineation can be implemented as follows:

- STEP 1. Soil and time-specific plant spatial information acquisition, pre-processing, interpretation, and interpolation
- STEP 2. Time-specific sub-field soil–plant modeling
- STEP 3. Time-specific MZ delineation with cluster analysis
- STEP 4. Evaluation/interpretation of time-specific MZ design

2.1.1. Soil and Time-Specific Plant Spatial Information

In STEP 1, high-resolution spatial measurements for target soil properties and in-season plant-canopy information, such as canopy reflectance, are acquired, pre-processed, interpreted, and interpolated. In-season plant-canopy reflectance is acquired and used as an indication of crop status. Soil spatial information is used to interpret the crop canopy measurements.

Soil sensor data acquisition should be carried out according to established protocols [50–52] to increase the accuracy and consistency of the survey across large areas. Attention should be paid to selecting those sensors which represent the spatial variability of soil properties known or believed to influence a crop at the site of interest [53,54]. For information on sensor data pre-processing (e.g., conversion of spatial coordinates, removals of outliers), readers are referred to the first protocol step of Córdoba, Bruno, Costa, Peralta, and Balzarini [49]. Subsequent to the sensor surveys, soil sampling should be carried out across the root zone (e.g., 0–1.2 m) to calibrate/interpret the sensor readings [50]. Sensor-directed sampling schemes can be used to minimize the number of sampling sites [55,56]. Exploratory analyses, such as correlation analysis and principal component analysis (PCA) should be carried out to investigate the relationships between soil sensor and laboratory soil analyses. The strength of the relationships between collocated soil sensor values and laboratory soil analyses should be investigated. If these relationships are moderately to very strong, the sensor data can be considered as an indicator of spatial variability of the target soil properties. Then, sensor data can be used to generate maps of the selected soil properties [57,58]. Soil maps should be generated only if acceptable prediction errors [59] are obtained. Alternatively, for weak to moderately strong relationships, the soil sensor maps should be used as qualitative indicator of soil spatial variability. Córdoba, Bruno, Costa, Peralta, and Balzarini [49] describe how to practically process interpolated data to obtain a raster of desired block support (e.g., of the same resolution chosen for the MZ design) in the second step of their MZ-delineation protocol.

Free (e.g., Sentinel 2 satellite) or inexpensive (e.g., from UAV) crop measurements are available throughout the growing season with moderately-high temporal resolution. For example, the Sentinel 2 satellite from the European Space Agency provides multi-spectral canopy reflectance at the 10×10-m resolution, with a 5-day revisit time, free of charge. Remote sensing of crop canopy data can be used to calculate vegetation indices [34]. Point measurements, such as those from tractor-mounted active spectrometers [30], should be pre-processed and interpolated similarly to soil sensing measurements. High-resolution raster data, such as that from satellite imagery, may need to be re-gridded and scaled to the selected block support.

2.1.2. Time-Specific Sub-Field Soil–Plant Modeling

In STEP 2, soil information from STEP 1 is used to interpret in-season (i.e., time-specific) measurements of crop status. The interpretation of crop canopy sensing measurements is not straightforward—as they are influenced by species $\times$ growth-stage $\times$ stress levels $\times$ soil background interactions [60–62].

Moving window spatial regression modeling, such as geographically-weighted regression (GWR) [63,64], can be used to understand the local (i.e., sub-field scale) variation in the plant–soil relationship. With GWR, a regression is run for each grid location, rather than for the whole study area [65]. Soil map(s) are used as the independent (explanatory) variable(s) and the in-season crop status maps are the dependent variable. When multiple explanatory variables are available, one should consider standardizing them. The GWR allows non-stationarity of the regression equation parameters a (e.g., intercept, slope), estimating their values at each location i. For a dependent variable y the equation reads:

$$y_i = a_{0i} + a_{1i}x_{1i} + \ldots + a_{ki}x_{ki} + \varepsilon_i \tag{1}$$

where ε, is a random error term, a_0 is the regression intercept, and a_1 to a_k are the regression coefficient(s) for each explanatory variable(s) x_1 to x_k. The spatial variability of the soil–plant relationship can be described with the maps of local a_1 to a_k coefficients (i.e., the regression slope[s]). Maps of GWR slope coefficient(s) show the spatial variability of the impact (i.e., sensitivity) of the explanatory variable on the regression [65].

In the GWR framework, spatial weighting is determined by incorporating all the dependent and explanatory variables falling within a geographical kernel of each target feature [65]. The values of the regression parameters and goodness-of-fit of the GWR depend on how the kernel size is chosen [66]. Maps of the estimated dependent variable, local coefficient of determination (R^2), and local Pearson correlation coefficient r can be generated with the GWR [65]. The GWR is available in commercial GIS software platforms (e.g., ArcMap's Spatial Statistics package [65], version 10.5.1; ESRI, Redlands, CA, USA), and freeware (e.g., spgwr package in R, version 3.4.1; the R Foundation for Statistical Computing Platform, Vienna, Austria).

2.1.3. Time-Specific MZ Delineation with Cluster Analysis

In STEP 3, the MZ delineation is carried out using the soil map(s) and the time-specific GWR slope maps from STEP 2 as ancillary variables. As indicated by Córdoba, Bruno, Costa, Peralta, and Balzarini [49], fuzzy c-means (also known as "k-means") unsupervised clustering algorithms [67] can be employed to classify the data into MZs. ArcMap's Grouping Analysis tool (e.g., [68]), the Management Zone Analyst software [69] or the EZZone online tool [70] can be used to delineate MZs. Several MZ designs can be tested. The optimum number of MZs can be identified using cluster validity functions, including: the Calinski–Harabasz criterion [71], the fuzziness performance index [67], the normalized classification entropy index [67,72], and the Jenks optimization method [73]. The Calinski–Harabasz criterion (*CHC*), also known as pseudo F-statistic, describes the ratio between within-MZ similarity and between-MZ differences. It is defined as:

$$CHC = \frac{BMZSS/(MZn - 1)}{WMZSS/(N - MZn)} \tag{2}$$

where N is the number of data points, MZn is the number of considered MZs, $BMZSS$ is the between-MZ sum of squares, and $WMZSS$ is the within group sum of squares. The larger the value of the CHC the higher are the within-MZ homogeneity and between-MZ differences. Finally, one may consider smoothing the fuzzy c-means clustering results to reduce zone fragmentation [49,70].

2.1.4. Time-Specific MZ-Design Quality Control and Interpretation

In STEP 4, the quality of the MZ design from STEP 3 should be checked. Each MZ should identify a unique combination of soil and soil–plant relationship characteristics. To infer differences across MZs, parametric analysis of variance or the Kruskal–Wallis (KW) rank test [74] can be used. The KW test is a nonparametric analysis assessing if samples originate from the same distribution. The test can be used as alternative to the standard analysis of variance when assumptions for parametric testing are not met.

2.2. Synthetic Data Example

The proposed workflow can be demonstrated using synthetic data for soil clay content (percentage) and normalized difference vegetation index (NDVI) [75].

A spatial random field for clay content was generated using the RFsimulate function in the RandomFields package in R. The function estimated a Gaussian random field (RPgauss function) having an exponential covariance function (RMexp function with variance = 25 and scale = 300), a nugget effect (RMnugget function with variance = 5), and a pure trend model with covariance 0 (RMtrend function with mean = 15). The simulations were carried out over a square of size 250 × 250 cells. The random field was then converted into a 250 × 250-pixel raster with pixel size = 1 m using the function raster from the raster package in R. The clay content raster was then exported from R to a text file using the writeRaster function. We refer to this raster as the true clay content (*TCC*) map.

Next, we mimicked a field procedure for generating a soil clay content map, using *TCC* as the true, unknown, high resolution clay content. Accurate approximations of true soil properties can be obtained when high-resolution covariates, such as those from proximal-soil sensing [76], are available [77]. Soil sensor surveys can be calibrated to estimate the spatial variability of a target soil property by using laboratory measurements on collocated soil cores [58]. Heggemann et al. [78] reported that gamma-ray sensor readings can be calibrated to predict texture values with high accuracy (up to ~95 percent of observed variance in soil texture). We mimicked a typical sensor survey (e.g., Lesch [55]) in which 1870 data points (average nearest neighbor distance = 3.2 m) were spread across 24 nearly-parallel transects (average nearest neighbor distance = 9.9 m). Values from the *TCC* map were extracted at the sensor survey locations. A random error having mean = 0 and variance equal to 5 percent of the extracted *TCC* values was added to the sensor data. This simulated a realistic sensor calibration with goodness-of-fit with a R^2 close to 0.95. The spatial autocorrelation of the calibrated sensor measurements was described with a spherical semivariogram having range = 50.3 m, and nugget equal to 67% of the total sill. This spatial structure was similar to those reported by other authors in a 5-ha clay loam field in Italy [79] and in a cluster of fields with contrasting soil properties (from clay to gravelly) in Germany [80]. These calibrated sensor measurements were then interpolated using simple kriging in with ArcMap's Geostatistical Analyst package. The resulting sensor-derived clay content (*SCC*) map was retained for further analyses.

NDVI is a vegetation index calculated from surface reflectance in the red (RED) and near-infrared (NIR) regions of the electromagnetic spectrum. It is defined as:

$$\text{NDVI} = \frac{(\text{NIR} - \text{RED})}{(\text{NIR} + \text{RED})} \tag{3}$$

NDVI ranges from -1 to $+1$. NDVI for agricultural crops usually ranges from ~0.1 to ~0.9, with lush vegetation generally having high NDVI [81,82]. The NDVI information was simulated with the understanding that, often, the spatial relationship between remote sensing canopy measurement and collocated soil properties has both deterministic and spatial random components [83–85]. The deterministic component of the relationship can be a linear model between the soil property and the available remote sensing plant information [84]. The spatial random component is often equal to the field of spatially correlated residuals from the deterministic linear model [57,83]. We simulated the NDVI as follow:

$$\text{NDVI} = S \times (TCC + SpERR) + O \tag{4}$$

where *SpERR* was a spatial error raster, *S* was a scaling factor = 0.01, and *O* was an offset coefficient = 0.4. *SpERR* was generated in R, using the RandomFields package. We estimated a Gaussian random field based on a model with exponential covariance function (variance = 7.5 and scale = 50) and a nugget effect (variance = 5).

A GWR with 30-m bandwidth size was used to model the spatial variability of the synthetic NDVI image using the *SCC* map as explanatory variable. Then, MZs were delineated with a fuzzy c-mean unsupervised cluster analysis. Clustering was carried out in ArcMap using the Grouping Analysis tool from the Spatial Statistics package. The SCC and the time-specific GWR slope maps were used for the cluster analysis. No spatial constraints were set for the cluster analysis: the covariate data points could be grouped according to their value, disregarding the values of their geographical coordinates (i.e., points did not need to be neighbors to be part of the same MZ). The classifications were carried out for scenarios outputting two, three, and four MZs. Then, the optimal number of MZs was identified using the *CHC*. Differences in clay (from the *SCC* map), NDVI, and GWR slope across MZs were investigated with the KW test using STATISTICA (version 12, StatSoft Inc., Tulsa, OK, USA).

2.3. Field Data Example

2.3.1. Site Description

Another example is given using experimental data from researchers at the University of Padua, Italy [86–88]. Here, we describe two time-specific MZ designs for a maize field based on remote sensing imagery acquired during the late kernel blister stage (R-2), in two consecutive growing seasons.

The study site (Figure 1) is a 21-ha field located at Chioggia, Venice, Italy, along the southern margin of the Venice Lagoon (45°10′7.0″ N, 12°13′55.0″ E). Previous studies [86–89] discussed the relationship between soil and maize yield at this site. The site lies below average sea level, and the groundwater level is kept fairly shallow (approximately between -0.6 to -1.5 m below ground level) by a pumping station [87] to promote sub-irrigation [88]. The soil in the area is classified as Molli-Gleyic Cambisol [90] with two coarsely-textured paleochannels crossing it with SW-NE direction (Figure 1). The field is affected by soil salinity due to its proximity to the Venice Lagoon [86].

Here, we analyzed the two maize growing seasons (April to September), 2010 and 2011. The two growing seasons differed greatly in terms of meteorology. Compared to the average April to September rainfall from 1993 to 2012, 2010 was rainy (in the upper third quartile, 540 mm) and 2011 was a drought year (in the first quartile, 200 mm) [88]. The daily average reference evapotranspiration was 4.01 mm day^{-1} in 2010 and 4.42 mm day^{-1} in 2011. For the entire growing season, the reference evapotranspiration was 569 mm in 2010 and 672 mm in 2011 [88]. Agronomic management (base-dressing of 64 kg N ha^{-1}, and 94 kg P_2O_5 ha^{-1} and urea top-dressing of 184 kg N ha^{-1}) and maize hybrid (PR32P26, Pioneer Hi-Bred Italia, Gadesco-Pieve Delmona, Italy) were the same in the two years [86].

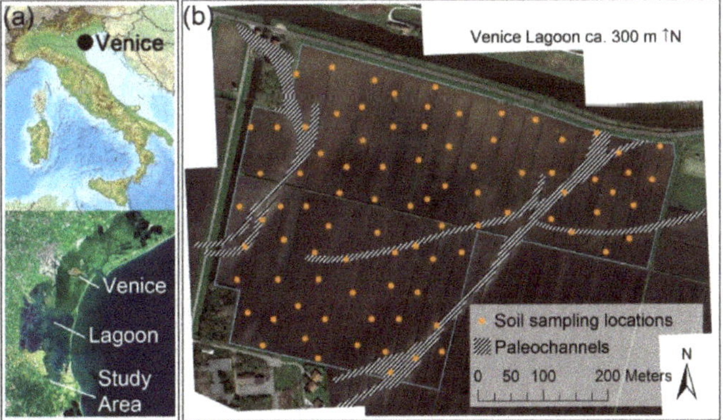

Figure 1. (**a**) Geographical location of the study area relatively to the Venice Lagoon, Italy, and (**b**) locations of soil sampling locations and coarsely textured paleochannels. Modified after Scudiero, Teatini, Corwin, Dal Ferro, Simonetti, and Morari [88].

2.3.2. Soil Maps and Plant Spatiotemporal Information

Geospatial measurements of soil apparent electrical conductivity (EC_a) were used as an indicator of soil spatial variability at the study site. EC_a over the 0–1.5 m soil profile (EC_a Deep) was measured in April of 2011 across the site with a frequency-domain electromagnetic induction sensor (CMD-1, GF Instruments, Brno, Czech Republic) at 20,470 geo-referenced locations. Soil samples were taken at 91 locations in May 2010 (Figure 1) down to 1.2 m with 0.3-m increments. Here we discuss salinity and texture for the 0–1.2-m soil profile. Texture was measured with a Mastersizer 2000 (Malvern Instruments Ltd., Great Malvern, UK) and salinity was measured as $EC_{1:2}$ (i.e., the electrical conductivity of a 1:2 soil–water extract) [91]. Principal component analysis (PCA) carried out with STATISTICA was used to visually assess similarities and differences between the EC_a Deep dataset and the selected soil properties [92]. The point measurements were preprocessed with a procedure comparable to that of Córdoba, Bruno, Costa, Peralta, and Balzarini [49] and spatially interpolated with ordinary kriging following Scudiero, Teatini, Corwin, Deiana, Berti, and Morari [86] on a 10 × 10-m block support—the desired support for MZ at the site.

Two remotely sensed measures of crop status were available. On 31 July 31 2010 (10:39:40 UTC) and 9 July 2011 (10:47:30 UTC) WorldView-2 (DigitalGlobe, Westminster, CO, USA) satellite scenes were acquired over the study area. For both years, acquisition dates corresponded to late R-2 [88]. Pre-processing procedures, including sensor calibration, atmospheric correction, and radiometric normalization, were applied according to Vicente-Serrano et al. [93]. Radiometric calibration was required to convert digital numbers to top-of-atmosphere radiances [W m^{-2} sr^{-1} μm^{-1}], using the absolute radiometric calibration factors and effective bandwidths for each band, according to the satellite data provider [94]. No topographic correction was applied to the images because the study area is relatively flat and the solar incident angles (28.11° in 2010; 23.58° in 2011) were quite similar at the two acquisition times. The top-of-atmosphere radiance was then transformed to surface reflectance through the *6S* code [95]. The values of aerosol optical thickness (AOT) at 550 nm collected from a nearby AERONET (Aerosol Robotic Network, https://aeronet.gsfc.nasa.gov/) [96] station (45°18′50.0″ N, 12°30′29.9″ E) were used as input for *6S*. The AOT measurements were obtained simultaneously to the satellite overpasses. WorldView-2 reflectance has 2 × 2-m spatial resolution over eight spectral bands, at wavelengths spanning from 400 to 1040 nm. The red band (RED, 630–690 nm) and the near-infrared band at 770–895 nm where used to calculate NDVI according to Equation (3). The NDVI maps were aggregated (i.e., the coarsened pixel is the average of the pixels within the aggregated cell) to the 10 × 10-m cell size, as suggested by Córdoba, Bruno, Costa, Peralta, and Balzarini [49].

2.3.3. Time-Specific Spatial Soil–Plant Modeling

The time-specific relationship between soil properties and in-season NDVI were described using geographically weighted regressions (GWRs). The GWRs were carried out using the NDVI maps as the dependent variable and spatial soil information (i.e., EC_a Deep) as the independent variable. GWRs were carried out with ArcMap using an adaptive kernel (i.e., moving window) of 70 neighbors. For locations with significant GWR, the observed-estimated NDVI relationship was used as indication of goodness-of-fit of the GWR models. The GWR slope map was selected as indicator of soil–plant relationship type.

2.3.4. Delineating MZs with Cluster Analysis

Clustering was carried out using the Grouping Analysis tool in ArcMap. Spatial soil information (i.e., EC_a Deep) and the time-specific GWR slope map were used for the cluster analysis. No spatial constraints were set for the cluster analysis. The classifications were carried out for scenarios outputting three, four, and five MZs. Then, the optimal number of MZs was identified using the *CHC*. To compare the proposed time-specific soil–plant based MZ delineation with a more traditional, static MZ design, EC_a Deep alone was used to select a static MZ design.

2.3.5. Evaluation and Interpretation of MZ Design

Differences across MZs in the two years were investigated using the KW test in STATISTICA. The following variables were tested for differences across MZs: EC_a Deep, GWR slope, GWR local r, $EC_{1:2}$, and sand and clay contents. Additionally, in-season potential yield estimations were used to evaluate the MZ designs. Appendix A reports on how yield maps (obtained from Scudiero, Teatini, Corwin, Dal Ferro, Simonetti, and Morari [88]) and NDVI measurements were used to calculate yield prediction maps.

3. Results and Discussion

3.1. Synthetic Data Example

Every time the crop status is mapped during the season (e.g., every week), a new set of MZ is delineated (i.e., time-specific MZ design). Clearly, this applies only for crop growth stages when crop canopy plays a relevant role in determining reflectance readings. For soil tillage, sowing, and early vegetation stages when surface reflectance is mostly determined by soil, a static MZ delineation approach based on the spatial variability of soil properties (e.g., [9]) may be more adequate.

Figure 2 outlines the proposed analytical protocol to delineate time-specific MZs from soil information and in-season crop information using the synthetic dataset. In Step 1 of the protocol, the soil and plant information were acquired and pre-processed. The clay content (percentage) map had mean = 18.6%, standard deviation = 4.5%, minimum = 9%, and maximum = 30.1%. The NDVI map had mean = 0.58, standard deviation = 0.04, minimum = 0.45, and maximum = 0.69. The NDVI values are typical of vegetative stages for crops such as maize [97] and soybean (*Glycine max* (L.) Merr.) [98]. A simple linear model between the two maps had $R^2 = 0.61$, with an intercept of 0.44 and a positive slope of 0.007. Similar goodness-of-fit values between soil properties and vegetation reflectance were reported by Gomez et al. [99] for clay and by Gomez et al. [100] for soil organic carbon. In Step 2, the clay map was used as the explanatory variable in a GWR with NDVI as dependent variable. Through the GWR analysis, clay explained 94.5% of the observed variance of NDVI. Such high goodness-of-fit can be found in real-world data: Scudiero, Corwin, Wienhold, Bosley, Shanahan, and Johnson [54] observed R^2 ranging between 0.83 and 0.94 for their GWR analyses between soil maps and remotely sensed winter wheat (*Triticum aestivum* L.) canopy reflectance. The GWR slope had mean = 0.009, standard deviation = 0.008, minimum = -0.018, and maximum = 0.038. In Step 3, the clay and the GWR slope maps were used to delineate MZs via unsupervised fuzzy c-means clustering. The *CHC* indicated that the best number of MZs was three.

In Step 4, the MZ-design was evaluated and interpreted. The three MZs identified unique combinations of clay content and GWR slope values. MZ 1 was characterized by low NDVI, low clay content, and large GWR slope values (indicating high sensitivity of NDVI to clay). MZ 2 was characterized by high NDVI, high clay, and moderately low GWR slope values. MZ 3 was characterized by low NDVI, low clay content, and the smallest GWR slope values.

Step 1: Soil and time-specific plant spatial information processing

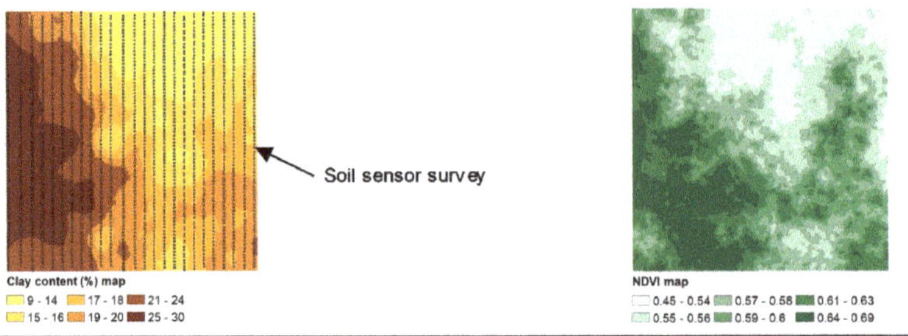

Step 2: Time-specific soil–plant modeling

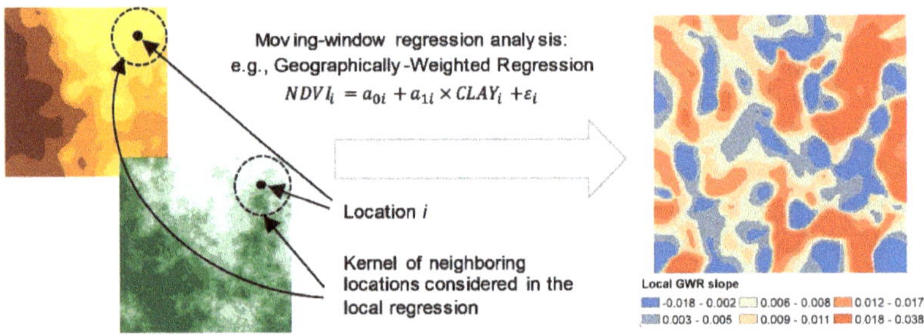

Step 3: Time-specific MZ delineation

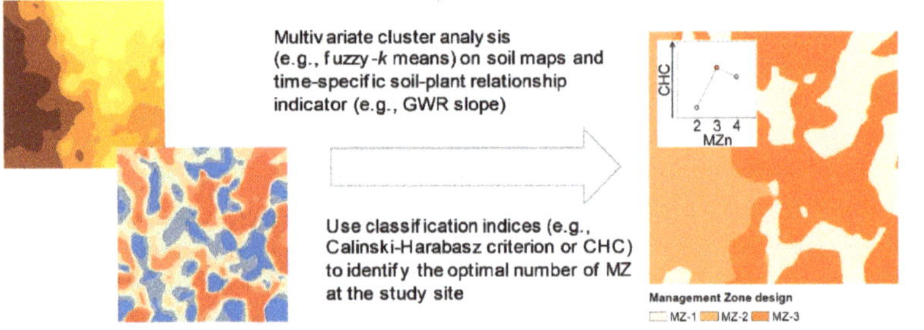

Step 4: Evaluation & interpretation of MZ-design

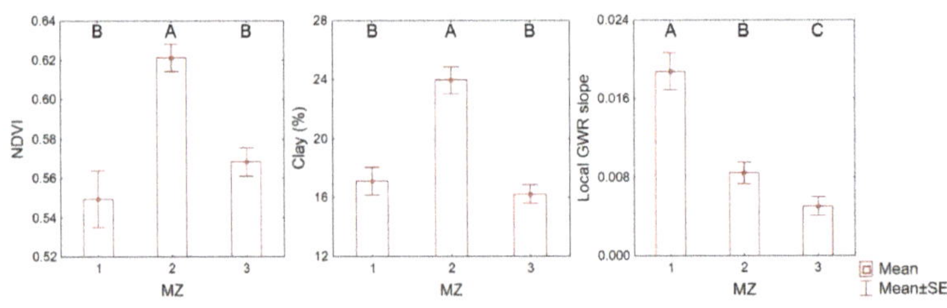

Figure 2. Diagram outlining the proposed analytical protocol to delineate time-specific management zones (MZs) from soil and (in-season) crop information. In Step 3, MZn refers to the number of considered management zones.

3.2. Field Data Example

3.2.1. Soil and Plant Information

Figure 3a shows an example of short-scale soil spatial variation at the site. Figure 3b depicts the frequency distribution of measured soil salinity. At this site, soils with $EC_{1:2}$ >0.79 dS m^{-1} should be considered salt affected, with $EC_{1:2}$ >1.4 dS m^{-1} indicating strongly saline soils [88]. Figure 3c reports the observed texture at the 91 soil sampling locations. Soil texture is predominantly in the loam and sandy loam classes. Few samples are in the clay loam and loamy sand class.

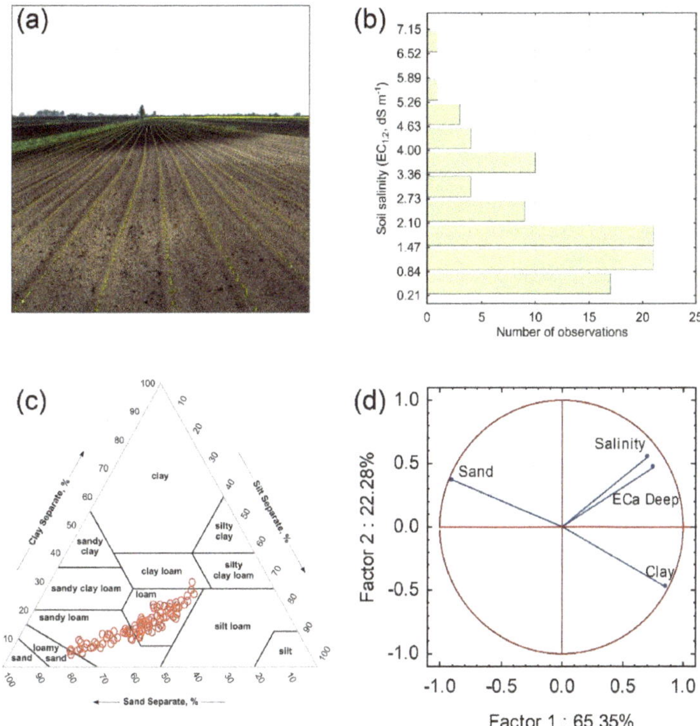

Figure 3. (**a**) Photograph showing the sharp textural change at the site (light vs. dark colors); (**b**) histogram for measured soil salinity (EC$_{1:2}$); (**c**) soil textural triangle; and (**d**) bi-plot of selected variables (soil apparent electrical conductivity [EC$_a$ Deep], salinity, clay, and sand) on the two larger factors in the principal components analysis.

The EC$_a$ Deep map is shown in Figure 4a. Descriptive statistics for the map are reported in Table 1. EC$_a$ Deep showed a Pearson correlation coefficient of 0.89 with salinity, −0.51 with sand, and 0.39 with clay. The PCA extracted two factors (with eigenvalues of 2.61 and 0.89) explaining 65.3% (first component) and 22.3% (second component) of the total variance in the dataset. The bi-plot in Figure 3d—see Abdi and Williams [92] for interpretation guidelines—indicates that the first component contrasts the positive contribution of clay content and salinity on the EC$_a$ Deep measurements with the negative correlation between sand content and EC$_a$ Deep. In Figure 3d, clay content was not clustered with salinity and EC$_a$ Deep, indicating that the three variables were not collinear, yet negatively correlated with sand content. According to the correlation and PCA analyses, high values of EC$_a$ Deep were mainly interpreted as an indication of high salinity and fine soil texture. EC$_a$ readings larger than 1–2 dS m^{-1} are most likely due to high soil salinity rather than other edaphic factors contributing to soil conductivity [101,102]. Conversely, lower EC_a Deep was interpreted as an indication of coarser soils with low salinity. Note that Scudiero, Teatini, Corwin, Deiana, Berti, and Morari [86] reported that the spatial patterns of EC_a Deep remained stable over time at the study area.

The two in-season NDVI maps for the late R-2 growth stage are reported in Figure 4b (2010) and Figure 4e (2011). Table 1 reports the descriptive statistics for the two NDVI maps. The 2010 NDVI was more heterogeneous (mean = 0.79, standard deviation = 0.046) than that of 2011 (mean = 0.83, standard deviation = 0.018). During the R-2 growth stage, environmental stressors (e.g., drought, heat, severe nutrient deficiency) may prevent kernels from developing properly (e.g., kernel abortion at the ear tip) [103,104].

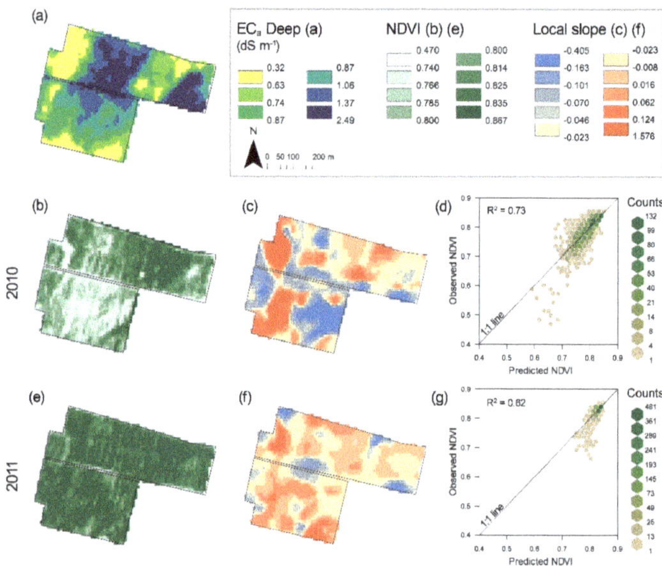

Figure 4. Maps of: (**a**) apparent electrical conductivity of the 0–1.5-m soil profile (EC_a Deep); Normalized Difference Vegetation Index (NDVI) for (**b**) 2010 and (**e**) 2011; local geographically-weighted regressions (GWRs) slope coefficient for (**c**) 2010 and (**f**) 2011; and observed versus estimated NDVI relationship for (**d**) 2010 and (**g**) 2011 on significant local GWRs.

Table 1. Descriptive statistics for the soil apparent electrical conductivity map (EC_a Deep), and the 2010 and 2011 Normalized Difference Vegetation Index (NDVI) maps.

Variable	Mean	Median	Minimum	Maximum	Standard Deviation
EC_a Deep (dS m^{-1})	0.96	0.86	0.32	2.49	0.36
2010 NDVI	0.790	0.80	0.470	0.867	0.046
2011 NDVI	0.829	0.83	0.664	0.858	0.018

3.2.2. Plant–Soil Relationship at Different Time Points

Figure 4 summarizes the GWR analysis between soil EC_a Deep and in-season NDVI for 2010 and 2011. For 2010, 1149 cells (62.5% of total) were characterized by a significant ($p < 0.05$) local Pearson correlation coefficient r. For 2011, 56.8% of the cells had a significant local r. At the locations with significant local r, the observed-estimated NDVI relationship was characterized by $R^2 = 0.73$ in 2010 (Figure 3d) and $R^2 = 0.62$ in 2011 (Figure 3g). There was a strong negative relationship in both years ($r = -0.36$ in 2010 and $r = -0.43$ in 2011) between the significant local r and EC_a Deep maps. The slope of this relationship was steeper in 2011 (-0.46 with standard error = 0.027) than in 2010

(−0.38 with standard error = 0.026). This may indicate that low EC_a Deep (e.g., indicating coarse texture) was a greater constrain in 2011 than 2010. Scudiero, Teatini, Corwin, Dal Ferro, Simonetti, and Morari [88] showed that water stress in areas with coarser soil texture is particularly limiting in dry years, such as 2011.

In areas where the significant local r values were consistently negative in both years, the EC_a Deep averaged 1.17 dS m^{-1} (standard deviation = 0.39 dS m^{-1}), whereas, in areas with negative local r only in 2010, it averaged 0.94 dS m^{-1} (standard deviation = 0.28 dS m^{-1}). This suggested that only very high EC_a Deep, which indicates high salinity, consistently limited crop growth in the two years. As discussed by other authors [105–107], the effects of high salinity on crops are generally stable throughout different growing seasons.

The GWR slope maps (Figure 4c for 2010, Figure 4f for 2011) changed remarkably between the two years, indicating a change in the sensitivity of plant NDVI to changes in soil properties. Changes of crop growth and yield spatial patterns are expected between seasons having widely different meteorology, as discussed by Maestrini and Basso [38] and McBratney, Whelan, Ancev, and Bouma [37].

3.2.3. Time-Specific MZ Delineation

The GWR slope and EC_a Deep maps were used to drive the delineation of in-season site-specific MZs. The *CHC* index indicated that a five-MZ configuration was optimal for 2010. The *CHC* was 1901.3 for three MZs, 1979.7 for four MZs, and 2002.7 for five-MZs. In 2011, the *CHC* index indicated that a four-MZ configuration was optimal. *CHC* was 1248.9 for three MZs, 1383.4 for four MZs, and 1374.0 for five MZs. The two MZ delineations are shown in Figure 5a (for 2010) and Figure 5b (for 2011).

Overall, 45.1% of the cells were classified in the same MZ at both times. The consistency rate was 94.3% for MZ I, 75.6% for MZ II, 20.1% for MZ III, and 48.5% for MZ IV. Conversely, changes in local GWR slope led to 54.9% of the cells being assigned to a different time-specific MZ. As reported by Maestrini and Basso [38], it is reasonable to expect portions of a field to have stable (e.g., consistently high yields) and unstable crop outputs. Management in the unstable areas should be addressed within each growing season to meet desired agronomic output goals [18,38]. The proposed methodology helps characterizing time-specific changes in the crop output according to soil spatial variability.

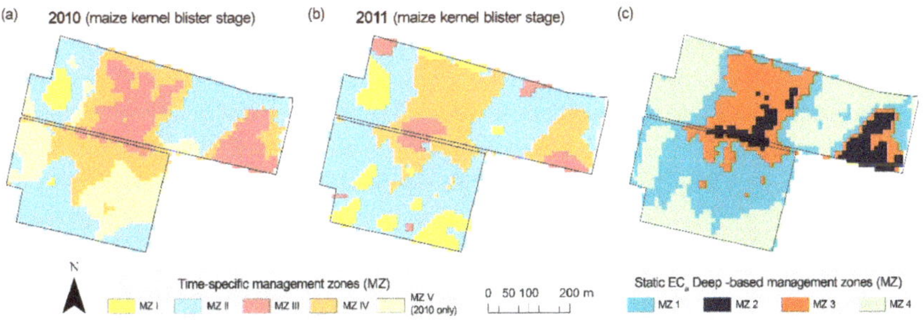

Figure 5. In-season delineation of time-specific management zones (MZs) for maize late kernel blister stage of (**a**) 2010 and (**b**) 2011. (**c**) Static MZ designs using soil apparent electrical conductivity for the 0–1.5 m soil profile (EC_a Deep) only.

3.2.4. Time-Specific MZ-Design Quality Control and Interpretation

In both years, each time-specific MZ was characterized by unique combinations of the variables used in MZ delineation. Moreover, the time-specific MZs differed greatly in terms of EC_a Deep, significant GWR r and slope, salinity, and texture (Table 2). Note that the local GWR r and slope, which can be used to interpret the EC_a Deep–NDVI relationship, remained fairly consistent over time at the different MZs.

Table 2. Count (n), mean, and standard deviation (Std. Dev.) of apparent electrical conductivity for the 0–1.5-m soil profile (EC_a Deep), local regression slope and Pearson correlation coefficient (r) obtained from significant geographically weighted regressions (GWR), soil salinity ($EC_{1:2}$), sand and clay contents, and potential yield estimations at each time-specific management zone (MZ) for 2010 and 2011 (n.a. = not available; NA = not assessed).

Variable (Unit)	MZ	2010				2011			
		n	Mean	Std. Dev.	KW [1]	n	Mean	Std. Dev.	KW [1]
	I	53	0.44	0.10	e	263	0.65	0.21	d
	II	779	0.68	0.12	d	1164	0.78	0.15	c
EC_a Deep (dS m^{-1})	III	402	1.54	0.19	a	146	1.24	0.45	a
	IV	513	1.09	0.12	b	574	1.38	0.23	b
	V	400	0.80	0.12	c	n.a.	n.a.	n.a.	n.a.
	I	53 (100.0% [2])	0.790	0.316	a	221 (84.0% [2])	0.073	0.040	a
	II	371 (47.6% [2])	0.065	0.113	b	488 (41.9% [2])	−0.024	0.028	b
GWR slope	III	334 (83.1% [2])	−0.069	0.069	c	146 (100.0% [2])	−0.132	0.050	c
	IV	294 (57.3% [2])	−0.048	0.055	c	364 (63.4% [2])	−0.018	0.029	b
	V	397 (99.3% [2])	−0.176	0.092	d	n.a.	n.a.	n.a.	n.a.
	I	53 (100.0% [2])	0.587	0.102	a	221 (84.0% [2])	0.433	0.126	a
	II	371 (47.6% [2])	0.028	0.411	b	488 (41.9% [2])	−0.279	0.285	b
GWR local r	III	334 (83.1% [2])	−0.439	0.239	c	146 (100.0% [2])	−0.597	0.190	d
	IV	294 (57.3% [2])	−0.399	0.281	c	364 (63.4% [2])	−0.308	0.349	c
	V	397 (99.3% [2])	−0.537	0.150	d	n.a.	n.a.	n.a.	n.a.
	I	1	0.21	NA	ab	11	1.73	1.33	b
	II	32	1.38	0.89	b	49	1.49	1.01	b
$EC_{1:2}$ (dS m^{-1})	III	17	3.67	1.46	a	3	3.16	1.32	ab
	IV	26	2.28	1.18	a	28	3.03	1.45	a
	V	15	1.35	0.85	b	n.a.	n.a.	n.a.	n.a.
	I	1	71.1	NA	ab	11	54.1	15.6	ab
	II	32	50.6	12.6	b	49	53.1	12.4	a
Sand (%)	III	17	42.8	10.9	b	3	37.5	13.6	ab
	IV	26	42.7	8.8	b	28	42.1	9.2	b
	V	15	63.8	8.5	a	n.a.	n.a.	n.a.	n.a.
	I	1	9.0	NA	ab	11	12.6	6.6	ab
	II	32	13.5	5.1	a	49	12.6	4.8	b
Clay (%)	III	17	17.4	5.7	a	3	20.8	9.3	ab
	IV	26	16.8	4.7	a	28	17.3	4.7	a
	V	15	8.7	3.2	b	n.a.	n.a.	n.a.	n.a.
	I	53	7.03	1.90	bc	263	10.0	1.7	b
Potential Yield	II	779	8.38	2.24	a	1164	10.5	1.2	a
(Mg ha^{-1})	III	402	7.61	1.75	b	146	9.3	2.1	c
	IV	513	7.19	2.01	c	574	10.3	1.3	b
	IV	400	7.08	1.90	c	n.a.	n.a.	n.a.	n.a.

[1] Different letters are significantly different between MZs at the $p < 0.05$ level according to the Kruskal–Wallis (KW) test. [2] Percentage of cells having significant GWR within the MZs.

In both years, MZ I consisted of soils with low EC_a Deep, low salinity, and coarse soil texture. MZ I had the highest positive GWR slope values, meaning NDVI would decrease with increasing sand content. Scudiero, Teatini, Corwin, Dal Ferro, Simonetti, and Morari [88] monitored crop water and salt stress at five soil–plant–water monitoring stations at the site. One of their stations ("Station E") was located in our MZ I. Scudiero, Teatini, Corwin, Dal Ferro, Simonetti, and Morari [88] indicated that at Station E maize was not stressed by salinity but was under water stress, particularly in 2011, when water stress was described as "severe".

In contrast, MZ II in both years had GWR slope and r distributions overlapping with 0, indicating little to no influence of soil on crop NDVI variability. This is perhaps the reason why MZ II was consistently characterized by the highest potential yield estimations (Appendix A). The very saline MZ III (highest measured average EC_a Deep and $EC_{1:2}$) was characterized by lower potential yield estimations for both years in comparison with the other MZs (Table 2), as expected for (unmanaged) very saline portions of farmlands [106]. Similarly, the moderately saline MZ IV (second highest average EC_a Deep), was characterized by the second lowest yield predictions in 2010 and 2011, together with MZ I. MZ V was only selected in 2010. It grouped soils with moderately high EC_a Deep (third highest MZ average) together with the strongest negative GWR slope values and low potential yield estimations. In 2011, 89.5% of the 2010 MZ V locations were classified into MZ II. Previous research focusing on spatiotemporal variability of yield patterns [38], indicates that fields generally have areas of stable and unstable yield patterns. For unstable areas, in-season NDVI is suggested as the best predictor for yield spatial distribution [38]. Our proposed approach further refines the NDVI information by interpreting its spatial variability as a function of soil spatial variability.

A static MZ design based on EC_a Deep only (Figure 5c) would identify four areas with significantly different soil properties (Table 3), but very heterogeneous and inconsistent in terms of GWR local r and slope (Table 4).

Table 3. Count (n), mean, and standard deviation (Std. Dev.) of apparent electrical conductivity for the 0–1.5-m soil profile (EC_a Deep), soil salinity ($EC_{1:2}$), and sand and clay contents for the static management zones (MZs) delineated using EC_a Deep only.

Variable (Unit)	Static MZ	n	Mean	Std. Dev.	KW [1]
EC_a Deep (dS m^{-1})	1	718	0.92	0.10	c
	2	181	1.71	0.17	a
	3	444	1.31	0.12	b
	4	804	0.63	0.10	d
$EC_{1:2}$ (dS m^{-1})	1	28	1.98	1.04	b
	2	9	4.07	1.51	ab
	3	21	2.82	1.25	a
	4	33	1.07	0.68	c
Sand (%)	1	28	50.3	11.9	ab
	2	9	41.0	9.3	b
	3	21	41.1	9.8	b
	4	33	55.9	12.9	a
Clay (%)	1	28	13.3	5.0	bc
	2	9	18.3	5.7	ab
	3	21	17.9	5.0	a
	4	33	11.8	4.9	c

[1] Different letters are significantly different between MZs at the $p < 0.05$ level according to the Kruskal–Wallis (KW) test.

Table 4. Mean, standard deviation (Std. Dev.), and count (n) for the correlation coefficient (r) and slope obtained from significant geographically weighted regressions (GWRs) for the static management zones (MZs) delineated using soil apparent electrical conductivity for the 0–1.5-m soil profile (EC_a Deep) only.

GWR Variable	Static MZ	2010 n	2010 Mean	2010 Std. Dev.	2011 n	2011 Mean	2011 Std. Dev.
Local r	1	490 (68.2% [1])	−0.374	0.351	349 (48.6% [1])	−0.156	0.383
	2	163 (90.1% [1])	−0.474	0.158	148 (81.8% [1])	−0.542	0.158
	3	313 (70.5% [1])	−0.408	0.281	289 (65.1% [1])	−0.303	0.408
	4	483 (60.1% [1])	−0.098	0.472	433 (53.9% [1])	−0.041	0.420
Slope	1	490 (68.2% [1])	−0.086	0.148	349 (48.6% [1])	−0.009	0.039
	2	163 (90.1% [1])	−0.077	0.065	148 (81.8% [1])	−0.074	0.068
	3	313 (70.5% [1])	−0.058	0.065	289 (65.1% [1])	−0.023	0.045
	4	483 (60.1% [1])	0.066	0.304	433 (53.9% [1])	−0.002	0.079

[1] Percentage of cells having significant GWR within the MZ.

4. Conclusions

Characterizing the temporal variability of the soil–plant relationship within the growing season and over years is very relevant to developing best crop management practices with VRM. Soil influences plants according to complex interactions between factors that may change in time, such as meteorological conditions, nutrient availability, and water content. Static MZs are not ideal when spatial patterns of soil–plant relationship change in time—because of changing meteorology and/or other transient factors. We presented a time-specific MZ-delineation workflow to address such spatiotemporal variability.

The proposed approach for time-specific MZs should be further tested and evaluated in the field. Future work should focus on multi-field and multi-year comparisons of time-specific and static MZs and their relative impact on crop yields and resource use efficiency (e.g., water, nutrients).

Author Contributions: Conceptualization, E.S. and T.H.S.; methodology, E.S.; software, E.S. and F.B.; formal analysis, investigation, resources, data curation, writing—original draft preparation, writing—review and editing, All Authors; visualization, E.S.; funding acquisition, F.M. and P.T.

Funding: This study was funded within the Research Program "GEO-RISKS: Geological, Morphological and Hydrological Processes: Monitoring, Modeling and Impact in North-Eastern Italy: Work package 4", the University of Padua, Italy.

Acknowledgments: We acknowledge the outstanding work of Nicola Dal Ferro, Davide Piragnolo, Diego Mengardo, and Gianluca Simonetti who helped in the laboratory analyses and/or field surveys. Special thanks to John Shanahan and George Vellidis for providing valuable comments and advice on early versions of this manuscript.

Conflicts of Interest: The authors declare no conflict of interest. The funding sponsors had no role in the design of the study, in the collection, analyses, or interpretation of data, in the writing of the manuscript, nor in the decision to publish the results.

Abbreviations

AOT, aerosol optical thickness; BMZSS, between-MZ sum of squares; *CHC*, Calinski–Harabasz criterion; $EC_{1:2}$, salinity measured as electrical conductivity of a 1:2 soil–water extract; EC_a, apparent electrical conductivity; EC_a Deep, apparent electrical conductivity measurements of the 0–1.5 m soil profile; GWR, geographically weighted regression; MZ, management zone; NDVI, normalized difference vegetation index; PCA, principal component analysis; *r*, Pearson's correlation coefficient; R^2, coefficient of determination; R-2, kernel blister stage of maize growth; *SCC*, (synthetic) sensor-derived clay content; *TCC*, (synthetic) true clay content; UAV, unmanned aerial vehicle; VRM, variable rate management; WMZSS, within-MZ sum of squares; WV2, WorldView-2 satellite.

Appendix A   Interpreting NDVI as Potential Yield Indicator

In-season potential yield predictions can be very useful for the producers to decide on the feasibility of VRM. Throughout the season NDVI can be used as potential yield indicator [33,35]. Estimations can be made at different growth stages [32], provided the NDVI measurements are calibrated accordingly [33]. This is particularly relevant when historical yield data is available, so that yield-predicting NDVI functions can be developed and used in following years.

According to [16], the in-season potential yield can be estimated by fitting the NDVI values to the 68th percentile (i.e., sum of average and one standard deviation) of yield. The NDVI and yield data from 2010 and 2011 were clumped together. The yield data was obtained from Scudiero, Teatini, Corwin, Dal Ferro, Simonetti, and Morari [88]. The dataset was divided into fifty quantiles according to NDVI. The within-quantile average NDVI was then used as explanatory variable to estimate the within-quintile 68th percentile of yield. The estimation was done using a second-degree locally-weighted least squares regression. The regression modeling was carried out through the loess function (with span = 0.45) from the stats package in R. The span parameter was selected through a leave-one-out cross-validation using the loess.wrapper function from the bisoreg package in R. Figure A1a shows the scatterplot between NDVI (i.e., at late R-2) and maize yield, all points were clustered on a uniform grid of hexagons using the hexbin package in R.

The model describing the potential yield (Mean absolute error = 0.26 Mg ha^{-1}, mean error = -0.027 Mg ha^{-1}, and $R^2 = 0.98$) is reported with a solid red line in Figure A1. The locally-weighted least squares regression estimations were then applied to the two NDVI maps (Figure 4b,e) to obtain potential yield estimations across the whole study area for 2010 (Figure A1b) and 2011 (Figure A1c).

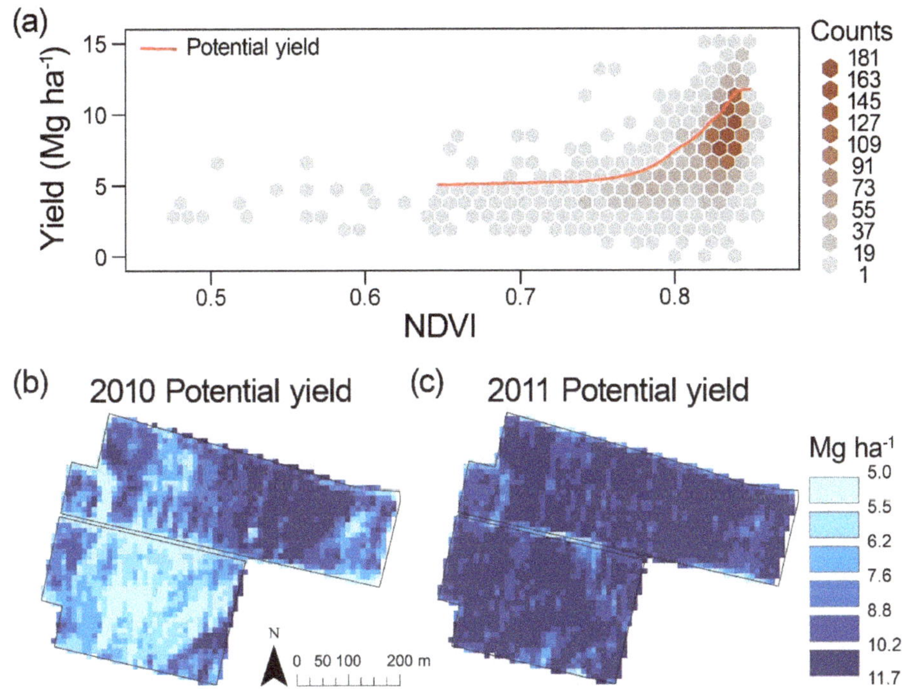

Figure A1. (**a**) Scatterplot between yield and WorldView-2 NDVI for both years. The modeled in-season potential yield is depicted with a solid red line. Potential yield maps calculated for (**b**) 2010 and (**c**) 2011.

References

1. Lobell, D.B.; Azzari, G. Satellite detection of rising maize yield heterogeneity in the US Midwest. *Environ. Res. Lett.* **2017**, *12*. [CrossRef]
2. Corwin, D.L.; Lesch, S.M.; Shouse, P.J.; Soppe, R.; Ayars, J.E. Identifying soil properties that influence cotton yield using soil sampling directed by apparent soil electrical conductivity. *Agron. J.* **2003**, *95*, 352–364. [CrossRef]
3. Kitchen, N.R.; Drummond, S.T.; Lund, E.D.; Sudduth, K.A.; Buchleiter, G.W. Soil Electrical Conductivity and Topography Related to Yield for Three Contrasting Soil–Crop Systems. *Agron. J.* **2003**, *95*, 483–495. [CrossRef]
4. Bobryk, C.W.; Myers, D.B.; Kitchen, N.R.; Shanahan, J.F.; Sudduth, K.A.; Drummond, S.T.; Gunzenhauser, B.; Gomez Raboteaux, N.N. Validating a Digital Soil Map with Corn Yield Data for Precision Agriculture Decision Support. *Agron. J.* **2016**, *108*, 957–965. [CrossRef]
5. Kaffka, S.R.; Lesch, S.M.; Bali, K.M.; Corwin, D.L. Site-specific management in salt-affected sugar beet fields using electromagnetic induction. *Comput. Electron. Agric.* **2005**, *46*, 329–350. [CrossRef]
6. Singh, G.; Williard, K.J.; Schoonover, J.E. Spatial Relation of Apparent Soil Electrical Conductivity with Crop Yields and Soil Properties at Different Topographic Positions in a Small Agricultural Watershed. *Agronomy* **2016**, *6*, 57. [CrossRef]
7. Zhang, N.Q.; Wang, M.H.; Wang, N. Precision agriculture—A worldwide overview. *Comput. Electron. Agric.* **2002**, *36*, 113–132. [CrossRef]
8. Fleming, K.L.; Westfall, D.G.; Wiens, D.W.; Brodahl, M.C. Evaluating Farmer Defined Management Zone Maps for Variable Rate Fertilizer Application. *Precis. Agric.* **2000**, *2*, 201–215. [CrossRef]

9. Fraisse, C.W.; Sudduth, K.A.; Kitchen, N.R. Delineation of site-specific management zones by unsupervised classification of topographic attributes and soil electrical conductivity. *Trans. ASAE* **2001**, *44*, 155–166. [CrossRef]

10. Moral, F.J.; Terrón, J.M.; Silva, J.R.M.D. Delineation of management zones using mobile measurements of soil apparent electrical conductivity and multivariate geostatistical techniques. *Soil Tillage Res.* **2010**, *106*, 335–343. [CrossRef]

11. Baveye, P.C.; Laba, M. Moving away from the geostatistical lamppost: Why, where, and how does the spatial heterogeneity of soils matter? *Ecol. Model.* **2015**, *298*, 24–38. [CrossRef]

12. Schepers, A.R.; Shanahan, J.F.; Liebig, M.A.; Schepers, J.S.; Johnson, S.H.; Luchiari, A. Appropriateness of management zones for characterizing spatial variability of soil properties and irrigated corn yields across years. *Agron. J.* **2004**, *96*, 195–203. [CrossRef]

13. Koch, B.; Khosla, R.; Frasier, W.M.; Westfall, D.G.; Inman, D. Economic Feasibility of Variable-Rate Nitrogen Application Utilizing Site-Specific Management Zones. *Agron. J.* **2004**, *96*, 1572–1580. [CrossRef]

14. Liang, X.; Liakos, V.; Wendroth, O.; Vellidis, G. Scheduling irrigation using an approach based on the van Genuchten model. *Agric. Water Manag.* **2016**, *176*, 170–179. [CrossRef]

15. Vellidis, G.; Liakos, V.; Porter, W.; Tucker, M.; Liang, X. A dynamic variable rate irrigation control system. In Proceedings of the 13th International Conference on Precision Agriculture, Monticello, IL, USA, 24–27 June 2018.

16. Raun, W.R.; Solie, J.B.; Johnson, G.V.; Stone, M.L.; Mullen, R.W.; Freeman, K.W.; Thomason, W.E.; Lukina, E.V. Improving nitrogen use efficiency in cereal grain production with optical sensing and variable rate application. *Agron. J.* **2002**, *94*, 815–820. [CrossRef]

17. Hedley, C.B.; Yule, I.J. Soil water status mapping and two variable-rate irrigation scenarios. *Precis. Agric.* **2009**, *10*, 342–355. [CrossRef]

18. Basso, B.; Ritchie, J.T.; Cammarano, D.; Sartori, L. A strategic and tactical management approach to select optimal N fertilizer rates for wheat in a spatially variable field. *Eur. J. Agron.* **2011**, *35*, 215–222. [CrossRef]

19. Cohen, Y.; Alchanatis, V.; Saranga, Y.; Rosenberg, O.; Sela, E.; Bosak, A. Mapping water status based on aerial thermal imagery: Comparison of methodologies for upscaling from a single leaf to commercial fields. *Precis. Agric.* **2017**, *18*, 801–822. [CrossRef]

20. Helman, D.; Bahat, I.; Netzer, Y.; Ben-Gal, A.; Alchanatis, V.; Peeters, A.; Cohen, Y. Using Time Series of High-Resolution Planet Satellite Images to Monitor Grapevine Stem Water Potential in Commercial Vineyards. *Remote Sens.* **2018**, *10*, 1615. [CrossRef]

21. Shanahan, J.F.; Kitchen, N.R.; Raun, W.R.; Schepers, J.S. Responsive in-season nitrogen management for cereals. *Comput. Electron. Agric.* **2008**, *61*, 51–62. [CrossRef]

22. Long, D.S.; Whitmus, J.D.; Engel, R.E.; Brester, G.W. Net Returns from Terrain-Based Variable-Rate Nitrogen Management on Dryland Spring Wheat in Northern Montana. *Agron. J.* **2015**, *107*, 1055–1067. [CrossRef]

23. Quebrajo, L.; Perez-Ruiz, M.; Pérez-Urrestarazu, L.; Martínez, G.; Egea, G. Linking thermal imaging and soil remote sensing to enhance irrigation management of sugar beet. *Biosyst. Eng.* **2018**, *165*, 77–87. [CrossRef]

24. Adamchuk, V.I.; Hummel, J.W.; Morgan, M.T.; Upadhyaya, S.K. On-the-go soil sensors for precision agriculture. *Comput. Electron. Agric.* **2004**, *44*, 71–91. [CrossRef]

25. De Lara, A.; Khosla, R.; Longchamps, L. Characterizing Spatial Variability in Soil Water Content for Precision Irrigation Management. *Agronomy* **2018**, *8*, 59. [CrossRef]

26. Badewa, E.; Unc, A.; Cheema, M.; Kavanagh, V.; Galagedara, L. Soil Moisture Mapping Using Multi-Frequency and Multi-Coil Electromagnetic Induction Sensors on Managed Podzols. *Agronomy* **2018**, *8*, 224. [CrossRef]

27. Gabriel, J.L.; Zarco-Tejada, P.J.; López-Herrera, P.J.; Pérez-Martín, E.; Alonso-Ayuso, M.; Quemada, M. Airborne and ground level sensors for monitoring nitrogen status in a maize crop. *Biosyst. Eng.* **2017**, *160*, 124–133. [CrossRef]

28. Irmak, S.; Haman, D.Z.; Bastug, R. Determination of Crop Water Stress Index for Irrigation Timing and Yield Estimation of Corn. *Agron. J.* **2000**, *92*, 1221–1227. [CrossRef]

29. Raun, W.R.; Solie, J.B.; Martin, K.L.; Freeman, K.W.; Stone, M.L.; Johnson, G.V.; Mullen, R.W. Growth stage, development, and spatial variability in corn evaluated using optical sensor readings. *J. Plant Nutr.* **2005**, *28*, 173–182. [CrossRef]

30. Solari, F.; Shanahan, J.; Ferguson, R.; Schepers, J.; Gitelson, A. Active sensor reflectance measurements of corn nitrogen status and yield potential. *Agron. J.* **2008**, *100*, 571–579. [CrossRef]

31. Franke, J.; Menz, G. Multi-temporal wheat disease detection by multi-spectral remote sensing. *Precis. Agric.* **2007**, *8*, 161–172. [CrossRef]

32. Shanahan, J.F.; Schepers, J.S.; Francis, D.D.; Varvel, G.E.; Wilhelm, W.W.; Tringe, J.M.; Schlemmer, M.R.; Major, D.J. Use of remote-sensing imagery to estimate corn grain yield. *Agron. J.* **2001**, *93*, 583–589. [CrossRef]

33. Teal, R.K.; Tubana, B.; Girma, K.; Freeman, K.W.; Arnall, D.B.; Walsh, O.; Raun, W.R. In-season prediction of corn grain yield potential using normalized difference vegetation index. *Agron. J.* **2006**, *98*, 1488–1494. [CrossRef]

34. Torino, M.S.; Ortiz, B.V.; Fulton, J.P.; Balkcom, K.S.; Wood, C.W. Evaluation of Vegetation Indices for Early Assessment of Corn Status and Yield Potential in the Southeastern United States. *Agron. J.* **2014**, *106*, 1389–1401. [CrossRef]

35. Tagarakis, A.C.; Ketterings, Q.M. In-Season Estimation of Corn Yield Potential Using Proximal Sensing. *Agron. J.* **2017**, *109*, 1323–1330. [CrossRef]

36. Blackmore, S.; Godwin, R.J.; Fountas, S. The analysis of spatial and temporal trends in yield map data over six years. *Biosyst. Eng.* **2003**, *84*, 455–466. [CrossRef]

37. McBratney, A.; Whelan, B.; Ancev, T.; Bouma, J. Future directions of precision agriculture. *Precis. Agric.* **2005**, *6*, 7–23. [CrossRef]

38. Maestrini, B.; Basso, B. Predicting spatial patterns of within-field crop yield variability. *Field Crop. Res.* **2018**, *219*, 106–112. [CrossRef]

39. Fullmer, D.; Chetty, V.; Warnick, S. How good is bad weather? In Proceedings of the 2014 American Control Conference, Portland, OR, USA, 4–6 June 2014; pp. 2711–2716.

40. Sadler, E.J.; Evans, R.G.; Stone, K.C.; Camp, C.R. Opportunities for conservation with precision irrigation. *J. Soil Water Conserv.* **2005**, *60*, 371–378.

41. Vellidis, G.; Snider, J.; Liakos, V.; Porter, W.; Perry, C. Dynamic Variable Rate Irrigation Management Using Soil Moisture and Canopy Temperature Sensors in the Southeastern USA. In Proceedings of the 2017 ASA, CSSA, and SSSA International Annual Meeting, Tampa, FL, USA, 22–25 October 2017.

42. Franzen, D.W. Profitable Use of Site-Specific Nutrient Management Technologies. In Proceedings of the 2017 ASA, CSSA, and SSSA International Annual Meeting, Tampa, FL, USA, 22–25 October 2017.

43. Scudiero, E.; Morari, F.; Skaggs, T.H.; Braga, F.; Teatini, P. Understanding spatiotemporal variability of soil–plant relationships in a heterogeneous coastal farmland in Northern Italy. In Proceedings of the 2016 ASA, CSSA, and SSSA International Annual Meeting, Phoenix, AZ, USA, 6–9 November 2016.

44. Liu, H.; Whiting, M.L.; Ustin, S.L.; Zarco-Tejada, P.J.; Huffman, T.; Zhang, X. Maximizing the relationship of yield to site-specific management zones with object-oriented segmentation of hyperspectral images. *Precis. Agric.* **2018**, *19*, 348–364. [CrossRef]

45. Myers, D.B. Measurements That Matter for Decision Agriculture. In Proceedings of the 2016 ASA, CSSA, and SSSA International Annual Meeting, Phoenix, AZ, USA, 6–9 November 2016.

46. Betzek, N.M.; Souza, E.G.D.; Bazzi, C.L.; Schenatto, K.; Gavioli, A. Rectification methods for optimization of management zones. *Comput. Electron. Agric.* **2018**, *146*, 1–11. [CrossRef]

47. Gavioli, A.; de Souza, E.G.; Bazzi, C.L.; Guedes, L.P.C.; Schenatto, K. Optimization of management zone delineation by using spatial principal components. *Comput. Electron. Agric.* **2016**, *127*, 302–310. [CrossRef]

48. Taylor, J.A.; McBratney, A.B.; Whelan, B.M. Establishing Management Classes for Broadacre Agricultural Production. *Agron. J.* **2007**, *99*, 1366–1376. [CrossRef]

49. Córdoba, M.A.; Bruno, C.I.; Costa, J.L.; Peralta, N.R.; Balzarini, M.G. Protocol for multivariate homogeneous zone delineation in precision agriculture. *Biosyst. Eng.* **2016**, *143*, 95–107. [CrossRef]

50. Corwin, D.L.; Lesch, S.M. Characterizing soil spatial variability with apparent soil electrical conductivity I. Survey protocols. *Comput. Electron. Agric.* **2005**, *46*, 103–133. [CrossRef]

51. Rossel, R.A.V.; Taylor, H.J.; McBratney, A.B. Multivariate calibration of hyperspectral γ-ray energy spectra for proximal soil sensing. *Eur. J. Soil Sci.* **2007**, *58*, 343–353. [CrossRef]

52. Minty, B.R.S. Fundamentals of airborne gamma-ray spectrometry. *AGSO J. Aust. Geol. Geophys.* **1997**, *17*, 39–50.

53. Priori, S.; Martini, E.; Andrenelli, M.C.; Magini, S.; Agnelli, A.E.; Bucelli, P.; Biagi, M.; Pellegrini, S.; Costantini, E.A.C. Improving Wine Quality through Harvest Zoning and Combined Use of Remote and Soil Proximal Sensing. *Soil Sci. Soc. Am. J.* **2013**, *77*, 1338–1348. [CrossRef]

54. Scudiero, E.; Corwin, D.L.; Wienhold, B.J.; Bosley, B.; Shanahan, J.F.; Johnson, C.K. Downscaling Landsat 7 canopy reflectance employing a multi-soil sensor platform. *Precis. Agric.* **2016**, *17*, 53–73. [CrossRef]

55. Lesch, S.M. Sensor-directed response surface sampling designs for characterizing spatial variation in soil properties. *Comput. Electron. Agric.* **2005**, *46*, 153–179. [CrossRef]

56. Van Groenigen, J.W.; Stein, A. Constrained optimization of spatial sampling using continuous simulated annealing. *J. Environ. Qual.* **1998**, *27*, 1078–1086. [CrossRef]

57. Hengl, T.; Heuvelink, G.B.M.; Stein, A. A generic framework for spatial prediction of soil variables based on regression-kriging. *Geoderma* **2004**, *120*, 75–93. [CrossRef]

58. Lesch, S.M.; Corwin, D.L. Prediction of spatial soil property information from ancillary sensor data using ordinary linear regression: Model derivations, residual assumptions and model validation tests. *Geoderma* **2008**, *148*, 130–140. [CrossRef]

59. Nelson, M.A.; Bishop, T.F.A.; Triantafilis, J.; Odeh, I.O.A. An error budget for different sources of error in digital soil mapping. *Eur. J. Soil Sci.* **2011**, *62*, 417–430. [CrossRef]

60. Gausman, H.W.; Allen, W.A. Optical parameters of leaves of 30 plant species. *Plant Physiol.* **1973**, *52*, 57–62. [CrossRef] [PubMed]

61. Huete, A.R.; Jackson, R.D.; Post, D.F. Spectral Response of a Plant Canopy with Different Soil Backgrounds. *Remote Sens. Environ.* **1985**, *17*, 37–53. [CrossRef]

62. Li, H.; Lascano, R.J.; Barnes, E.M.; Booker, J.; Wilson, L.T.; Bronson, K.F.; Segarra, E. Multispectral Reflectance of Cotton Related to Plant Growth, Soil Water and Texture, and Site Elevation. *Agron. J.* **2001**, *93*, 1327–1337. [CrossRef]

63. Brunsdon, C.; Fotheringham, S.; Charlton, M. Geographically weighted regression—Modelling spatial non-stationarity. *J. R. Stat. Soc. Ser. D-Stat.* **1998**, *47*, 431–443. [CrossRef]

64. Brunsdon, C.; Fotheringham, A.S.; Charlton, M.E. Geographically Weighted Regression: A Method for Exploring Spatial Nonstationarity. *Geogr. Anal.* **1996**, *28*, 281–298. [CrossRef]

65. Mitchell, A. *The ESRI Guide to GIS Analysis. Vol. II: Spatial Measurements and Statistics*; ESRI Press: Redlands, CA, USA, 2005.

66. Fotheringham, A.S.; Brunsdon, C.; Charlton, M.E. *Geographically Weighted Regression: The Analysis of Spatially Varying Relationships*; John Wiley: Hoboken, NJ, USA, 2002.

67. Odeh, I.O.A.; McBratney, A.B.; Chittleborough, D.J. Soil pattern recognition with fuzzy-c-means: Application to classification and soil-landform interrelationships. *Soil Sci. Soc. Am. J.* **1992**, *56*, 505–516. [CrossRef]

68. Venkatramanan, S.; Chung, S.Y.; Rajesh, R.; Lee, S.Y.; Ramkumar, T.; Prasanna, M.V. Comprehensive studies of hydrogeochemical processes and quality status of groundwater with tools of cluster, grouping analysis, and fuzzy set method using GIS platform: A case study of Dalcheon in Ulsan City, Korea. *Environ. Sci. Pollut. Res.* **2015**, *22*, 11209–11223. [CrossRef] [PubMed]

69. Fridgen, J.J.; Kitchen, N.R.; Sudduth, K.A.; Drummond, S.T.; Wiebold, W.J.; Fraisse, C.W. Management Zone Analyst (MZA): Software for subfield management zone delineation. *Agron. J.* **2004**, *96*, 100–108. [CrossRef]

70. Lowrance, C.; Fountas, S.; Liakos, V.; Vellidis, G. EZZone–An Online Tool for Delineating Management Zones. In Proceedings of the 13th International Conference on Precision Agriculture, St. Louis, MI, USA, 24–27 June 2018.

71. Caliński, T.; Harabasz, J. A dendrite method for cluster analysis. *Commun. Stat.* **1974**, *3*, 1–27. [CrossRef]

72. Bezdek, J.C. *Pattern Recognition with Fuzzy Objective Function Algorithms*; Plenum Press: New York, NY, USA, 1981.

73. Jenks, G.F. The data model concept in statistical mapping. *Int. Yearb. Cartogr.* **1967**, *7*, 186–190.

74. Kruskal, W.H.; Wallis, W.A. Use of ranks in one-criterion variance analysis. *J. Am. Stat. Assoc.* **1952**, *47*, 583–621. [CrossRef]

75. Rouse, J.; Haas, R.; Schell, J.; Deering, D. Monitoring vegetation systems in the Great Plains with ERTS. In Proceedings of the Third Earth Resources Technology Satellite Symposium, Washington, DC, USA, 10–14 December 1973; pp. 309–317.

76. Viscarra Rossel, R.A.; Adamchuk, V.I.; Sudduth, K.A.; McKenzie, N.J.; Lobsey, C. Chapter Five—Proximal Soil Sensing: An Effective Approach for Soil Measurements in Space and Time. In *Advances in Agronomy*; Sparks, D.L., Ed.; Academic Press: Cambridge, MA, USA, 2011; Volume 113, pp. 243–291.

77. McBratney, A.B.; Mendonça Santos, M.L.; Minasny, B. On digital soil mapping. *Geoderma* **2003**, *117*, 3–52. [CrossRef]

78. Heggemann, T.; Welp, G.; Amelung, W.; Angst, G.; Franz, S.O.; Koszinski, S.; Schmidt, K.; Pätzold, S. Proximal gamma-ray spectrometry for site-independent in situ prediction of soil texture on ten heterogeneous fields in Germany using support vector machines. *Soil Tillage Res.* **2017**, *168*, 99–109. [CrossRef]

79. Morari, F.; Castrignano, A.; Pagliarin, C. Application of multivariate geostatistics in delineating management zones within a gravelly vineyard using geo-electrical sensors. *Comput. Electron. Agric.* **2009**, *68*, 97–107. [CrossRef]

80. Weller, U.; Zipprich, M.; Sommer, M.; Castell, W.Z.; Wehrhan, M. Mapping Clay Content across Boundaries at the Landscape Scale with Electromagnetic Induction. *Soil Sci. Soc. Am. J.* **2007**, *71*, 1740–1747. [CrossRef]

81. Wardlow, B.D.; Egbert, S.L. Large-area crop mapping using time-series MODIS 250 m NDVI data: An assessment for the U.S. Central Great Plains. *Remote Sens. Environ.* **2008**, *112*, 1096–1116. [CrossRef]

82. Masialeti, I.; Egbert, S.; Wardlow, B.D. A Comparative Analysis of Phenological Curves for Major Crops in Kansas. *GIScience Remote Sens.* **2010**, *47*, 241–259. [CrossRef]

83. Xu, Y.; Smith, S.E.; Grunwald, S.; Abd-Elrahman, A.; Wani, S.P.; Nair, V.D. Estimating soil total nitrogen in smallholder farm settings using remote sensing spectral indices and regression kriging. *CATENA* **2018**, *163*, 111–122. [CrossRef]

84. Samuel-Rosa, A.; Heuvelink, G.B.M.; Vasques, G.M.; Anjos, L.H.C. Do more detailed environmental covariates deliver more accurate soil maps? *Geoderma* **2015**, *243*, 214–227. [CrossRef]

85. Hengl, T.; Heuvelink, G.B.M.; Rossiter, D.G. About regression-kriging: From equations to case studies. *Comput. Geosci.* **2007**, *33*, 1301–1315. [CrossRef]

86. Scudiero, E.; Teatini, P.; Corwin, D.L.; Deiana, R.; Berti, A.; Morari, F. Delineation of site-specific management units in a saline region at the Venice Lagoon margin, Italy, using soil reflectance and apparent electrical conductivity. *Comput. Electron. Agric.* **2013**, *99*, 54–64. [CrossRef]

87. Manoli, G.; Bonetti, S.; Scudiero, E.; Morari, F.; Putti, M.; Teatini, P. Modeling Soil–Plant Dynamics: Assessing Simulation Accuracy by Comparison with Spatially Distributed Crop Yield Measurements. *Vadose Zone J.* **2015**, *14*. [CrossRef]

88. Scudiero, E.; Teatini, P.; Corwin, D.L.; Dal Ferro, N.; Simonetti, G.; Morari, F. Spatiotemporal Response of Maize Yield to Edaphic and Meteorological Conditions in a Saline Farmland. *Agron. J.* **2014**, *106*, 2163–2174. [CrossRef]

89. Grosso, C.; Manoli, G.; Martello, M.; Chemin, Y.; Pons, D.; Teatini, P.; Piccoli, I.; Morari, F. Mapping Maize Evapotranspiration at Field Scale Using SEBAL: A Comparison with the FAO Method and Soil–Plant Model Simulations. *Remote Sens.* **2018**, *10*, 1452. [CrossRef]

90. FAO-UNESCO. *Soil Map of the World, Revised Legend*; FAO: Rome, Italy, 1989.

91. Rhoades, J.; Chanduvi, F.; Lesch, S.M. *Soil Salinity Assessment: Methods and Interpretation of Electrical Conductivity Measurements*; FAO: Rome, Italy, 1999; Volume 57.

92. Abdi, H.; Williams, L.J. Principal component analysis. *Wiley Interdiscip. Rev. Comput. Stat.* **2010**, *2*, 433–459. [CrossRef]

93. Vicente-Serrano, S.M.; Perez-Cabello, F.; Lasanta, T. Assessment of radiometric correction techniques in analyzing vegetation variability and change using time series of Landsat images. *Remote Sens. Environ.* **2008**, *112*, 3916–3934. [CrossRef]

94. Updike, T.; Comp, C. *Radiometric Use of WorldView-2 Imagery*; DigitalGlobe, Inc.: Longmont, CO, USA, 2010; pp. 1–17.

95. Vermote, E.F.; Tanre, D.; Deuze, J.L.; Herman, M.; Morcrette, J.J. Second Simulation of the Satellite Signal in the Solar Spectrum, 6S: An overview. *IEEE Trans. Geosci. Remote Sens.* **1997**, *35*, 675–686. [CrossRef]

96. Holben, B.N.; Eck, T.F.; Slutsker, I.; Tanré, D.; Buis, J.P.; Setzer, A.; Vermote, E.; Reagan, J.A.; Kaufman, Y.J.; Nakajima, T.; et al. AERONET—A Federated Instrument Network and Data Archive for Aerosol Characterization. *Remote Sens. Environ.* **1998**, *66*, 1–16. [CrossRef]

97. Xia, T.; Miao, Y.; Wu, D.; Shao, H.; Khosla, R.; Mi, G. Active Optical Sensing of Spring Maize for In-Season Diagnosis of Nitrogen Status Based on Nitrogen Nutrition Index. *Remote Sens.* **2016**, *8*, 605. [CrossRef]
98. Peng, Y.; Nguy-Robertson, A.; Arkebauer, T.; Gitelson, A. Assessment of Canopy Chlorophyll Content Retrieval in Maize and Soybean: Implications of Hysteresis on the Development of Generic Algorithms. *Remote Sens.* **2017**, *9*, 226. [CrossRef]
99. Gomez, C.; Oltra-Carrio, R.; Bacha, S.; Lagacherie, P.; Briottet, X. Evaluating the sensitivity of clay content prediction to atmospheric effects and degradation of image spatial resolution using Hyperspectral VNIR/SWIR imagery. *Remote Sens. Environ.* **2015**, *164*, 1–15. [CrossRef]
100. Gomez, C.; Viscarra Rossel, R.A.; McBratney, A.B. Soil organic carbon prediction by hyperspectral remote sensing and field vis-NIR spectroscopy: An Australian case study. *Geoderma* **2008**, *146*, 403–411. [CrossRef]
101. Corwin, D.L.; Lesch, S.M. Protocols and Guidelines for Field-scale Measurement of Soil Salinity Distribution with ECa-Directed Soil Sampling. *J. Environ. Eng. Geophys.* **2013**, *18*, 1–25. [CrossRef]
102. Corwin, D.L. Field-scale monitoring of the long-term impact and sustainability of drainage water reuse on the west side of California's San Joaquin Valley. *J. Environ. Monit.* **2012**, *14*, 1576–1596. [CrossRef] [PubMed]
103. Ransom, J.; Endres, G.J.; McWilliams, D.A. Corn Growth and Management Quick Guide A1173. Available online: www.ag.ndsu.edu/pubs/plantsci/crops/a1173.pdf (accessed on 20 October 2018).
104. Nielsen, R.L. Effects of Severe Stress During Grain Filling in Corn. Available online: http://www.kingcorn.org/news/timeless/GrainFillStress.html (accessed on 22 October 2018).
105. Lobell, D.B.; Lesch, S.M.; Corwin, D.L.; Ulmer, M.G.; Anderson, K.A.; Potts, D.J.; Doolittle, J.A.; Matos, M.R.; Baltes, M.J. Regional-scale Assessment of Soil Salinity in the Red River Valley Using Multi-year MODIS EVI and NDVI. *J. Environ. Qual.* **2010**, *39*, 35–41. [CrossRef] [PubMed]
106. Lobell, D.B.; Ortiz-Monasterio, J.I.; Gurrola, F.C.; Valenzuela, L. Identification of saline soils with multiyear remote sensing of crop yields. *Soil Sci. Soc. Am. J.* **2007**, *71*, 777–783. [CrossRef]
107. Madrigal, L.P.; Wiegand, C.L.; Meraz, J.G.; Rubio, B.D.R.; Estrada, X.C.; Ramirez, O.L. Soil salinity and its effect on crop yield—A study using satellite imagery in three irrigation districts. *Ing. Hidraul. En Mex.* **2003**, *18*, 83–97.

Article

Prediction of Sugarcane Yield Based on NDVI and Concentration of Leaf-Tissue Nutrients in Fields Managed with Straw Removal

Izaias Pinheiro Lisboa, Júnior Melo Damian *, Maurício Roberto Cherubin, Pedro Paulo Silva Barros, Peterson Ricardo Fiorio, Carlos Clemente Cerri and Carlos Eduardo Pellegrino Cerri

"Luiz de Queiroz" College of Agriculture, University of São Paulo, 11 Pádua Dias Avenue, Piracicaba, SP 13418-900, Brazil; iplisboaa@gmail.com (I.P.L.); cherubin@usp.br (M.R.C.); pedropaulo@usp.br (P.P.S.B.); fiorio@usp.br (P.R.F.); cerri@cena.usp.br (C.C.C.); cepcerri@usp.br (C.E.P.C.)
* Correspondence: damianjrm@usp.br; Tel.: +55-559-996-90524

Received: 29 July 2018; Accepted: 13 September 2018; Published: 19 September 2018

Abstract: The total or partial removal of sugarcane (*Saccharum* spp. L.) straw for bioenergy production may deplete soil quality and consequently affect negatively crop yield. Plants with lower yield potential may present lower concentration of leaf-tissue nutrients, which in turn changes light reflectance of canopy in different wavelengths. Therefore, vegetation indexes, such as the normalized difference vegetation index (*NDVI*) associated with concentration of leaf-tissue nutrients could be a useful tool for monitoring potential sugarcane yield changes under straw management. Two sites in São Paulo state, Brazil were utilized to evaluate the potential of *NDVI* for monitoring sugarcane yield changes imposed by different straw removal rates. The treatments were established with 0%, 25%, 50%, and 100% straw removal. The data used for the *NDVI* calculation was obtained using satellite images (CBERS-4) and hyperspectral sensor (FieldSpec Spectroradiometer, Malvern Panalytical, Almelo, Netherlands). Besides sugarcane yield, the concentration of the leaf-tissue nutrients (N, P, K, Ca, and S) were also determined. The *NDVI* efficiently predicted sugarcane yield under different rates of straw removal, with the highest performance achieved with *NDVI* derived from satellite images than hyperspectral sensor. In addition, leaf-tissue N and P concentrations were also important parameters to compose the prediction models of sugarcane yield. A prediction model approach based on data of *NDVI* and leaf-tissue nutrient concentrations may help the Brazilian sugarcane sector to monitor crop yield changes in areas intensively managed for bioenergy production.

Keywords: crop residue management; remote sensing; satellite images; hyperspectral sensor; vegetation index; yield monitoring

---

1. Introduction

Brazil is the largest sugarcane producer in the world, representing 40% of global production [1]. Sugarcane area extends for approximately 9 million ha, with an ethanol production at around 27 billion liters and about 40 million tonnes of sugar [2]. Although Brazil is a major player in the global biofuel market, growing internal and external demands have been stimulated an increase in ethanol production. Considering the current global production of 98.6 billion liters [3], it will be necessary to double ethanol production to meet the estimated global demand of 200 billion liters in 2021 [4]. In addition, bioelectricity represents an important proportion of the revenues in the industry, and it is projected that about 17% of the domestic electric energy production in Brazil will be provided by sugarcane biomass by 2023 [5].

An alternative to increase bioenergy production, without expanding area, is to use lignocellulosic raw materials, such as bagasse and straw. In this process, crop residues, rich in carbohydrates, are converted into simple sugars using enzymatic hydrolysis, followed by yeast fermentation to produce ethanol, or crop residues are burnt to generate bioelectricity [6].

More recently, sugarcane straw has been considered as an alternative raw material to address the increasing demand for bioenergy production [7]. Since adoption of mechanized harvesting of sugarcane, about 10 to 20 Mg ha^{-1} (as dry mass) remains on the field annually [8]. However, integral or partially straw retention on the soil surface is essential to support several soil functions [9,10], such as thermal regulation [11], control of soil compaction [12], water retention [13], carbon storage [14], nutrient cycling [15], biological activities [16], and control of erosion [17]. Consequently, changes in soil quality can affect, directly or indirectly, crop production [14,18,19]. Therefore, monitoring the yield of sugarcane could be a strategy to evaluate the effects of varying the amount of straw in the soil–plant system.

Crop yield can be monitored by vegetation indexes. Indeed, Tucker et al. [20] reported that up to 64% of the grain yield variation was explained by spectral data. The applications of *NDVI* has been broadly increased, not only because *NDVI* can be derived from data of either hyperspectral sensors or satellite images [21], but also because it is closely correlated with yield of different crops [22], such as maize (*Zea mays* L.) [23], soybean (*Glycine max* L.) [24], and sugarcane [25]. Additionally, deficiency and excess of nutrients in the leaves can be diagnosed through spectral responses obtained by sensors and, consequently, by vegetation indexes (e.g., *NDVI*) [26]). The variation in the concentration of nutrients in the leaves indicates physiological stress, which can be caused by diverse limiting factors that occurring in the production environment [27].

Although *NDVI* has been used for different purposes, as previously stated, in our knowledge, there are no studies in the literature in which *NDVI* and concentration of leaf-tissue nutrients under sugarcane straw removal management were used to develop predictive models for stalk yield. Thus, the following hypotheses were established: (i) the *NDVI* and concentration of leaf-tissue nutrients are efficient parameters to predict sugarcane yield under different straw removal rates; (ii) the *NDVI* obtained using data from satellite images or hyperspectral sensors have similar performance for predicting sugarcane yield; and (iii) the nutritional status of the crop (evaluated by the leaf-tissue nutrient concentrations) is affected by the different straw removal rates. To test the hypotheses, a study was conducted in areas managed with straw removal in southeastern Brazil to predict sugarcane yield based on *NDVI* and plant nutritional status. In addition, the efficiency of the *NDVI* data for predicting sugarcane yield, obtained using remote sensing imagery from various platforms, was compared.

2. Materials and Methods

2.1. Study Sites

The field study was conducted at two sites in southeastern Brazil, which represents the largest sugarcane-producing region of the country. These areas are located in São Paulo state, near to Capivari, at the Bom Retiro mill (Area 1) (Lat. 22°59′42″ S; Long. 47°30′34″ W) and Valparaiso, at the Univalem mill (Area 2) (Lat. 21°14′48″ S; Long. 50°47′04″ W) (Figure 1).

Soil is classified as Rhodic Kandiudox [28] with a sandy clay loam texture in Area 1, whereas in Area 2, the soil is classified as Kanhaplic Haplustults [28] with a sandy loam texture.

The regional climate for the Area 1 is humid subtropical-Cwa type (Köppen classification) characterized by dry winters and hot summers, with a mean annual temperature of 21.8 °C and annual precipitation of 1289 mm (Figure 2A). In the Area 2, the climate is tropical-Aw type, characterized by dry winters, with a mean annual temperature of 23.4 °C and annual precipitation of 1241 mm (Figure 2B). Rainfall at both sites is concentrated in the spring and summer (October to April), while the dry season is in the autumn and winter (May to September).

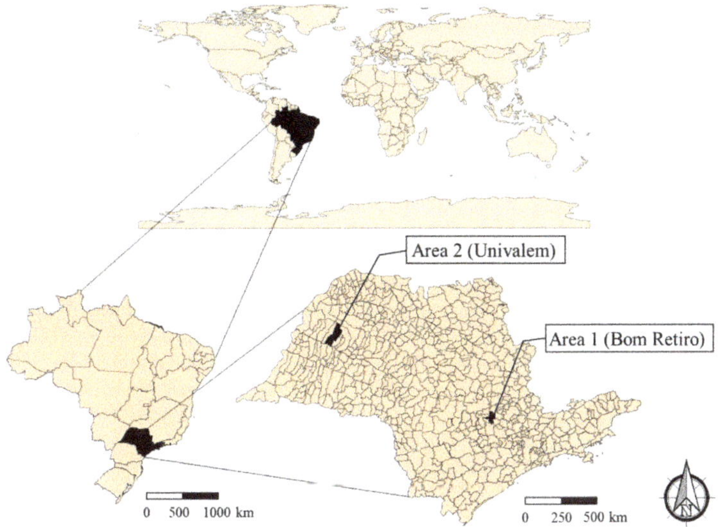

Figure 1. Geographical location of the experimental areas used in the study.

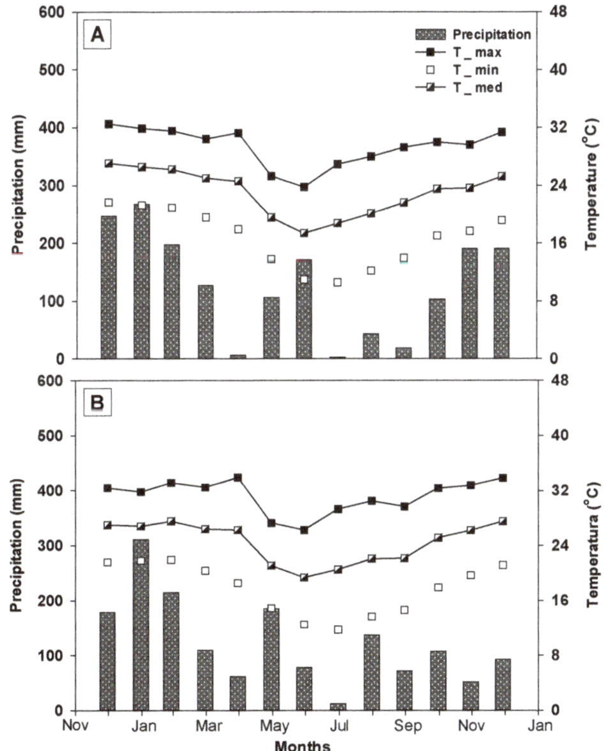

Figure 2. Mean monthly temperatures (maximum, mean, and minimum) (°C) and monthly precipitation (mm) in Area 1 (Capivari, SP) (**A**) and Area 2 (Valparaiso, SP) (**B**) from 1 December 2015 to 31 December 2016. Sources: CEPAGRI [29] and ESALQ [30].

2.2. Experimental Design

Within each site, sugarcane was planted in 15 February 2013, using the varieties CTC 14 and RB 867515 in Area 1 and Area 2, respectively. These varieties are broadly cultivated in this sugarcane-producing region of Brazil. Before the experiment was installed, the areas were sampled in order to obtain a chemical and physical characterization of the soils (Table 1), as previously presented by Satiro et al. [12].

Table 1. Initial characterization of the soils in the experimental areas.

Soil Layer (cm)	pH H$_2$O	C g kg^{-1}	P mg dm^{-3}	K	Ca	Mg	Base Saturation	Al Saturation	Clay
				mmol$_c$ dm^{-3}			%		
				Area 1 (Bom Retiro)					
0–10	5.2	11.3	29.3	9.3	26.1	7.7	68.8	0.8	33
10–20	4.8	11.0	24.9	5.1	19.0	5.9	54.7	3.5	33
20–30	4.5	9.4	22.1	3.3	12.5	2.95	36.8	4.2	34
				Area 2 (Univalem)					
0–10	5.2	6.1	17.4	3.3	9.3	2.9	51.1	2.4	11
10–20	4.8	5.5	14.1	2.6	4.8	1.5	34.8	5.6	11
20–30	4.5	4.9	12.7	2.1	3.6	1.0	27.5	7.4	12

The cane plant (i.e., first cycle) was cultivated for 20 months, thereafter the harvested was performed in 15 October 2014 and the experiments with different rates of straw removal were established. After 14 months, in 17 December 2015, the areas were harvested again and similar experiments were conducted. Leaf sampling and sensor data collection were done during the second sugarcane ratoon season (Figure 3). The experiments were mechanically established with harvester, which was set up with different angular velocities on the primary extractor fan while the secondary extractor fan was switched on or off to remove different quantities of sugarcane straw. Details on how extractors were set up are described in Lisboa et al. [7].

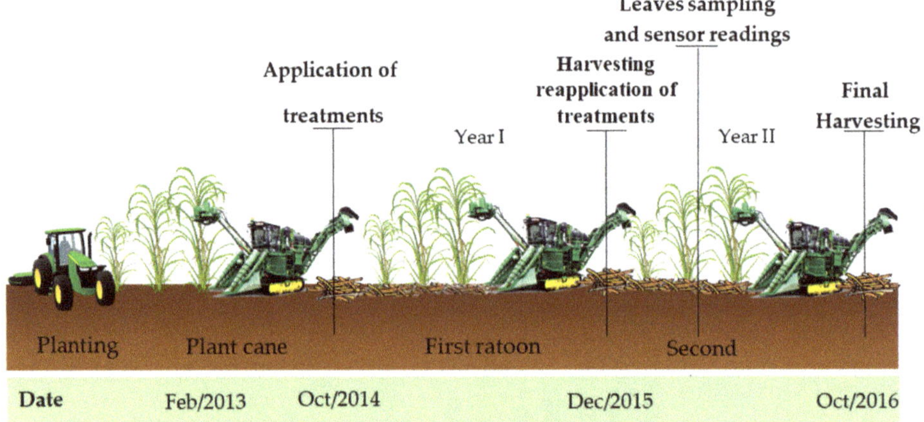

Figure 3. Timeline of the experiments establishment (at harvesting) and conduction over two years in southeast of Brazil. Source: adapted from Lisboa et al. [19].

The experimental design was randomized blocks with four treatments (i.e., straw removal rates), and four replications (four plots of ~50 × 25 m). Initially, the aim was to remove the amount of straw proportional to 0%, 25%, 50%, and 100% of the straw yield in each area. However, under field conditions, the exact proportion was hard to be achieved, but the removal rates were very close to those proposed (Table 2).

Table 2. Amount of sugarcane straw left on soil surface after each treatment in the first and second years in each experimental area.

Straw Removal Rate (%)	Area 1 (Bom Retiro)		Area 2 (Univalem)	
	Year I	Year II	Year I	Year II
	Amount of straw left on the soil surface (Mg ha^{-1}) [#]			
100	0.0 ± 0.0	0.0 ± 0.0	0.0 ± 0.0	0.0 ± 0.0
50	8.7 ± 0.9	5.5 ± 1.0	10.2 ± 0.9	6.8 ± 0.4
25	15.1 ± 1.5	10.5 ± 0.1	12.5 ± 0.9	11.6 ± 0.5
0	18.9 ± 1.6	13.6 ± 2.0	16.4 ± 1.4	13.7 ± 1.1

[#] Dry mass; Source: adapted from Lisboa et al. [19].

At the start of each experiment period, the sugarcane straw was sampled to determine its elemental composition (Table 3).

Table 3. Carbon (C) and macronutrient (nitrogen—N, phosphorus—P, potassium—K, calcium—Ca, magnesium—Mg, sulphur—S) content of the sugarcane straw used in each experiment.

Sites	C	N	P	K	Ca	Mg	S	C:N Ratio
				g kg^{-1}				
Area 1 [I] (Bom Retiro)	479	2.58	0.39	1.66	1.96	1.38	0.45	177
Area 1 [II] (Bom Retiro)	470	3.10	0.34	0.56	2.44	1.24	0.35	152
Area 2 [I] (Univalem)	467	4.02	0.38	2.45	2.44	1.54	0.74	116
Area 2 [II] (Univalem)	422	6.04	0.58	1.30	8.55	2.55	0.95	73

[I] and [II] denote the first and second sugarcane ratoon, respectively. Source: adapted from Lisboa et al. [19].

Overall, this study frames the development of models based on *NDVI* and plant nutritional status to predict sugarcane yield in field managed with straw removal. Plant tillering, growth, and stalk yield under the same straw management were discussed in detail by Lisboa et al. [19]. Thus, part of the primary data (i.e., amount of straw, characterization and stalk yield) presented on this study (mainly in Material and Methods) were originally shown in Lisboa et al. [19].

2.3. Calculation of the NDVI

The data were obtained in the field using the FieldSpec Spectroradiometer (ASD–Analytical Spectral Devices Inc., Boulder, CO, USA) hyperspectral sensor. This equipment is classified as a passive hyperspectral sensor that uses the visible and infrared wavelengths (325 to 1075 nm), with a spectral resolution of 3 nm and a view angle of 25°. Before the canopy reading was initiated, the spectroradiometer was calibrated using the standard Lambertian plate that accompanies the apparatus. This calibration was repeated after reading approximately four plots or whenever a change of light intensity was observed in the field. The *NDVI* was calculated using Equation (1).

$$NDVI = \frac{(NIR - R)}{(NIR + R)} \tag{1}$$

where *NDVI* is the Normalized Difference Vegetation Index (unitless); NIR is the near infrared band and R is the red band; NIR represents the reflected light at the near infrared band and R represents the reflected light at the red band.

The field evaluations with the hyperspectral sensor were carried out on 28 April 2016 and 4 March 2016 for Areas 1 and 2, respectively, four months after harvesting. This period coincides with the recommended period for foliar diagnosis in sugarcane crop [31]. The evaluations were always carried out on sunny days, between 10:00 a.m. and 2:00 p.m., with the sensor positioned 1 m above the crop canopy and a field of view of 25°, which allowed the evaluation of a circular area of approximately 0.25 m² (Figure 4).

Figure 4. Representation of the hyperspectral sensor in the field (**a**) and Lambertini plate used for calibration (**b**).

The satellite images were obtained from the satellite CBERS-4 (China Brazil Earth Resources Satellite) on the 17 March 2016 and the 4 May 2016 for Areas 1 and 2, respectively. The images presented a spatial resolution of 5 m, radiometric resolution of 8 bits and wavelength of 0.50–0.840 μm. As the panchromatic band of CBERS-4 was supplied in digital numbers without atmospheric calibration, the images were submitted to atmospheric correction according to Vermote et al. [32]. To perform *NDVI* calculations, the image was converted to surface reflectance according to the method described by Carlotto [33].

2.4. Plant Parameters

After measurements to calculate the *NDVI*, 50 leaves (i.e., the diagnose leaf) were randomly collected from each plot for the determination of the leaf-tissue nutrient content. The third leaf from top to bottom of the stalk with clearly visible dewlap, is the most photosynthetically active and is used for monitoring the nutritional status of sugarcane crop [31].

At the end of the ratoon cycle, approximately one year after previous harvesting, the sugarcane yield was quantified. The fresh stalk mass was mechanically harvested from the five central rows (500 m long covering an area of 525 m^2) of each plot, weighed in the field using a wagon coupled to a balance and the results expressed in Mg ha^{-1}.

2.5. Data Analysis

The data were initially subjected to exploratory analysis (descriptive statistics), aiming to verify the position and dispersion of the data using the Statistical Analysis System (SAS) 9.4 Software Package for Windows 8 (SAS Inc, Cary, NC, USA).

In order to select the explaining variables of stalk yield, a stepwise multiple regression ($Y = a + b_1x^1 + b_2x^2 + ... + b_nx^n$) was applied. This functions by the systematic addition or removal of variables in the regression, based on a statistical test of significance for each variable ($p < 0.05$). The final model includes only those variables that have a decisive influence on the dependent variable. The models were evaluated according to the coefficient of determination, residual standard error and the Durbin Watson test. The Durbin–Watson test ranges from 0 to 4, where values closer to 2 indicate optimal values [34], i.e., the absence of autocorrelation in the data. The means of variables selected by stepwise multiple regression were compared according to Tukey's test ($p < 0.05$). In addition, the overall model that pools data from both studied areas was tested through sensitivity analysis, where position measurements (maximum, minimum, and mean) of each selected variable were inserted independently in the model, aiming to evaluate the resulting variations in sugarcane yield.

Finally, a model validation process was carried out with 30% of the data that were not used to generate the models. The relationship between the observed and predicted values were evaluated by the root mean square error (RMSE). The Pearson correlation matrix was also used to evaluate the relationship between observed and predicted productivity of sugarcane with variables selected with the stepwise models.

3. Results

3.1. NDVI Values

The *NDVI* obtained using the data from the hyperspectral sensor and the satellite images presented differences with respect to the distribution and magnitude of the values between the experimental areas (Figure 5). In Area 1, the *NDVI* values using the sensor and the satellite images ranged between 0.60 to 0.80 and 0.40 to 0.60, respectively. However, in Area 2 the *NDVI* values were smaller, ranging between 0.40 to 0.70 and 0.20 to 0.50 for the sensor and the satellite images, respectively. On average, *NDVI* values (from both methodologies) were 34% higher in Area 1 than in Area 2.

In both experimental areas, only the *NDVI* values derived from hyperspectral sensor followed a normal distribution (Figure 5). The largest differences were found in the *NDVI* values derived from the satellite images, especially when the removal rates were 25% and 50% for Area 1 and Area 2, which did not follow a normal data distribution. Although there are variations between *NDVI* derived by these two methods, the results showed a quite similar pattern in the variations of the *NDVI* data due to straw removal rates.

3.2. Sugarcane Leaf-Tissue Nutrient Concentration

The nutrient concentration in the sugarcane leaves varied according to the straw removal rate at both sites (Figures 6 and 7). The greatest variation was observed in the elements K (5 to 15 g/kg^{-1}) and Ca (2 to 6 g/kg^{-1}), whereas that the variation was less intense for the other elements. Despite that, for the majority of cases, nutrient data did not followed a normal distribution (Figures 6 and 7).

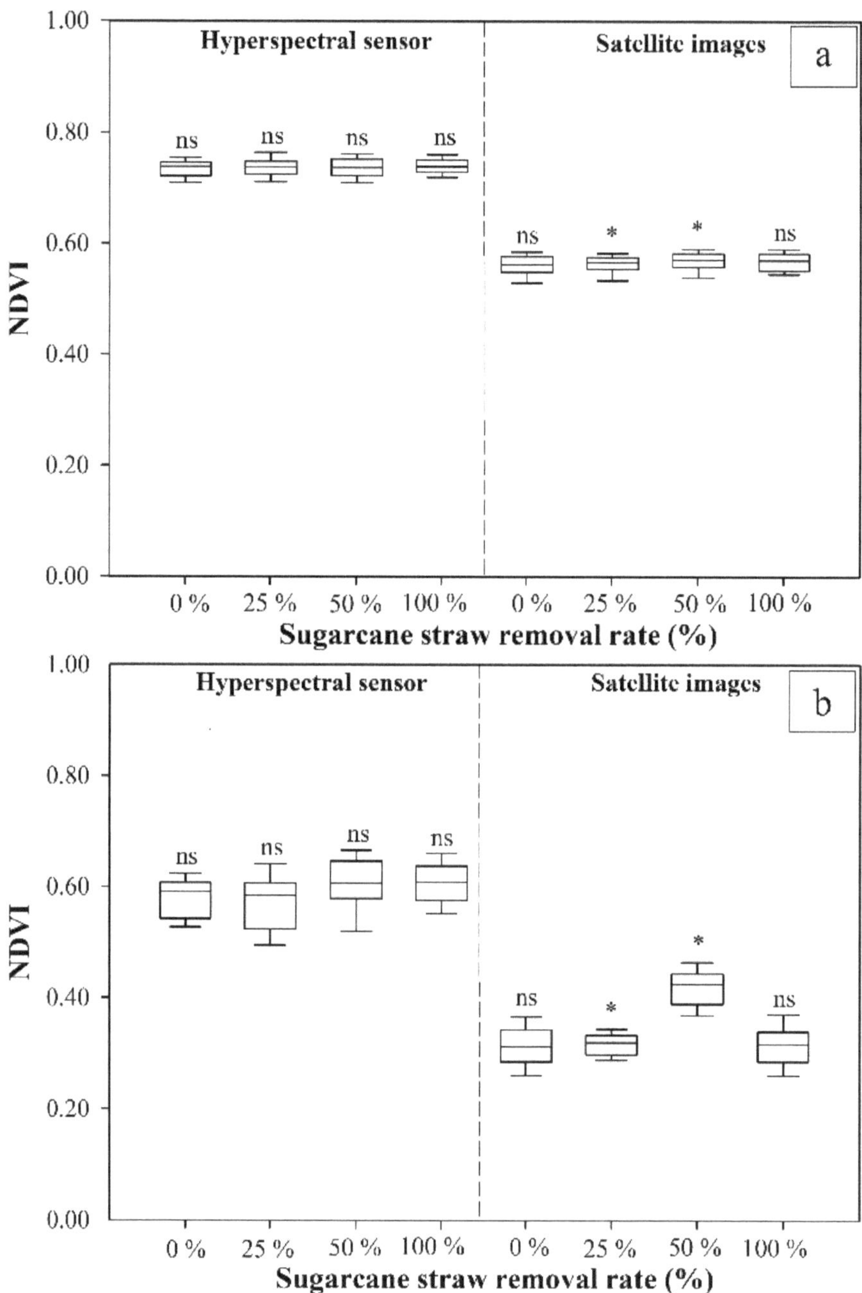

Figure 5. Values of *NDVI* for Areas 1 (**a**) and 2 (**b**) at the different rates of straw removal. Shapiro–Wilk Test for normal distribution, where: * significant ($p < 0.05$) and ns not significant. When significant, it indicates that the hypothesis for normal distribution is rejected.

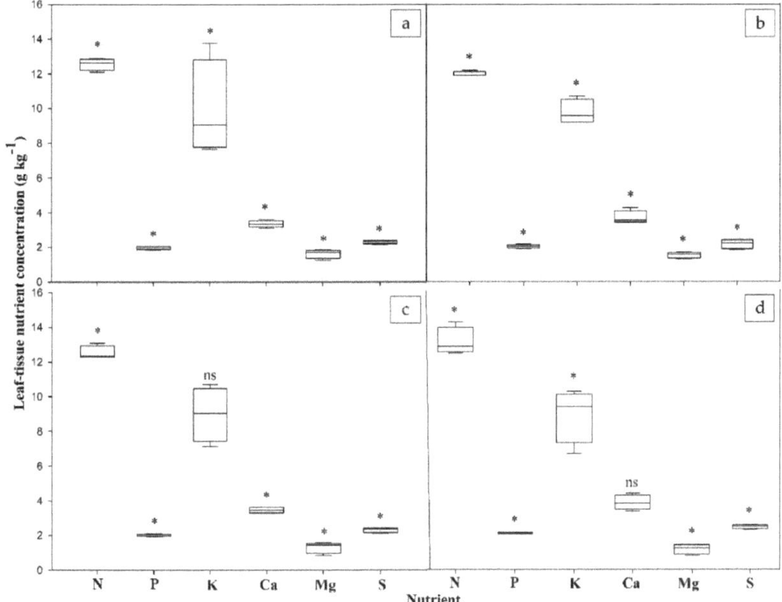

Figure 6. Leaf-tissue concentrations of N, P, K, Ca, Mg, and S in sugarcane grown under straw removal rates of 0 (**a**), 25 (**b**), 50 (**c**), and 100% (**d**) in Area 1. Shapiro–Wilk Test for normal distribution, where: * significant ($p < 0.05$) and ns not significant. When significant, it indicates that the hypothesis for normal distribution is rejected.

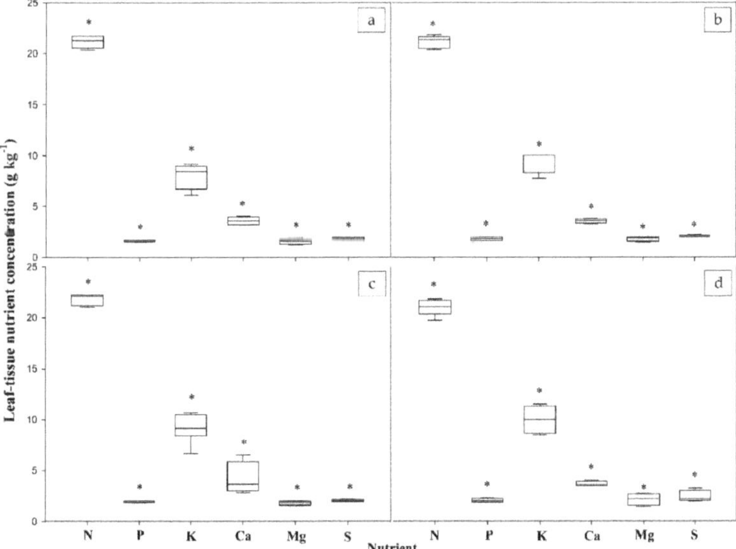

Figure 7. Leaf-tissue concentration of N, P, K, Ca, Mg, and S in sugarcane grown under straw removal rates of 0 (**a**), 25 (**b**), 50 (**c**), and 100% (**d**) in Area 2. Shapiro–Wilk Test for normal distribution, where: * significant ($p < 0.05$) and ns not significant. When significant, it indicates that the hypothesis for normal distribution is rejected.

3.3. Sugarcane Yield

Sugarcane yield varied from 60 to 85 Mg ha^{-1} and from 10 to 30 Mg ha^{-1} for the Areas 1 and 2, respectively. On average, sugarcane yield was three times higher in Area 1 than in Area 2 (Figure 8). Our findings showed that straw removal rates did not linearly affect sugarcane yield, but it depends on the site-specific characteristics related to soil, climate, and variety.

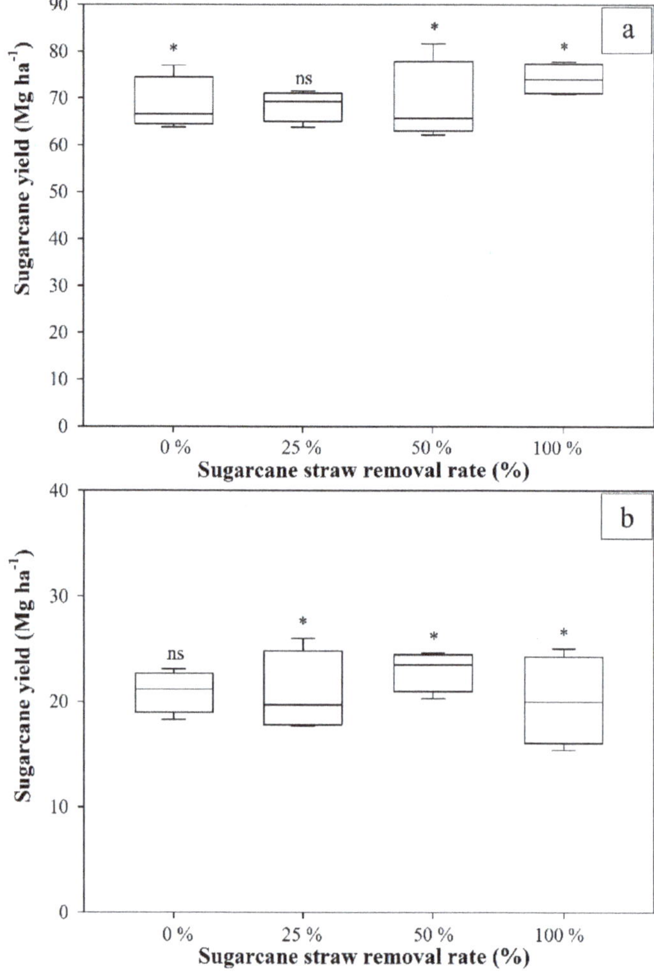

Figure 8. Sugarcane yield (Mg ha^{-1}) changes induced by straw removal rates in Areas 1 (**a**) and 2 (**b**). Shapiro–Wilk Test for normal distribution, where: * significant ($p < 0.05$) and ns not significant. When significant, it indicates that the hypothesis for normal distribution is rejected.

The data did not follow a normal distribution for the majority of the treatments (Figure 8), except for data from 25% and 0% straw removal plots in Area 1 and 2, respectively.

3.4. Modelling Sugarcane Yields by Stepwise Analysis

Explaining variable of sugarcane yield were selected for each straw removal rate using stepwise analysis (Table 4). The *NDVI* obtained from satellite images was more efficient to predict sugarcane yield compared to *NDVI* derived from hyperspectral sensor, and therefore, this parameter was included in all the models.

Table 4. Stepwise models for estimating the components of sugarcane yield under straw removal rates in Areas 1 and 2.

Straw Removal Rate (%)	Models	r^2 *	SRE	DW
	Area 1			
0	$y = 205.38 - 48.66\,P - 0.83\,K - 58.36\,NDVI_{SI}$	0.94	1.76	1.78
25	$y = 546.65 - 58.07\,P - 30.31\,N + 7.18\,NDVI_{SI}$	0.91	1.31	1.52
50	$y = 322.62 - 108.24\,P - 6.33\,N + 60.36\,NDVI_{SI}$	0.88	4.07	1.48
100	$y = 193.25 - 56.54\,P + 4.25\,NDVI_{SI} + 0.27\,N$	0.85	2.10	1.65
Overall (Area 1)	$y = 136.70 + 42.84\,NDVI_{SI} - 51.19\,P + 1.61\,N$	0.97	2.56	1.56
	Area 2			
0	$y = -13.20 + 1.75\,N - 13.76\,NDVI_{SI}$	0.87	6.64	1.44
25	$y = -166.31 + 25.60\,P + 6.51\,N + 22.76\,NDVI_{SI}$	0.91	8.65	1.65
50	$y = 42.25 - 12.41\,P + 22.23\,NDVI_{SI}$	0.76	4.50	1.59
100	$y = 134.25 - 20.90\,Ca - 20.51\,P + 31.89\,NDVI_{SI}$	0.94	4.29	1.56
Overall (Area 2)	$y = -127.21 + 3.80\,NDVI_{SI} - 0.26\,N + 54.88\,P$	0.88	3.86	1.61
Overall (Area 1 and Area 2)	$y = -150.028 + 79.11\,NDVI_{SI} - 3.65\,N + 115.28\,P$	0.89	1.79	1.72

SRE: standard residual error (%); DW: Durbin–Watson test. N: nitrogen; P: phosphorus; K: potassium; Ca: calcium; and $NDVI_{SI}$: *NDVI* obtained by satellite images. * correlation coefficient.

The nutrients that most influenced crop yield were P and N, being included in the model for at least one straw removal rate in both experimental areas; while the K and Ca were only included in the models for the removal rates of 0% and 25% for Area 1 and 100% for Area 2. Overall models for Area 1 and 2, separately and together, included the parameters such as *NDVI* obtained from the satellite images, leaf-tissue concentration of P and N.

The models presented efficient performance to predict the sugarcane yield (Table 4). In Area 1, the r^2 values varied between 0.85 and 0.94 with the highest SRE encountered under the 50% removal rate (4.07%). The r^2 values varied between 0.76 and 0.94 for Area 2, with the highest SRE observed under the 25% removal rate (8.65%). Although SRE of 8.65% was the highest value observed in the two areas it was considered acceptable by the validation of the model. In the overall models, the r^2 and SRE values for Areas 1 and 2 were 0.97% and 2.56%, and 0.88% and 3.86%, respectively, and for the general model representing both areas the r^2 and SRE values were 0.89% and 1.79%, respectively. For the models generated individually for each straw removal rate in each of the two areas, for the general models for each area and for the general model representing both areas, the Durbin–Watson test values were similar, with values ranging between 1.44 and 1.78. These values were close to 2 (ideal threshold), showing that the data do not present autocorrelation [34]. The sensitivity analysis of the overall model for Areas 1 and 2, showed that yield ranged 11% (i.e., 39.2 to 52.5 Mg ha^{-1}) when the selected parameter changed from the minimum to maximum value. Individually, variation on *NDVI*, P, and N impacts were 55.84%, 55.03%, and 53.10% in the sugarcane yield.

3.5. Effect of Sugarcane Straw Removal on Selected Yield-Explaining Variables

The comparison of the means of the selected variables by the stepwise model indicated that sugarcane yield was not altered by straw removal management within the study period (Figure 9a). In relation to the variables that explained the sugarcane yield, it was observed that 25% straw removal rate induced a reduction in the *NDVI*, using the satellite image data (Figure 9b), and the leaf-tissue P content (Figure 9d) in Area 1. However, 50% straw removal rate induced an increase in the *NDVI*, using the satellite image data (Figure 9b) and the leaf-tissue N (Figure 9c) and P (Figure 9d) contents in Area 2.

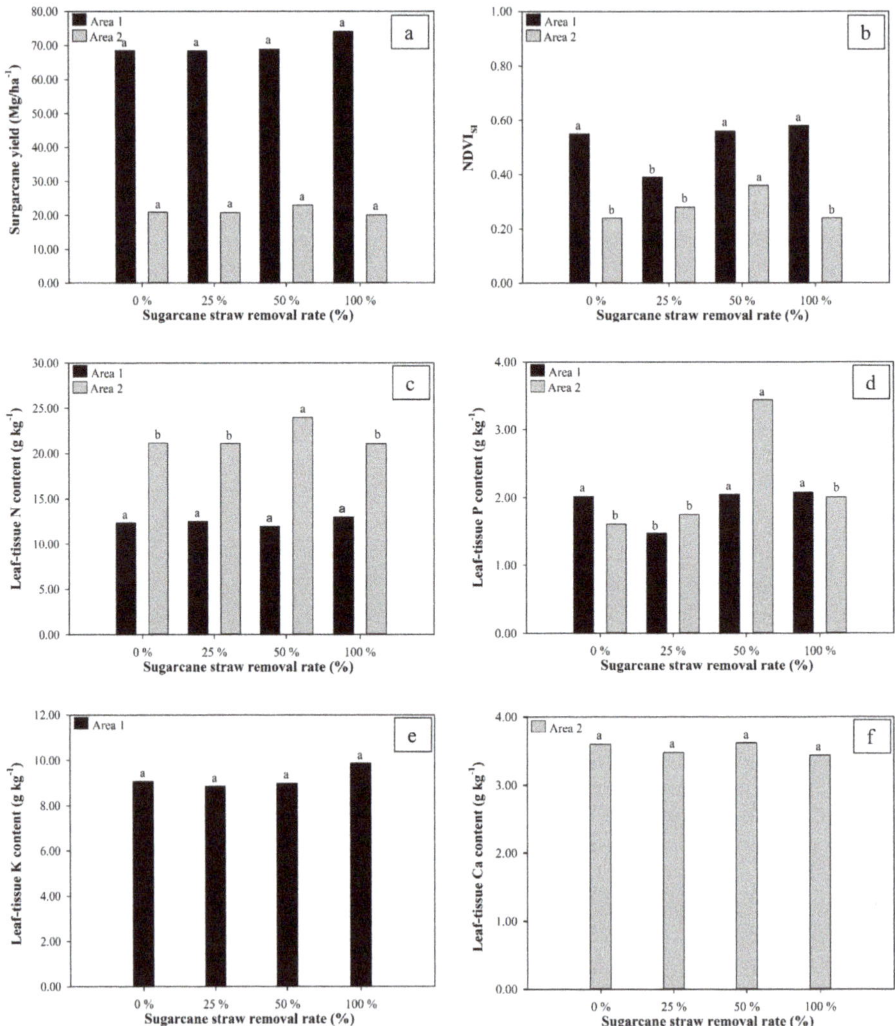

Figure 9. Comparison of the sugarcane straw removal rates with the variables selected by the stepwise model means. Sugarcane yield (**a**), the *NDVI* obtained from the satellite image data (**b**), leaf-tissue N (**c**), P (**d**), K (**e**), and Ca (**f**) contents in Areas 1 and 2. Means followed by the same letter are not statistically different by the Tukey test ($p \leq 0.05$).

3.6. Validation of Models

The models were validated based on relationship between observed and predicted values (Figure 10). In general, the results revealed increases of estimate errors (i.e., RMSE values) with the increase in the straw removal rate (0% to 100% straw removal rate). The highest values were observed for the rates of 50 (10.34 Mg ha^{-1}) and 100% (4.15 Mg ha^{-1}) for Areas 1 and 2 respectively. The overall model, including Areas 1 and 2, was efficient to predict the sugarcane yield, presenting RMSE of 0.77 Mg ha^{-1}.

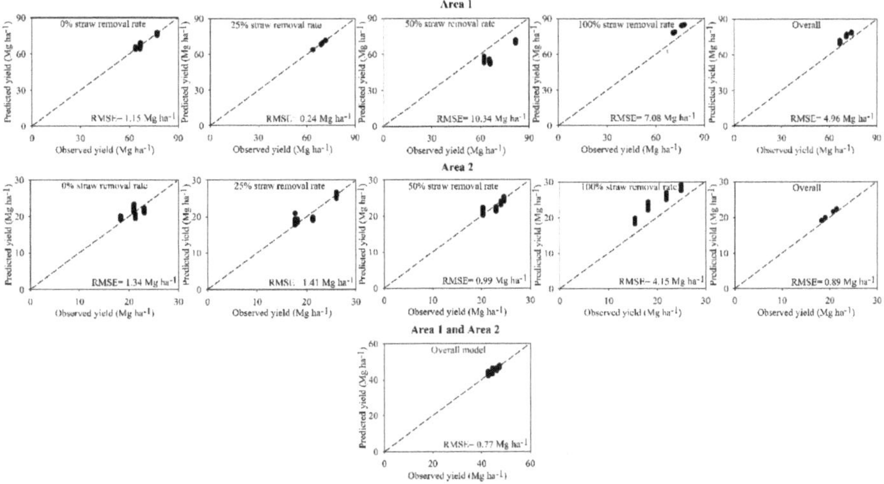

Figure 10. Comparison of sugarcane yield observed with predicted values with Stepwise models. The diagonal dashed line is the 1:1 line, whereas RMSE is the root mean square error.

Observed (OY) and predicted (PY) yield data presented satisfactory performance when compared to the variables selected through the stepwise models (Table 5). Leaf-tissue P concentration presented negative correlation with OY and PY yield data for the Area 1 subjected to all rates of straw removal (i.e., 0%, 25%, 50%, and 100%). Similar pattern was also observed to the general model for the same area. While for Area 2, P presented negative correlation with PY and OY data only under 50% of straw removal. In addition, N presented positive and significant correlations with OY and PY data from Area 1 under 50% and 100% of straw removal as well as for general model for this area, coefficients of correlation (r) ranged from 0.33 to 0.68. The correlations among $NDVI_{SI}$, OY, and PY data were positive for all rates of straw removal and overall model for the area 1. Same pattern was verified for all rates of straw removal in the Area 2, except for the rate of 25% of straw removal; coefficients of correlation ranged from −0.40 to 0.93 in Area 2.

It taking into account overall model established for both areas, leaf-tissue concentrations of P and N were significantly correlated with stalk yield, but in different order, as follow: P negatively correlated with OY ($r = -0.45$) and PY ($r = -0.50$) data, while N was positively correlated with OY ($r - 0.55$) and PY ($r = 0.58$) data. Similarly, $NDVI_{SI}$ correlated positively with OY ($r = 0.60$, $p < 0.05$) and PY ($r = 0.68$, $p < 0.05$) data.

Table 5. Correlation matrix between observed and predicted sugarcane yield values with selected variables with stepwise models.

0%

Area 1	P	K	$NDVI_{SI}$	OY	PY
P	-				
K	0.19	-			
$NDVI_{SI}$	−0.23	0.22	-		
OY	−0.84 **	0.60 **	0.71 **	-	
PY	−0.86 **	0.62 **	0.74 **	0.97 **	-

Area 2	N	$NDVI_{SI}$	OY	PY
N	-			
$NDVI_{SI}$	−0.14	-		
OY	0.54 *	0.37 *	-	
PY	0.87 **	0.60 **	0.75 **	-

25%

Area 1	P	N	$NDVI_{SI}$	OY	PY
P	-				
N	−0.85 **	-			
$NDVI_{SI}$	0.01	0.38	-		
OY	−0.62 *	0.25	0.50 *	-	
PY	−0.71 **	0.23	0.45 *	0.98 **	-

Area 2	P	N	$NDVI_{SI}$	OY	PY
P	-				
N	−0.49 *	-			
$NDVI_{SI}$	−0.56 *	0.02	-		
OY	0.62 **	0.23	−0.37 *	-	
PY	0.68 **	0.26	−0.40 *	0.90 **	-

50%

Area 1	P	N	$NDVI_{SI}$	OY	PY
P	-				
N	0.15	-			
$NDVI_{SI}$	−0.16	0.46 *	-		
OY	−0.90 **	0.33 *	0.84 **	-	
PY	−0.96 **	0.35 *	0.81 **	0.93 **	-

Area 2	P	$NDVI_{SI}$	OY	PY
P	-			
$NDVI_{SI}$	−0.41 *	-		
OY	−0.73 **	0.60 **	-	
PY	−0.88 **	0.76 **	0.89 **	-

100%

Area 1	P	N	$NDVI_{SI}$	OY	PY
P	-				
N	−0.62 **	-			
$NDVI_{SI}$	−0.39 *	0.25	-		
OY	−0.96 **	0.56 **	0.73 **	-	
PY	−0.91 **	0.64 **	0.61 **	0.98 **	-

Area 2	P	Ca	$NDVI_{SI}$	OY	PY
P	-				
Ca	−0.54 *	-			
$NDVI_{SI}$	0.23	0.30	-		
OY	−0.22	0.64 **	0.30	-	
PY	−0.14	0.72 **	0.29	0.97 **	-

Overall

Area 1	P	N	$NDVI_{SI}$	OY	PY
P	-				
N	0.95 **	-			
$NDVI_{SI}$	−0.54 *	−0.28	-		
OY	−0.84 **	0.68 **	0.86 **	-	
PY	−0.81 **	0.61 **	0.93 **	0.96 **	-

Area 2	P	N	$NDVI_{SI}$	OY	PY
P	-				
N	0.84 **	-			
$NDVI_{SI}$	−0.24	0.12	-		
OY	−0.28	−0.26	0.61 **	-	
PY	−0.24	−0.28	0.69 **	0.97 **	-

Table 5. *Cont.*

Area 1						Area 2
	Overall (Area 1 and Area 2)					
	P	N	$NDVI_{SI}$	OY	PY	
P	-					
N	0.96 **	-				
$NDVI_{SI}$	−0.08	0.08	-			
OY	−0.45 *	0.55 *	0.60 **	-		
PY	−0.50 *	0.58 *	0.68 **	0.90 **	-	

*, ** Significant at $p < 0.05$ and $p < 0.01$, respectively. OY: observed yield; PY: predicted yield; N: nitrogen; P: phosphorous; K: potassium; Ca: calcium and $NDVI_{SI}$: NDVI obtained by satellite images.

4. Discussion

4.1. Effect of Straw Removal on the Leaf-Tissue Nutrient Concentration and Sugarcane Yield

Sugarcane straw removal had little or even no impacts on leaf-tissue nutrient content (Figures 6 and 7). Nutrient cycling in the soil was minimally affected by the straw removal rates in the short term [12], and therefore, plant nutrition was little affected. Among the evaluated elements, K and Ca presented the largest variations in both experimental areas. The sugarcane straw can recycle about 80–100 kg ha^{-1} of K [35,36]. According to Rosolem et al. [37], more than 80% of the K is present in a soluble form in plant tissues, which makes this element easily leached from plant tissues [9] for later absorption by the plant root system. Unbalanced absorption of K may also change Ca content in plant tissues, since both elements interfere in the physiological process of transporting sugars into the plant [38].

Although K and Ca were the most variable in the plant tissues as a function of straw removal, N and P were the most sensitive nutrients to the straw removal rates (Figure 9c,d). When compared to other nutrients—such as P, K, Ca, Mg, and S—N presented the largest reduction in the soil with straw removal [39]. On the other hand, straw removal led to small reductions in the level of soil P, since the content of this element in straw is low (Table 3). However, the majority of P is highly adsorbed in the soil particles (especially Fe and Al oxides) [40], thus, small inputs of P in organic forms via straw decomposition can be important to increase plant-availability P in the soils [41]. In addition, the organic acids released during straw decomposition may compete for adsorption sites and increase the availability of phosphate ions in the soil solution [42].

The sugarcane yield varied three times between the two areas evaluated (Figure 8). This difference may be related to the greater soil fertility of Area 1, in which soil P, K, Ca, and Mg contents and base saturation percentage were on the average 42%, 55%, 69%, 67%, and 29%, respectively higher than Area 2. Thus, higher soil fertility in Area 1 likely allowed that plants accumulate higher levels of these nutrients in the leaf tissue, which is directly related to the increase of the sugarcane yield [43]. Sugarcane presents high nutritional needs once it a crop that remains in the field for several years (~5 to 8 years), with high nutrient removal in each annual harvesting cycle. Therefore, improper replenishment of nutrients is one of the reasons for yield loss during the annual harvesting cycles [44,45].

The sugarcane straw removal did not affect biomass production (Figure 9a). These results are in agreement with other studies reporting that the removal of up to 50% of the straw produced did not impact crop yield [18,19,46]. In addition, through an extensive literature review, Carvalho et al. [9] suggested that at least 7 Mg ha^{-1} of straw should remain on the soil surface to sustain suitable soil functioning and plant yield over time. However, as the crop tends to lose the productive potential (stalk and straw) over successive harvesting cycles [19,47], the amount of straw that should be removed after each harvesting cycle may be different.

Over the past few years, studies have brought very relevant understanding about the impacts of straw removal management on soil quality [12,14,36,48], SOC accumulation [49,50], greenhouse gas emissions [51–53], plant growth and stalk yield [18,19,46,54–56]. Despite this fact, future studies are essential to understand the long-term implications of straw removal management on soil and stalk yield. So far, we only have prediction performed by DayCent model of sugarcane straw removal impacts on SOC accumulation across the different soil types within the main core occupied with the crop in Brazil (i.e., central-south region) [49,50]. However, these estimates performed by Carvalho et al. [49] and Oliveira et al. [50] still have to be validated by long-term experiment, still inexistent in Brazil. The same statement is true for the yield prediction models developed in this present study, which gave us relevant insights related to the potentialities of NVDI and leaf-tissue nutrient concentration to predict sugarcane yield in fields impacted by straw removal management.

4.2. NDVI Acquisition by Satellite Image vs Hyperspectral Sensor

The quality of the data used to calculate the *NDVI* is affected by interference from diverse factors, depending on the form of data acquisition. When satellite images are used, although the data is collected with minimum cloud cover, the interference of these few clouds at the moment of data acquisition could alter the quality of the incident and reflected radiation [57]. Besides this, the capture of satellite images tends to suffer a greater influence from the soil than that experienced using the hyperspectral sensor, where readings are taken above the canopy and rows with minimal soil influence. Thus, the *NDVI* values using satellite images are normally lower [58]. These interferences explain the difference of 34% observed between the *NDVI* using the hyperspectral sensor those obtained using satellite images for the sugarcane in the two experimental areas (Figure 5). Although several studies have shown a high correlation between the *NDVI* obtained using hyperspectral sensor and satellite image data [59,60], we suggest caution when choosing the spectral data acquisition platform to calculate the *NDVI*.

The *NDVI* from the satellite images data was included in all the models generated (Table 4) and was efficient for detecting the effects of straw removal rates on sugarcane yield in Areas 1 and 2 (Figure 9b). In spite of limitations attributed to *NDVI* data derived from satellite images, the results showed that acquired data from satellite images better represented the variation in sugarcane yield [61]. This type of data acquisition has the advantage that it can be applied for larger areas in relation to the data acquired using hyperespectral sensors, which are generally limited to smaller areas of the crop canopy.

4.3. Sugarcane Yield Prediction Using the NVDI and Leaf-Tissue Nutrient Concentration

In all the stepwise models established to estimate sugarcane yield as a function of each straw removal rate, the *NDVI* was included (Table 4). This inclusion can be explained by the fact that most of models used to estimate sugarcane yield are based on the estimated biomass production [62]. This would explain, at least in part, the efficiency of the *NDVI* to detect stalk yield changes as a function of the straw removal rates, as this index has a high correlation with crop biomass [63].

Among the nutrients included in the site-specific models (i.e., N, P, and K for Area 1; and N, P, and Ca for Area 2), only N and P were included in the overall model for the two areas (Table 4). Nutrients such as N and P are present in larger quantities in younger plants, being more sensitive to changes caused by different management systems [64,65]. In addition, the relationship of these nutrients with the *NDVI* in the models was expected due to the influence of these elements on the photosynthetic process, which in turn, change the spectral responses by the plant, influencing the *NDVI* [66,67]. In this approach, other studies can be performed to identify the spectral signatures of the nutrients in the leaves so that the model can only be applied using spectral data.

The performance of the models in the validation process were also satisfactory when observed RMSE values between the rates of 0% to 100%, ranged from 0.24 to 10.34 Mg ha^{-1} for the area 1 and 0.99 to 4.15 Mg ha^{-1} for Area 2 (Figure 10). These values agree with those reported by Fernandes et al. [68], which found RMSE values of predicted sugarcane yields ranged from 7.20 to 11.00 Mg ha^{-1}. In addition, the sensitivity analysis showed that overall model for Areas 1 and 2 provides a satisfactory margin for the prediction of sugarcane yield with similar soil and climate conditions of this study. Therefore, the overall model performance showed *NDVI* from satellite images and leaf-tissue nutrient concentrations can be useful and alternative tool to improve the sugarcane yield predictions.

5. Conclusions

The normalized difference vegetation index (*NDVI*) can be a useful tool for predicting sugarcane yield in fields managed with straw removal management in Brazil. Between the methods of data acquisition to calculate *NDVI*, satellite image was more efficient than hyperspectral sensor for detecting straw removal effects on sugarcane yield. In addition, concentrations of leaf-tissue N and P were also important parameters to develop the prediction models of sugarcane yield.

Prediction model approach based on data of *NDVI* and concentration of leaf-tissue nutrients collected in early stages of crop growth can help the Brazilian sugarcane sector to predict crop yield in fields intensively managed for bioenergy production. In addition, these models can be used for monitoring spatio-temporal crop yield changes induced by straw removal and supporting decision making towards a more sustainable crop residue management in Brazilian sugarcane fields. The sugarcane yield prediction models developed in this study have to be further tested and validated using data from long-term straw removal experiments, so far, still inexistent in Brazil.

Author Contributions: I.P.L.: field sampling, writing, review; J.M.D.: data analyses, writing, review; M.R.C.: writing and review; P.P.S.B.: field sampling and writing; P.R.F.: financial support and writing; C.C.C.: this paper is in his memory; C.E.P.C.: financial support, writing and review.

Funding: I.P.L. thanks The Brazilian Federal Agency for Support and Evaluation of Graduate Education (CAPES process #88881.134605/2016-01) and CNPq (processes #141459/2015-8 and #201207/2017-6) for providing his Ph.D. scholarships in Brazil and the United States. M.R.C. thanks the "Fundação de Estudos Agrários Luiz de Queiroz" (project #67555) for providing his post-doctoral fellowship. Also, we want to thank the Brazilian Development Bank–BNDES and the Raízen Energia S/A for funding our research (project #14.2.0773.1).

Acknowledgments: We are grateful to Lilian Duarte, Admilson Margato, Ralf Araújo, Sandra Nicolete, Dagmar Vasca and Eleusa Basse for the strong assistance in the experiment conduction and samples analyses in the laboratory.

Conflicts of Interest: The authors declare no conflict of interest.

References

1. FAO—Food and Agriculture Organization. Available online: http://faostat.fao.org/ (accessed on 25 July 2018).
2. Companhia Nacional de Abastecimento-Conab. Available online: https://www.conab.gov.br/ (accessed on 25 July 2018).
3. REN21-Highlights of the REN21 Renewables 2017 Global Status Report in Perspective. Available online: http://www.ren21.net/wp-content/uploads/2017/06/GSR2017_Highlights_FINAL.pdf (accessed on 15 June 2018).
4. Goldemberg, J.; Mello, F.F.; Cerri, C.E.; Davies, C.A.; Cerri, C.C. Meeting the global demand for biofuels in 2021 through sustainable land use change policy. *Energy Policy* **2014**, *69*, 14–18. [CrossRef]
5. de Expansão de Energia 2024. Available online: http://www.epe.gov.br (accessed on 1 May 2018).
6. Cunha, F.M.; Badino, A.C.; Farinas, C.S. Effect of a novel method for in-house cellulase production on 2G ethanol yields. *Biocatal. Agric. Biotechnol.* **2017**, *9*, 224–229. [CrossRef]
7. Lisboa, I.P.; Cherubin, M.R.; Cerri, C.C.; Cerri, D.G.P.; Cerri, C.E.P. Guidelines for the recovery of sugarcane straw from the field during harvesting. *Biomass Bioenergy* **2017**, *96*, 69–74. [CrossRef]
8. Leal, M.R.L.V.; Galdos, M.V.; Scarpare, F.V.; Seabra, J.E.A.; Walter, A.; Oliveira, C.O.F. Sugarcane straw availability, quality, recovery and energy use: A literature review. *Biomass Bioenergy* **2013**, *53*, 11–19. [CrossRef]
9. Carvalho, J.L.N.; Nogueirol, R.C.; Menandro, L.M.S.; Bordonal, R.D.O.; Borges, C.D.; Cantarella, H.; Franco, H.C.J. Agronomic and environmental implications of sugarcane straw removal: A major review. *Glob. Chang. Biol. Bioenergy* **2017**, *9*, 1181–1195. [CrossRef]
10. Cherubin, M.R.; Oliveira, D.M.S.; Feigl, B.J.; Pimentel, L.G.; Lisboa, I.P.; Gmach, M.R.; Varanda, L.L.; Moraes, M.C.; Satiro, L.S.; Popin, G.V.; et al. Crop residue harvest for bioenergy production and its implications on soil functioning and plant growth: A review. *Sci. Agric.* **2018**, *75*, 255–272. [CrossRef]
11. Corrêa, S.T.R.; Carvalho, J.L.N.; Hernandes, T.A.D.; Barbosa, L.C.; Menandro, L.M.S.; Leal, M.R.L.V. Assessing the effects of different amountsof sugarcane straw on temporal variability of soil moisture content and temperature. In Proceedings of the 25th European Biomass Conference and Exhibition, Stockholm, Sweden, 12–15 June 2017; pp. 12–15.
12. Satiro, L.S.; Cherubin, M.R.; Safanelli, J.L.; Lisboa, I.P.; Junior, P.R.R.; Cerri, C.E.E.; Cerri, C.C. Sugarcane straw removal effects on Ultisols and Oxisols in south-central Brazil. *Geod. Reg.* **2017**, *11*, 86–95. [CrossRef]

13. Dos Anjos, J.C.R.; Júnior, A.S.A.; Bastos, E.A.; Noleto, D.H.; Melo, F.B.; De Brito, R.R. Water storage in a Plinthaqualf cultivated with sugarcane under straw levels. *Pesqui. Agropecu. Brasil.* **2017**, *52*, 464–473. [CrossRef]

14. Bordonal, R.O.; Menandro, L.M.S.; Barbosa, L.C.; Lal, R.; Milori, D.M.B.P.; Kolln, O.T.; Franco, H.C.J.; Carvalho, J.L.N. Sugarcane yield and soil carbon response to straw removal in south-central Brazil. *Geoderma* **2018**, *328*, 79–90. [CrossRef]

15. Fortes, C.; Vitti, A.C.; Otto, R.; Ferreira, D.A.; Franco, H.C.J.; Trivelin, P.C.O. Contribution of nitrogen from sugarcane harvest residues and urea for crop nutrition. *Sci. Agric.* **2013**, *70*, 313–320. [CrossRef]

16. Paredes Junior, F.P.; Portilho, I.I.R.; Mercante, F.M. Atributos microbiológicos de um latossolo sob cultivo de cana-de-açúcar com e sem queima da palhada. *Semin. Ciênc. Agrár.* **2015**, *36*, 151–164. [CrossRef]

17. Valim, W.C.; Panachuki, E.; Pavei, D.S.; Sobrinho, T.A.; Almeida, W.S. Effect of sugarcane waste in the control of interrill erosion. *Semin. Ciênc. Agrár.* **2016**, *37*, 1155–1164. [CrossRef]

18. Aquino, G.S.; Medina, C.C.; Costa, D.C.; Shahaba, M.; Santiago, A.D.; Cunha, A.C.B.; Kussaba, D.A.O.; Carvalho, J.B.; Moreira, A. Does straw mulch partial-removal from soil interfere in yield and industrial quality sugarcane? A long term study. *Ind. Crops Prod.* **2018**, *111*, 573–578. [CrossRef]

19. Lisboa, I.P.; Cherubin, M.R.; Lima, R.P.; Cerri, C.C.; Satiro, L.S.; Wienhol, B.J.; Schmer, M.R.; Jin, V.L.; Cerri, C.E.P. Sugarcane straw removal effects on plant growth and stalk yield. *Ind. Crops Prod.* **2018**, *111*, 794–806. [CrossRef]

20. Tucker, C.J.; Holben, B.N.; Elgin, J.H.; Mcmurtrey, J.E. Relationship of spectral data to grain yield variation. *Photogramm. Eng. Remote Sens.* **1980**, *46*, 657–666.

21. Tarnavsky, E.; Garrigues, S.; Brown, M.E. Multiscale geostatistical analysis of AVHRR, SPOT-VGT, and MODIS global *NDVI* products. *Remote Sens. Environ.* **2008**, *112*, 535–549. [CrossRef]

22. Sanches, G.M.; Duft, D.G.; Kölln, O.T.; Luciano, A.C.; De Castro, S.G.; Okuno, F.M.; Franco, H.C. The potential for RGB images obtained using unmanned aerial vehicle to assess and predict yield in sugarcane fields. *Int. J. Remote Sens.* **2018**, *15*, 1–3. [CrossRef]

23. Peralta, N.R.; Assefa, Y.; Du, J.; Barden, C.J.; Ciampitti, I.A. Mid-season high-resolution satellite imagery for forecasting site-specific corn yield. *Remote Sens.* **2016**, *8*, 1–16. [CrossRef]

24. Damian, J.M.; Santi, A.L.; Fornari, M.; Da Ros, C.O.; Eschner, V.L. Monitoring variability in cash-crop yield caused by previous cultivation of a cover crop under a no-tillage system. *Comput. Electron. Agric.* **2017**, *142*, 607–621. [CrossRef]

25. Lofton, J.; Tubana, B.S.; Kanke, Y.; Teboh, J.; Viator, H.; Dalen, M. Estimating sugarcane yield potential using an in-season determination of normalized difference vegetative index. *Sensors* **2012**, *12*, 7529–7547. [CrossRef] [PubMed]

26. Yang, H.; Yang, X.; Heskel, M.; Sun, S.; Tang, J. Seasonal variations of leaf and canopy properties tracked by ground-based *NDVI* imagery in a temperate forest. *Sci. Rep.* **2017**, *1267*, 1–10. [CrossRef] [PubMed]

27. Sardans, J.; Grau, O.; Chen, H.Y.H.; Janssens, I.A.; Ciais, P.; Piao, S.; Peñuelas, J. Changes in nutrient concentrations of leaves and roots in response to global change factors. *Glob. Chang. Biol.* **2017**, *23*, 3849–3856. [CrossRef]

28. USDA-United States Department of Agriculture. Available online: https://www.usda.gov/ (accessed on 18 June 2018).

29. CEPAGRI—Tempo e Clima Unicamp. Available online: https://www.cpa.unicamp.br (accessed on 2 February 2017).

30. Posto Meteorológico "Professor Jesus Marden dos Santos" ESALQ—USP. Available online: http://www.leb.esalq.usp.br/posto/ (accessed on 2 February 2017).

31. Raij, B.; Cantarella, H.; Guaggio, J.A.; Furlani, A.M.C. *Recomendações De Adubação E Calagem Para O Estado De São Paulo*, 1st ed.; Instituto Agronômico and Fundacão IAC: Campinas, Brazil, 1997; pp. 6–13.

32. Vermote, E.F.; Tanre, D.; Deulze, J.L.; Herman, M.; Morcrette, J.J. Second Simulation of the Satellite Signal in the Solar Spectrum, 6S: An overview. *IEEE Trans. Geosci. Remote* **1997**, *35*, 675–686. [CrossRef]

33. Carlotto, M.J. Reducing the effects of space-varying, wavelength-dependent scattering in multispectral imagery. *Int. J. Remote Sens.* **1999**, *20*, 3333–3344. [CrossRef]

34. Neter, J.; Wasserman, W.; Kutner, M.H. *Applied Linear Statistical Methods*, 5th ed.; Irwin Professional Publishing: Chicago, IL, USA, 1985; pp. 50–78.

35. Franco, H.C.J.; Pimenta, M.T.B.; Carvalho, J.L.N.; Magalhães, P.S.G.; Rossell, C.E.V.; Braunbeck, O.A.; Vitti, A.C.; Kolln, O.T.; Neto, J.R. Assessment of sugarcane trash for agronomic and energy purposes in Brazil. *Sci. Agric.* **2013**, *70*, 305–312. [CrossRef]

36. Trivelin, P.C.O.; Franco, H.C.J.; Otto, R.; Ferreira, D.A.; Vitti, A.C.; Fortes, C.; Faroni, C.E.; Oliveira, E.C.A.; Cantarella, H. Impact of sugarcane trash on fertilizer requirements for São Paulo, Brazil. *Sci. Agric.* **2013**, *70*, 345–352. [CrossRef]

37. Rosolem, C.A.; Calonego, J.C.; Foloni, J.S.S. Lixiviação de potássio da palha de espécies de cobertura de solo de acordo com a quantidade de chuva aplicada. *Rev. Bras. Cienc. Sol.* **2003**, *27*, 355–362. [CrossRef]

38. Endres, L.; Cruz, S.J.S.; Vilela, R.D.; Santos, J.M.; Barbosa, G.V.S.; Silva, J.A.C. Foliar applications of calcium reduce and delay sugarcane flowering. *Bioenerg. Res.* **2016**, *9*, 98–108. [CrossRef]

39. Vitti, G.; Luz, P.; Otto, R. *Agrícola Ouro Verde*, 10th ed.; Lençóis Paulista: Lencóis Paulista, Brazil, 2008; pp. 1–19.

40. Campos, M.; Antonangelo, J.A.; Alleoni, L.R.F. Phosphorus sorption index in humid tropical soils. *Soil Tillage Res.* **2016**, *156*, 110–118. [CrossRef]

41. Damon, P.M.; Bowden, B.; Rose, T.; Rengel, Z. Crop residue contributions to phosphorus pools in agricultural soils: A review. *Soil Biol. Biochem.* **2014**, *74*, 127–137. [CrossRef]

42. Pavinato, P.S.; Rosolem, C.A. Disponibilidade de nutrientes no solo—Decomposição e liberação de compostos orgânicos de resíduos vegetais. *Rev. Bras. Cienc. Sol.* **2008**, *32*, 911–920. [CrossRef]

43. Leite, J.M.; Ciampitti, I.A.; Mariano, E.; Vieira-Megda, M.X.; Trivelin, P.C. Nutrient, partitioning and stoichiometry in unburnt sugarcane ratoon at varying yield levels. *Front. Plant Sci.* **2016**, *7*, 1–14. [CrossRef] [PubMed]

44. Paul, G.C.; Bokhtiar, S.M.; Rehman, H.; Kabiraj, R.C.; Rahman, A.B.M.M. Efficacies of some organic fertilizers on sustainable sugarcane production in old Himalayan piedmont plain soil of Bangladesh. *Pak. Sugar J.* **2005**, *20*, 2–5.

45. Dotaniya, M.L.; Datta, S.C.; Biswas, D.R.; Dotaniya, C.K.; Meena, B.L.; Rajendiran, S.; Regar, K.L.; Lata, M. Use of sugarcane industrial by-products for improving sugarcane productivity and soil health. *Int. J. Recycl. Org. Waste Agric.* **2016**, *5*, 185–194. [CrossRef]

46. Aquino, G.S.; Medina, C.C.; Costa, D.C.; Shahaba, M.; Santiago, A.D. Sugarcane straw management and its impact on production and development of ratoons. *Ind. Crops Prod.* **2017**, *102*, 58–64. [CrossRef]

47. Singh, S.N.; Singh, A.K.; Malik, J.P.S.; Kumar, R.; Sharma, M.L. Cultural-practice packages and trash management effects on sugarcane ratoons under sub-tropical climatic conditions of India. *J. Agric. Sci.* **2012**, *150*, 237–247. [CrossRef]

48. Castioni, G.A.; Cherubin, M.R.; Menandro, L.M.S.; Sanches, G.M.; de Oliveira Bordonal, R.; Barbosa, L.C.; Carvalho, J.L.N. Soil physical quality response to sugarcane straw removal in Brazil: A multi-approach assessment. *Soil Tillage Res.* **2018**, *184*, 301–309. [CrossRef]

49. Carvalho, J.L.; Hudiburg, T.W.; Franco, H.C.; DeLucia, E.H. Contribution of above-and belowground bioenergy crop residues to soil carbon. *Glob. Chang. Biol. Bioenerg.* **2017**, *9*, 1333–1343. [CrossRef]

50. Oliveira, D.M.; Williams, S.; Cerri, C.E.; Paustian, K. Predicting soil C changes over sugarcane expansion in Brazil using the DayCent model. *Glob. Chang. Biol. Bioenerg.* **2017**, *9*, 1436–1446. [CrossRef]

51. Pitombo, L.M.; Cantarella, H.; Packer, A.P.C.; Ramos, N.P.; do Carmo, J.B. Straw preservation reduced total N2O emissions from a sugarcane field. *Soil Use Manag.* **2017**, *33*, 583–594. [CrossRef]

52. Vasconcelos, A.L.S.; Cherubin, M.R.; Feigl, B.J.; Cerri, C.E.; Gmach, M.R.; Siqueira-Neto, M. Greenhouse gas emission responses to sugarcane straw removal. *Biomass Bioenergy* **2018**, *113*, 15–21. [CrossRef]

53. Tavares, R.L.M.; Spokas, K.; Hall, K.; Colosky, E.; Souza, Z.M.D.; Scala, N.L. Sugarcane residue management impact soil greenhouse gas. *Ciênc. Agrotec.* **2018**, *42*, 195–203. [CrossRef]

54. da Silva, J. Sistemas de manejo da palhada influenciam acúmulo de biomassa e produtividade da cana-de-açúcar (var. RB855453). *Act. Sci. Agron.* **2010**, *32*, 345–350. [CrossRef]

55. Aquino, G.S.; de Conti Medina, C.; Junior, A.D.O.M.; Pasini, A.; Brito, O.R.; Cunha, A.C.B.; Almeida, L.F. Impact of harvesting with burning and management of straw on the industrial quality and productivity of sugarcane. *Afr. J. Agric. Res.* **2016**, *11*, 2462–2468. [CrossRef]

56. Aquino, G.S.; de Conti Medina, C.; Silvestre, D.A.; Gomes, E.C.; Benitez, A.C.; Cunha, D.A.O.K.; Santiago, A.D. Straw removal of sugarcane from soil and its impacts on yield and industrial quality ratoons. *Sci. Agric.* **2018**, *75*, 526–529. [CrossRef]

57. Wang, Q.; Tenhunen, J.; Dinh, N.Q.; Reichstein, M.; Vesala, T.; Keronen, P. Similarities in ground- and satellite-based *NDVI* time series and their relationship to physiological activity of a scots pine forest in finland. *Remote Sens. Environ.* **2004**, *93*, 225–237. [CrossRef]

58. Balzarolo, M.; Anderson, K.; Nichol, C.; Rossini, M.; Vescovo, L.; Arriga, N.; Wohlfahrt, G.; Calvet, J.C.; Carrara, A.; Cerasoli, S.; et al. Ground-based optical measurements at european flux sites: A review of methods, instruments and current controversies. *Sensors* **2011**, *11*, 7954–7981. [CrossRef] [PubMed]

59. Genc, L.; Turhan, H.; Asar, B.; Smith, S.E. Comparison of spectral indices from QUICKBIRD and ground based hyper-spectral data for winter wheat. *World Appl. Sci. J.* **2009**, *7*, 756–762. [CrossRef]

60. Caturegli, L.; Casucci, M.; Lulli, F.; Grossi, N.; Gaetani, M.; Magni, S.; Bonari, E.; Volterrani, M. GeoEye-1 satellite versus ground-based multispectral data for estimating nitrogen status of turfgrasses. *Int. J. Remote Sens.* **2015**, *36*, 2238–2251. [CrossRef]

61. Bu, H.; Sharma, L.K.; Denton, A.; Franzen, D.W. Comparison of Satellite Imagery and Ground-Based Active Optical Sensors as Yield Predictors in Sugar Beet, Spring Wheat, Corn, and Sunflower. *Agron. J.* **2017**, *109*, 299–308. [CrossRef]

62. Morel, J.; Bégué, A.; Todoroff, P.; Martiné, J.F.; Lebourgeois, V.; Petit, M. Coupling a sugarcane crop model with the remotely sensed time series of fIPAR to optimise the yield estimation. *Eur. J. Agron.* **2014**, *61*, 60–68. [CrossRef]

63. Bégué, A.; Lebourgeois, V.; Bappel, E.; Todoroff, P.; Pellegrino, A.; Baillarin, F.; Siegmund, B. Spatio-temporal variability of sugarcane fields and recommendations for yield forecast using *NDVI*. *J. Remote Sens.* **2010**, *31*, 5391–5407. [CrossRef]

64. Jarrell, W.M.; Beverly, R.B. The dilution effect in plant nutrition studies. *Adv. Agron.* **1981**, *34*, 197–224. [CrossRef]

65. Muchovej, R.M.; Newman, P.R.; Luo, Y. Sugarcane leaf nutrient concentrations: With or without midrib tissue. *J. Plant. Nutr.* **2005**, *28*, 1271–1286. [CrossRef]

66. Al-Abbas, A.H.; Barr, R.; Hall, J.D.; Crane, F.L.; Baumgardner, M.F. Spectra of normal and nutrient-deficient maize leaves. *Agron. J.* **1974**, *66*, 16–20. [CrossRef]

67. Raper, T.B.; Varco, J.J.; Hubbard, K.J. Canopy-Based normalized difference vegetation index sensors for monitoring cotton nitrogen status. *Agron. J.* **2013**, *105*, 1345–1354. [CrossRef]

68. Fernandes, J.L.; Ebecken, N.F.F.; Esquerdo, J.C.D.N. Sugarcane yield prediction in Brazil using *NDVI* time series and neural networks ensemble. *ISPRS J. Photogramm. Remote Sens.* **2017**, *38*, 4631–4644. [CrossRef]

MDPI

St Alban-Anlage 66

4052 Basel

Switzerland

Tel. +41 61 683 77 34

Fax +41 61 302 89 18

www.mdpi.com

Agronomy Editorial Office

E-mail: agronomy@mdpi.com

www.mdpi.com/journal/agronomy